Fundamentals of Turbomachinery

Fundamentals of Turbomachinery simplifies complex turbomachinery concepts, helping engineering students understand the basics step by step to prepare for more advanced studies and real-world applications. It provides the theoretical background needed to explain the principles underlying turbomachinery behavior with a strong emphasis on a solid mathematical foundation.

Designed to guide readers through an understanding of machines in terms of increasing complexity, this textbook focuses on core principles instead of advanced approximations or computational tools. It covers dimensional analysis and pump systems before progressing into axial compressors, axial turbines, centrifugal compressors, and centripetal turbines. The chapters transition from theoretical and mathematical concepts to practical applications and feature numerous end-of-chapter problems and examples.

This book is intended for senior undergraduate engineering students taking turbomachinery and fluid mechanics courses.

Instructors will be able to utilize a Solutions Manual and Figure Slides for their course.

Dr. Ernesto Benini is a Full Professor in the Department of Industrial Engineering of the University of Padova, Italy, and leads the "TEMPEST" research group. He earned his PhD in Energy Science and Engineering (2000) from the University of Padova, Italy. He has more than 25 years of experience in the research and development of advanced methods for the design and analysis of high-speed fluid machines, with an emphasis on super-hypersonic propulsion and transonic machines. He is a Principal Investigator of numerous private and European research projects, as well as a Scientific Advisor to major international companies in the propulsion sector. He is also a Fellow of the American Society of Mechanical Engineers (ASME), an Associate Fellow of the American Institute of Aeronautics and Astronautics (AIAA), and a member of the Advisory Council for Aviation Research and Innovation in Europe (ACARE). He serves as the Chair of the AIAA Gas Turbine Engines Technical Committee – Education. He is the author or co-author of more than 350 scientific publications concerning fluid dynamics of machines and has also been granted several international patents.

Fundamentals of Turbomachinery

Ernesto Benini

CRC Press
Taylor & Francis Group
Boca Raton London New York

CRC Press is an imprint of the
Taylor & Francis Group, an **informa** business

Designed cover image: Ernesto Benini – The fan of the Eurojet EJ200 engine.

First edition published 2026
by CRC Press
2385 NW Executive Center Drive, Suite 320, Boca Raton FL 33431

and by CRC Press
4 Park Square, Milton Park, Abingdon, Oxon, OX14 4RN

CRC Press is an imprint of Taylor & Francis Group, LLC

Library of Congress Cataloging-in-Publication Data
Names: Benini, Ernesto author
Title: Fundamentals of turbomachinery / Ernesto Benini.
Description: First edition. | Boca Raton, FL : CRC Press, 2026. |
Includes bibliographical references and index.
Identifiers: LCCN 2025051820 (print) | LCCN 2025051821 (ebook) | ISBN 9781041025665 hbk | ISBN 9781041025702 pbk | ISBN 9781003619956 ebk
Subjects: LCSH: Turbomachines–Fluid dynamics–Textbooks | Turbomachines–Aerodynamics–Textbooks | LCGFT: Textbooks
Classification: LCC TJ267 .B46 2026 (print) | LCC TJ267 (ebook)
LC record available at https://lccn.loc.gov/2025051820
LC ebook record available at https://lccn.loc.gov/2025051821

ISBN: 978-1-041-02566-5 (hbk)
ISBN: 978-1-041-02570-2 (pbk)
ISBN: 978-1-003-61995-6 (ebk)

DOI: 10.1201/9781003619956

Typeset in Times
by codeMantra

Access the Support Material: www.routledge.com/9781041025665

To my family,
for their constant presence and support along the way.

To my students,
in recognition of their hard work and commitment to understanding,
with the hope that this book may support their growth and development.

Contents

List of Figures

List of Tables

Preface

This book, *Fundamentals of Turbomachinery*, is intended for undergraduate and graduate students enrolled in BSc and MSc engineering programs. It provides a structured and progressively developed introduction to turbomachinery, with the aim of supporting students in building a solid understanding of the subject.

The topics have been arranged to gradually enhance the students' confidence and comprehension as they progress through the material. The early chapters establish a robust foundation for the theory of turbomachines by revisiting fundamental concepts from fundamental disciplines such as Thermodynamics and Fluid Mechanics. In these sections, the emphasis is on the preservation of mathematical rigor, which enables students to establish a connection with the material that was previously covered in courses such as Fluid Mechanics and Theoretical/Applied Mechanics. In light of this emphasis, this book restricts its focus to semi-empirical methods, reserving them for more advanced studies. The objective is to motivate students to acquire a comprehensive understanding of theoretical concepts prior to utilizing these methods in a more mature and deliberate manner. Computer modeling is also not addressed for comparable reasons, which permits the emphasis to remain on the development of a theoretical comprehension.

The content becomes more practical as this book progresses, with later chapters transitioning from a mathematical approach to more application-oriented discussions. This change is intended to ensure that the learning process is in accordance with the practical requirements that students will face as they transition into professional environments.

Ernesto Benini
15 April 2026

About the Author

Dr. Ernesto Benini graduated with honors in Mechanical Engineering from the University of Padova, Italy, receiving the prestigious A. Sarpi Gold Medal as the top graduate of his class. He subsequently earned his Ph.D. in Energy Science and Engineering from the same institution.

After completing a postdoctoral fellowship at Cranfield University (UK), he joined the University of Padova on a permanent basis, where he is currently a Full Professor of Fluid Machines in the Department of Industrial Engineering and leads the "TEMPEST" research group.

His expertise lies in the research and development of advanced methods for the design, analysis, and optimization of high-speed fluid machines, with a particular focus on super-hypersonic propulsion applications. He serves as a Principal Investigator for numerous privately and European-funded research projects. He also acts as a Scientific Advisor to major international companies in the aeroengine sector.

He is a Fellow of the American Society of Mechanical Engineers (ASME), an Associate Fellow of the American Institute of Aeronautics and Astronautics (AIAA), and a Member of the Advisory Council for Aviation Research and Innovation in Europe (ACARE). He has authored or co-authored more than 350 international scientific publications on the fluid dynamics of machines and holds several international patents in the field of turbomachinery.

His academic and scientific accomplishments have earned him multiple honors, including best paper awards and a nomination from the New York Academy of Sciences for the Outstanding Young Researchers award. According to the Stanford University global ranking, he is recognized among the World's Top 2% Most-Cited Scientists.

In addition to his research contributions, he is a dedicated educator in undergraduate and graduate engineering programs. He supervises numerous doctoral students and mentors young researchers in the field of turbomachinery fluid dynamics.

For more information, visit the author's Google Scholar Profile.

List of Symbols

LATIN SYMBOLS

A	Cross-sectional area
$\mathbf{B}(x)$	Bézier polynomial
b	Linear dimension, e.g. width
c	Absolute velocity
c_p	Specific heat at constant pressure
c_v	Specific heat at constant volume
Ca	Cavitation number
C_d	Discharge coefficient
C_D	Drag coefficient
C_L	Lift coefficient
C_{L_0}	Lift coefficient of an isolated airfoil
C_p	Pressure coefficient
$C_{p,r}, C'_{p,r}$	Pressure recovery coefficient
C_q	Heat capacity
d	Distance
D	Diameter or Drag force
D_h	Hydraulic diameter
D_s	Specific diameter
DF	Diffusion factor
$\dot{m}$	Mass flow rate
$\dot{m}_c$	Corrected mass flow rate or Cooling mass flow rate
$\dot{m}_s$	Mass flow rate per unit area
$\dot{q}$	Heat transfer rate per unit area
e_c	Compressor's polytropic efficiency
e_t	Turbine's polytropic efficiency
E	Aerodynamic efficiency or Energy
f	Function or Friction loss factor or Frequency
$\mathbf{f}$	Body force per unit volume
$\boldsymbol{f}_\mu$	Vector of viscous forces per unit mass
F	Force
F_{AR}	Blade aspect ratio factor
g	Acceleration due to gravity
h	Specific enthalpy or Head
h_r	Head loss
h_t	Theoretical head
H	Linear dimension, e.g. height, or Wake shape factor
i	Incidence angle
I	Rothalpy
k	Thermal conductivity or Stiffness
K	Type number or Loss coefficient or Factor
K_p	Permeability
l	Length
ℓ	Chord
L	Specific work
m	Meridional or Variable or Carter's parameter or Mass
m'	Radius-normalized meridional coordinate
Ma	Mach number

M	Momentum
n	Polytropic exponent or Number of elements or Generic exponent
n_d	Nodal diameter
N	Rotations per minute
N_c	Corrected rotational speed
N_s	Shape number
Nu	Nusselt number
p	Pressure
p_v	Vapor pressure
P	Power or Perimeter
$\mathbf{P}$	Control point
P_f	Power factor
Pr	Prandtl number
Q	Volumetric flow rate or Specific heat transfer
r	Radial coordinate or Recovery factor
R	Specific gas constant or Radius of curvature
Re	Reynolds number
s	Specific entropy or Variable or Electromagnetic slip or Spacing
S	Curve length
SM	Stall margin
t	Time or Variable or Thickness
t_b	Thickness of compressor base profile
t_t	Thickness of turbine profile
Th	Thoma number
T	Temperature or Period
u	Blade velocity
U_∞	Freestream velocity
v	Specific volume or Local velocity
$\mathcal{V}$	Seepage velocity
V	Volume or Velocity of cooling air
w	Relative velocity
W_{cycle}	Work per cycle
x	Coordinate or Variable
$\mathbf{x}$	Vector quantity
y	Coordinate or Variable
y_c	Camber line ordinate
Y	Total pressure loss coefficient in compressible flow
z	Axial or Elevation coordinate
Z	Number of blades
Z_s	Number of stages
Zw	Zweifel number

GREEK SYMBOLS

α	Angle of absolute velocity or Convective heat transfer coefficient
α'	Stator vane angle
β	Angle of relative velocity
β'	Rotor blade angle
γ	Heat capacity ratio
Γ	Circulation
δ	Angular flow deviation or Boundary layer thickness
δ^*	Boundary layer displacement thickness
δ_{AERO}	Aerodynamic damping coefficient
δ_{corr}	Pressure correction ratio

δ_{ij}	Kronecker delta
ε	Roughness or Tip clearance
ε_{ij}	Rate of strain tensor
ε_r	Degree of reaction
ε^*	Howell's nominal deflection
ζ	Total pressure loss coefficient
η	Efficiency or Local coordinate
η_x	Film-cooling effectiveness
θ	Azimuthal angle or Generic angle
θ^*	Momentum thickness
θ_c	Camber angle
θ_{corr}	Temperature correction ratio
θ_i	Blade angle at leading edge relative to chord direction
θ_u	Blade angle at trailing edge relative to chord direction
ι	$\tan^{-1}$(1/Aerodynamic efficiency)
κ	Local streamline inclination angle
$\varkappa$	Blades' blockage
λ	Bulk viscosity or Loss coefficient or Eigenvalues
λ_w	Work done factor
Λ	Power coefficient
μ	Dynamic viscosity
ν	Kinematic viscosity or Hub-to-tip radius ratio or Coolant flow ratio
ξ	Lean angle or Stagger angle or Blockage factor or Enthalpy loss coefficient
Ξ	Inter-blade phase angle (IBPA)
π_c	Compression ratio
π_e	Expansion ratio
π_i	I-th non-dimensional group
π_N	Total pressure loss in nozzle vanes
π_s	Stage pressure ratio
Π	Overall pressure ratio
ρ	Density
ρ_r	Stage density ratio
σ	Cascade solidity or Entropy generation
σ_{ij}	Total stress tensor
τ	Torque or Total temperature ratio
τ_{adm}	Allowable shear stress
τ_{ij}	Viscous stress tensor
ϕ	Flow coefficient or Scalar flow quantity
φ	Momentum ratio between coolant and hot gas
Φ	Rate of energy dissipation per unit volume or Potential function
χ	Contraction coefficient
ψ	Head/Pressure coefficient or Work coefficient
ψ_s	Isentropic work coefficient
Ψ	Stream function
ω	Angular velocity or Vibration frequency
ω_s	Specific speed
ϖ	Vibration frequency
Ω	Vorticity or Diffuser total pressure loss coefficient or Excitation frequency
Ω_c	Critical rotational speed

APICES

$-$	Mean value
0	Stagnation quantity
$*$	Referred to critical state or Design quantity

PEDICES

–	Referred to lower flow rate
0	Reference state
1	Station location (e.g. inlet)
2	Station location (e.g. outlet)
3	Station location (e.g. outlet)
∞	Referred to cascade mean quantities
atm	Atmospheric
E	Eulerian
f	Referred to body forces
hyd	Hydraulic
i, j, k	Indices
id	Ideal
in	Inlet
irr	Referred to a irreversible process
is	Isentropic
l	Length
mech	Mechanical and windage
n	Normal, perpendicular
out	Outlet, exit
R	Referred to rotor
r, u, z	Radial, Tangential, Axial
s	Referred to shear stresses or Referred to an isentropic process or Referred to a stage
S	Referred to stator
pol	Polytropic
rev	Referred to a reversible process
θ	Tangential
tip	Referred to the tip section
ts	Total-to-static
TT	Total-to-total
v	Referred to a volume or Volute
v	Volumetric
x, y, z	Along x, y and z direction

MATHEMATICAL OPERATORS

$\cdot$	Dot product
∇	Differential operator
$\times$	Cross product
:	Double contraction (or Frobenius inner product) between two second-order tensors

OTHER SYMBOLS

$[\mathbf{X}]$	Matrix X
$\hat{\dots}$	Unit vector
$\langle x \rangle$	Time averaged value of x
$\mathfrak{L}$	Area-averaged specific work

1 Basic Concepts

The fundamental principles of fluid mechanics and thermodynamics in turbomachinery are rooted in classical physics and form a cohesive and unified framework. The relationship between forces and changes in momentum in circulating fluids is described by Newton's equations of motion, while the concept of continuity guarantees mass conservation within control volumes. At the same time, the laws of thermodynamics govern the conservation and transformation of energy throughout these processes. These principles are axiomatic foundations, or *principia*—from the Latin *princeps-cĭpis*, meaning «first» or «chief»—serving as essential pillars for both theoretical understanding and practical analysis in the study of turbomachines.

1.1 SELECTED TOPICS FROM FLUID MECHANICS

1.1.1 CONSTITUTIVE EQUATIONS FOR NEWTONIAN FLUIDS

Newtonian fluids are identified by a linear relationship between the stress tensor and the rate of strain tensor. The second-rank total Cauchy stress tensor $\boldsymbol{\sigma}$ can be decomposed into two components: the isotropic stress tensor and the viscous stress tensor.

The isotropic stress is associated with the hydrostatic pressure p, which acts equally in all directions:

$$\boldsymbol{\sigma}^{\text{iso}} = -p\mathbf{I} \tag{1.1}$$

where $\mathbf{I}$ is the identity tensor (i.e. the identity matrix in our case).

The viscous, or deviatoric, stress tensor $\boldsymbol{\tau}_w$ is proportional to the rate of strain tensor and accounts for internal fluid friction:

$$\boldsymbol{\tau}_w = 2\mu\boldsymbol{\varepsilon} + \lambda(\nabla\cdot\mathbf{c})\mathbf{I} \tag{1.2}$$

where the rate of strain tensor $\boldsymbol{\varepsilon}$ is defined as:

$$\boldsymbol{\varepsilon} = \frac{1}{2}\left(\nabla\mathbf{c} + (\nabla\mathbf{c})^T\right) \tag{1.3}$$

while μ is the dynamic viscosity, λ the bulk viscosity, $\nabla\mathbf{c}$ the velocity gradient tensor, and $\nabla\cdot\mathbf{c}$ the divergence of the velocity field.

In general, the bulk viscosity λ is treated as an independent coefficient of the dynamic viscosity μ. However, for many Newtonian fluids, it is common to adopt Stokes' hypothesis, which assumes:

$$\lambda = -\frac{2}{3}\mu \tag{1.4}$$

DOI: 10.1201/9781003619956-1

Substituting this into equation 1.2, the total Cauchy stress tensor becomes:

$$\boldsymbol{\sigma} = \boldsymbol{\sigma}^{\text{iso}} + \boldsymbol{\tau}_w = -p\mathbf{I} + 2\mu\boldsymbol{\varepsilon} + \lambda(\nabla\cdot\mathbf{c})\mathbf{I} = -p\mathbf{I} + 2\mu\left(\boldsymbol{\varepsilon} - \frac{1}{3}(\nabla\cdot\mathbf{c})\mathbf{I}\right) \quad (1.5)$$

1.1.2 VISCOUS DISSIPATION

The rate of mechanical energy dissipation per unit volume, known as viscous dissipation, denoted by Φ, quantifies the irreversible conversion of kinetic energy into internal energy due to the action of viscous stresses within a deforming fluid. The mechanical power per unit volume delivered by internal stresses is given by the double contraction[1]:

$$\boldsymbol{\sigma} : \nabla\mathbf{c} = -p(\nabla\cdot\mathbf{c}) + \boldsymbol{\tau}_w : \nabla\mathbf{c} \quad (1.6)$$

The second term represents the viscous dissipation:

$$\Phi = \boldsymbol{\tau}_w : \nabla\mathbf{c} \quad (1.7)$$

which, for a Newtonian fluid under Stokes' hypothesis, becomes:

$$\Phi = \left(2\mu\boldsymbol{\varepsilon} - \frac{2}{3}\mu(\nabla\cdot\mathbf{c})\mathbf{I}\right) : \nabla\mathbf{c} \quad (1.8)$$

Upon decomposing the velocity gradient into symmetric and antisymmetric parts:

$$\nabla\mathbf{c} = \boldsymbol{\varepsilon} + \boldsymbol{\Omega}, \quad \text{where} \quad \boldsymbol{\Omega} = \frac{1}{2}\left(\nabla\mathbf{c} - (\nabla\mathbf{c})^T\right) \quad (1.9)$$

and since $\boldsymbol{\varepsilon} : \boldsymbol{\Omega} = 0$ and $\mathbf{I} : \boldsymbol{\Omega} = 0$, we have:

$$\boldsymbol{\varepsilon} : \nabla\mathbf{c} = \boldsymbol{\varepsilon} : \boldsymbol{\varepsilon}, \quad \text{and} \quad \mathbf{I} : \nabla\mathbf{c} = \nabla\cdot\mathbf{c} \quad (1.10)$$

Using (1.10) in (1.8):

$$\Phi = 2\mu\left(\varepsilon : \varepsilon - \frac{1}{3}(\nabla\cdot\mathbf{c})^2\right) \quad (1.11)$$

1.1.3 MASS FLOW AND VOLUMETRIC FLOW, THE CONTINUITY EQUATION

The elementary mass and volumetric flow rate $d\dot{m}$ and dQ through an infinitesimal element $d\mathbf{A}$ with normal unit vector $\hat{\mathbf{n}}$ and velocity vector $\mathbf{c}$ are given by:

$$d\dot{m} = \rho\mathbf{c}\cdot d\mathbf{A}, dQ = \mathbf{c}\cdot d\mathbf{A} \quad (1.12)$$

where ρ is the density of the fluid, $\mathbf{c}$ is the velocity vector of the fluid, $d\mathbf{A}$ is the infinitesimal area vector with magnitude dA and direction $\hat{\mathbf{n}}$ (unit vector). The dot product $\mathbf{c}\cdot d\mathbf{A}$ represents the component of the velocity vector $\mathbf{c}$ that is perpendicular to the area vector $d\mathbf{A}$, hence determining the rate of mass flow and volume flow through the infinitesimal area dA, respectively.

The principle of mass conservation, expressed by the continuity equation, stipulates that the mass flow rate remains constant throughout a steady-state arbitrary flow. In other words, the principle states that the mass flow entering a control volume must equal the mass flow leaving it, thus ensuring continuity of the fluid. In differential form, the continuity equation reads as follows:

$$\nabla \cdot (\rho \mathbf{c}) = 0 \tag{1.13}$$

where $\nabla \cdot (\rho \mathbf{c})$ represents the divergence of the mass flux vector $\rho \mathbf{c}$.

In incompressible flows, e.g. flows involving liquids, the fluid density remains constant under normal operating conditions, a fact that simplifies the calculations of mass flow rates. Similar to compressible flows, the conservation of mass principle applies, which leads to the conservation of volumetric flux:

$$\nabla \cdot \mathbf{c} = 0 \tag{1.14}$$

indicating that the divergence of the velocity vector $\mathbf{c}$ is zero, which implies that the volume flow rate through any closed surface remains constant over time.

To derive the integral form of the continuity equation for a steady flow, we integrate equation 1.13 over a fixed control volume V enclosed within a boundary surface ∂V:

$$\iiint_V \nabla \cdot (\rho \mathbf{c})\, dV = 0 \tag{1.15}$$

Applying the Green–Gauss divergence theorem to convert the volume integral into a surface integral yields:

$$\oiint_{\partial V} \rho \mathbf{c} \cdot \hat{\mathbf{n}}\, dA = 0 \tag{1.16}$$

where, by convention, $\hat{\mathbf{n}}$ points outward from the control volume.

When the control volume V has two open surface boundaries for inlet and outlet mass flows, A_1 and A_2, the preceding equation leads to:

$$\oiint_{\partial V} \rho \mathbf{c} \cdot \hat{\mathbf{n}}\, dA = 0 \implies \iint_{A_1} \rho \mathbf{c} \cdot \hat{\mathbf{n}}\, dA + \iint_{A_2} \rho \mathbf{c} \cdot \hat{\mathbf{n}}\, dA = 0 \tag{1.17}$$

which ultimately reads as:

$$\iint_{A_2} \rho c_{n2}\, dA - \iint_{A_1} \rho c_{n1}\, dA = 0 \tag{1.18}$$

c_{n2} and c_{n1} being the velocity components normal to their pertinent area, or, more simply:

$$\dot{m}_1 = \rho_1 c_{n1} A_1 = \dot{m}_2 = \rho_2 c_{n2} A_2 \tag{1.19}$$

which states that in steady flow regime, whatever the nature of the fluid involved, the mass flow rate is conserved between two arbitrary stations along the flow path.

In the special case of steady incompressible flow, therefore, the volumetric flow rate remains constant:

$$Q_1 = c_{n1}A_1 = Q_2 = c_{n2}A_2 \tag{1.20}$$

1.1.4 MOMENTUM EQUATION

The differential form of the compressible momentum equation in steady state is derived from the Navier–Stokes equations, which themselves originate from the application of Newton's second law of motion to a viscous fluid:

$$(\rho \mathbf{c} \cdot \nabla)\mathbf{c} = -\nabla p + \nabla \cdot \boldsymbol{\tau}_w + \mathbf{f} \tag{1.21}$$

where $\boldsymbol{\tau}_w$ is the viscous stress tensor and $\mathbf{f}$ is a body force per unit volume, e.g. gravitational forces ($\mathbf{f} = -\rho g \hat{\mathbf{k}}$).

The integral form of the steady-state compressible momentum equation over a control volume V and a boundary surface ∂V is as follows:

$$\oiint_{\partial V} \rho \mathbf{c}(\mathbf{c} \cdot \hat{\mathbf{n}})\, dA = -\oiint_{\partial V} p\hat{\mathbf{n}}\, dA + \oiint_{\partial V} \boldsymbol{\tau}_w \cdot \hat{\mathbf{n}}\, dA + \iiint_V \mathbf{f}\, dV \tag{1.22}$$

In the case where body forces act throughout the volume V, and the control surface has one inlet and one outlet area on which only pressure forces are exerted, the integral form of the momentum equation becomes:

$$\begin{aligned} &\iint_{A_{\text{out}}} \rho \mathbf{c}(\mathbf{c} \cdot \hat{\mathbf{n}})\, dA + \iint_{A_{\text{in}}} \rho \mathbf{c}(\mathbf{c} \cdot \hat{\mathbf{n}})\, dA \\ &= -\iint_{A_{\text{out}}} p\hat{\mathbf{n}}\, dA - \iint_{A_{\text{in}}} p\hat{\mathbf{n}}\, dA + \iint_{A_w} \boldsymbol{\tau}_w \cdot \hat{\mathbf{n}}\, dA + \iiint_V \mathbf{f}\, dV \end{aligned} \tag{1.23}$$

In the special case of steady incompressible and inviscid flow ($\boldsymbol{\tau}_w \to \mathbf{0}$, ρ = const.), equation 1.21 becomes:

$$(\mathbf{c} \cdot \nabla)\mathbf{c} = -\frac{1}{\rho}\nabla p + \frac{1}{\rho}\mathbf{f} \tag{1.24}$$

which can be easily integrated along a streamline from points 1 to 2, leading to the well-known Bernoulli equation:

$$p + \frac{1}{2}\rho c^2 + \rho g z = \text{constant along a streamline} \tag{1.25}$$

1.1.5 MOMENT OF MOMENTUM EQUATION

The rotational effects in the fluid flow are described using the differential form of the steady-state moment of momentum equation, which is derived from the momentum equation. The equation is directly derived from the principle of conservation of angular momentum, which is a direct result of the application of Newton's second law of motion to rotational motion.

If $\mathbf{r}$ is the position vector from an arbitrary reference point O, the moment of momentum equation reads as:

$$\mathbf{r} \times [(\rho\mathbf{c} \cdot \nabla)\mathbf{c}] = \mathbf{r} \times [-\nabla p + \nabla \cdot \boldsymbol{\tau}_w + \mathbf{f}] \tag{1.26}$$

This gives:

$$(\rho\mathbf{c} \cdot \nabla)(\mathbf{r} \times \mathbf{c}) - \rho\mathbf{c} \times \mathbf{c} = \mathbf{r} \times (-\nabla p) + \mathbf{r} \times (\nabla \cdot \boldsymbol{\tau}_w) + \mathbf{r} \times \mathbf{f} \tag{1.27}$$

Since $\mathbf{c} \times \mathbf{c} = 0$ and $\mathbf{r} \times (-\nabla p) = -\mathbf{r} \times \nabla p$, the latter becomes:

$$(\rho\mathbf{c} \cdot \nabla)(\mathbf{r} \times \mathbf{c}) = -\mathbf{r} \times \nabla p + \mathbf{r} \times (\nabla \cdot \boldsymbol{\tau}_w) + \mathbf{r} \times \mathbf{f} \tag{1.28}$$

Applying the divergence theorem to equation 1.28 over a steady control volume V with outward unit normal $\hat{\mathbf{n}}$, one obtains:

$$\underbrace{\oiint_A \rho\,(\mathbf{r} \times \mathbf{c})(\mathbf{c} \cdot \hat{\mathbf{n}})\,dA}_{\text{net flux of angular momentum}} = \underbrace{\oiint_A \mathbf{r} \times \left(-p\hat{\mathbf{n}} + \boldsymbol{\tau}_w \cdot \hat{\mathbf{n}}\right) dA}_{\text{surface torque}} + \underbrace{\iiint_V \mathbf{r} \times \mathbf{f}\,dV}_{\text{body-force torque}} \tag{1.29}$$

For a control volume with both open boundaries (inlets and outlets) and solid walls, the control surface A can be decomposed as $A = A_{\text{open}} \cup A_{\text{walls}}$. The moment of momentum balance then becomes:

$$\begin{aligned}\iint_{A_{\text{open}}} \rho\,(\mathbf{r} \times \mathbf{c})(\mathbf{c} \cdot \hat{\mathbf{n}})\,dA = &\underbrace{\iint_{A_{\text{open}}} \mathbf{r} \times (-p\hat{\mathbf{n}})\,dA}_{\mathbf{M_p}} \\ &+ \underbrace{\iint_{A_{\text{walls}}} \mathbf{r} \times \left(-p\hat{\mathbf{n}} + \boldsymbol{\tau}_w \cdot \hat{\mathbf{n}}\right) dA}_{\mathbf{M_w}} + \underbrace{\iiint_V \mathbf{r} \times \mathbf{f}\,dV}_{\mathbf{M_f}}\end{aligned} \tag{1.30}$$

$\mathbf{M}_\text{p}$ denotes the net moment of the pressure forces acting on the open boundaries, $\mathbf{M}_\text{w}$ the net torque from wall pressure and viscous stresses, and $\mathbf{M}_\text{f}$ the torque due to body forces.

Therefore, the *net moment* exchanged between a steady flow and the rotating equipment that encloses it is obtained by summing the angular–momentum fluxes across all open boundaries and subtracting the contribution of the associated pressure moments. In compact form:

$$\mathbf{M}_{\text{net}} = \mathbf{M}_\text{w} + \mathbf{M}_\text{f} = \iint_{A_{\text{open}}} \rho\,(\mathbf{r} \times \mathbf{c})\,(\mathbf{c} \cdot \hat{\mathbf{n}})\,dA - \mathbf{M}_\text{p} \tag{1.31}$$

An alternative physical interpretation of this equation is that the net torque—which directly corresponds to the actual moment of the forces acting on the walls or blades inside the rotor—can be evaluated solely from the flow properties at the open boundaries. This result represents a fundamental advantage of the moment–of–momentum theorem, since it allows one to compute the shaft torque without explicitly resolving all forces and moments on the individual blades inside the rotor, a task that would otherwise be extremely challenging or practically impossible.

1.1.5.1 Turbomachinery

In turbomachines, the computation of torque about the rotational axis, usually the z axis in cylindrical coordinates, is essential to evaluate performance and power transfer. If the velocity vector in cylindrical coordinates is:

$$\mathbf{c} = c_r \hat{\mathbf{i}}_r + c_u \hat{\mathbf{i}}_\theta + c_z \hat{\mathbf{i}}_z \tag{1.32}$$

and the position vector from the axis of rotation is:

$$\mathbf{r} = r\hat{\mathbf{i}}_r \tag{1.33}$$

the angular momentum per unit mass about the z-axis is given by the z-component of $\mathbf{r} \times \mathbf{c}$ and $\mathbf{r} \times (p\hat{\mathbf{n}})$:

$$\mathbf{r} \times \mathbf{c} = r(c_u \hat{\mathbf{i}}_z - c_z \hat{\mathbf{i}}_\theta) \implies \hat{\mathbf{i}}_z \cdot (\mathbf{r} \times \mathbf{c}) = rc_u \tag{1.34}$$

in such a way that:

$$\mathbf{r} \times (p\hat{\mathbf{n}}) = r\hat{\mathbf{i}}_r \times (p\hat{\mathbf{n}}) \implies \hat{\mathbf{i}}_z \cdot [\mathbf{r} \times (p\hat{\mathbf{n}})] = r\hat{\mathbf{i}}_z \cdot [\hat{\mathbf{i}}_r \times (p\hat{\mathbf{n}})] \tag{1.35}$$

The latter term depends on the geometry and orientation of $\hat{\mathbf{n}}$. In turbomachinery, where axis-symmetric control volumes are dealt with, the pressure torque vanishes as $\hat{\mathbf{i}}_z \perp (\hat{\mathbf{i}}_r \times \hat{\mathbf{n}})$.

Furthermore, it is worth noting that, in almost all practical turbomachinery applications, the contribution of body-force torque $\mathbf{M}_\mathrm{f}$ is negligible when compared with both the pressure moment $\mathbf{M}_\mathrm{p}$ and the wall moment $\mathbf{M}_\mathrm{w}$.

As a direct consequence of the preceding discussion, the resulting net torque about the z-axis can be expressed as:

$$M_z = \tau = \iint_{A_\mathrm{out}} \rho r c_u (\mathbf{c} \cdot \hat{\mathbf{n}}) \, dA + \iint_{A_\mathrm{in}} \rho r c_u (\mathbf{c} \cdot \hat{\mathbf{n}}) \, dA \tag{1.36}$$

1.1.6 ONE-DIMENSIONAL FLOW

In a simplified representation, the properties of the fluid are subject to variation in a single spatial direction, resulting in a one-dimensional flow. When fluid behavior is examined along a single coordinate axis, this idealization is frequently employed, disregarding variations in other directions. The fundamental conservation laws of mass, momentum, and energy continue to be observed in one-dimensional flows. The singular coordinate axis, x, is used to express these laws in differential form.

In the x direction, the continuity equation in one-dimensional steady-state flow states is plainly stated:

$$\dot{m}_1 = \rho_1 c_{x1} A_1 = \dot{m}_2 = \rho_2 c_{x2} A_2 \tag{1.37}$$

Under the same hypothesis, the scalar form of the momentum equation along the x-direction is:

$$\sum F_x = \dot{m}(c_{x2} - c_{x1}) \tag{1.38}$$

where $\sum F_x$ represents the sum of all external forces acting in the x direction on the system under consideration. We will derive a one-dimensional form of the moment of momentum equation later on.

The one-dimensional flow assumption serves as a simplification tool for engineering calculations and design processes to handle complex fluid dynamics problems.[2] Engineers use this assumption to rapidly determine flow characteristics, which helps them make their first design choices. Relevant examples include: (i) the study of long pipes and channels works because radial changes are smaller than axial changes; (ii) aerodynamic analyses of thin airfoils and diffusers and nozzles use simplified methods; (iii) the Eulerian theory of turbomachinery, which represents a simplified approach to modeling turbomachinery flows.

1.2 SELECTED TOPICS FROM THERMODYNAMICS

1.2.1 CONSERVATION OF ENERGY – FIRST LAW OF THERMODYNAMICS

The principle of energy conservation, also known as the first law of thermodynamics, states that energy cannot be created or destroyed in an isolated system. Instead, energy can only be transformed from one form to another, but the total amount of energy remains constant over time.

It is well known that the first law of thermodynamics for any kind of substance flowing in an open system following a one-dimensional model is given in differential form as follows:

$$dh + d(c^2/2) + g\,dz = dQ - dL \tag{1.39}$$

where

- h is the specific enthalpy, which quantifies the total energy content per unit mass as a result of both internal energy and the energy associated with pressure-volume work:

  $$h = u + \frac{p}{\rho}$$

  u being the specific internal energy, p the pressure, and ρ the density;
- c denotes the fluid flow velocity. The term $c^2/2$ refers to the specific kinetic energy, which is the kinetic energy per unit of mass;
- z is the elevation or height above a reference level. The term $g\,dz$ specifies the change in gravitational potential energy per unit mass, where g is the acceleration due to gravity;

- dQ is the differential amount of heat added to the system per unit mass. It denotes the energy transfer into the system as a result of heat;
- dL represents the differential work of the system per unit mass. This is in accordance with the specific work that the system exchanges with the external environment, such as the shaft work of a turbomachine.

An alternative form of this law is:

$$Tds = dh - dp/\rho \tag{1.40}$$

By casting the two expressions 1.39 and 1.40 the following law is obtained:

$$Tds + dp/\rho + d(c^2/2) + gdz = dQ - dL \tag{1.41}$$

1.2.2 IDEAL AND REAL INCOMPRESSIBLE FLOWS

Regardless of variations in pressure, an incompressible fluid is a fluid substance whose density remains constant. The analysis of fluid flow is simplified by this assumption, which is frequently employed in fluid mechanics for liquids. This is due to the fact that, compared to gases, the density of liquids typically changes very little under pressure variations. After this, an incompressible flow is a flow in which the density of a fluid element remains constant during its motion.[3]

In turbomachinery dealing with liquids, e.g. water, the energy exchange process is virtually always adiabatic, so that $dQ = 0$. From equation 1.41, it follows that:

$$|dL| = Tds + dp/\rho + d(c^2/2) + gdz \tag{1.42}$$

If the energy exchange were internally reversible, the entropy rise of the liquid would be null, and the preceding equation would lead to:

$$|dL_{id}| = dp/\rho + d(c^2/2) + gdz \tag{1.43}$$

Real processes do not allow internal reversibility of energy exchange, so the increase in entropy of the liquid needs to be included. The main factors that lead to irreversibility in liquid-based processes include heat transfer across nonzero temperature gradients, viscous dissipation, rapid mixing, chemical reactions, and uncontrolled expansion into a vacuum or pressure-less environment.

The entropy change for a generic substance in a process requires the following thermodynamic definition:

$$ds = \frac{dQ_{\text{rev}}}{T} \tag{1.44}$$

where dQ_{rev} is the infinitesimal amount of heat added to the system in a reversible process, at a given absolute temperature T. For a liquid, the entropy can be computed using this definition if, during the reversible heat exchange process, the heat capacity $C_q(T)$ is known over a temperature range:

$$ds = \frac{C_q(T)}{T} dT \tag{1.45}$$

from which the absolute entropy s of a liquid at a temperature T can be found by integrating from absolute zero (0 K) to the temperature T:

$$s(T) = \int_0^T \frac{C_q(T)}{T} dT + s_0 \tag{1.46}$$

where s_0 is the entropy at absolute zero, which is typically set to 0 for a perfectly ordered substance, often referred to as a perfect crystal, according to the third law of thermodynamics. The calculation of entropy in the case where the heat capacity is constant with temperature is immediate.[4]

Calculating the entropy change ds from heat exchange becomes complicated when dQ is provided or extracted during an irreversible liquid process. The state function nature of entropy allows its change to depend solely on the initial and final states rather than the chosen path.

The actual heat dQ_{irrev} exchanged in the irreversible process differs from the heat exchanged in the hypothetical reversible process. The second law of thermodynamics requires that the total entropy change of the universe (system + surroundings) increase during an irreversible process. The system's entropy change Δs maintains its reversible path calculation, but the surroundings generate additional entropy due to internal irreversibilities.

The total entropy change Δs_{total} for an irreversible process consists of the generation of entropy $\sigma_{irr} \geq 0$ that results from irreversibilities:

$$\Delta s_{total} = \sigma_{irr} = \Delta s_{system} + \Delta s_{surroundings} \geq 0 \tag{1.47}$$

The system does not exchange heat with its environment during an adiabatic process. The change in entropy in the environment $\Delta s_{surroundings}$ is equal to zero because the heat transfer in the environment $dQ_{surroundings}$ is zero. The first law expression for liquids in adiabatic systems exchanging work can be written as follows:

$$|dL| = dp/\rho + d(c^2/2) + gdz + d\sigma_{irr} \tag{1.48}$$

which is sometimes referred to as the *generalized Bernoulli equation.*

1.2.3 COMPRESSIBLE FLOWS

A compressible fluid is a fluid whose density changes significantly with variations in pressure and temperature, i.e. a gas. Compressible flow occurs when the effects of fluid compressibility significantly alter flow behavior compared to the incompressible case.

In addition, in most turbomachines that deal with gases, the process of energy exchange is virtually adiabatic, so that $dQ = 0$.[5]

Furthermore, in gases, potential energy changes due to elevation are negligible, i.e. $dz = 0$.

Under these assumptions, from equation 1.39, it follows that:

$$|dL| = dh + d(c^2/2) \tag{1.49}$$

By introducing the stagnation (or total) enthalpy, denoted as h^0:

$$h^0 = h + \frac{c^2}{2} \tag{1.50}$$

it is found that:

$$|dL| = dh^0 \tag{1.51}$$

1.2.3.1 Work Done in Compression

During compression, work $dL_{\text{compression}}$ is done on the fluid:

$$dL_{\text{compression}} = dh^0_{\text{compression}} \tag{1.52}$$

Since h^0 increases during compression:

$$L_{\text{compression}} = h^0_{\text{outlet}} - h^0_{\text{inlet}} \tag{1.53}$$

where h^0_{outlet} is the stagnation enthalpy at the outlet and h^0_{inlet} is the stagnation enthalpy at the inlet.

1.2.3.2 Work Done in Expansion

During expansion, work $dL_{\text{expansion}}$ is done by the fluid:

$$dL_{\text{expansion}} = -dh^0_{\text{expansion}} \tag{1.54}$$

Since h^0 decreases during expansion:

$$L_{\text{expansion}} = h^0_{\text{inlet}} - h^0_{\text{outlet}} \tag{1.55}$$

where h^0_{inlet} is the stagnation enthalpy at the inlet and h^0_{outlet} is the stagnation enthalpy at the outlet.

From now on, the subscripts *compression* and *expansion* are omitted in the expressions for work, making it clear from context whether work is done on the fluid (compression) or by the fluid (expansion).

1.2.4 ENTROPY GENERATION DUE TO FRICTION AND VISCOUS DISSIPATION

As previously demonstrated, the viscous dissipation rate per unit volume in a simple shear flow can be expressed in terms of the velocity gradients within the fluid. In fact, for a steady, incompressible Newtonian fluid:

$$\Phi = \mu \left(\frac{\partial u}{\partial y} \right)^2 \tag{1.56}$$

The local rate of entropy generation per unit volume, σ_v, is related to the rate of energy dissipation Φ by the temperature T at which the process occurs:

$$\sigma_v = \frac{\dot{Q}_{\text{irr}}}{T} = \frac{\Phi}{T} \tag{1.57}$$

Substituting the previous expression for Φ into equation 1.57, we get:

$$\sigma_v = \frac{\Phi}{T} = \frac{\mu\left(\frac{\partial u}{\partial y}\right)^2}{T} \tag{1.58}$$

This equation represents the local rate of entropy generation due to viscous effects in a liquid.

In the case of compressible flows, the full expression for viscous dissipation provided in equation 1.11 must be used instead.

In addition to σ_v, if there is heat transfer with a finite temperature difference during the process, the entropy generation due to heat transfer σ_q can be calculated as:

$$\sigma_q = \frac{Q}{T_{\text{system}}} - \frac{Q}{T_{\text{surroundings}}}$$

where Q is the rate of heat transfer, T_{system} the temperature of the system and $T_{\text{surroundings}}$ the temperature of the surroundings.

Once σ_v (+ σ_q if present) has been computed, the total entropy generation σ_{irr} in the process can be obtained by integrating the local entropy generation over the entire volume of the liquid and summing the contributions from different sources of irreversibility:

$$\sigma_{\text{irr}} = \iiint_V \sigma_v \, dV + \iint_A \sigma_q \, dA \tag{1.59}$$

Example: Laminar Flow Between Parallel Plates (Couette Flow). Consider a simple case of laminar flow between two parallel plates, where the top plate moves with a constant velocity U and the bottom plate is stationary. The fluid velocity profile $u(y)$ is linear:

$$u(y) = \frac{U}{b}y$$

where b is the distance between the plates and y the vertical position. The velocity gradient is computed as:

$$\frac{\partial u}{\partial y} = \frac{U}{b}$$

such that:

$$\Phi = \mu\left(\frac{U}{b}\right)^2$$

and

$$\sigma_v = \frac{\mu}{T}\left(\frac{U}{b}\right)^2$$

For more complex flow fields, the local entropy generation rate σ_v needs to be integrated over the entire volume of the liquid to obtain the total entropy generation σ:

$$\sigma_{\text{irr}} = \iiint_V \sigma_v \, dV = \iiint_V \frac{\mu}{T}\left(\frac{\partial c_i}{\partial x_j}\right)^2 dV \tag{1.60}$$

where the indices i and j run over the spatial coordinates and V is the volume of the fluid.

Example:Turbulent Flows. The calculation becomes more complex in turbulent flows because of the chaotic nature of the velocity field. The dissipation function must be able to account for turbulent eddies, and statistical methods are typically employed. The entropy generation in turbulent flows is often higher than in laminar flows because of the increased rate of energy dissipation.

1.3 SELECTED TOPICS FROM HYDRAULICS

1.3.1 ENERGY BALANCES

The field of hydraulics investigates incompressible fluid flows, with particular emphasis on water. The examination of steady hydrodynamics provides the most suitable approach for our needs because it studies incompressible fluid movement through reservoirs and pipes of different diameters, valves, pumps, and other similar components. The study of hydrodynamics relies on two fundamental principles, which include the generalized Bernoulli equation 1.43 and the continuity equation 1.20. The principles control the behavior of the fluid motion while explaining pressure and velocity and elevation variations throughout the flow. The following relationships exist for any two points on a generic streamtube within a steady, incompressible flow:

$$\frac{p_1}{\rho} + \frac{c_1^2}{2} + gz_1 = \frac{p_2}{\rho} + \frac{c_2^2}{2} + gz_2 + gh_r + gh_{1,2} \tag{1.61}$$

$$Q_1 = Q_2 \tag{1.62}$$

Here, p_1 and p_2 are the pressures at points 1 and 2, c_1 and c_2 the flow velocities, z_1 and z_2 the elevations, h_r is the head loss due to friction and other factors, $gh_{1,2}$ representing the energy exchanged between an external device and the fluid between stations 1 and 2, ρ the fluid density, and Q_1 and Q_2 the volumetric flow rates at points 1 and 2, respectively. When the device is a pump, mechanical work is transferred to the fluid, so the energy in 2 is higher compared to 1. The opposite occurs for a

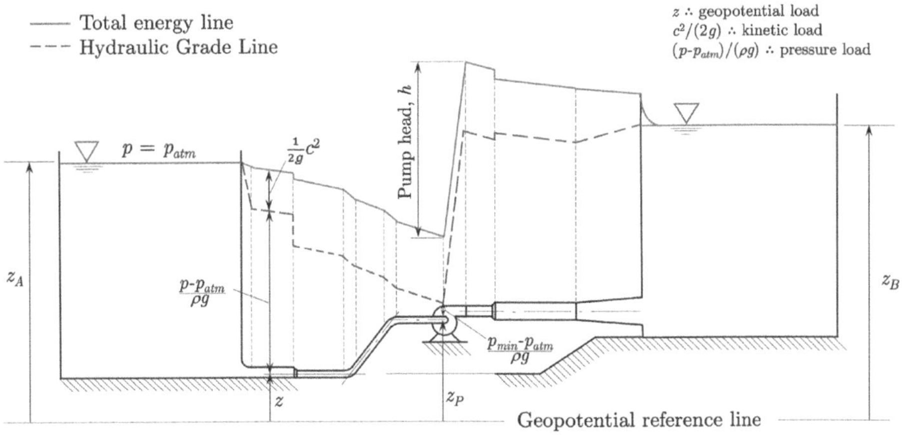

Figure 1.1 Pump circuit.

hydraulic turbine. Expressing energy per unit of weight, or in equivalent meters, the generalized Bernoulli equation is:

$$\frac{p_1}{\rho g}+\frac{c_1^2}{2g}+z_1+h_{1,2}=\frac{p_2}{\rho g}+\frac{c_2^2}{2g}+z_2+h_r \text{ (pump)} \tag{1.63}$$

$$\frac{p_1}{\rho g}+\frac{c_1^2}{2g}+z_1-h_{1,2}=\frac{p_2}{\rho g}+\frac{c_2^2}{2g}+z_2+h_r \text{ (turbine)} \tag{1.64}$$

where $h_{1,2}$, or simply h, is termed *head*. If we subtract the atmospheric pressure, $\frac{p_{\text{atm}}}{\rho g}$, from both sides the validity of equation is not altered, and we get:

$$\frac{p_1-p_{\text{atm}}}{\rho g}+\frac{c_1^2}{2g}+z_1+h_{1,2}=\frac{p_2-p_{\text{atm}}}{\rho g}+\frac{c_2^2}{2g}+z_2+h_r \text{ (pump)} \tag{1.65}$$

$$\frac{p_1-p_{\text{atm}}}{\rho g}+\frac{c_1^2}{2g}+z_1-h_{1,2}=\frac{p_2-p_{\text{atm}}}{\rho g}+\frac{c_2^2}{2g}+z_2+h_r \text{ (turbine)} \tag{1.66}$$

where p_1-p_{atm} is named *gauge pressure*, that is the pressure relative to the atmospheric one.

A graphic representation of equation 1.65 is given in Figure 1.1, where the point-by-point specific energy $\frac{p-p_{\text{atm}}}{\rho g}+\frac{c^2}{2g}+z$ distribution along a pumping circuit is shown by the solid line (the *total energy line*). The slope of this line indicates the rate of energy loss.

Since local flow velocity obeys the continuity $c=Q/A$, if Q is given and the geometric details of the plant specified, the kinetic load $c^2/(2g)$ is also known at each station. Subtracting the kinetic load from the total energy line gives the distribution of the *hydraulic grade line*, $\frac{p-p_{\text{atm}}}{\rho g}+z$, represented as a dashed line in Figure 1.1. The value of the hydraulic grade line at a certain location is, in fact, the level that water would reach in a hypothetical inspection sink or an open tank at that location. Equations 1.65 hold for an arbitrarily chosen couple of points along the flow path. If these points are selected along this flow path with no work-exchanging device in between, the generalized Bernoulli equation stipulates that the total energy of the fluid always decreases because head losses occur from the first point to the second point. On the other hand, when a pump is installed between these points, the fluid energy increases between the inlet and exit of the machine due to the pump's head $h_{1,2} = h$ (head h must not be confused with enthalpy). This head does not coincide with the work spent by the pump, as this work must account for the head losses inside the pump itself. The hydraulic work of the pump is often referred to as the *theoretical head* h_t and is related to h via:

$$h_t = \frac{h}{\eta_{\text{hyd}}} \text{ (pump)} \tag{1.67}$$

where η_{hyd} is the hydraulic efficiency of the pump.

1.3.2 HEAD LOSSES IN A HYDRAULIC CIRCUIT

Hydraulic circuits experience head losses in closed conduits due to continuous and concentrated factors that affect fluid flow.

The Darcy-Weisbach equation defines continuous head losses as follows:

$$h_{r,c} = f \cdot \frac{l}{D_h} \cdot \frac{c^2}{2g} \tag{1.68}$$

where f is the Darcy-Weisbach friction factor, l the length of the pipe, D_h the hydraulic diameter of the pipe.[6] Factor f depends on the flow regime (laminar or turbulent) and the relative roughness of the pipe and can be determined from empirical correlations or from the Moody chart (Figure 1.2). In particular, for laminar flows ($\text{Re} < 2000$), the friction factor is inversely proportional to the Reynolds number:

$$f = \frac{64}{\text{Re}} \tag{1.69}$$

In turbulent Flows ($\text{Re} > 4000$), the friction factor is calculated using the Colebrook-White equation, which implicitly relates f to Re and the relative roughness of the pipe ε/D:

$$\frac{1}{\sqrt{f}} = -2\log\left(\frac{\varepsilon/D}{3.7} + \frac{2.51}{\text{Re}\sqrt{f}}\right) \tag{1.70}$$

In the transitional regime ($2000 < \text{Re} < 4000$), the friction factor is less predictable and often requires empirical data or interpolation between laminar and turbulent values.

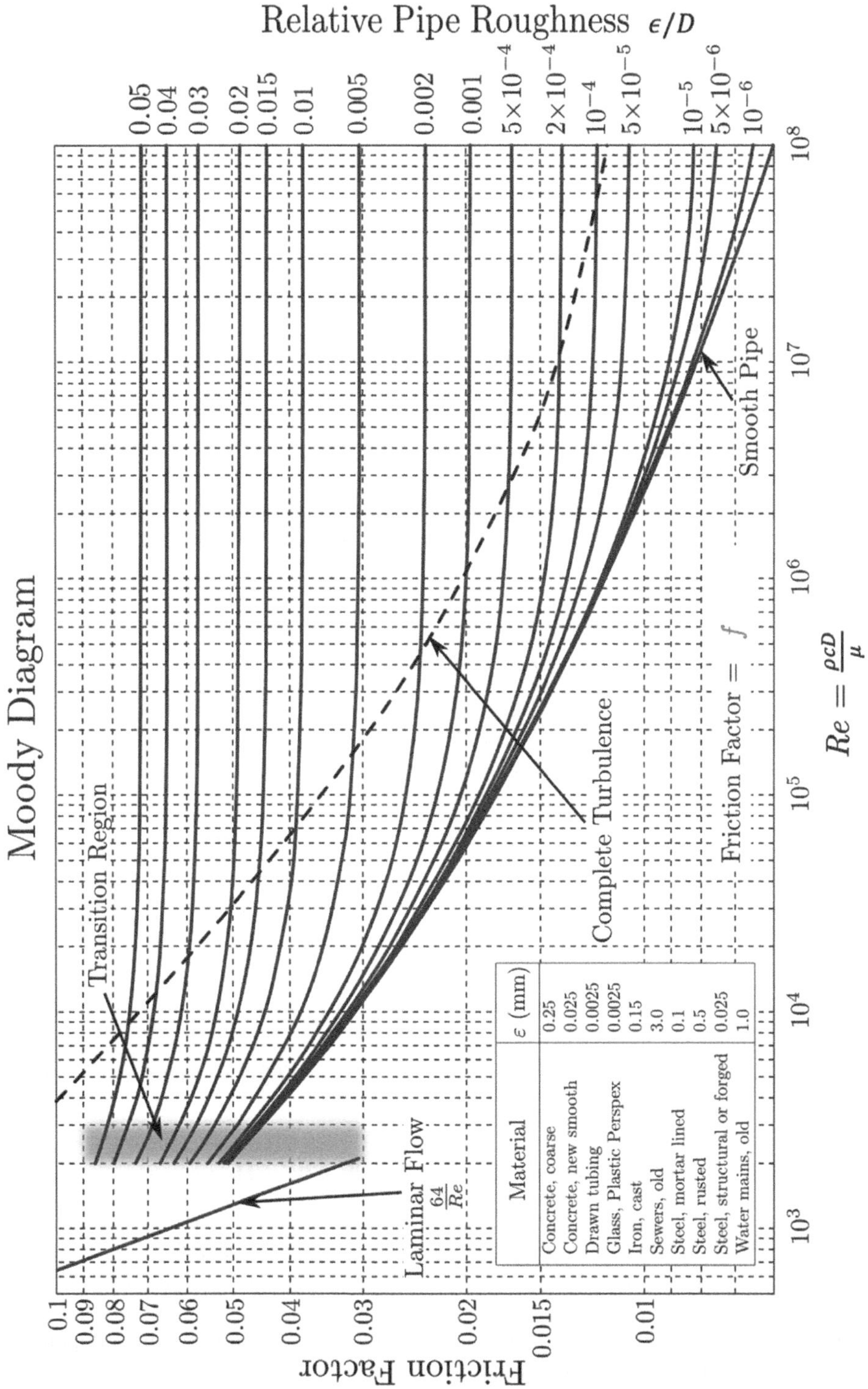

Material	ε (mm)
Concrete, coarse	0.25
Concrete, new smooth	0.025
Drawn tubing	0.0025
Glass, Plastic Perspex	0.0025
Iron, cast	0.15
Sewers, old	3.0
Steel, mortar lined	0.1
Steel, rusted	0.5
Steel, structural or forged	0.025
Water mains, old	1.0

Figure 1.2 Moody diagram.

The second type of head loss is characterized by sudden changes in the flow pattern, such as those occurring in elbows, valves, or fittings. Also, sudden changes in the diameter of the pipe cause localized head losses.

In elbows, these losses can be estimated using the following formula.

$$h_e = K_e \cdot \frac{c^2}{2g} \tag{1.71}$$

where K_e is a loss coefficient that is dependent upon the Reynolds number, the radius of curvature relative to the pipe diameter and the angle of the bend. The typical values for a standard 90° elbow are between 0.3 and 0.9.

Localized head losses in valves can be estimated using the previous formula, where the K_e value is subject to significant changes depending on the type (e.g. globe, gate, and ball) and the extent of opening. For example, a partially open globe valve may have a K_e value that exceeds 10, whereas a fully open gate valve may have a K_e value as low as 0.2.

As mentioned above, flow separations and vortex formation occur when the cross-sectional area suddenly changes through expansions or contractions, resulting in localized head losses. The Borda-Carnot formula serves to quantify these concentrated losses by enabling the calculation of head loss from disruptions in the flow path:

$$h_B = \frac{(c_1 - c_2)^2}{2g} \tag{1.72}$$

where c_1 and c_2 are the velocities of fluid before and after the area change, respectively. A special case of Borda loss is associated with a free-flow outlet in a reservoir, where $c_2 = 0$:

$$h_{\text{outlet}} = \frac{c_1^2}{2g} \tag{1.73}$$

Inlet loss occurs when fluid accelerates from zero velocity to c_1 downstream of the *vena contracta*. The head loss can be calculated using:

$$h_{\text{inlet}} = \frac{K_{\text{inlet}} c_1^2}{2g} \tag{1.74}$$

where K_{inlet} is the inlet loss coefficient (typically $K_{\text{inlet}} \approx 0.5$ for a sharp-edged entrance).

1.3.3 CHARACTERISTIC CURVE OF A PUMPING CIRCUIT

The total energy levels of a pumping circuit are illustrated by its characteristic curve, which describes the cumulative effect of static and dynamic phenomena. The former are encapsulated in the pressure and elevation characteristics of inlet and outlet reservoirs and, as such, do not depend on the flow rate (static phenomenon). The latter include the characteristics of overall head losses in the circuit, which depend on the flow rate (dynamic phenomenon).

The end-to-end total energy balance (e.g. the one in Figure 1.1) is expressed mathematically through the generalized Bernoulli equation, which applies from level A to level B while ignoring the pump's head. The relationship exists only between the circuit characteristics without any influence from the pump.

$$h_{\text{circuit}} = \frac{p_B - p_A}{\rho g} + (z_B - z_A) + \sum_i h_{(\text{r,c})_i} + \sum_j h_{\text{r}_j} \tag{1.75}$$

where $\sum_i h_{\text{r,c},i}$ represents the sum of continuous frictional head losses along the length of the $i-th$ pipe and $\sum_j h_{\text{r},j}$ is the sum of concentrated head losses. In the example of Figure 1.1, since $p_B - p_A = 0$, the characteristic curve of the circuit is:

$$h_{\text{circuit}} = (z_B - z_A) + \sum_i f_i \frac{l_i}{D_i} \frac{c_i^2}{2g} + \sum_j K_j \frac{c_j^2}{2g} \tag{1.76}$$

or:

$$h_{\text{circuit}} = (z_B - z_A) + \sum_i f_i \frac{l_i}{D_i} \frac{Q^2}{2gA_i^2} + \sum_j K_j \frac{Q^2}{2gA_j^2} = (z_B - z_A) + f(Q^2) \tag{1.77}$$

A sample plot of equation 1.77 is given in Figure 1.3 for parametric values of overall loss counts.

1.4 SELECTED TOPICS FROM COMPRESSIBLE FLUID DYNAMICS

The fluid dynamics of compressible flows, which is also known as gas dynamics, is extensively based on ideal gases as their fundamental component. The field of gas dynamics deals with many problems that involve gases under high velocity conditions,

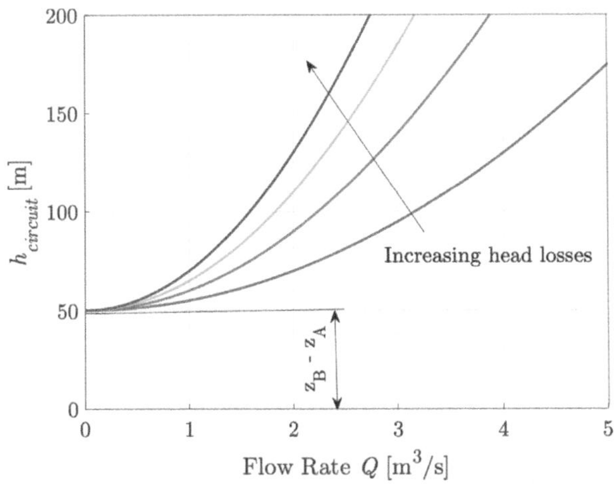

Figure 1.3 Circuit head curves h_{circuit} versus flow rate Q.

where density changes become important and need to be considered. The treatment of compressible flows differs from that of incompressible flows because it needs to include thermodynamic relationships.

1.4.1 IDEAL GAS LAW

The ideal gas law relates pressure, density, and temperature as follows:

$$p = \rho RT \tag{1.78}$$

where p is the pressure, ρ the density, R the specific gas constant and T the absolute temperature.

In ideal gases, the specific heat at constant pressure c_p relates the heat added to the gas at constant pressure to the resulting temperature change:

$$c_p = \frac{\gamma R}{\gamma - 1} \tag{1.79}$$

where γ is the *adiabatic index* (otherwise referred to as *specific heat ratio*).

Specific heat at constant volume c_v describes the heat added to the gas at constant volume to produce a temperature change.

$$c_v = \frac{R}{\gamma - 1} \tag{1.80}$$

c_p and c_v are linked by:

$$c_p = c_v + R; \; c_p/c_v = \gamma \tag{1.81}$$

The specific heats c_p and c_v of ideal gases are constants and do not depend on the temperature. Ideal gases do not change phase and do not have intermolecular forces between particles except during collisions.

1.4.2 SPEED OF SOUND AND MACH NUMBER FOR IDEAL GASES

The speed of sound in an ideal gas is the speed at which small pressure disturbances (sound, or Mach, waves) propagate through it. It can be calculated as[7]:

$$a = \sqrt{\gamma RT} \tag{1.82}$$

The Mach number is a dimensionless quantity that expresses the ratio of the speed of a fluid element to the speed of sound in the surrounding medium:

$$\text{Ma} = \frac{c}{a} \tag{1.83}$$

In subsonic flow $\text{Ma} < 1$ the Mach number must be less than 1 for this flow regime to occur. The pressure disturbances in this flow regime move at a speed that exceeds the speed of the fluid particles. A transonic flow $\text{Ma} \approx 1$ exists near Mach 1 and displays both subsonic and supersonic flow areas within the same field. Supersonic flow occurs when the Mach number exceeds 1.

1.4.3 STAGNATION, OR TOTAL, QUANTITIES FOR IDEAL GASES

Stagnation – often referred to as *total* – quantities are defined as the state variables of a fluid when it is brought to rest isentropically (i.e. without heat transfer or work done and with no entropy change) from its actual state.

Total enthalpy is defined as:

$$h^0 = h + \frac{c^2}{2} \tag{1.84}$$

The total temperature can be derived from the total enthalpy in an ideal gas:

$$T^0 = T + \frac{c^2}{2c_p} \tag{1.85}$$

By substituting $c = \mathrm{Ma} \cdot a$ into the preceding equation, we get:

$$T^0 = T + \frac{(a\mathrm{Ma})^2}{2c_p} \rightarrow T^0 = T + \frac{\gamma R T \mathrm{Ma}^2}{2c_p} \tag{1.86}$$

and finally:

$$T^0 = T\left(1 + \frac{\gamma - 1}{2}\mathrm{Ma}^2\right) \tag{1.87}$$

Stagnation pressure is given by[8]:

$$p^0 = p\left(1 + \frac{\gamma - 1}{2}\mathrm{Ma}^2\right)^{\frac{\gamma}{\gamma - 1}} \tag{1.88}$$

Stagnation density is obtained from $p^0 = \rho^0 R T^0$ as:

$$\rho^0 = \rho\left(1 + \frac{\gamma - 1}{2}\mathrm{Ma}^2\right)^{\frac{1}{\gamma - 1}} \tag{1.89}$$

The stagnation speed of sound is given by:

$$a^0 = \sqrt{\gamma R T^0} \tag{1.90}$$

1.4.4 CRITICAL STATE

The Mach number of fluid flow equals 1 when the flow velocity c matches the local speed of sound a in the medium. The flow reaches its sonic condition when the Mach number is equal to 1. The sonic condition also goes by the name *critical state*. Critical quantities are denoted with *:

$$\text{Critical speed of sound } a^*: \quad a^* = \sqrt{\gamma \frac{p^*}{\rho^*}} \tag{1.91}$$

$$\text{Critical temperature } T^*: \quad T^* = T^0 / \left(\frac{\gamma+1}{2}\right) \tag{1.92}$$

$$\text{Critical pressure } p^*: \quad p^* = p^0 / \left(\frac{\gamma+1}{2}\right)^{\frac{\gamma}{\gamma-1}} \tag{1.93}$$

$$\text{Critical density } \rho^*: \quad \rho^* = \rho^0 / \left(\frac{\gamma+1}{2}\right)^{\frac{1}{\gamma-1}} \tag{1.94}$$

1.4.5 ISENTROPIC FLOW FOR IDEAL GASES

An isentropic flow is a flow process that is both adiabatic and reversible. For an ideal gas undergoing isentropic flow, the following relationships hold:

$$\frac{T_2}{T_1} = \left(\frac{p_2}{p_1}\right)^{\frac{\gamma-1}{\gamma}}; \frac{T_2}{T_1} = \left(\frac{\rho_2}{\rho_1}\right)^{\gamma-1} \tag{1.95}$$

from which:

$$\frac{T_2^0}{T_1^0} = \left(\frac{p_2^0}{p_1^0}\right)^{\frac{\gamma-1}{\gamma}}; \frac{T_2^0}{T_1^0} = \left(\frac{\rho_2^0}{\rho_1^0}\right)^{\gamma-1} \tag{1.96}$$

The mass flow rate can be easily computed in an isentropic flow as a function of the initial stagnation quantities, local cross-sectional area of a streamtube, and local Mach number. Starting from the definition:

$$\dot{m} = \rho c A \tag{1.97}$$

let us use the stagnation density relationship:

$$\rho = \rho^0 \left(1 + \frac{\gamma-1}{2}\text{Ma}^2\right)^{-\frac{1}{\gamma-1}} = \frac{p^0}{RT^0}\left(1 + \frac{\gamma-1}{2}\text{Ma}^2\right)^{-\frac{1}{\gamma-1}}$$

and the local velocity c as a function of the Mach number Ma and the speed of sound a:

$$c = \text{Ma} \cdot a = \text{Ma}\sqrt{\gamma RT}$$

Using the isentropic relations for total temperature and total pressure, we arrive at the following law:

$$\dot{m} = \frac{p^0 A\sqrt{\gamma}}{\sqrt{RT^0}}\text{Ma}\left(1 + \frac{\gamma-1}{2}\text{Ma}^2\right)^{-\frac{\gamma+1}{2(\gamma-1)}} \tag{1.98}$$

The specific mass flow can be defined when describing isentropic flow in a generic duct:

$$\dot{m}_s = \frac{\dot{m}}{A} = \frac{p^0}{\sqrt{T^0}} \cdot \left(\frac{\text{Ma}\sqrt{\gamma}}{\sqrt{R}}\right)\left(1 + \frac{\gamma - 1}{2}\text{Ma}^2\right)^{-\frac{\gamma+1}{2(\gamma-1)}} \tag{1.99}$$

which, for a given gas, can be related to a known, or given, initial state 1, leading to the *corrected specific mass flow rate*:

$$\frac{\dot{m}_s\sqrt{T_1^0}}{p_1^0} = \text{const} \cdot \text{Ma}\left(1 + \frac{\gamma - 1}{2}\text{Ma}^2\right)^{-\frac{\gamma+1}{2(\gamma-1)}} \tag{1.100}$$

The corrected specific mass flow rate in equation 1.100 is shown in Figure 1.4 as a function of the Mach number. The figure shows that the maximum mass flow rate in a duct occurs at a cross-sectional area A where the flow is sonic (Ma $= 1$) for a given inlet stagnation state (p_1^0 and T_1^0). At this point, the duct is said to be *choked* as no further increase in the mass flow is possible. Also, in this condition, the mass flow rate becomes independent of the downstream pressure. To prove this, let us express the specific mass flow rate as a function of the pressure ratio $\frac{p}{p_1^0}$. First, let us work on ρ:

$$\rho = \rho^0\left(\frac{p}{p^0}\right)^{\frac{1}{\gamma}} = \frac{p^0}{RT^0}\left(\frac{p}{p^0}\right)^{\frac{1}{\gamma}} \tag{1.101}$$

and then let us use the energy equation to work on c:

$$c^2 = \frac{2\gamma R}{\gamma - 1}(T^0 - T) = \frac{2\gamma R}{\gamma - 1}\left(T^0 - T^0\left(\frac{p}{p^0}\right)^{\frac{\gamma-1}{\gamma}}\right) = \frac{2\gamma RT^0}{\gamma - 1}\left(1 - \left(\frac{p}{p^0}\right)^{\frac{\gamma-1}{\gamma}}\right) \tag{1.102}$$

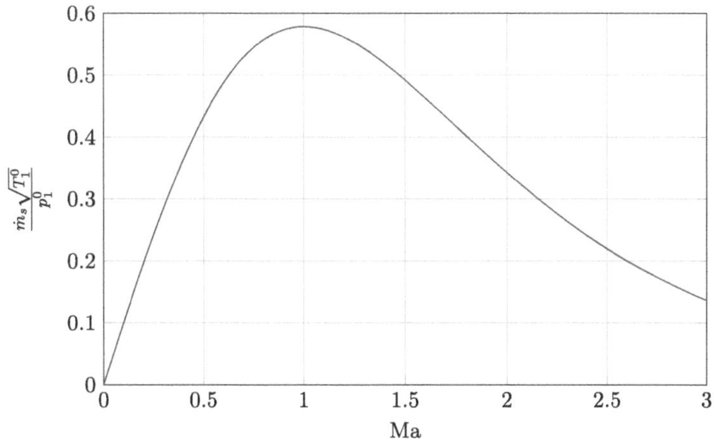

Figure 1.4 Corrected specific mass flow as a function of Mach number (const. = 1; $\gamma = 1.4$).

Now the specific mass flow as a function of the pressure ratio $\frac{p}{p_1^0}$ becomes:

$$\frac{\dot{m}}{A} = \rho c = \frac{p_1^0}{RT_1^0}\left(\frac{p}{p_1^0}\right)^{\frac{1}{\gamma}}\sqrt{\frac{2\gamma}{\gamma-1}RT_1^0\left[1-\left(\frac{p}{p_1^0}\right)^{\frac{\gamma-1}{\gamma}}\right]} \tag{1.103}$$

and the corresponding corrected specific mass flow:

$$\frac{\dot{m}_s\sqrt{T_1^0}}{p_1^0} = \text{const.}\sqrt{\left(\frac{p}{p_1^0}\right)^{\frac{2}{\gamma}}\left[1-\left(\frac{p}{p_1^0}\right)^{\frac{\gamma-1}{\gamma}}\right]} \tag{1.104}$$

Equation 1.104 is plotted in Figure 1.5 as a function of $\frac{p}{p_1^0}$.

Calculus allows us to discover that equation 1.104 has its maximum when $p/p_1^0 = p^*/p_1^0$, i.e. at the critical state, where $\frac{\dot{m}}{A} = (\frac{\dot{m}}{A})_{\max}$:

$$\left(\frac{\dot{m}}{A}\right)_{\max} = \frac{p_0}{\sqrt{T_0}}\sqrt{\frac{\gamma}{R}}\left(\frac{2}{\gamma+1}\right)^{\frac{\gamma+1}{2(\gamma-1)}} \tag{1.105}$$

Despite mathematical findings, however, for $\frac{p}{p_1^0} < \frac{p^*}{p_1^0}$ the choking condition still applies to a generic critical cross section of a duct, such that the flow rate maintains its maximum value. According to the Hugoniot theorem, a critical cross-section must be a throat.[9]

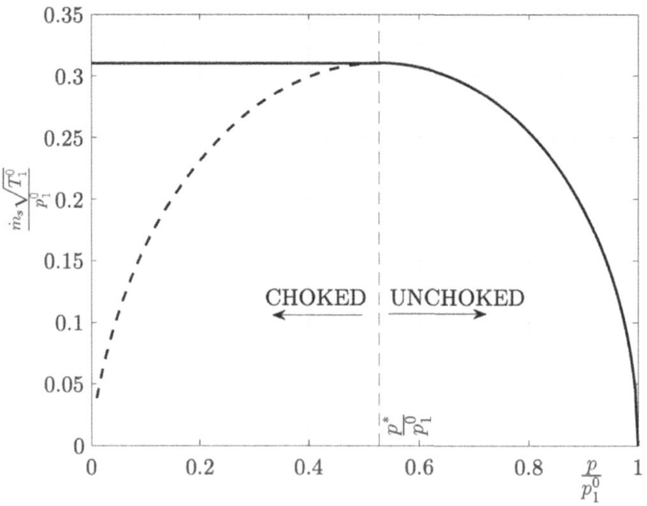

Figure 1.5 Corrected specific mass flow rate versus pressure ratio $\frac{p}{p_1^0}$ in variable area duct.

1.4.6 THE NAVIER-STOKES EQUATIONS IN CYLINDRICAL COORDINATES

The analysis of turbomachinery flows uses cylindrical coordinates because these systems have specific geometric shapes. The Navier-Stokes equations in vector-tensor form appear as follows when body forces are neglected:

$$\frac{D\mathbf{c}}{Dt} = \frac{\partial \mathbf{c}}{\partial t} + (\mathbf{c} \cdot \nabla)\mathbf{c} = -\frac{1}{\rho}\nabla p + \boldsymbol{f}_\mu \tag{1.106}$$

Here, all variables are functions of the cylindrical coordinates and time (r, θ, z, t), e.g. $\mathbf{c} = c_r \hat{\mathbf{i}}_r + c_\theta \hat{\mathbf{i}}_\theta + c_z \hat{\mathbf{i}}_z$. ∇ represents the formal differential operator in cylindrical coordinates and is given by:

$$\nabla = \hat{\mathbf{i}}_r \frac{\partial}{\partial r} + \hat{\mathbf{i}}_\theta \frac{1}{r}\frac{\partial}{\partial \theta} + \hat{\mathbf{i}}_z \frac{\partial}{\partial z}$$

and $\boldsymbol{f}_\mu$ denotes the contribution from viscous forces.

Expanding the vector equation into its components, we have:

$$\begin{aligned}
&\text{Radial:} && \frac{\partial c_r}{\partial t} + c_r \frac{\partial c_r}{\partial r} + \frac{c_\theta}{r}\frac{\partial c_r}{\partial \theta} - \frac{c_\theta^2}{r} + c_z \frac{\partial c_r}{\partial z} = -\frac{1}{\rho}\frac{\partial p}{\partial r} + f_{\mu,r} \\
&\text{Tangential:} && \frac{\partial c_\theta}{\partial t} + c_r \frac{\partial c_\theta}{\partial r} + \frac{c_\theta}{r}\frac{\partial c_\theta}{\partial \theta} + \frac{c_r c_\theta}{r} + c_z \frac{\partial c_\theta}{\partial z} = -\frac{1}{\rho r}\frac{\partial p}{\partial \theta} + f_{\mu,\theta} \\
&\text{Axial:} && \frac{\partial c_z}{\partial t} + c_r \frac{\partial c_z}{\partial r} + \frac{c_\theta}{r}\frac{\partial c_z}{\partial \theta} + c_z \frac{\partial c_z}{\partial z} = -\frac{1}{\rho}\frac{\partial p}{\partial z} + f_{\mu,z}
\end{aligned} \tag{1.107}$$

where terms $-\frac{c_\theta^2}{r}$ and $\frac{c_r c_\theta}{r}$ appear because the cylindrical basis vectors change with θ.[10]

1.4.7 THE NAVIER–STOKES EQUATIONS IN CYLINDRICAL COORDINATES FOR A ROTATING FRAME

To derive the Navier–Stokes equations in a reference frame rotating with angular velocity $\boldsymbol{\omega} = \omega \hat{\mathbf{i}}_z$ about the z-axis, we begin by defining the relationship between the absolute velocity $\mathbf{c}$, the relative velocity $\mathbf{w}$ and the rotational velocity $\mathbf{u}$:

$$\mathbf{c} = \mathbf{w} + \mathbf{u} = \mathbf{w} + \boldsymbol{\omega} \times \mathbf{r} \tag{1.108}$$

Taking the total derivative of $\mathbf{c}$ with respect to time in the inertial frame yields:

$$\frac{D\mathbf{c}}{Dt} = \frac{D}{Dt}(\mathbf{w} + \boldsymbol{\omega} \times \boldsymbol{r}) = \frac{D\mathbf{w}}{Dt} + \frac{D}{Dt}(\boldsymbol{\omega} \times \boldsymbol{r}) \tag{1.109}$$

Since $\boldsymbol{\omega}$ does not vary with time and $\frac{D\boldsymbol{r}}{Dt} = \mathbf{c} = \mathbf{w} + \boldsymbol{\omega} \times \boldsymbol{r}$, we get:

$$\frac{D\mathbf{c}}{Dt} = \frac{D\mathbf{w}}{Dt} + \boldsymbol{\omega} \times \mathbf{w} + \boldsymbol{\omega} \times (\boldsymbol{\omega} \times \boldsymbol{r}) \tag{1.110}$$

Now we substitute the rotating frame form of acceleration in the inertial form given in equation 1.106:

$$\left(\frac{D\mathbf{w}}{Dt}\right)_{\text{rotating}} + 2\boldsymbol{\omega} \times \mathbf{w} + \boldsymbol{\omega} \times (\boldsymbol{\omega} \times \boldsymbol{r}) = -\frac{1}{\rho}\nabla p + \boldsymbol{f}_{\mu}$$

and move the fictitious forces to the right-hand side while taking the total derivative of $\mathbf{w}$ with respect to time:

$$\frac{\partial \mathbf{w}}{\partial t} + (\mathbf{w} \cdot \nabla)\mathbf{w} = -\frac{1}{\rho}\nabla p + \boldsymbol{f}_{\mu} - 2\boldsymbol{\omega} \times \mathbf{w} - \boldsymbol{\omega} \times (\boldsymbol{\omega} \times \boldsymbol{r}) \tag{1.111}$$

Equation 1.111 can be expanded into its r, θ and z components in cylindrical coordinates:

$$\begin{aligned}
\text{Radial:} \quad & \frac{\partial w_r}{\partial t} + w_r \frac{\partial w_r}{\partial r} + \frac{w_\theta}{r}\frac{\partial w_r}{\partial \theta} - \frac{w_\theta^2}{r} + w_z \frac{\partial w_r}{\partial z} = -\frac{1}{\rho}\frac{\partial p}{\partial r} + f_{\mu,r} + 2\omega w_\theta + \omega^2 r \\
\text{Tangential:} \quad & \frac{\partial w_\theta}{\partial t} + w_r \frac{\partial w_\theta}{\partial r} + \frac{w_\theta}{r}\frac{\partial w_\theta}{\partial \theta} + \frac{w_r w_\theta}{r} + w_z \frac{\partial w_\theta}{\partial z} = -\frac{1}{\rho r}\frac{\partial p}{\partial \theta} + f_{\mu,\theta} - 2\omega w_r \\
\text{Axial:} \quad & \frac{\partial w_z}{\partial t} + w_r \frac{\partial w_z}{\partial r} + \frac{w_\theta}{r}\frac{\partial w_z}{\partial \theta} + w_z \frac{\partial w_z}{\partial z} = -\frac{1}{\rho}\frac{\partial p}{\partial z} + f_{\mu,z}
\end{aligned} \tag{1.112}$$

The Coriolis force $-2\boldsymbol{\omega} \times \mathbf{w}$ affects both the tangential and the radial momentum equations, but the centrifugal force $\boldsymbol{\omega} \times (\boldsymbol{\omega} \times \boldsymbol{r}) = \omega^2 r \hat{\mathbf{i}}_r$ appears only in the radial equation.

1.4.7.1 Special Cases: Non-axisymmetrical Radial Flow

The flow type occurs in radial turbomachines where $w_z = 0$, $w_\theta \neq 0$ and $\frac{\partial(\cdot)}{\partial z} = 0$. The conditions apply specifically to purely centrifugal and centripetal machines. The Navier–Stokes equations are presented for steady inviscid (Eulerian) relative flow:

$$(\mathbf{w} \cdot \nabla)\mathbf{w} = -\frac{1}{\rho}\nabla p - 2\boldsymbol{\omega} \times \mathbf{w} + \boldsymbol{\omega} \times (\boldsymbol{\omega} \times \mathbf{r}) \tag{1.113}$$

Given that $w_z = 0$, the velocity $\mathbf{w}$ simplifies to $\mathbf{w} = w_r \hat{\mathbf{i}}_r + w_\theta \hat{\mathbf{i}}_\theta$, such that the Coriolis force expression becomes:

$$\mathbf{F}_{\text{Coriolis}} = -2\omega \hat{\mathbf{i}}_z \times (w_r \hat{\mathbf{i}}_r + w_\theta \hat{\mathbf{i}}_\theta) = -2\omega(w_r \hat{\mathbf{i}}_\theta - w_\theta \hat{\mathbf{i}}_r) \tag{1.114}$$

since: $\hat{\mathbf{i}}_z \times \hat{\mathbf{i}}_r = \hat{\mathbf{i}}_\theta, \hat{\mathbf{i}}_z \times \hat{\mathbf{i}}_\theta = -\hat{\mathbf{i}}_r$

The centrifugal force is given by:

$$\mathbf{F}_{\text{Centrifugal}} = \boldsymbol{\omega} \times (\boldsymbol{\omega} \times \mathbf{r}) = \boldsymbol{\omega} \times (\omega r \hat{\mathbf{i}}_\theta) = \omega r \hat{\mathbf{i}}_z \times \hat{\mathbf{i}}_\theta = -\omega^2 r \hat{\mathbf{i}}_r \tag{1.115}$$

An interesting finding regarding this type of flow relates to the blade-to-blade relative velocity gradient. This can be computed from the tangential flow represented in equation 1.112:

$$w_r \frac{\partial w_\theta}{\partial r} + \frac{w_\theta}{r}\frac{\partial w_\theta}{\partial \theta} + \frac{w_r w_\theta}{r} = -\frac{1}{\rho r}\frac{\partial p}{\partial \theta} - 2\omega w_r \tag{1.116}$$

Assuming $w_\theta = w_\theta(r)$ is constant with respect to θ, so $\frac{\partial w_\theta}{\partial \theta} = 0$, we simplify to:

$$\int_{\theta_1}^{\theta_2} \left(w_r \frac{\partial w_\theta}{\partial r} + \frac{w_r w_\theta}{r} \right) d\theta = \int_{\theta_1}^{\theta_2} \left(-\frac{1}{\rho r}\frac{\partial p}{\partial \theta} - 2\omega w_r \right) d\theta \tag{1.117}$$

where, since w_θ is constant with respect to θ:

$$\int_{\theta_1}^{\theta_2} \frac{w_r w_\theta}{r} d\theta = \frac{w_\theta}{r} \left[w_r \right]_{\theta_1}^{\theta_2} \tag{1.118}$$

while for the right-hand side of equation 1.117:

$$\int_{\theta_1}^{\theta_2} \left(-\frac{1}{\rho r}\frac{\partial p}{\partial \theta} - 2\omega w_r \right) d\theta = -\frac{1}{\rho r} \left[p \right]_{\theta_1}^{\theta_2} - 2\omega w_r (\theta_2 - \theta_1) \tag{1.119}$$

Combining these results:

$$\frac{w_\theta}{r} \left[w_r \right]_{\theta_1}^{\theta_2} = -\frac{1}{\rho r} \left[p \right]_{\theta_1}^{\theta_2} - 2\omega w_r (\theta_2 - \theta_1) \tag{1.120}$$

or:

$$\left(\frac{w_\theta}{r} + 2\omega \right) (w_{r2} - w_{r1}) = -\frac{1}{\rho r} \left(p(\theta_2) - p(\theta_1) \right) \tag{1.121}$$

From the equation, it is straightforward to prove that if $p(\theta_1) > p(\theta_2)$ — as is typically the case for the pressure side of a bladed turbomachine compared to the suction side at the same radius — the difference $w_{r2} - w_{r1}$ must be positive. Given that $w_\theta = w_\theta(r)$ is constant, this implies that the radial component of the relative velocity is minimized on the pressure side and maximized on the suction side at the same radius.

The relative vorticity vector $\mathbf{\Omega_w}$ is defined as the curl of the relative velocity field $\mathbf{w}$:

$$\mathbf{\Omega_w} = \nabla \times \mathbf{w} \tag{1.122}$$

For the cylindrical coordinate system (r, θ, z), the vorticity components $\Omega_{w,r}$, $\Omega_{w,\theta}$ and $\Omega_{w,z}$ are given by:

$$\Omega_{w,r} = \frac{1}{r} \left(\frac{\partial w_z}{\partial \theta} - \frac{\partial (r w_\theta)}{\partial z} \right) \tag{1.123}$$

$$\Omega_{w,\theta} = \frac{\partial w_r}{\partial z} - \frac{\partial w_z}{\partial r} \tag{1.124}$$

$$\Omega_{w,z} = \frac{1}{r} \left(\frac{\partial (r w_\theta)}{\partial r} - \frac{\partial w_r}{\partial \theta} \right) \tag{1.125}$$

For a non-axisymmetrical radial flow, the components simplify to:

$$\Omega_{w,r} = 0; \, \Omega_{w,\theta} = 0; \, \Omega_{w,z} = \frac{1}{r} \left(\frac{\partial (r w_\theta)}{\partial r} - \frac{\partial w_r}{\partial \theta} \right) \tag{1.126}$$

This implies that, in general, as the flow enters the impeller, it begins to acquire a vortical character in the plane normal to the shaft axis and, as a result, can no longer be considered irrotational.

1.4.7.2 Special Cases: Non-axisymmetrical Axial Flow

The flow type occurs in an axial flow rotor when $w_r = 0$. The velocity field reduces to the following form under this condition:

$$\mathbf{w} = w_\theta \hat{\mathbf{i}}_\theta + w_z \hat{\mathbf{i}}_z \tag{1.127}$$

and the Coriolis force becomes:

$$\boldsymbol{F}_{\text{Coriolis}} = -2\omega \hat{\mathbf{i}}_z \times (w_\theta \hat{\mathbf{i}}_\theta + w_z \hat{\mathbf{i}}_z) = -2\omega\,(w_\theta(-\hat{\mathbf{i}}_r) + w_z(0)) = 2\omega w_\theta \hat{\mathbf{i}}_r \tag{1.128}$$

owing to $\hat{\mathbf{i}}_z \times \hat{\mathbf{i}}_\theta = -\hat{\mathbf{i}}_r$ and $\hat{\mathbf{i}}_z \times \hat{\mathbf{i}}_z = 0$.

On the other hand, the centrifugal force is once again given by:

$$\mathbf{F}_{\text{Centrifugal}} = \boldsymbol{\omega} \times (\boldsymbol{\omega} \times \mathbf{r}) = -\omega^2 r \hat{\mathbf{i}}_r \tag{1.129}$$

The Coriolis force contains two components in a purely radial flow because it acts in both the tangential and radial directions through $-2\omega w_r \hat{\mathbf{i}}_\theta$ and $2\omega w_\theta \hat{\mathbf{i}}_r$. The Coriolis force shows one radial component of $2\omega w_\theta \hat{\mathbf{i}}_r$ which operates exclusively in the radial direction during purely axial flow.

The centrifugal force maintains its identical form in both the discussed cases through the expression $-\omega^2 r \hat{\mathbf{i}}_r$ which operates only in the radial direction. The rotational effects impact radial flows more strongly than they affect axial flows. The increased sensitivity leads to greater deviations in the relative velocity direction, which is often referred to as *velocity slip.*

The radial pressure gradient emerges from the radial component of the governing equation 1.113 as a significant result for axial flow:

$$\frac{\partial p}{\partial r} = \rho \omega^2 r \tag{1.130}$$

which is well-known as the *simple radial equilibrium equation.*

In non-axisymmetrical axial flow, the components of vorticity are:

$$\Omega_{w,r} = \frac{1}{r}\left(\frac{\partial w_z}{\partial \theta} - r\frac{\partial w_\theta}{\partial z}\right); \Omega_{w,\theta} = -\frac{\partial w_z}{\partial r}; \Omega_{w,z} = \frac{1}{r}\frac{\partial (r w_\theta)}{\partial r} \tag{1.131}$$

Assuming a *free vortex law*, where $r w_\theta = \text{const.}$ along an axisymmetrical flow path $\left(\frac{\partial}{\partial \theta} = 0\right)$, in Chapter 7 it will be shown, based on the principle of energy conservation, that $\frac{\partial w_z}{\partial r} = 0$. Consequently:

$$\boldsymbol{\Omega}_w = \mathbf{0} \tag{1.132}$$

from which it follows that, in this type of flow, the relative velocity field is irrotational. For all other types of vortex distributions in non-axisymmetrical axial flow where $r w_\theta \neq \text{const.}$ the velocity field is rotational.

1.5 PROBLEMS

1. A tapered horizontal pipe carries water. The inlet (station 1) has an area $A_1 = 0.2\,\text{m}^2$ and velocity $c_1 = 2\,\text{m/s}$, while the outlet (station 2) has an area $A_2 = 0.05\,\text{m}^2$. The pressure at the inlet is $p_1 = 300\,\text{kPa}$. Assuming incompressible flow, determine the pressure p_2 at the outlet. Neglect frictional losses.

   [270 kPa]

2. A velocity field in three-dimensional flow is given by $\mathbf{c} = (2xy\,\hat{\imath} + y^2\,\hat{\jmath} - z\hat{k})\,\text{m/s}$. Calculate the volumetric flow rate Q through a square surface in the y-z plane, defined by $y = 0$ to $y = 2\,\text{m}$ and $z = 0$ to $z = 1\,\text{m}$. Assume $x = 1\,\text{m}$.

   [$4\ \text{m}^3/\text{s}$]

3. Water flows through a stationary curved duct. The inlet and outlet velocities are $c_1 = 10\,\text{m/s}$ and $c_2 = 6\,\text{m/s}$ and the cross-sectional areas are $v_1 = 0.05\,\text{m}^2$ and $v_2 = 0.08\,\text{m}^2$, respectively. The flow is steady and incompressible with $\rho = 1000\,\text{kg/m}^3$. Determine the horizontal force exerted by the fluid on the duct walls.

   [2 kN]

4. Water flows between two parallel plates separated by $b = 1\,\text{cm}$. The top plate moves with a velocity $U = 1\,\text{m/s}$, while the bottom plate is stationary. If $\mu = 0.001\,\text{Pa·s}$, calculate the rate of viscous dissipation per unit volume.

   [$10\ \text{W/m}^3$]

5. Using the dissipation function from Problem 4 ($\Phi = 10\,\text{W/m}^3$), calculate the entropy generation rate per unit volume if the flow occurs at $T = 300\,\text{K}$.

   [$0.0333\ \text{W/(m}^3\text{K)}$]

6. A turbine operates with air entering at a pressure of $p_1 = 8\,\text{MPa}$ and temperature $T_1 = 500\,\text{K}$ and exiting at $p_2 = 0.1\,\text{MPa}$ and $T_2 = 350\,\text{K}$. The mass flow rate is $\dot{m} = 3\,\text{kg/s}$. Assume the process is adiabatic and $c_1 = c_2$. Calculate:
   a. The change in entropy of the air per unit mass.
   b. The power generated by the turbine.

   [0.901 kJ/(kg·K), 452.25 kW]

7. An ideal gas with $c_p = 1.0\,\text{kJ/(kg·K)}$ and $R = 0.287\,\text{kJ/(kg·K)}$ expands isentropically from $p_1 = 2\,\text{MPa}$, $T_1 = 500\,\text{K}$ to $p_2 = 0.2\,\text{MPa}$. Calculate the final temperature T_2 and the work done per unit mass by the gas during the expansion.

   [315.5 K, 461.25 kJ/kg]

8. A gas undergoes isothermal compression from an initial state of $p_1 = 500\,\text{kPa}$ and $T_1 = 400\,\text{K}$ to $p_2 = 1000\,\text{kPa}$. Calculate the entropy change per unit mass of the gas. Use $R = 0.287\,\text{kJ/(kg·K)}$.

   [0.199 kJ/(kg·K)]

9. A water pipeline with a diameter $D_1 = 0.5\,\text{m}$ at the inlet and $D_2 = 1.0\text{m}$ at the outlet experiences a sudden expansion. The flow rate is $Q = 0.5\,\text{m}^3/\text{s}$ and the water enters the pipe with a velocity $c_1 = 2\,\text{m/s}$. The inlet pressure is $p_1 = 500\,\text{kPa}$ and the outlet pressure is $p_2 = 450\,\text{kPa}$. Calculate the head loss due to the sudden expansion.

[51.875 m]

10. Water flows through a 200-meter-long horizontal pipe with a diameter of $D = 0.4\,\text{m}$ and a flow rate of $Q = 0.2\,\text{m}^3/\text{s}$. The pipe material has a roughness coefficient $\varepsilon = 0.01\,\text{m}$. Calculate the head loss due to friction.

[1.16 m]

11. A pump lifts water through a 300-meter-long horizontal pipe with a diameter of $D = 0.2\,\text{m}$. The flow rate is $Q = 0.03\,\text{m}^3/\text{s}$ and the total head loss is due to both friction and concentrated losses, $h_f = 10\,\text{m}$. Calculate the required pump power.

[2.943 kW]

12. A pump is installed in a pipeline between two reservoirs at different elevations. The first reservoir is located at an elevation of $z_1 = 20\,\text{m}$ and the second reservoir is located at $z_2 = 10\,\text{m}$. The pipeline connecting the reservoirs experiences a sudden contraction and several bends. The diameter of the pipe at the inlet is $D_1 = 0.5\,\text{m}$ and at the outlet it is $D_2 = 0.25\,\text{m}$. The length of the pipe is $l = 1000\,\text{m}$ and the flow rate through the system is $Q = 0.25\,\text{m}^3/\text{s}$. The water entering the pipe has a velocity of $c_1 = 2\,\text{m/s}$ and the inlet pressure is $p_1 = 500\,\text{kPa}$. The pressure at the outlet is $p_2 = 450\,\text{kPa}$. The pipe is made of steel, with a roughness coefficient of $\varepsilon = 0.02\,\text{m}$. The flow is turbulent and minor losses are attributed to a sudden contraction with a loss coefficient $K_{\text{contraction}} = 0.5$ and two elbows, each with a loss coefficient $K_{\text{elbow}} = 0.75$. Calculate the required pump power to overcome head losses and deliver water from one reservoir to another.

[8.54 kW]

13. A pumping circuit is used to transfer water through a hydraulic system, governed by the characteristic curves of the pump and the circuit. The characteristic curve of the pump head is described by the equation $h_p(Q) = h_0 - kQ^2$, where $h_p(Q)$ is the head of the pump (in meters), h_0 is the shut-off head (in meters), k is the pump constant (in m^3/s and Q is the flow rate (in m^3/s. The energy losses of the hydraulic circuit are described by the relationship $h_r(Q) = k_f Q^2$, where $h_r(Q)$ represents the loss of head due to friction and other factors (in meters) and k_f is the friction factor (in $\text{m}^{-1} \cdot \text{s}^2$. For the given system, the pump has a shut-off head of $h_0 = 25\,\text{m}$, the pump constant is $k = 5$, and the friction factor is $k_f = 3$. Determine the operating point of the system by finding the flow rate Q and pump head h_p. Furthermore, analyze the system under the following conditions: (i) the flow rate is reduced by 50% and (ii) the friction factor increases to $k_f = 5$. For each condition, calculate the flow rate Q and the pump head h_p.

[1st point: $Q = 1.77\text{m}^3/\text{s}$, $h_p = 9.4\text{m}$; 2nd point: not in equilibrium; 3rd point: $Q = 1.58\text{m}^3/\text{s}$, $h_p = 12.5\text{m}$]

14. Air enters a compressor at a stagnation enthalpy $h_1^0 = 320\,\text{kJ/kg}$ and an inlet velocity $c_1 = 100\,\text{m/s}$. The compressor delivers a specific work input of $|dL| = 90\,\text{kJ/kg}$ and the outlet stagnation enthalpy is h_2^0. Assume the outlet velocity is $c_2 = \alpha c_1$, where α is a scaling factor. Determine h_2^0 and the value of α if the outlet static enthalpy is $h_2 = 370\,\text{kJ/kg}$.

[410 kJ/kg, 0.0894]

15. A multi-stage compressor operates with air entering at a stagnation enthalpy $h_1^0 = 300\,\text{kJ/kg}$ and exiting at $h_3^0 = 500\,\text{kJ/kg}$. The intermediate stage exits with $h_2^0 = 400\,\text{kJ/kg}$. The velocity at the inlet is $c_1 = 50\,\text{m/s}$, at the intermediate stage $c_2 = 100\,\text{m/s}$ and at the outlet $c_3 = 150\,\text{m/s}$. Determine the work input for each stage.

[100 kJ/kg]

16. A convergent nozzle is used to accelerate air from a stagnation state to sonic conditions at the throat $\text{Ma} = 1$. The air has a stagnation pressure $p^0 = 300\,\text{kPa}$, stagnation temperature $T^0 = 500\,\text{K}$ and specific heat ratio $\gamma = 1.4$. The gas constant for air is $R = 287\,\text{J/(kg·K)}$. Determine the specific mass flow rate $\dot{m}/A^*$ through the nozzle as a function of the stagnation pressure p^0 and the specific mass flow rate for the given $p^0 = 300\,\text{kPa}$.

[540 kg/(s m^2)]

17. In a cylindrical coordinate system (r, θ, z), consider non-axis-symmetric axial flow through an axial flow rotor. The flow occurs in a rotating reference frame where the Navier-Stokes equation for the axial velocity w_z is given by:

$$\frac{\partial w_z}{\partial t} + w_\theta \frac{\partial w_z}{r\partial \theta} + w_z \frac{\partial w_z}{\partial z} = -\frac{1}{\rho}\frac{\partial p}{\partial z},$$

where
- w_z is the axial velocity component,
- $w_\theta = r\omega$ is the tangential velocity component (solid body rotation),
- ρ is the density,
- p is the pressure,
- ω is the angular velocity of the rotor.

Assume steady-state flow $\frac{\partial w_z}{\partial t} = 0$ and constant axial pressure gradient $\frac{\partial p}{\partial z} = -G$, where $G>0$.

Simplify the equation for w_z under the given assumptions. Solve for w_z as a function of θ and z, assuming $w_z(\theta, z) = f(\theta) + g(z)$. Discuss how the tangential velocity w_θ influences the axial flow w_z.

NOTES

1. The double dot product

$$\mathbf{A} : \mathbf{B} = \sum_{i,j} A_{ij} B_{ij}$$

is a scalar inner product between two second-order tensors. Also known as the Frobenius inner product, it generalizes the dot product for vectors. In general,

$$\mathbf{A} : \mathbf{B} = \mathrm{tr}\left(\mathbf{A}^T \mathbf{B}\right),$$

and for symmetric tensors this reduces to

$$\mathbf{A} : \mathbf{B} = \mathrm{tr}(\mathbf{AB}).$$

2. The one-dimensional flow model remains useful for many applications, yet fails to capture essential three-dimensional phenomena including boundary layer formation, secondary flow patterns, and complex flow interactions. The model remains restricted to situations where such effects remain minimal or can be handled independently.
3. This does not imply that the fluid is incompressible in a thermodynamic sense (e.g. real fluids such as water are slightly compressible); rather, it indicates that density changes are irrelevant for the purposes of flow analysis.
4. As an example, suppose we want to calculate the entropy change when heating a liquid from temperature T_1 to T_2. When the heat capacity C_q is constant in the temperature range, the entropy change is:

$$\Delta s = C_q \ln\left(\frac{T_2}{T_1}\right)$$

   The integral form must be used when C_q is not constant because it requires numerical methods or specific heat capacity functions.
5. The process of energy exchange in turbomachines that operate with gases, such as compressors, turbines, and fans, is frequently thought to be adiabatic, which implies that there is no heat transfer to or from the environment. Nevertheless, the process is not adiabatic in a number of instances. Most of these non-adiabatic cases involve heat transfer as a result of a variety of factors. As an illustration, gas turbine blades are frequently cooled to preserve their structural integrity at elevated temperatures.
6. The hydraulic diameter D_h is a characteristic length used to describe the flow of fluid even in non-circular conduits. It is defined as:

$$D_h = \frac{4A}{P} \tag{1.133}$$

   where A is the cross-sectional area of the flow and P the wetted perimeter of the conduit. For instance, for a rectangular duct with width b and height H, the hydraulic diameter is given by:

$$D_h = \frac{2bH}{b+H} \tag{1.134}$$

   For flow between two concentric cylinders with inner diameter D_{in} and outer diameter D_{out}, the hydraulic diameter is:

$$D_h = D_{\text{out}} - D_{\text{in}} \tag{1.135}$$

   For a circular pipe, the hydraulic diameter is simply the actual diameter of the pipe:

$$D_h = D \tag{1.136}$$

7. For an ideal gas, the speed of sound a is given by:

$$a = \sqrt{\left(\frac{\partial p}{\partial \rho}\right)_s}$$

where the subscript s indicates the derivative is taken at constant entropy (isentropic process). In an isentropic process:

$$p\rho^{-\gamma} = \text{constant}$$

which takes the natural logarithm of both sides of the isentropic condition:

$$\ln(p) - \gamma \ln(\rho) = \text{constant}$$

and differentiating with respect to ρ becomes:

$$\frac{1}{p}\frac{dp}{d\rho} - \gamma\frac{1}{\rho} = 0 \Longrightarrow \frac{dp}{d\rho} = \gamma\frac{p}{\rho} = \frac{dp}{d\rho} = \gamma\frac{p}{\rho} = \gamma\frac{\rho RT}{\rho} = \gamma RT$$

from which:

$$a = \sqrt{\left(\frac{\partial p}{\partial \rho}\right)_s} = \sqrt{\gamma RT}$$

8. Total pressure p^0 can be approximated by $p^0 \approx p + \frac{\rho c^2}{2}$ for small Mach numbers. In fact, when Ma $\ll 1$, the binomial series expansion for $(1+x)^n$:

$$(1+x)^n \approx 1 + nx + \frac{n(n-1)}{2}x^2 + \cdots$$

can be used. Here, $x = \frac{\gamma-1}{2}\left(\frac{c}{a}\right)^2$ and $n = \frac{\gamma}{\gamma-1}$. Thus:

$$\left(1 + \frac{\gamma-1}{2}\left(\frac{c}{a}\right)^2\right)^{\frac{\gamma}{\gamma-1}} \approx 1 + \frac{\gamma}{\gamma-1} \cdot \frac{\gamma-1}{2}\left(\frac{c}{a}\right)^2 = 1 + \frac{\gamma}{2}\left(\frac{c}{a}\right)^2$$

Multiplying by the static pressure p:

$$p^0 \approx p\left(1 + \frac{\gamma}{2}\left(\frac{c}{a}\right)^2\right)$$

The speed of sound a in a gas is related to the density ρ and pressure p by:

$$a^2 = \frac{\gamma p}{\rho}$$

Therefore:

$$p^0 \approx p + \frac{p}{p} \cdot \frac{\gamma}{2} \cdot \frac{c^2 \rho}{\gamma} = p^0 \approx p + \frac{1}{2} \cdot c^2 \rho = p^0 \approx p + \frac{\rho c^2}{2}$$

9. Choking in compressors and turbines occurs at the narrowest cross-sectional area of the flow path, often at or close to the leading edge of rotor blades in compressors and near the trailing edge of nozzle vanes in turbines.
10. Since unit vectors vary with θ, in a standard right-handed cylindrical frame $\hat{\mathbf{i}}_\theta$ points tangentially, in the direction of increasing θ (i.e. counterclockwise):

$$\frac{\partial \hat{\mathbf{i}}_r}{\partial \theta} = \frac{d}{d\theta}(\cos\theta)\,\hat{\mathbf{i}} + \frac{d}{d\theta}(\sin\theta)\,\hat{\ } = -\sin\theta\,\hat{\mathbf{i}} + \cos\theta\,\hat{\ } = \hat{\mathbf{i}}_\theta$$

$$\frac{\partial \hat{\mathbf{i}}_\theta}{\partial \theta} = \frac{d}{d\theta}(-\sin\theta)\,\hat{\mathbf{i}} + \frac{d}{d\theta}(\cos\theta)\,\hat{\ } = -\cos\theta\,\hat{\mathbf{i}} - \sin\theta\,\hat{\ } = -\hat{\mathbf{i}}_r$$

$$\frac{\partial \hat{\mathbf{i}}_z}{\partial \theta} = 0$$

Next, we compute:

$$\frac{c_\theta}{r}\frac{\partial}{\partial \theta}(c_\theta \hat{\mathbf{i}}_\theta) = \frac{c_\theta}{r}\left(\frac{\partial c_\theta}{\partial \theta}\hat{\mathbf{i}}_\theta + c_\theta \frac{\partial \hat{\mathbf{i}}_\theta}{\partial \theta}\right) = \frac{c_\theta}{r}\left(\frac{\partial c_\theta}{\partial \theta}\hat{\mathbf{i}}_\theta - c_\theta \hat{\mathbf{i}}_r\right)$$

$$\frac{c_\theta}{r}\frac{\partial}{\partial \theta}(c_r \hat{\mathbf{i}}_r) = \frac{c_\theta}{r}\left(\frac{\partial c_r}{\partial \theta}\hat{\mathbf{i}}_r + c_r \frac{\partial \hat{\mathbf{i}}_r}{\partial \theta}\right) = \frac{c_\theta}{r}\left(\frac{\partial c_r}{\partial \theta}\hat{\mathbf{i}}_r + c_r \hat{\mathbf{i}}_\theta\right)$$

2 Introduction to Turbomachines

A turbomachine is a mechanical device that enables continuous energy transfer between fluids and rotating elements, called *rotors*. Rotors have multiple blades arranged in one or more rows that provide energy transfer through dynamic interaction (momentum exchange) with a *working fluid*. For this reason, a turbomachine is referred to as a *rotordynamic machine*.

2.1 CLASSIFICATION OF TURBOMACHINES

The classification of turbomachines follows multiple criteria that facilitate systematic understanding. The criteria incorporate functional principles, type and characteristics of the working fluid, and the mechanisms governing fundamental energy exchange. No single classification criterion is universally comprehensive, as turbomachines often operate across overlapping domains featuring complex interactions between fluid dynamics, thermodynamics, and mechanical design. A commonly accepted classification is shown in Figure 2.1.

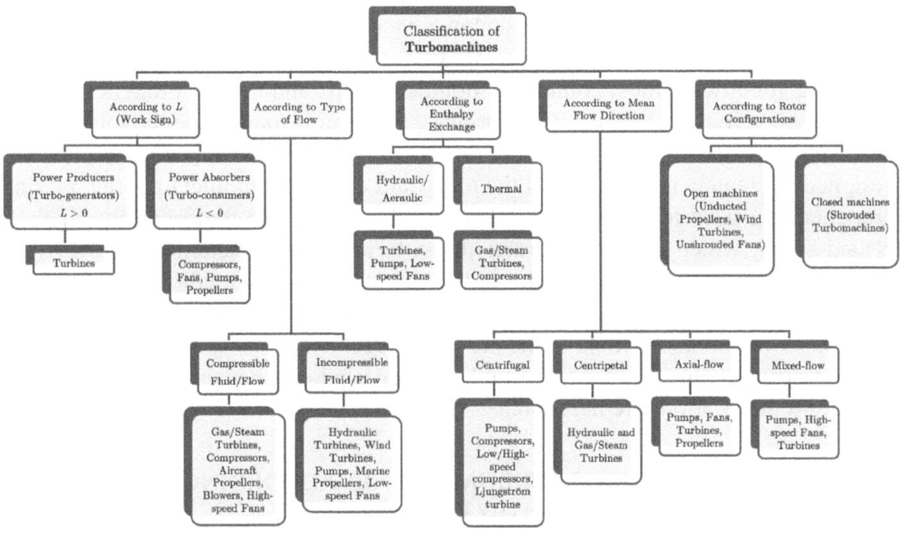

Figure 2.1 Classification of turbomachines.

DOI: 10.1201/9781003619956-2

2.1.1 CLASSIFICATION OF TURBOMACHINES ACCORDING TO SIGN OF L

The sign of work exchange between the fluid and the machine determines the first, fundamental distinction between *power producers* and *power absorbers*.

Power producers, also known as turbo generators, provide work to the environment by converting fluid energy into mechanical work ($L>0$). These devices are generally named *turbines*.

When $L < 0$, machines absorb mechanical work to increase the energy of the working fluid. These include *turbo-pumps*, *turbo-compressors*, and others.[1]

2.1.2 CLASSIFICATION OF TURBOMACHINES ACCORDING TO TYPE OF FLOW

Turbomachines can also be classified according to the nature of the flow, either compressible or incompressible.

Compressible flow turbomachines operate with working fluids whose density varies significantly during the process of energy exchange. Such density variation influences the performance significantly and can by no means be neglected. Among the most notable examples are gas and steam turbines, gas compressors, high-speed propulsive fans, and aircraft propellers. In addition, helicopter rotors, commonly referred to as rotary-wing machines, are in fact a special type of high-speed fan. Instead, low-speed fans and blowers, although handling a compressible fluid, are used primarily for air circulation, not for significant compression, and do not fall under the present category. In fact, blowers achieve only a modest increase in total pressure and, as a result, cause only a small rise in the fluid density that leads engineers to treat them as incompressible flow machines.

Following the reasoning given just above, *incompressible flow turbomachines* are designed to operate with fluids whose density remains nearly constant under typical operating conditions, most commonly liquids. Pumps and marine propellers are an essential class in this category. Their primary function is to increase the total energy of a liquid, predominantly through a rise in pressure. However, while pumps are designed to create a net increase in the total fluid energy from inlet to outlet, marine propellers are used to increase the fluid pressure locally, in the proximity of their bladed screw, which in turn generates thrust to propel a vehicle. Another relevant example of an incompressible flow turbomachine is the hydraulic turbine, which converts the total energy of water into mechanical work. Wind turbines, although operating with air, also belong to the present category. This is justified because wind turbines generally operate at Mach number regimes where changes in air density due to pressure fluctuations are negligible.

2.1.3 CLASSIFICATION OF TURBOMACHINES ACCORDING TO THE ROLE OF ENTHALPY EXCHANGE

Turbomachines can also be broadly categorized according to the predominant mode of energy transfer, specifically as *hydraulic/aeraulic machines* and *thermal machines*.

Hydraulic and aeraulic machines primarily transfer energy through mechanical interaction with fluids that are incompressible or nearly incompressible, such as liquids and low-speed gases. In these machines, the energy exchange is largely mechanical, with negligible thermal effects. Typical examples include hydraulic turbines and wind turbines as well as pumps, fans, blowers, and marine propellers.

Thermal machines transfer energy through thermodynamic processes that involve significant changes in the enthalpy of the working fluid. The operation of these machines relies on the conversion of thermal energy into mechanical work, or vice versa, and is governed by the principles of thermodynamics. Gas and steam turbines are essential examples as they convert the enthalpy of high-temperature, high-pressure gases or steam into mechanical output. Compressors and high-speed fans, and even helicopter rotors—as an extension—also belong to this class.

2.1.4 CLASSIFICATION OF TURBOMACHINES ACCORDING TO PRIMARY FLOW DIRECTION

Turbomachines can also be classified on the basis of the direction of the primary flow within the machine. The main categories include *centrifugal*, *centripetal*, *mixed-flow*, and *axial-flow* turbomachines. For the purpose of this classification, therefore, it is essential to distinguish between the primary flow, which in a turbomachine is the *meridional flow*, and the *tangential flow*.

The meridional flow refers to the principal direction of fluid motion as visualized on a plane passing through the axis of rotation. The tangential flow is visualized in surfaces orthogonal to the previously defined plane, and is primarily defined by the circumferential component of fluid motion caused by the interaction of the flow with either rotating blades or stationary vanes.

Centrifugal turbomachines are characterized by a meridional flow that moves radially outwards from the center of rotation. These machines are especially suited for applications that demand substantial increases in fluid pressure, and include centrifugal pumps, fans, and compressors.

Centripetal turbomachines exhibit a meridional flow directed radially inward toward the center of rotation. An example is the inward flow radial turbine, which converts the energy of a fluid moving radially inward into mechanical work.

Mixed flow turbomachines combine both radial and axial components in the meridional flow path. This configuration allows the machines to exploit the advantages of both radial and axial designs in order to achieve a desired balance between pressure rise and flow rate.

Finally, axial flow turbomachines are distinguished by a meridional flow that is parallel to the axis of rotation. These machines are typically used when high flow rates and relatively low to moderate energy transfer are required.

2.1.5 CLASSIFICATION ACCORDING TO ROTOR CONFIGURATION

Turbomachines may also be classified according to the geometric configuration of their rotor(s), which is(are) typically distinguished as *shrouded* (or ducted) or *un-*

shrouded (or unducted). This distinction is based on whether the rotating component is enclosed within a casing, which contains and directs the fluid flow, or whether it operates in an open environment.

Shrouded rotors (Figure 2.2) are commonly employed in machines where a physical division between internal and external environments is essential to the purpose of the desired operation. The internal environment coincides with the flow domain, where the fluid interacts with the rotors, while the external environment may either itself contain a fluid or not. In any case, external flow does not interact with the shrouded rotors, as happens regularly in centrifugal, centripetal, and mixed-flow machines. In the realm of shrouded rotors, a typical subclass includes those shapes used in centrifugal or centripetal machines: these are called *impellers*, which appear in closed, semi-open, or open configurations. Closed impellers (Figure 2.3) have enclosed vanes on both sides of the hub and shroud, thus offering effective guidance to meridional flow and appropriate mechanical strength, particularly in high-pressure applications. Open impellers have open blades on both sides without any protective shroud and are secured to a hub that provides the rotation: they are preferred to shrouded designs when smaller, more inexpensive configurations, often handling slurry fluids, are required. Semi-open impellers have a back-wall shroud that keeps the other side open while giving the vanes more mechanical strength: in terms of gen-

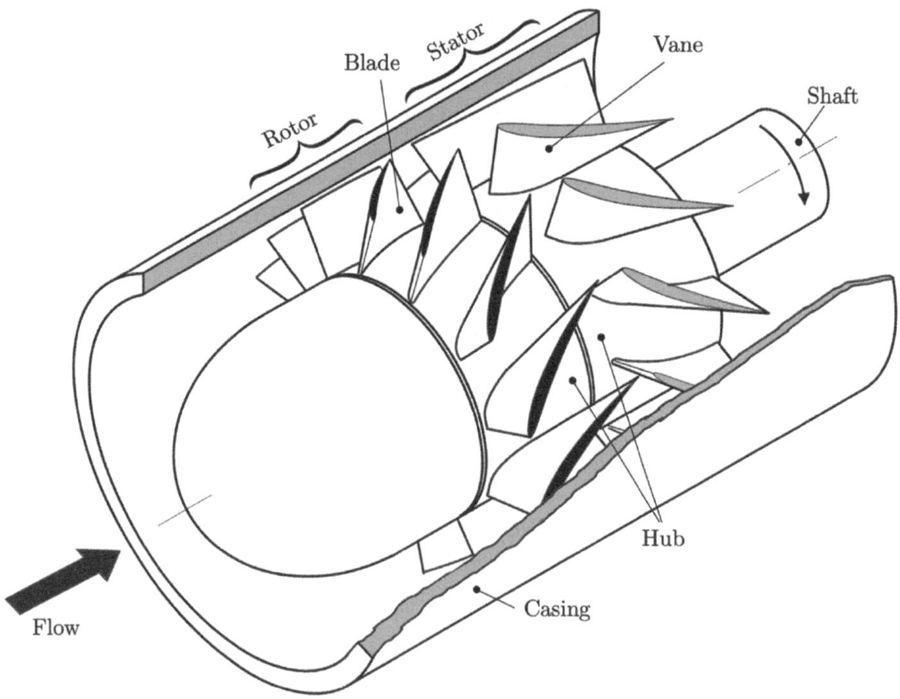

Figure 2.2 Schematic cutaway view of a shrouded turbomachine.

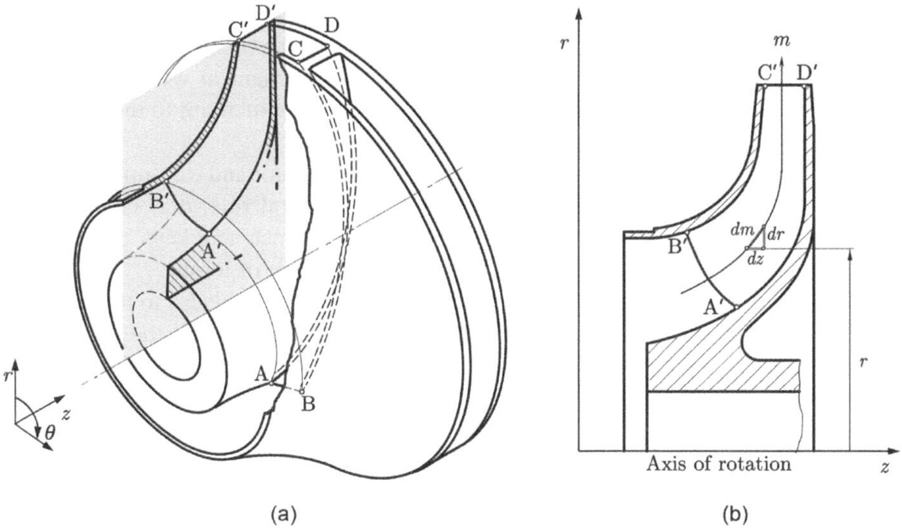

Figure 2.3 Meridional view of a turbomachine flow passage. (a) 3D visualization. (b) 2D schematic of the meridional plane illustrating the main flow path.

eral characteristics, they fall somewhere between open and closed impellers, which makes them appropriate for medium-sized designs where the fluid has a mild degree of impurity.

Unshrouded rotors operate without a surrounding casing, where the flow is far less constrained and large volumetric flow rate or direct interaction with the environment is desired. Typical examples include open aircraft- or marine propellers, open wind turbines, and helicopter rotors, which all accommodate an axial-flow. The unshrouded design allows for simpler construction and a deliberate need for interaction with the open environment, albeit with less control over the fluid path compared to shrouded configurations.

2.2 CONSTRUCTIVE ELEMENTS OF TURBOMACHINES

A turbomachine is made up of several essential components, each serving a specific function to ensure efficient energy transfer between the fluid and the bladed system. The key constructive elements are briefly described below (Figure 2.2).

2.2.1 ROTOR

As already stated, the rotor is the central rotating component of a turbomachine and is responsible for direct energy transfer with the working fluid. Typically, the rotor is composed of one or more sets of blades fastened to a central hub, the latter being mounted on a rotating shaft.

The blades are the specific elements that enable energy exchange between the fluid and the machine. In turbines, the working fluid imparts angular momentum to the blades, causing the rotor to rotate and generate mechanical work. In contrast, in compressors and pumps, the blades transfer angular momentum to the fluid, thus resulting in an increase in the total energy of the fluid.

The shaft serves as the central axis of rotation of the rotor and transmits rotational motion either to or from the turbomachine. As a general rule, it is mechanically coupled to the hub by means of removable links (for instance, keys or keyways) providing the necessary structural coupling for the transfer of torque.

The hub acts as a central support structure, connecting the blades to the shaft and distributing the mechanical loads imposed by both the blades and the fluid forces and moments encountered during operation. It is designed to withstand centrifugal and aerodynamic forces and ensure the needed structural integrity and stiffness of the rotor assembly.

2.2.2 STATOR

The stator is the stationary part of the turbomachine, which often includes guide vanes or stationary blades. In centrifugal machines, the stator may include a volute, i.e. a spiral casing that converts the kinetic energy of the flow into pressure energy (in pumps and compressors) or vice versa (in turbines).

Guide vanes direct the flow to or out of the rotors. In turbines, guide vanes are typically placed upstream of the rotor to direct the flow at a desired angle for energy transfer, while in compressors and pumps, they are generally used downstream of the rotor to recover flow static pressure from its kinetic energy. Guide vanes are usually secured to the casing and, in cases where the hub is fixed, they can also be attached to the hub itself (Figure 2.2), which ensures structural support and supplementary flow guidance.

The *casing* constitutes the outer enclosure that supports and contains all the internal components of the turbomachine. In centrifugal pumps and compressors, the volute is actually a part of the casing.

2.2.3 ADDITIONAL ELEMENTS

Turbomachines incorporate several auxiliary elements in addition to the above mentioned components.

Bearings support the rotating shaft ensuring both precise dynamic alignment and structural support, smooth rotation, and effective reaction to both radial and axial loads encountered during service.

Seals are employed to prevent leakage of the working fluid between the high-pressure and low-pressure regions within the turbomachine. In doing this, they help preserve operational efficiency and protect the machine against fluid loss or contamination.

Inlet and outlet structures are designed to guide the working fluid as it enters and exits the turbomachine. These elements may include nozzles and diffusers and serve also as matching components between the machine and the plant where it operates.

2.3 GEOMETRIC REPRESENTATION OF TURBOMACHINES

To obtain an accurate and comprehensible representation of a turbomachine, it is necessary to utilize two distinct families of planes: the *meridional plane* and a number of *blade-to-blade planes*.

The meridional plane provides a view of the machine's cross section along the axis of rotation; as such, it makes it possible to detail the path of the primary flow and visualize the fundamental contours of the bladed/vaned rows from the hub to the shroud.

The blade-to-blade planes are obtained from two-dimensional unwrapped views of the curved surfaces generated by revolution about the rotational axis. The generatrix, which can be selected to mimic the trajectory of the primary flow in the meridional plane, defines the locus where the blade-to-blade planes are constructed. At each point along this generatrix, the blade-to-blade plane is geometrically orthogonal to the meridional plane. As such, these planes slice through the blade rows and reveal the geometry of blades' profile.

2.3.1 THE MERIDIONAL PLANE

The meridional plane represents an equivalent cross-sectional view of the structural elements of the turbomachine, extending from the hub to the shroud. This plane contains the axis of rotation and captures the radial and axial dimensions of the turbomachine. Therefore, it provides a longitudinal section that illustrates the path through which the working fluid travels.

The meridional plane does not coincide with a literal cross-section of the turbomachine obtained by slicing the machine along a plane that passes through the rotational axis. In fact, the blade view displayed on the meridional plane is obtained by means of a circumferential projection of the actual blade shapes onto this plane, thus providing a simplified yet comprehensive view of the blade arrangements and flow path (the gray-colored plane in Figure 2.3). The elements visible on the meridional plane of a turbomachine include the hub contour, the shroud contour, the leading and trailing edges of rotor blades and stator vanes, if present.

2.3.2 THE BLADE-TO-BLADE FAMILY OF PLANES

The blade-to-blade family of planes is a geometric representation used to analyze the flow characteristics between successive blades within a blade row. The location and number of these planes depend on the type of turbomachine, as detailed in the following.

- **Radial machines with purely radial, single-curvature blades.** A centrifugal machine of purely radial single-curvature bladed arrangement needs just one blade-to-blade plane to be uniquely described (Figure 2.4). This plane is positioned perpendicular to the axis of rotation and slices through the impeller capturing the cross-sectional profile of the blades at a single z coordinate. More specifically, a single-curvature rotor refers to a design in which the blade surfaces are

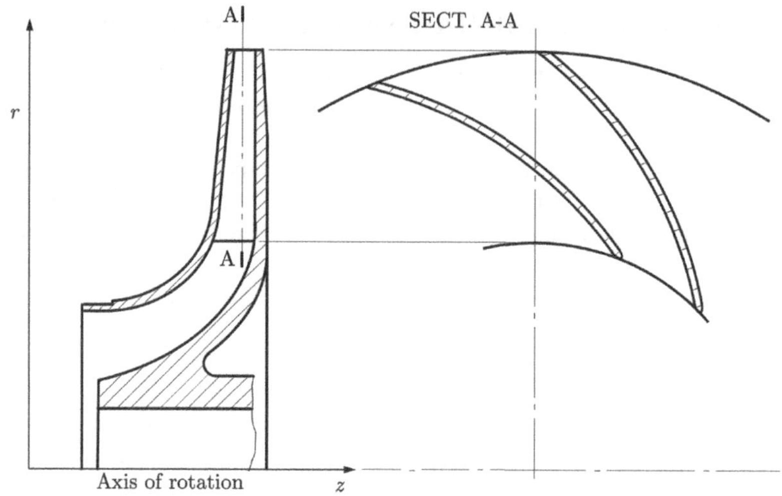

Figure 2.4 Meridional and blade-to-blade views of a purely radial turbomachine.

shaped such that their cross-sectional profile exhibits no variation in blade curvature along the span from the hub to the shroud.

Single blade-to-blade plane at $z = z_0$: $\perp$ to the axis of rotation

- **Radial machines with double-curvature blades.** When the blade configuration of the impeller transitions from a purely radial to a double-curvature geometry, the blade shapes become three-dimensional. This requires multiple blade-to-blade planes to fully capture the blade geometry. The conformal geometric transformations of the coordinates are used for this purpose (Figure 2.7). This approach introduces the conformal plane, where the geometry of the blade profiles is mathematically related to the actual coordinates measured on a closed surface of rotation whose generatrix is a proper curve – often an approximate, educated guess of a meridional streamline m – drawn in the meridional plane (Figure 2.3b).

Conformal transformation: $(r, \theta, z) \to (m', \theta)$

- **Axial flow machines.** In axial flow machines, the use of cylindrical sections of the blade at various radii is customary. These cylindrical sections are then unwrapped onto a plane and provide a clear, detailed view of the blade profiles at different heights along the span.

Cylindrical Sections taken at radii r_i for $i = 1, 2, \ldots, n$

2.3.3 THE CONFORMAL TRANSFORMATION OF COORDINATES

The conformal transformation of coordinates maps the three-dimensional cylindrical coordinates (r, θ, z) of a generic point within the flow domain onto a conformal

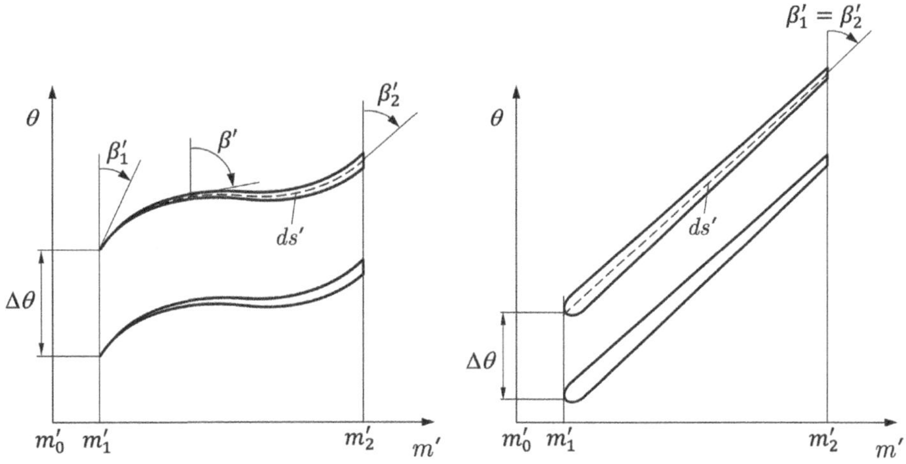

Figure 2.5 Blade-to-blade view in the conformal (m', θ) plane showing blade passage geometry.

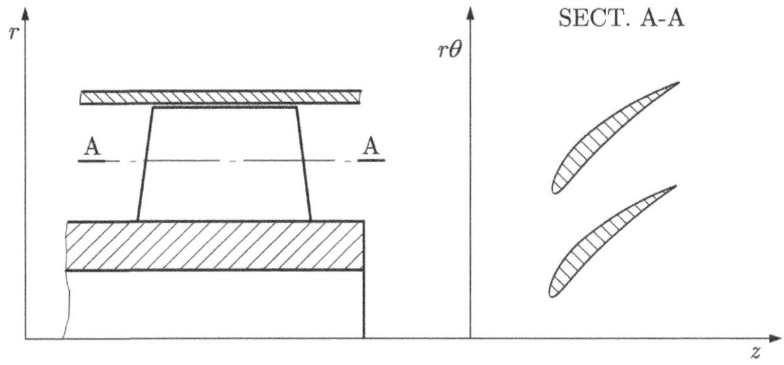

Figure 2.6 (Left) Meridional view of an axial flow turbomachine. (Right) Cross-section in the $(r\theta, z)$ plane at cylindrical section A–A.

plane (m', θ). The conformal plane coincides with a streamsurface if m' is defined by the local radial projection normalized with respect to r of a 3D streamline onto the meridional plane (Figure 2.5). The transformation preserves local angles and key shape characteristics of the original blade profile to some extent. In particular, m' represents the normalized arc length increment along the stream surface and θ is the circumferential angle.

The transformation equations are defined as follows (Figure 2.7):

$$m' = m_0' + \int \frac{\sqrt{dr^2 + dz^2}}{r} \tag{2.1}$$

$$\theta = \theta \tag{2.2}$$

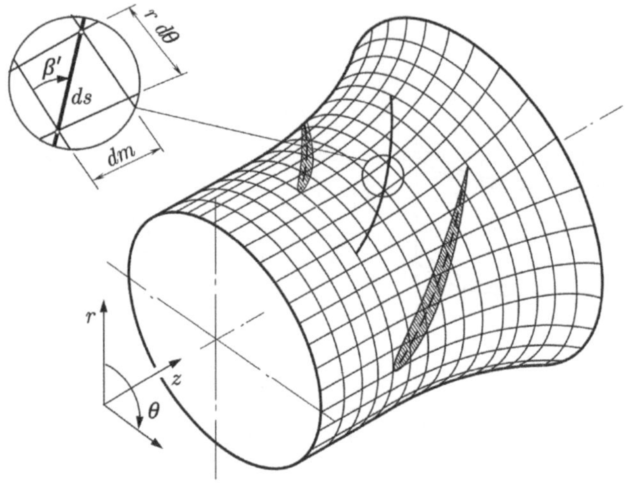

Figure 2.7 Three-dimensional representation of a turbomachine blade passage using a conformal coordinate grid.

where dr and dz represent infinitesimal changes in the radial and axial directions, respectively, while m_0' is an arbitrary integration constant. Following this, the total arc length increment is:

$$ds = \sqrt{dr^2 + dz^2 + (r\,d\theta)^2} = \sqrt{dm'^2 + (r\,d\theta)^2} = r\sqrt{dm'^2 + d\theta^2} = r\,ds' \quad (2.3)$$

The local blade angle (Figure 2.7) can be computed as follows:

$$\beta' = \tan^{-1}\left(\frac{dm'}{d\theta}\right) \quad (2.4)$$

and the blade circumferential wrap angle, $\Delta\theta$ (Figure 2.5), is given by:

$$\Delta\theta = \int_{r_1}^{r_2} \left(\frac{\partial\theta}{\partial r}\right)_z dr \quad (2.5)$$

For 2-D cascades used in axial flow machines, the conformal plane coincides with a cylindrical surface unwrapped on a plane, while r denotes the specific radius at which the cylindrical section is taken:

$$m' = \frac{z}{r} \quad \text{(axial cascade at } r = \text{const.)} \quad (2.6)$$

For a purely radial cascade, e.g. a purely radial outflow pump or a purely radial inflow turbine:

$$m' = m_0' + \int \frac{dr}{r} = m_0' + \ln r \quad \text{(radial outflow cascade)} \quad (2.7)$$

$$m' = m'_0 + \int \frac{-dr}{r} = m'_0 + \ln\frac{1}{r} \quad \text{(radial inflow cascade)} \tag{2.8}$$

The superposition of several conformal planes and the adjustment of the value of m_0 facilitate the visual inspection and alignment of the blade profiles.

The inverse, or anti-conformal, transformation is obtained by solving for r and z in terms of m' and θ. Since $ds = r\,ds'$, the axial coordinate z is calculated as:

$$dz = ds'\frac{z}{\sqrt{dm'^2 + d\theta^2}} \implies z(\theta) = z_0 + \int \frac{ds'}{\sqrt{dm'^2 + d\theta^2}} \tag{2.9}$$

The radial coordinate r is derived from $m' = \int \frac{\sqrt{dr^2 + dz^2}}{r}$:

$$dr = dm'\frac{r}{\sqrt{dm'^2 + d\theta^2}} \implies r(\theta) = r_0 + \int \frac{dm'}{\sqrt{dm'^2 + d\theta^2}} \tag{2.10}$$

Finally, the circumferential coordinate θ is simply given by:

$$\theta = \theta \tag{2.11}$$

2.4 ANALYTIC GEOMETRY OF TURBOMACHINE ELEMENTS

The geometry of turbomachine elements is defined using analytic methods that enable precise control over shape and ensure smooth transitions across component surfaces. Parametric and explicit curves are employed to describe key features such as the camber line, thickness distribution, and spatial orientation of the blade. Polynomials and parametric curves, in particular Bézier curves, provide a flexible framework for modeling smooth profiles with straightforward control over curvature and slope.

Camber lines are often constructed in a conformal plane to simplify aerodynamic analysis and alignment of blades with the flow. Thickness distributions are defined as mathematical functions along the chord and ensure consistency in profile generation. Blade lean introduces controlled variation of blade orientation along the span, a method to link geometric design to aerodynamic performance and structural response.

Conformal mapping allows bi-univocal correspondence between coordinates on construction planes (meridional and blade-to-blade) and those in the three-dimensional cylindrical frame.

2.4.1 BÉZIER CURVES

Bézier curves provide a robust method for representing the hub and shroud contours, as well as for generating blade profiles.

A cubic Bézier curve in 2D is given by:

$$\mathbf{B}(t) = (1-t)^3\mathbf{P}_0 + 3(1-t)^2 t\mathbf{P}_1 + 3(1-t)t^2\mathbf{P}_2 + t^3\mathbf{P}_3 \tag{2.12}$$

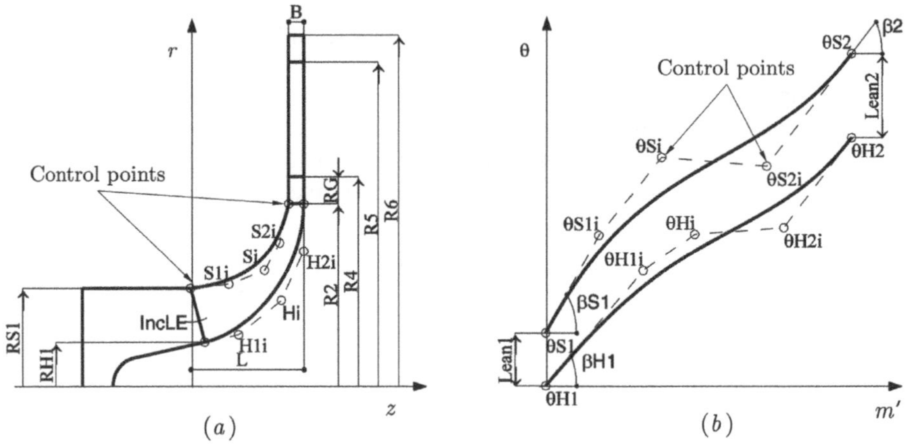

Figure 2.8 Blade geometry parameterization using Bézier curves of a centrifugal compressor. (a) Meridional view (r, z) showing the blade profile shaped by Bézier control points along the hub and shroud. (b) Conformal view of the impeller hub and shroud camber lines.

where $\mathbf{P}_0, \mathbf{P}_1, \mathbf{P}_2, \mathbf{P}_3$ are 2×1 vectors containing the coordinates of control points. These curves can be used to define the shapes of the hub and shroud contours, respectively (Figure 2.8a).

Intermediate contours from hub and shroud ones, often referred to as *layers*, are generated by interpolating between the hub and shroud curves. For a given spanwise position s, the blade profile $\mathbf{B}(t,s)$ of the layer is:

$$\mathbf{B}(t,s) = (1-s)\mathbf{H}(t) + s\mathbf{S}(t) \tag{2.13}$$

where $\mathbf{H}(t)$ and $\mathbf{S}(t)$ are Bézier curves defining the hub and the shroud, respectively.

Bézier curves are widely used for the representation of blade section geometry as well (Figures 2.8b and 2.9).

2.4.2 CAMBER LINES IN THE CONFORMAL PLANE

Other than Bézier curves, camber lines on the conformal plane can be represented using polynomial equations:

$$y_c(x) = \sum_{i=0}^{n} a_i x^i \tag{2.14}$$

where a_i are coefficients determined based on the desired shape of the camber line and n is the degree of the polynomial.

2.4.3 EXAMPLES OF CAMBER LINES

- **NACA 4-Digit camber line for axial flow machines.** NACA (National Advisory Committee for Aeronautics) airfoil camber lines are defined by specific polynomial equations. For a NACA 4-digit airfoil, the camber line $y_c(x)$ is typically given

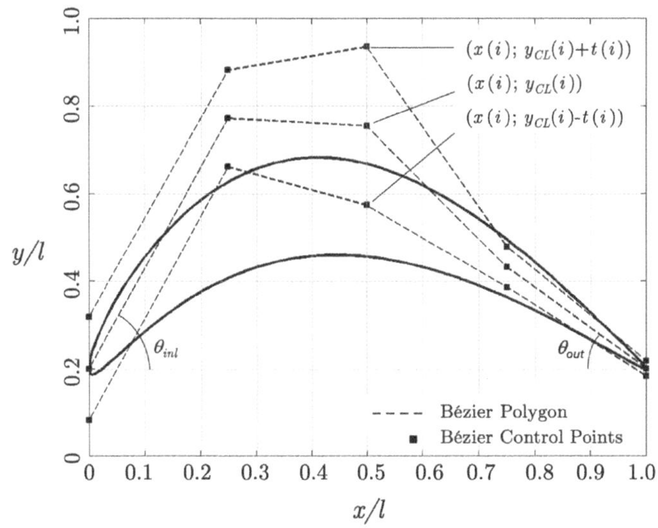

Figure 2.9 Geometry parameterization of an airfoil using Bézier curves. Squares represent control points.

by:

$$y_c(x) = m\left(\frac{x}{\ell}\right)^k\left[1-\left(\frac{x}{\ell}\right)^n\right] \tag{2.15}$$

where x is the chordwise position, ℓ is the chord length and m, k, n are parameters defining the camber distribution. An example of a 4-digit airfoil with a parabolic camber line is given in Figure 2.10.

- **Circular, multi-circular, or logarithmic camber line for radial machines**. Radial machines, such as centrifugal pumps often use circular, multi-circular or logarithmic camber lines. For example:
  - Circular camber line (Figure 2.11a):

$$y_c(x) = R_c - \sqrt{R_c^2 - x^2} \tag{2.16}$$

where R_c is the radius of the circular arc.
  - Multi-circular camber line (Figure 2.11b and c):

$$y_c(x) = \sum_{i=1}^{N}(R_{ci} - \sqrt{R_{ci}^2 - x^2}) \tag{2.17}$$

where R_{ci} is the radius of i-th individual circular arc. Multi-circular camber lines are widely used in the aerodynamic design of radial pumps, as well as of axial flow compressors and axial flow turbine blades to fine-tune the pressure distribution and control the flow characteristics.
  - Logarithmic camber line (Figure 2.11d):

$$y_c(x) = A\ln(x) + B \tag{2.18}$$

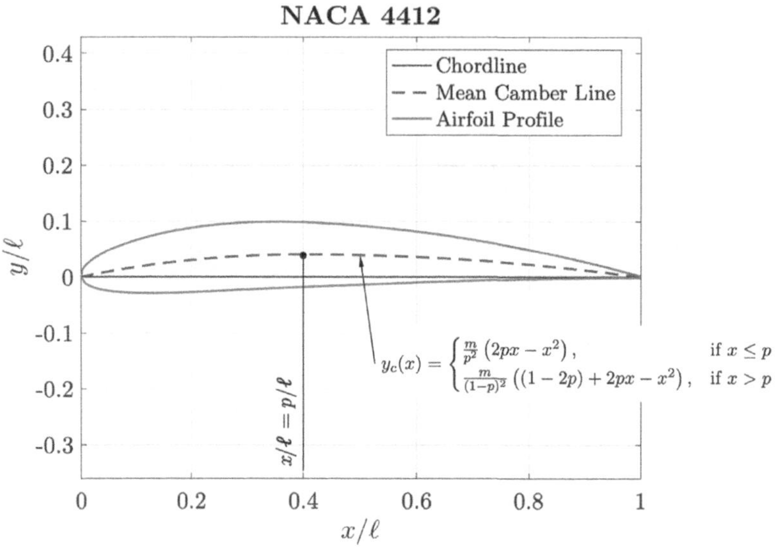

Figure 2.10 Geometric construction of the NACA 4412 airfoil.

where A and B are constants defining the logarithmic spiral. Logarithmic camber lines have been adopted in the design of centrifugal compressor and pump blades to achieve favorable flow guidance and reduced flow separation.

- **Higher-order polynomial camber line for centrifugal compressors and centripetal turbines**: Higher-order polynomial equations are frequently used, which provides great flexibility in tailoring the blade profile for optimal aerodynamic performance. A general form of such a camber line is given by:

$$y_c(x) = \sum_{i=0}^{N} a_i x^i \tag{2.19}$$

where a_i are coefficients that determine the specific camber line shape. For example, fifth-order polynomials have demonstrated their usefulness in the design optimization of the camber line of radial flow machines.

2.4.4 THICKNESS DISTRIBUTION

Thickness distributions can be represented using polynomial equations on the $m - r\theta$ plane. These equations define the variation in thickness along the m direction, or equivalently along the chord direction x of the blade sections.

For a NACA 4-digit airfoil, the thickness distribution $t(x)$ is typically given by:

$$t(x) = t_c \left[0.2969\sqrt{\frac{x}{\ell}} - 0.1260\left(\frac{x}{\ell}\right) - 0.3516\left(\frac{x}{\ell}\right)^2 + 0.2843\left(\frac{x}{\ell}\right)^3 - 0.1015\left(\frac{x}{\ell}\right)^4\right] \tag{2.20}$$

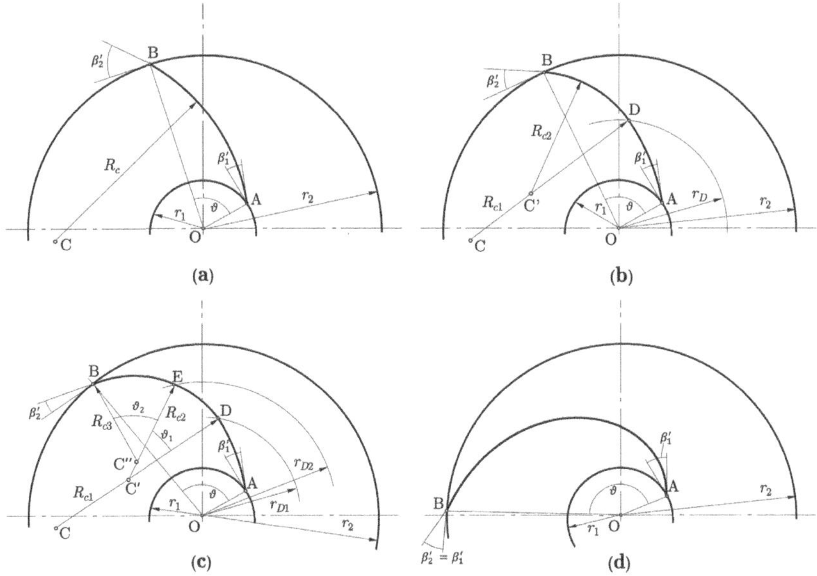

Figure 2.11 Geometric constructions of blade camber lines using circular arcs. (a) Single-arc camber with radius R_c. (b) Two-arc camber split at point D with radii R_{c1} and R_{c2}. (c) Three-arc camber. (d) Logarithmic camber.

where x is the chordwise position, ℓ the chord length and t_c the maximum thickness of the airfoil.

A constant thickness distribution $t(x) = t_0$ is frequently used in radial flow machines like centripetal turbines and centrifugal compressors since it makes manufacturing easier and guarantees structural integrity under high centrifugal forces. Under high rotational speeds, a consistent thickness preserves balance and lessens stress concentrations. Furthermore, general blade shape and flow guidance have a greater impact on aerodynamic performance in these machines than precise airfoil profiles.

2.4.5 BLADE LEAN

Lean angles are defined as:

$$\tan \xi_r = r\left(\frac{\partial \theta}{\partial r}\right)_z; \quad \tan \xi_z = r\left(\frac{\partial \theta}{\partial r}\right)_r \tag{2.21}$$

The angle ξ_r is zero for blades made entirely of radial fibers, and is typically small for the rows of the axial blade rotors, as well as for the leading edges of centrifugal impellers (Figure 2.8b) or the trailing edges of radial turbines. ξ_z corresponds to the blade camber angle in an axial blade element or near the leading edge of a typical radial impeller and the trailing edge of a radial turbine impeller. In an axial blade row, ξ_z is referred to as the blade inlet angle at the entry and the blade outlet angle at the exit.

It should be noted that there is a relationship among the lean angles, the local blade angle β' and the local streamline inclination angle, $\kappa = dr/dz = (dr/dm)(dm/dz)$:

$$\tan\beta' = r\frac{d\theta}{dm} = r\left(\frac{\partial\theta}{\partial r}\right)_z \frac{dr}{dm} + r\left(\frac{\partial\theta}{\partial z}\right)_r \frac{dz}{dm} \tag{2.22}$$

or:

$$\tan\beta' = \tan\xi_r \sin\kappa + \tan\xi_z \cos\kappa \tag{2.23}$$

Blade lean can improve aerodynamic performance while also allowing for better structural load management and vibrational behavior. However, excessive or poorly designed lean may introduce undesirable three-dimensional flow effects, increase mechanical complexity, and potentially lead to performance penalties due to off-design operation or manufacturing challenges.

2.5 PROBLEMS

1. The hub and shroud contours of a compressor rotor are defined as follows: hub contour starts at $z = 0.0\,\text{m}, r = 0.05\,\text{m}$ and ends at $z = 0.3\,\text{m}, r = 0.08\,\text{m}$; the shroud contour starts at $z = 0.0\,\text{m}, r = 0.1\,\text{m}$ and ends at $z = 0.3\,\text{m}, r = 0.12\,\text{m}$. Compute the arc length of the hub contour.

   [0.3015 m]

2. In a mixed-flow pump, the hub starts at $r = 0.1\,\text{m}, z = 0.0\,\text{m}$ and ends at $r = 0.2\,\text{m}, z = 0.3\,\text{m}$. The shroud contour starts at $r = 0.15\,\text{m}, z = 0.0\,\text{m}$ and ends at $r = 0.25\,\text{m}, z = 0.3\,\text{m}$. Compute the conformal coordinate (m') for the hub at $z = 0.3$.

   [≈ 2.19]

3. The blade angle for a turbine is defined by:

   $$\tan\beta' = 0.3 + 0.2m', \quad m' \in [0, 0.5]$$

   Calculate the wrap angle ($\Delta\theta$).

   [0.175 rad]

4. The hub contour of an inflow turbine is described in the meridional plane by the parametric equations:

   $$r(z) = 0.05 + 0.02z^2, \quad z \in [0, 0.3]$$

   Using the trapezoidal rule, numerically compute the conformal coordinate m' at $z = 0.3$, starting from $m'_0 = 0$.

   [5.93]

5. The hub contour of a radial inflow turbine is described in the meridional plane by a constant axial coordinate ($z = 0.2\,\text{m}$) and a radial coordinate r varying from $r_1 = 0.3\,\text{m}$ to $r_2 = 0.1\,\text{m}$. Assume $m'_0 = 0$, the blade angle at r_1 is $\beta'_1 = 90^\circ$ and

the angular wrap is $\Delta\theta = \frac{\pi}{3}$ rad $= 60°$. Compute the conformal coordinate m' at the end of the cascade (r_2) and blade angle $\beta_2'(r_2)$.

[2.30, 46.37°]

6. In an axial flow pump rotor, blade angles vary linearly along the camber line with the following values $\beta_1' = 65°$(at $z_1 = 0.2$ m), $\beta_2' = 50°$(at $z_2 = 0.4$ m). Assume the blade lies at a constant radius $r = 0.5$ m and blade angle β' varies linearly with z. Compute the blade camber line length (L) assuming a linear variation of β' along z. Next, assuming a blade wrap angle of $\Delta\theta = 10°$ and a mean radius of 0.15 m, adjust the camber line length to incorporate the wrap angle.

   [0.379 m, 0.380 m]

7. The leading edge of a centrifugal compressor blade is defined using a cubic Bézier curve. The control points for the leading edge in the radial-axial plane r-z are $\mathbf{P}_0 = (10,0)$ mm (hub radius at the leading edge), $\mathbf{P}_1 = (15,5)$ mm, $\mathbf{P}_2 = (20,8)$ mm, $\mathbf{P}_3 = (25,10)$ mm (shroud radius at the leading edge). Compute the coordinates of the leading edge at $t = 0.25, 0.5, 0.75$.

   [(13.75, 3.4) mm, (17.5, 6.2) mm, (21.25, 8.3) mm]

8. The hub and shroud contours of an impeller are defined as cubic Bézier curves in the z-r plane. The control points for the hub and shroud contours are as follows: $\mathbf{P}_h = [(0,10),(10,10),(15,16),(15,30)]$ mm; $\mathbf{P}_s = [(0,20),(7,20),(11,26),(11,30)]$ mm. Plot the hub and shroud contours over $t \in [0,1]$. Calculate the coordinates of the point on the intermediate layer located at span $s = 0.5$ and $t = 0.5$.

   [(9.7, 19.13) mm]

9. A purely radial cascade extends from $r_1 = 0.1$ m to $r_2 = 0.5$ m, with blades arranged at a constant axial position z = const. The camber line follows a logarithmic shape, i.e. the blade angles at the inlet and outlet are constant: $\beta_1' = \beta_2' = 30°$. The actual blade thickness $t = 0.01$ m is constant in the (θ, m) plane. Plot the camber line both in cylindrical (r, θ) and in conformal coordinates (m', θ). Compute the blade angular thickness $\Delta\theta$ at $r_1, r = 0.2$ m, $r = 0.3$ m, and r_2.

   [$\Delta\theta(r_1) = 0.1$ rad, $\Delta\theta(r = 0.2) = 0.05$ rad, $\Delta\theta(r = 0.3) \approx 0.033$ rad, $\Delta\theta(r_2) = 0.02$ rad]

NOTE

1. From this point onward, we shall omit the prefix *turbo-* and use simply the words *pumps* and *compressors* when referring to power absorber turbomachines.

3 One-Dimensional Theory of Turbomachines

The one-dimensional (1D), or *Eulerian*, theory stands as a fundamental framework for analyzing turbomachine performance. It provides a simplified yet effective analytical framework to study fluid flow and energy transfer in these devices. The theory bases its analysis on the assumption that fluid properties, such as velocity, pressure, or density, change only along the direction of the primary flow, wherein the principles laid out in Chapter 1 are applied.

3.1 DEFINITIONS

Assume we define a local system of coordinates (m,n) along a meridional flow trajectory with an origin at point P, where m denotes the coordinate along the flow direction and n is the coordinate normal to m. Let ϕ be a generic scalar flow quantity, such as a velocity component, pressure or temperature of the fluid, assumed as inviscid. Also assume that gravitational forces do not produce significant effects. The one-dimensional theory stipulates that the partial derivative of ϕ with respect to n be zero along the entire flow direction m, i.e.:

$$\frac{\partial \phi}{\partial n} = 0 \quad \text{for all } m \tag{3.1}$$

If we add a coordinate in the $r\theta$ direction to represent the direction normal to both m and n, forming a right-handed Cartesian coordinate system, the one-dimensional theory further stipulates that the partial derivative of ϕ with respect to $r\theta$ be zero along m:

$$\frac{\partial \phi}{r\partial \theta} = 0 \quad \text{for all } m \tag{3.2}$$

Therefore, for a given m, the flow properties are uniformly distributed along n and $r\theta$, and specifically at the inlet and outlet of a generic bladed system. This is only possible if the fluid is inviscid and the local direction of flow always matches the local direction of the blade angle, that is ($\beta_{\text{flow}} = \beta_{\text{blade}} = \beta'$) in the case of rotors and ($\alpha_{\text{flow}} = \alpha_{\text{blade}} = \alpha'$) in stators.

3.2 VELOCITY DIAGRAMS

A velocity diagram, often referred to as a *velocity triangle* in rotors, is a graphic representation of the relationship between the absolute velocity **c**, the relative velocity **w** and the peripheral velocity **u** at a given point in the flow domain.

DOI: 10.1201/9781003619956-3

3.2.1 DEFINITIONS

The *absolute velocity* **c** refers to the velocity of the fluid measured relative to an inertial (stationary) frame of reference. The *relative velocity* **w** is the velocity of the fluid observed from the rotating frame. The *peripheral velocity* (also called the *blade velocity*), denoted by **u**, represents the tangential speed of the rotating blade at a given radius from the axis of rotation, i.e. the velocity of the rotating frame relative to the inertial frame.

3.2.2 MATHEMATICAL AND PHYSICAL EXPLANATION

The basic laws of relative motion from classical mechanics establish that the absolute velocity vector **c** is always the vector sum of the relative velocity vector **w** and the peripheral velocity vector **u**. This relationship can be expressed as:

$$\mathbf{c} = \mathbf{w} + \mathbf{u} \tag{3.3}$$

or, in other terms, $\mathbf{c} = \mathbf{w} + \boldsymbol{\omega} \times \mathbf{r}$, where the absolute value of peripheral velocity **u** is given by:

$$|\mathbf{u}| = \omega r \tag{3.4}$$

ω being the angular velocity of the rotor and r the radius at the point of interest.

The velocity vectors can be represented using a triangular relationship at any given point within the flow domain. Specifically, at the inlet of the rotor $\mathbf{c}_1 = \mathbf{w}_1 + \mathbf{u}_1$. Similarly, at the rotor outlet $\mathbf{c}_2 = \mathbf{w}_2 + \mathbf{u}_2$. These relationships must hold at every point along the rotor inlet and outlet according to the assumptions of the one-dimensional theory.

A velocity triangle can be sketched as in Figure 3.1.

3.3 CALCULATION OF FLOW RATES

The mass and volumetric flow rates are calculated using the concept of meridional velocity c_m at the corresponding cross-sectional area A for an arbitrary point along the flow path:

$$\dot{m} = \rho A c_m, \quad Q = A c_m \tag{3.5}$$

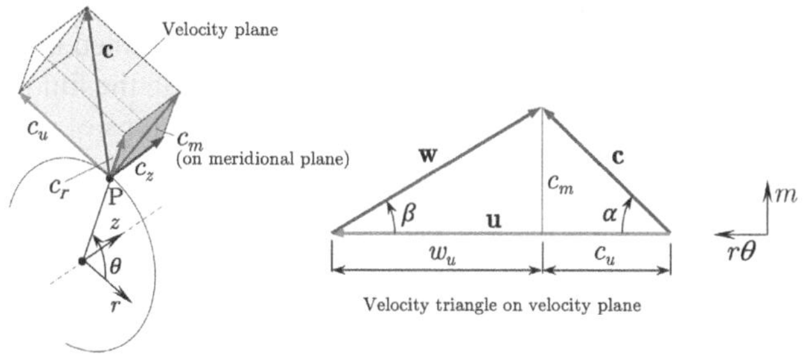

Figure 3.1 Definition of velocity plane and velocity components.

3.3.1 PURELY CENTRIFUGAL MACHINE

In a purely centrifugal machine, the flow is radial at inlet 1 and outlet 2. The cross-sectional area can be calculated as the lateral surface area of the cylindrical annulus defined between the hub and the shroud:

$$A_1 = 2\pi r_1 b_1, \quad A_2 = 2\pi r_2 b_2 \tag{3.6}$$

where r_{s1} and r_{s2}, the radii of the shroud at the inlet and outlet, equal r_{h1} and r_{h2}, the radii of the hub at the inlet and outlet, respectively; b_1 and b_2 are the blade heights at the inlet and outlet, respectively. Therefore, the continuity equation between inlet 1 and outlet 2 reads:

$$\dot{m} = \rho_1 A_1 c_{m1} = \rho_2 A_2 c_{m2}, \quad Q = A_1 c_{m1} = A_2 c_{m2} \tag{3.7}$$

3.3.2 MIXED FLOW MACHINE

In a mixed flow machine, the flow has both radial and axial components. The calculation of the cross-sectional area involves the calculation of the apothem b of a trunk conical surface (a slant height) connecting the edges of the two circular bases along the lateral surface. Therefore:

$$A_1 = 2\pi r_1 b_1, \quad A_2 = 2\pi r_2 b_2 \tag{3.8}$$

3.3.3 AXIAL FLOW MACHINE

In an axial flow machine, the flow is primarily in the axial direction. The cross-sectional area is typically constant along the flow path:

$$A_1 = A_2 = \pi(r_s^2 - r_h^2) \tag{3.9}$$

where r_s is the shroud radius and r_h the hub radius.

3.4 ENERGY EXCHANGE THROUGH A ROTOR

Work exchange between the fluid and the rotor can be determined by evaluating the torque τ around the axis of rotation resulting from changes in the fluid's angular momentum, as described by equation 1.36. Under the assumptions of one-dimensional theory, both gravitational forces and wall shear stresses are neglected, which simplifies the analysis to inertial effects alone.

Therefore, in case of a steady mass flow rate passing through a rotor having inlet station 1 and outlet station 2, the torque vector exchanged between the blades and the flow is derived (Figure 3.2[1]):

$$\boldsymbol{\tau} = \dot{m}\left[(\mathbf{r}_2 \times \mathbf{c}_2) - (\mathbf{r}_1 \times \mathbf{c}_1)\right] \tag{3.10}$$

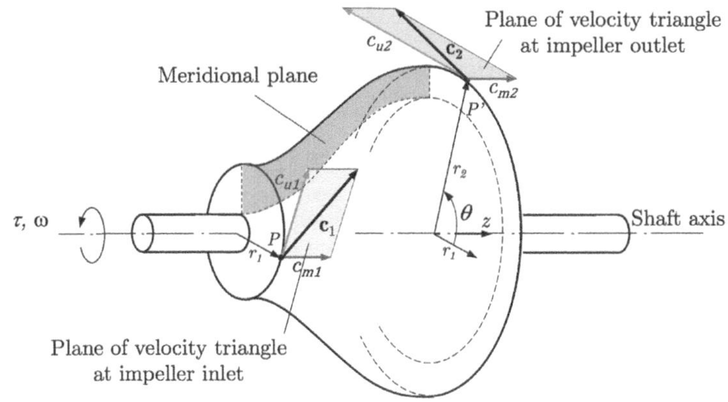

Figure 3.2 Schematic representation of flow through a rotating impeller showing the velocity triangles at the inlet and outlet stations.

If we name the tangential components of the velocity vectors as c_{u1} and c_{u2}, the z–component of torque then simplifies to:

$$\tau = \dot{m}\,(r_2 c_{u2} - r_1 c_{u1}) \tag{3.11}$$

The axis power P is related to the torque and the angular velocity:

$$P = \tau \cdot \omega \tag{3.12}$$

Substituting equation 3.11 into equation 3.12, we get:

$$P = \dot{m}\,(r_2 c_{u2} - r_1 c_{u1}) \cdot \omega \tag{3.13}$$

from which the specific work L done by the rotor on the fluid becomes:

$$L = \frac{P}{\dot{m}} = (u_2 c_{u2} - u_1 c_{u1}) \tag{3.14}$$

For turbines, the work is done by the fluid on the rotor:

$$L = (u_1 c_{u1} - u_2 c_{u2}) \tag{3.15}$$

These expressions of the specific work are often referred to as *Euler equations for turbomachinery*.[2] An example of velocity triangles at the inlet and outlet stations of the blade passages in a centrifugal impeller, drawn according to the one-dimensional theory, is given in Figure 3.3.

As a final statement pertaining to Euler's theory, we assert that the specific work of the rotors always equals the specific energy change in the flow. For instance, in a water pump, according to this theory, the work done by the rotor corresponds to the head acquired by the fluid. This implies that the efficiency η_{hyd} of the energy exchange in a pump rotor due to the blade-fluid interaction is 1. Similarly, in a water

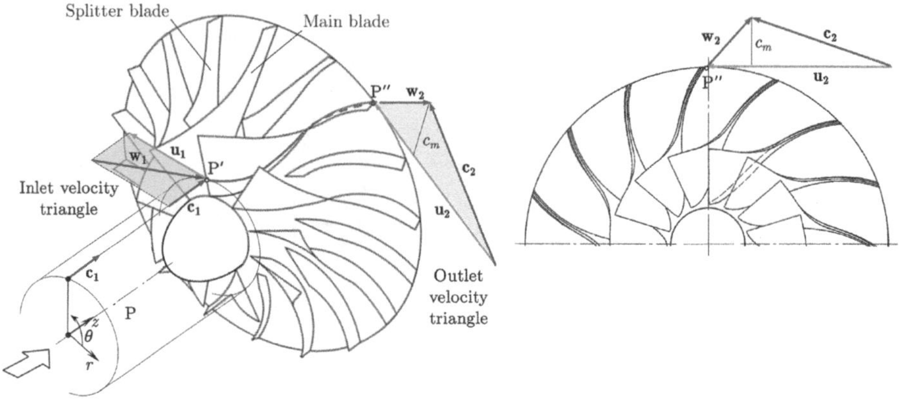

Figure 3.3 Inlet and outlet velocity triangles in a centrifugal impeller.

turbine, the energy drop of the fluid across the machine corresponds to the work generated by the rotor.

Mathematically, we express this as follows:

$$L_{\mathrm{E,rotor}} = gh \tag{3.16}$$

where – in the case of a pump – $L_{\mathrm{E,rotor}}$ represents the specific work done by the pump rotor and h is the specific head increase of the fluid, whose unit is $[m]$, while–in the case of a hydraulic turbine – these quantities represent the specific work done by the turbine rotor and the specific head decrease of the fluid, respectively. The suffix "E" in the previous equation emphasizes that it holds true only when a 1-D flow is postulated.

3.5 CORRECTIONS TO EULER'S THEORY

The one-dimensional theory requires corrections and enhancements to better reflect real-world flow physics and accurately predict performance.

First of all, in real flows, viscous effects cannot be neglected, especially near solid surfaces where boundary layers develop. In addition, viscous losses contribute to energy dissipation, which affects overall efficiency and power output. Next, the no-slip irrotational condition, where the fluid velocity at the blade surface matches the blade velocity exactly, is impractical in real flows. In fact, due to the finite number of blades of real rotors, actual blade shapes, and rotational effects, a deviation in the direction of the flow velocity occurs with respect to the direction of the blades at any arbitrary point in the flow domain.

3.5.1 CORRECTIONS TO EULER'S THEORY

The actual flow conditions at the rotor exit differ from the idealized one-dimensional flow predictions. One major consequence arising from this difference is that a flow deviation, or *slip*, appears at the exit station of a rotor which alters the exit velocity triangle and, therefore, the energy transfer between the rotor and the fluid compared to the Eulerian theory.

In radial flow rotors, the slip is primarily influenced by two factors:

1. vorticity generation and
2. finite pressure blade loading.

In axial flow rotors, vorticity is very limited compared to radial flow ones, so that slip (more often called simply *deviation*) is primarily influenced by finite pressure loading.

3.5.1.1 Rotational Effects

The spinning of a rotor causes vorticity to be generated in the flow, which affects the local direction of the flow and ultimately influences energy exchange (see Figure 3.4). For a steady-state motion, Euler's differential equations of fluid motion, which can be derived from the Navier-Stokes equations 1.21 in a rotating frame of reference by neglecting the effects of viscosity, include additional terms to account for the rotation of the frame:

$$(\mathbf{w}\cdot\nabla)\mathbf{w} = -\frac{1}{\rho}\nabla p + \mathbf{g} - 2\boldsymbol{\omega}\times\mathbf{w} - \boldsymbol{\omega}\times(\boldsymbol{\omega}\times\mathbf{r}) \tag{3.17}$$

Consider a rotor with a constant clockwise angular velocity $\boldsymbol{\omega} = -\omega\hat{\mathbf{i}}_z$ in a right-handed cylindrical frame (r,θ,z) and assume that the rotor motion only induces a tangential velocity in the flow while the rotor delivers no net flow (i.e. $Q=0$). In such a case, $\mathbf{u} = \boldsymbol{\omega}\times\boldsymbol{r} = (-\omega\hat{\mathbf{i}}_z)\times(r\hat{\mathbf{i}}_r) = -\omega r(\hat{\mathbf{i}}_z\times\hat{\mathbf{i}}_r) = -\omega r\hat{\mathbf{i}}_\theta$, i.e. the only fluid motion is the apparent one seen from the rotating blades, while, in the inertial frame, the fluid does not move at all. This means that:

$$\mathbf{c} = \mathbf{0} \implies \mathbf{w} = -\mathbf{u} \implies w_\theta = +\omega r \tag{3.18}$$

which shows that in the rotating frame, the fluid appears to swirl in the counterclockwise direction (Figure 3.4).

We can compute vorticity $\boldsymbol{\Omega}_w$ in the rotating frame as $\boldsymbol{\Omega}_w = \nabla\times\mathbf{w}$ using the curl operator:

$$\boldsymbol{\Omega}_w = \nabla\times\mathbf{w} = \begin{vmatrix} \hat{\mathbf{i}}_r & \hat{\mathbf{i}}_\theta & \hat{\mathbf{i}}_z \\ \dfrac{\partial}{\partial r} & \dfrac{\partial}{r\partial\theta} & \dfrac{\partial}{\partial z} \\ w_r & w_\theta & w_z \end{vmatrix}$$

yielding:

$$\boldsymbol{\Omega}_w = \left(\frac{1}{r}\frac{\partial w_z}{\partial\theta} - \frac{\partial w_\theta}{\partial z}\right)\hat{\mathbf{i}}_r + \left(\frac{\partial w_r}{\partial z} - \frac{\partial w_z}{\partial r}\right)\hat{\mathbf{i}}_\theta + \frac{1}{r}\left[\frac{\partial}{\partial r}(r w_\theta) - \frac{\partial w_r}{\partial\theta}\right]\hat{\mathbf{i}}_z \tag{3.19}$$

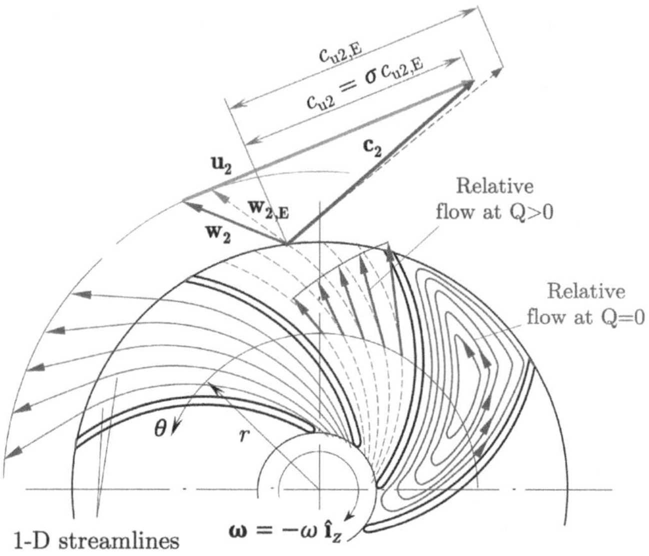

Figure 3.4 Corrections to the one-dimensional theory for a centrifugal impeller.

Since $w_\theta = \omega r$, we get:

$$\mathbf{\Omega}_w = \left(0, 0, \frac{1}{r}\frac{\partial(r\omega r)}{\partial r}\right) = (0, 0, +2\omega) \tag{3.20}$$

which states that, even for an inviscid fluid, the relative flow is not irrotational; rather, it possesses a constant vorticity of 2ω.

As already seen, the non-zero vorticity component $\Omega_z = 2\omega$ is compatible with a solid-body rotational vortex in the relative frame that has angular velocity $\omega\hat{\mathbf{i}}_z$ (equation 3.18). In fact, the presence of this vortex in the relative flow is the primary source of slip at the rotor's outlet section: when the volumetric flow rate is higher than 0 ($Q>0$), this vortex is superimposed onto the irrotational through-flow and results in the observed slip phenomenon. (Figure 3.4).

The slip factor σ is defined as the ratio of the actual tangential, vorticity-induced, component of the absolute velocity c_θ to the ideal, Eulerian, tangential component $c_{\theta,\text{ideal}}$:

$$\sigma_{\text{slip}} = \frac{c_{\theta 2}}{c_{\theta 2,\text{ideal}}} = \frac{c_{u2}}{c_{u2,\text{E}}} \tag{3.21}$$

Both theoretical and empirical models have been proposed to estimate the slip factor in centrifugal impellers. Stodola (1927)[3] attributed the slip to the solid body rotating motion of a cylinder, inscribed within one out of the total Z blade channels at the outlet of the impeller, rotating with ω. He computed the slip velocity as:

$$c_{\theta 2s} = \frac{\pi u_2 \sin\beta_2'}{Z} \tag{3.22}$$

and finally the slip factor as:

$$\sigma_{\text{slip}} = 1 - \left(\frac{\pi}{Z}\right) \frac{\sin\beta_2'}{1 - \frac{c_{m2}}{u_2}\cot\beta_2'} \tag{3.23}$$

In addition to Stodola's correlations, one of the most widely accepted corrections is attributed to Wiesner (1967)[4]:

$$\sigma_{\text{slip}} = 1 - \frac{\sqrt{\sin\beta_2'}}{Z^{0.70}(1 - c_{m2}/u_2\cot\beta_2')} \tag{3.24}$$

if $\varepsilon_{\text{limit}} = r_1/r_2 < \frac{1}{\ln^{-1}(8.16\sin\beta_2'/Z)}$, or:

$$\sigma_{\text{slip}} = \left(1 - \frac{\sqrt{\sin\beta_2'}}{Z^{0.70}(1 - c_{m2}/u_2\cot\beta_2')}\right) \cdot \left[1 - \left(\frac{r_1/r_2 - \varepsilon_{\text{limit}}}{1 - \varepsilon_{\text{limit}}}\right)^3\right] \tag{3.25}$$

otherwise. In the preceding two equations, β_2' is defined relative to the tangential direction.

3.5.1.2 Finite Blade Pressure Loading

In real turbomachines, the finite number of blades causes the momentum transfer between the blades and the fluid to create a pressure difference between the pressure and suction sides of each blade at a given radius r. In addition, for the same reason, the pressure varies along the length of the blade, from the leading edge to the trailing edge. This nonuniform pressure distribution, both tangentially and along the blade chord, results in a finite pressure loading on the blade surfaces.

Such loading causes a non-uniform flow field around the blade, such that the fluid near the pressure side moves more slowly due to higher pressure, while the fluid near the suction side moves faster. Eventually, this results in a skewed velocity profile at the blade exit (Figure 3.4), where the actual exit flow angle deviates from the design angle.[5]

The effect of finite-pressure loading is included in the slip factor in radial impellers.

3.5.1.3 The Special Case of Axial Flow Machines

The rotational effects linked to the vorticity produced by the rotor typically have a negligible impact on the overall flow behavior in axial flow machines. This can be measured by taking into account the inertial to Coriolis force ratio, which is commonly represented by the Rossby number and indicates the importance of rotational effects. This ratio is typically high ($\text{Ro} \gg 1$) in axial rotors, which indicates that inertial forces significantly dominate rotational effects. As a result, the pressure loading across the blades is the main cause of exit flow variation. This also accounts for the use of very similar empirical correlations for both rotor and stator blade rows in axial cascades, a principle often embodied by Carter's rule:

$$\delta_{\text{rot}} = \beta_2 - \beta_2' = \frac{m\theta_{c,\text{rot}}}{\sigma_R^n} \tag{3.26}$$

$$\delta_{\text{stat}} = \alpha_2 - \alpha_2' = \frac{m\theta_{c,\text{stat}}}{\sigma_S^n} \tag{3.27}$$

where δ_{rot} and δ_{stat} are the angular flow deviations for rotors and stators respectively, $\beta_2, \beta_2', \alpha_2, \alpha_2'$ the flow and blade angles measured with respect to the axial direction, $\theta_{c,\text{rot}}, \theta_{c,\text{stat}}$ the profile camber angles specific to rotors and stators, σ_R and σ_S the cascade solidity, i.e. the ratio between the profile chord length and the blade spacing at a given radius r. In compressor cascades $n = 0.5$, while in turbine cascades $n = 1$.

3.6 ENERGY EXCHANGE THROUGH A ROTOR AFTER CORRECTION

After considering the aforementioned corrections, equations 3.11–3.15 still hold if the quantities at station 2 are the actual ones instead of the Eulerian ones.

3.6.1 PUMP/COMPRESSOR ROTOR

To better identify the role of the corrections and to derive equations useful for preliminary blade row design, we can express the specific work of a generic power absorber rotor as:

$$L = u_2 c_{u2} - u_1 c_{u1} = u_2 \sigma_{\text{slip}} c_{u2,\text{E}} - u_1 c_{u1} \tag{3.28}$$

A useful expression is obtained when $c_{u1} = 0$, i.e. when no prewhirled flow at the impeller inlet is considered, a common scenario in single-stage pumps and compressors:

$$L = u_2 c_{u2} = u_2 \sigma_{\text{slip}} c_{u2,\text{E}} = u_2 \sigma_{\text{slip}} (u_2 - c_{m2} \cot \beta_2') \tag{3.29}$$

which is usually given in non-dimensional form:

$$\psi = \sigma_{\text{slip}} \left(1 - \phi \cot \beta_2'\right) \tag{3.30}$$

where $\psi = \frac{L}{u_2^2}$ is called the *work coefficient*, or *theoretical head coefficient*, and $\phi = \frac{c_{m2}}{u_2}$ the *flow coefficient*. while the corresponding Eulerian work coefficient is obtained for a non-slip condition, i.e. σ_{slip}=1:

$$\psi_E = 1 - \phi \cdot \cot \beta_2' \tag{3.31}$$

3.6.2 TURBINE ROTOR

Assuming that the slip only affects outlet flow in a centripetal turbine rotor, the specific work is given by:

$$L = u_1 c_{u1} - u_2 c_{u2} = u_1 c_{m1} \cos \alpha_1 - u_2 \frac{c_{u2,E}}{\sigma_{\text{slip}}} \tag{3.32}$$

therefore:

$$L = u_1 c_{m1} \cos\alpha_1 - u_2 c_{u2,E} \cdot \frac{1}{\sigma_{\text{slip}}} = u_1 c_{m1} \cos\alpha_1 - u_2 \frac{1}{\sigma_{\text{slip}}} (c_{m2} \cdot \cot\beta_2' - u_2) \quad (3.33)$$

and finally:

$$\psi = \phi \cos\alpha_1 - \frac{u_2}{u_1}\frac{1}{\sigma_{\text{slip}}}\left(\frac{c_{m2}}{u_1}\cot\beta_2' - \frac{u_2}{u_1}\right) = \phi\cos\alpha_1 - \frac{r_2 c_{m2}\cot\beta_2'}{\sigma_{\text{slip}} r_1^2} + \frac{r_2^2}{\sigma_{\text{slip}} r_1^2} \quad (3.34)$$

If we assume $c_{m2} = c_{m1}$, it further simplifies to (Figure 3.5):

$$\psi = \phi\left(\cos\alpha_1 - \frac{u_2}{u_1}\frac{1}{\sigma_{\text{slip}}}\cot(\beta_2')\right) + \frac{u_2^2}{u_1^2}\frac{1}{\sigma_{\text{slip}}} \quad (3.35)$$

while the Eulerian counterpart is obtained for a non-slip condition, $\sigma_{\text{slip}}=1$:

$$\psi_E = \phi\left(\cos\alpha_1 - \frac{u_2}{u_1}\cot(\beta_2')\right) + \frac{u_2^2}{u_1^2} \quad (3.36)$$

Equations 3.30 and 3.35 are named as the *non-dimensional theoretical characteristic curves* of power absorber- and power producer-turbomachines, respectively. These curves are fundamental as they are used to analyze and predict the performance of these devices across different operating conditions.

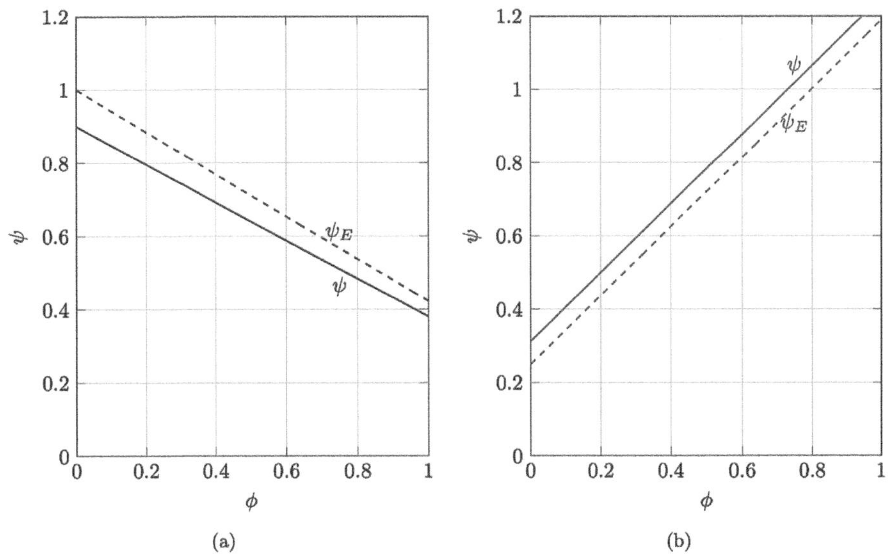

Figure 3.5 Plots of (a) ψ_E and ψ for a centrifugal pump/compressor rotor and (b) ψ_E and ψ for a centripetal turbine rotor, both as functions of ϕ.

3.7 ALTERNATIVE EXPRESSIONS OF ENERGY EXCHANGE THROUGH A ROTOR; ROTHALPY

The expressions for specific work, previously derived using the first law of thermodynamics and the principle of moment of momentum variation can be combined to make them applicable from the perspective of a moving observer. This results in an effective dualism between the two formulations (absolute and relative), as well as an alternative expression for specific work based on a balance of kinetic energies. The following discussion refers to the case of a power absorber machine but can easily be extended to a turbine by reversing the sign of the work and with small adjustments on the mathematical notation.

By equating the expressions for specific work derived from the first law of thermodynamics and Euler's formulation, for a generic compressor rotor, we obtain:

$$L = h_2^0 - h_1^0 = h_2 - h_1 + \frac{c_2^2 - c_1^2}{2} = u_2 c_{u2} - u_1 c_{u1} \tag{3.37}$$

from which, based on simple geometric considerations:

$$u c_u = u\,(u - w\cos\beta) \tag{3.38}$$

and, according to the law of cosines:

$$c^2 = w^2 + u^2 - 2uw\cos\beta \tag{3.39}$$

we obtain:

$$h_2 + \frac{w_2^2 - u_2^2}{2} = h_1 + \frac{w_1^2 - u_1^2}{2} \implies I_1 = I_2 \tag{3.40}$$

where $I = h + \frac{w^2 - u^2}{2}$ is a state function named *specific total enthalpy in the relative frame*, or more commonly *rothalpy*. As established by equation 3.40, in an adiabatic transformation, rothalpy is always conserved within moving ducts, in a manner analogous to the conservation of total enthalpy in stationary ducts. It is worth highlighting that, as long as it occurs without significant heat exchange with the surroundings, equation 3.40 is always valid, regardless of whether the thermodynamic transformation between the inlet and the outlet of the rotating ducts is reversible or not.

In moving ducts where liquid flows (ρ = const.), equation 3.40 must be written as:

$$\frac{p_2}{\rho} + \frac{w_2^2 - u_2^2}{2} = \frac{p_1}{\rho} + \frac{w_1^2 - u_1^2}{2} + gh_{\text{irr}} \tag{3.41}$$

which implies that, unless the flow is assumed to be inviscid, the total energy in the relative frame is not conserved for incompressible flow machines as a consequence of head losses gh_{irr}.

By combining equations 3.37 and 3.40, after removing the difference between the static enthalpies, an alternative expression for the specific work is easily obtained, and named *Euler's equation in energy form*. In this expression, specific work is given as a function of the differences in absolute, relative, and blade-specific kinetic energies at the inlet and outlet of a moving duct:

$$L = \frac{c_2^2 - c_1^2}{2} + \frac{w_1^2 - w_2^2}{2} + \frac{u_2^2 - u_1^2}{2} \tag{3.42}$$

which is of general validity in a rotor of a power absorber machine. As can be seen from the latter equation, work exchange implies: (i) *absolute flow acceleration* $\frac{c_2^2-c_1^2}{2}>0$, (ii) *relative flow deceleration* $\frac{w_2^2-w_1^2}{2} < 0$ and (iii) a *centrifugal effect* $\frac{u_2^2-u_1^2}{2}>0$.

These kinetic energy changes are easily realized in centrifugal rotors. In fact, as the flow moves radially outward from the center towards the periphery, it is subject to an increase in its tangential velocity according to the progressive accumulation of angular momentum. This results in a significant increase in the blade kinetic energy, i.e. $\frac{u_2^2-u_1^2}{2}>0$. In addition, the shape of the rotor blades is designed to divert the relative flow tangentially, as in a skewed rotating diffuser. This fact, combined with the requirement that the flow obey the continuity equation, results in a continuous decrease in the magnitude of the relative velocity within the rotor, i.e. $\frac{w_2^2-w_1^2}{2} < 0$.

The change in kinetic energy of the absolute flow is a direct consequence of these two effects, as $\mathbf{c} = \mathbf{w} + \mathbf{u}$, i.e. any changes in $\mathbf{w}$ and $\mathbf{u}$ will inherently affect $\mathbf{c}$.

The primary requirement for blade design is to ensure that the relative flow decelerates such that:

(i) the tangential component of the absolute flow, c_u, steadily increases from the inlet to the outlet of the rotor, i.e. $|c_{u2}|>|c_{u1}|$, as dictated by the Euler equation, and
(ii) there is a net increase in the kinetic energy of the absolute flow from the inlet to the outlet of the rotor, i.e. $\frac{c_2^2-c_1^2}{2}>0$.

In axial machines, implementing changes in the kinetic energy components is more challenging due to the absence of the centrifugal effect. In fact, in these machines, the flow moves parallel to the axis of rotation without a significant radial component. This means that there is no centrifugal force to aid in the acceleration of the flow, and, as a result, any changes in kinetic energy must be achieved primarily through a careful geometrical conformation of the blades, a method which is inherently less efficient than the centrifugal mechanism, as it always involves flow/blade interaction, which is affected by head losses. Furthermore, this limitation is linked to the fact that axial machines typically exchange less work with the fluid compared to centrifugal ones for the same blade velocity, making them less effective in applications that require significant energy transfer in relatively compact sizes. However, because the blade passages in axial machines are significantly larger than those in centrifugal configurations for the same rotor diameter, axial flow machines can accommodate larger flow rates, which makes them the most suitable choice when high specific speed is required.

Balancing the contributions from the different components of kinetic energy yields better results in terms of the efficiency of work exchange in a rotor. This concept will be further clarified with the formalization of the concept of degree of reaction in Section 3.9.

The corresponding equations for a turbine rotor may be derived by reversing the subscripts 1 and 2, thus reflecting the opposite direction of energy transfer. Moreover, in turbine rotors, the centrifugal effect is replaced by the centripetal one, as $\frac{u_1^2-u_2^2}{2}>0$.

3.8 EFFICIENCY

A fundamental element in evaluating the performance of turbomachines is the concept of efficiency, which serves as an effective measure to quantify the quality of the energy exchange process.

In general terms, efficiency is defined as the ratio of useful energy output to the energy input and, as such, it provides a clear metric for assessing how well the machine converts energy from one form to another. This parameter is crucial not only for determining the operational effectiveness of the turbomachine but also for optimizing its design, improving energy conservation, and reducing operational costs.

The importance of efficiency extends beyond the theoretical domain, impacting practical considerations such as fuel consumption, emission levels, and overall environmental footprint.

3.8.1 HYDRAULIC EFFICIENCY

In the context of incompressible flow machines, the hydraulic efficiency η_{hyd} is introduced.

In pumps, hydraulic efficiency is the ratio of the head h to the theoretical head, the latter corresponding to the specific work performed by the impeller. Mathematically, this is expressed as:

$$\eta_{\mathrm{hyd}}=\frac{gh}{|L|} \tag{3.43}$$

where both the head and the specific work are derived from the generalized Bernoulli equation:

$$gh=\frac{\Delta p_{i-e}}{\rho}+\Delta\left(\frac{c_{i-e}^2}{2}\right)+g\Delta z_{i-e} \tag{3.44}$$

$$|L|=\frac{\Delta p_{i-e}}{\rho}+\Delta\left(\frac{c_{i-e}^2}{2}\right)+g\Delta z_{i-e}+\sigma_{\mathrm{irr},i-e} \tag{3.45}$$

Δp_{i-e} being the pressure difference across the pump (from inlet i to exit e), $\Delta\left(\frac{c_{i-e}^2}{2}\right)$ the change in kinetic energy, $g\Delta z_{i-e}$ the change in potential energy, and $\sigma_{\mathrm{irr},i-e}$ the irreversible losses (or head losses) within the pump along the path from the inlet to the discharge sections.

Equation 3.44 can be rewritten also as:

$$\eta_{\mathrm{hyd}}=1-\frac{\sigma_{\mathrm{irr},i-e}}{|L|} \tag{3.46}$$

or as:

$$\eta_{\text{hyd}} = \frac{gh}{u_2 c_{u2} - u_1 c_{u1}} = 1 - \frac{\sigma_{\text{irr},i-e}}{u_2 c_{u2} - u_1 c_{u1}} \tag{3.47}$$

in view of the Euler equation, where 1 and 2 denote the inlet and outlet sections of the impeller, respectively.

In hydraulic turbines, hydraulic efficiency is similarly defined as the ratio of the actual specific work extracted by the rotor to the head drop of the flow from the inlet to the outlet of the machine:

$$\eta_{\text{hyd}} = \frac{L}{gh} \tag{3.48}$$

or equivalently as:

$$\eta_{\text{hyd}} = \frac{u_2 c_{u2} - u_1 c_{u1}}{gh} = 1 - \frac{\sigma_{\text{irr},i-e}}{gh} \tag{3.49}$$

in view of the fact that:

$$L = \frac{\Delta p_{i-e}}{\rho} + \Delta\left(\frac{c_{i-e}^2}{2}\right) + g\Delta z_{i-e} - \sigma_{\text{irr},i-e} \tag{3.50}$$

3.8.2 ISENTROPIC EFFICIENCY

In compressible flow machines, we define isentropic efficiency η_{is} to characterize the quality of energy exchange. Isentropic efficiency is the ratio of the actual work to the corresponding work that would be exchanged in an ideal isentropic process for the same static pressure change. For compressors, it is defined as:

$$\eta_{\text{is,c}} = \frac{h_{es}^0 - h_i^0}{h_e^0 - h_i^0} \tag{3.51}$$

while for turbines:

$$\eta_{\text{is,t}} = \frac{h_i^0 - h_e^0}{h_i^0 - h_{es}^0} \tag{3.52}$$

where h_i^0 is the total enthalpy at the inlet of the machine, h_e^0 is the total enthalpy at the outlet and h_{es}^0 is the total enthalpy at the outlet for a corresponding isentropic process reaching the same outlet pressure.

As already specified in Chapter 1, in the realm of ideal gases, the total enthalpy h^0 is related to the specific heat at constant pressure and total temperature $dh^0 = c_p dT^0$:

$$h^0 = h + \frac{c^2}{2} = c_p T + \frac{c^2}{2} \tag{3.53}$$

Since for an isentropic process involving an ideal gas with constant specific heats, the relationship between total temperatures and total pressures is given by:

$$\left(\frac{T_{e,s}^0}{T_i^0}\right) = \left(\frac{p_{e,s}^0}{p_i^0}\right)^{\frac{\gamma-1}{\gamma}} \simeq \left(\frac{p_e^0}{p_i^0}\right)^{\frac{\gamma-1}{\gamma}} \tag{3.54}$$

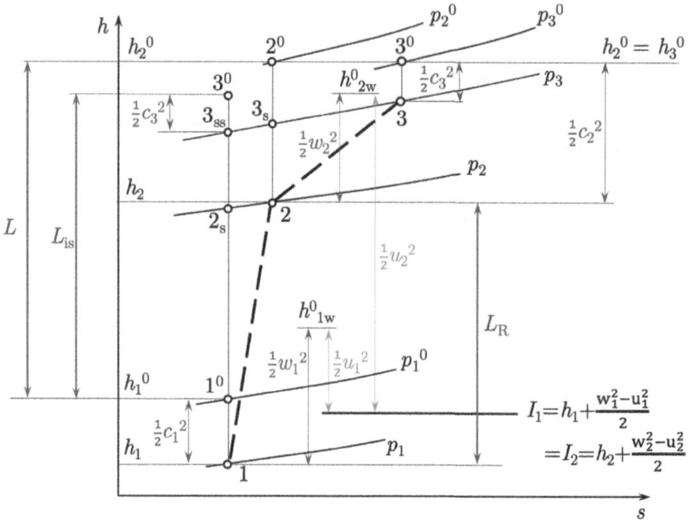

Figure 3.6 Compression transformation in $h-s$.

we can derive the following fundamental relations.
For compressors:

$$\eta_{\text{is,c}} = \frac{T_i^0\left[\left(\frac{p_e^0}{p_i^0}\right)^{\frac{\gamma-1}{\gamma}}-1\right]}{T_i^0\left[\left(\frac{T_e^0}{T_i^0}\right)-1\right]} = \frac{\left(\frac{p_e^0}{p_i^0}\right)^{\frac{\gamma-1}{\gamma}}-1}{\left(\frac{T_e^0}{T_i^0}\right)-1} \tag{3.55}$$

For turbines:

$$\eta_{\text{is,t}} = \frac{\left(\frac{T_i^0}{T_e^0}\right)-1}{\left(\frac{p_i^0}{p_e^0}\right)^{\frac{\gamma-1}{\gamma}}-1} \tag{3.56}$$

Compression and expansion processes through generic compressor and turbine stages (impeller and diffuser in a compressor, nozzle, and rotor in a turbine) are displayed in Figures 3.6 and 3.7.

3.8.3 POLYTROPIC EFFICIENCY

It is known that the slope of isobaric curves in a $T-s$ diagram, $\left(\frac{dT}{ds}\right)_p$, increases with T. Consequently, for a given initial state 1, the isentropic efficiency of a generic adiabatic process involving work exchange will depend on the temperature of the final state 2. It follows that two distinct isobaric curves have the same slope at points where the temperature is also equal. Since in real compression or expansion processes the final temperature will be, respectively, greater or lower than that of the

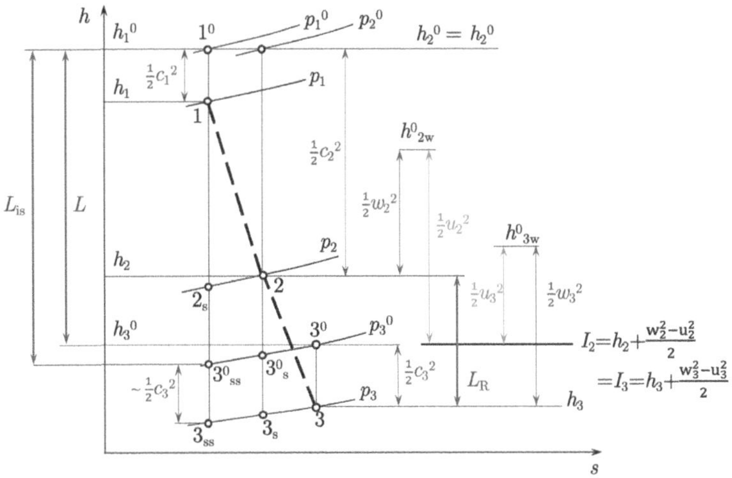

Figure 3.7 Expansion transformation in $h - s$.

corresponding ideal process, it follows that at the endpoint of the process, the slope of the curves will be such that the isentropic efficiency will be lower as the ratio of the initial to final pressures (and thus temperatures) increases. This means that, under the same conditions, isentropic efficiency will vary with the pressure ratio between states 1 and 2.

This dependence complicates comparisons between different machines with varying pressure ratios as the quality of energy exchange relies not only on the machine's design but also on the thermodynamic states of the fluid at the inlet and outlet. For example, consider two air compressors operating under the same inlet conditions. If the first compressor has a higher pressure ratio than the second, it will exhibit lower isentropic efficiency despite having similar design characteristics.

Therefore, comparing the isentropic efficiency of two machines with significantly different pressure ratios is meaningless, as the one characterized by a higher pressure ratio is either penalized or advantaged (in the case of a compressor or a turbine, respectively) by an intrinsic thermodynamic effect rather than by design choices.

To address this challenge, we need a new concept to quantify the thermodynamic efficiency of a machine that is independent of the pressure ratio. This concept is known as *polytropic efficiency*.

3.8.3.1 Compression with Work Exchange

The polytropic efficiency for an infinitesimal compression involving work exchange, $\eta_{pol,c}$, is defined in differential form as:

$$\eta_{pol,c} \equiv \frac{dh^0_{is}}{dh^0} = \frac{dL_{is}}{dL} \tag{3.57}$$

From the first law of thermodynamics in stagnation form[6]:

$$T^0 ds = dh^0 - \frac{dp^0}{\rho^0} \tag{3.58}$$

and, for an isentropic process (where $ds = 0$):

$$dh^0_{\mathrm{is}} = \frac{dp^0}{\rho^0} \tag{3.59}$$

If we then substitute dh^0_{is} from equation 3.59 into the definition of polytropic efficiency (equation 3.57):

$$\eta_{\mathrm{pol,c}} = \frac{dh^0_{\mathrm{is}}}{dh^0} = \frac{\frac{dp^0}{\rho^0}}{dh^0} \tag{3.60}$$

and express dh^0 in terms of specific heat c_p and temperature change for an ideal gas:

$$dh^0 = c_p dT^0 \tag{3.61}$$

we get:

$$\eta_{\mathrm{pol,c}} = \frac{\frac{dp^0}{\rho^0}}{c_p dT^0} \tag{3.62}$$

Next, since $R = \frac{p^0}{\rho^0 T^0}$:

$$\eta_{\mathrm{pol,c}} = \frac{\frac{dp^0}{\frac{p^0}{RT^0}}}{c_p dT^0} = \frac{RT^0 dp^0}{p^0 c_p dT^0} = \frac{dp^0}{p^0} \cdot \frac{RT^0}{c_p dT^0} \tag{3.63}$$

and $\frac{R}{c_p} = 1 - \frac{1}{\gamma}$:

$$\eta_{\mathrm{pol},c} = \frac{dp^0}{p^0} \frac{RT^0}{c_p dT^0} = \frac{dp^0}{p^0} \left(1 - \frac{1}{\gamma}\right) \frac{T^0}{dT^0} \tag{3.64}$$

which, upon separation of variables, leads to:

$$\frac{dp^0}{p^0} = \frac{\gamma \eta_{\mathrm{pol,c}}}{\gamma - 1} \frac{dT^0}{T^0} \tag{3.65}$$

We then integrate both sides from state 1 to state 2 by assuming constant γ and imposing $\eta_{\mathrm{pol,c}}$ = constant:

$$\int_{p_1^0}^{p_2^0} \frac{dp^0}{p^0} = \frac{\eta_{\mathrm{pol,c}} \gamma}{\gamma - 1} \int_{T_1^0}^{T_2^0} \frac{dT^0}{T^0} \tag{3.66}$$

leading to:

$$\ln\left(\frac{p_2^0}{p_1^0}\right) = \frac{\gamma \eta_{\mathrm{pol,c}}}{\gamma - 1} \ln\left(\frac{T_2^0}{T_1^0}\right) \Longrightarrow \frac{p_2^0}{p_1^0} = \left(\frac{T_2^0}{T_1^0}\right)^{\frac{\gamma \eta_{\mathrm{pol,c}}}{\gamma - 1}} \tag{3.67}$$

or, equivalently:

$$\pi_c = \frac{p_2^0}{p_1^0} = \left(\frac{T_2^0}{T_1^0}\right)^{\frac{\gamma\eta_{\text{pol,c}}}{\gamma-1}} \tag{3.68}$$

where $\pi_c = \frac{p_2^0}{p_1^0}$ is the total-to-total compression ratio.

From the integrated form:

$$\ln(\pi_c) = \frac{\gamma\eta_{\text{pol,c}}}{\gamma-1}\ln\left(\frac{T_2^0}{T_1^0}\right) \tag{3.69}$$

solving for $\eta_{\text{pol,c}}$, we obtain:

$$\eta_{\text{pol,c}} = \frac{\gamma-1}{\gamma}\frac{\ln\left(\frac{p_2^0}{p_1^0}\right)}{\ln\left(\frac{T_2^0}{T_1^0}\right)} \tag{3.70}$$

3.8.3.2 Relationship between Isentropic and Polytropic Efficiency for a Compression

Given initial and final thermodynamic states, the relationship between isentropic and polytropic efficiency for a compression can be established as follows:

$$\eta_{\text{is,c}} = \frac{T_{2,\text{is}}^0 - T_1^0}{T_2^0 - T_1^0} = \frac{\left(\frac{p_2^0}{p_1^0}\right)^{\frac{\gamma-1}{\gamma}} - 1}{\left(\frac{T_2^0}{T_1^0} - 1\right)} \tag{3.71}$$

And since:

$$\frac{T_2^0}{T_1^0} = \left(\frac{p_2^0}{p_1^0}\right)^{\frac{(\gamma-1)}{\gamma\eta_{\text{pol,c}}}} \tag{3.72}$$

we get:

$$\eta_{\text{is,c}} = \frac{\left(\frac{p_2^0}{p_1^0}\right)^{\frac{\gamma-1}{\gamma}} - 1}{\left(\frac{p_2^0}{p_1^0}\right)^{\frac{(\gamma-1)}{\gamma\eta_{\text{pol,c}}}} - 1} \tag{3.73}$$

As the compression ratio increases, the reduction in isentropic efficiency becomes more pronounced as the polytropic efficiency decreases. Consequently, aerodynamic losses increasingly impact the isentropic efficiency of the machine as the compression ratio increases. Furthermore, for a given value of polytropic efficiency, the slope of the isentropic efficiency curves decreases in absolute value as the compression ratio increases. Therefore, if a machine consists of multiple stages, it is advantageous for the initial stages to operate with high aerodynamic efficiencies; otherwise, the overall isentropic efficiency will be significantly impaired.

3.8.3.3 Expansion with Work Exchange

In the case of an infinitesimal expansion involving work exchange, the polytropic efficiency takes the form:

$$\eta_{\text{pol,exp}} \equiv \frac{dh^0}{dh^0_{\text{is}}} \tag{3.74}$$

By taking the equivalent steps previously shown for a compression, we get the expansion ratio:

$$\pi_e = \frac{p_1^0}{p_2^0} = \left(\frac{T_1^0}{T_2^0}\right)^{\frac{\gamma}{(\gamma-1)\eta_{\text{pol,exp}}}} \tag{3.75}$$

and the corresponding polytropic efficiency:

$$\eta_{\text{pol,exp}} = \frac{\gamma}{\gamma-1}\frac{\ln\left(\frac{T_2^0}{T_1^0}\right)}{\ln\left(\frac{p_2^0}{p_1^0}\right)} \tag{3.76}$$

3.8.3.4 Relationship between Isentropic and Polytropic Efficiency for an Expansion

Given the initial and final thermodynamic states, the relationship between isentropic and polytropic efficiency for an expansion can be established as follows:

$$\eta_{\text{is,exp}} = \frac{T_1^0 - T_2^0}{T_1^0 - T_{2,\text{is}}^0} = \frac{1 - \frac{T_2^0}{T_1^0}}{1 - \left(\frac{p_2^0}{p_1^0}\right)^{\frac{\gamma-1}{\gamma}}} \tag{3.77}$$

and since:

$$\frac{T_2^0}{T_1^0} = \left(\frac{p_2^0}{p_1^0}\right)^{\frac{(\gamma-1)\eta_{\text{pol,exp}}}{\gamma}} \tag{3.78}$$

we get:

$$\eta_{\text{is,exp}} = \frac{\left(\frac{p_2^0}{p_1^0}\right)^{\frac{(\gamma-1)\eta_{\text{pol,exp}}}{\gamma}} - 1}{\left(\frac{p_2^0}{p_1^0}\right)^{\frac{\gamma-1}{\gamma}} - 1} \tag{3.79}$$

The losses that occur during the expansion process in a stage increase the fluid's temperature, and this thermal energy can be partially utilized, according to the second law of thermodynamics, to produce work in subsequent stages. The efficiency with which these losses are converted back into work is maximized in the initial stages (in fact, the slope of the polytropic efficiency curves is greatest for expansion ratios close to one), while it becomes virtually negligible in the final stages. Consequently, in a turbine, it is crucial that the last stages are designed with particular care.

3.8.3.5 Final Remarks on Polytropic Efficiency

Equations 3.72 and 3.78 suggest that both actual compression or expansion transformations can be analytically described by the following law:

$$\frac{T_2^0}{T_1^0} = \left(\frac{p_2^0}{p_1^0}\right)^{\frac{(n-1)}{n}} \tag{3.80}$$

where:

$$\frac{(n-1)}{n} = \frac{(\gamma-1)\eta_{\text{pol,compr}}}{\gamma} \text{ (compression)} \tag{3.81}$$

and:

$$\frac{(n-1)}{n} = \frac{\gamma-1}{\gamma\eta_{\text{pol,exp}}} \text{ (expansion)} \tag{3.82}$$

thus suggesting that:

$$\frac{p}{\rho^n} = \text{const. (polytropic law)} \tag{3.83}$$

$$\eta_{\text{pol,compr}} = e_c = \frac{n-1}{n} \cdot \frac{\gamma}{\gamma-1} \quad \text{(compression)} \tag{3.84}$$

$$\eta_{\text{pol,exp}} = e_t = \frac{\gamma-1}{\gamma} \cdot \frac{n}{n-1} \text{ (expansion)} \tag{3.85}$$

Since the real 1-2 transformation occurs with an increase in entropy, the equivalent polytropic transformation is only possible by assuming a reversible heat exchange with the ambient at temperature T, i.e. $dQ_{\text{rev}} = Tds$, which leads to:

$$Q_{\text{rev, 1-2}} = \int_1^2 Tds \tag{3.86}$$

It should be noted, finally, that the definition of polytropic efficiency is equivalent to that obtained by comparing the actual work with that of a reversible polytropic transformation between the same initial and final states. This equivalence is of fundamental importance because it allows us to understand the true meaning of the polytropic efficiency and its relevance for the design of a turbomachine.

For the sake of clarity, let us consider a compression process. From the generalized Bernoulli equation, we have that, in the case of a reversible compression (i.e. without internal losses), the work exchanged is equal to:

$$L_{\text{rev}} = \int_1^2 vdp \tag{3.87}$$

Therefore, if such a transformation is a reversible polytropic one that follows $\frac{p}{\rho^n}$, from the previous equation we immediately obtain:

$$L_{\text{rev}} = \int_1^2 vdp = \int_1^2 \left(\frac{p_1 v_1^n}{p}\right)^{\frac{1}{n}} dp = (p_1 v_1^n)^{\frac{1}{n}} \int_1^2 p^{-\frac{1}{n}} dp \tag{3.88}$$

and:

$$L_{\text{rev}} = (p_1 v_1^n)^{\frac{1}{n}} \left[\frac{n}{n-1} p^{1-\frac{1}{n}} \right]_{p_1}^{p_2} \tag{3.89}$$

Substituting the limits p_1 and p_2, we get:

$$L_{\text{rev}} = \frac{n}{n-1} (p_1 v_1^n)^{\frac{1}{n}} \left(p_2^{1-\frac{1}{n}} - p_1^{1-\frac{1}{n}} \right) \tag{3.90}$$

Using the relation $(p_1 v_1^n)^{\frac{1}{n}} = p_1^{\frac{1}{n}} v_1$ and the equation of state for ideal gases, the expression becomes:

$$L_{\text{rev}} = \frac{n}{n-1} p_1^{\frac{1}{n}} v_1 \left(p_2^{1-\frac{1}{n}} - p_1^{1-\frac{1}{n}} \right) = \frac{nR}{n-1} T_1 \left[\left(\frac{p_2}{p_1} \right)^{\frac{n-1}{n}} - 1 \right] \tag{3.91}$$

Ultimately, by computing L_{rev}/L in a compression, we obtain:

$$\frac{L_{\text{rev}}}{L} = \frac{n-1}{n} \cdot \frac{\gamma}{\gamma - 1} = e_c \quad \text{(compression)} \tag{3.92}$$

This means that the polytropic efficiency can also be viewed, in the case of compression, as the ratio between the reversible work along the polytropic and the corresponding actual work between the initial and final states. Since the reversible polytropic work is, in fact, computed using the Bernoulli equation, it isolates the component of mechanical loss (due to wall friction, wakes, secondary flows, etc.) from the part degraded into heat (which is merely the consequence) that appears in the expression of the actual work. Therefore, we can assert that the polytropic efficiency is the true efficiency by which the quality of the aerodynamic design of a turbomachine can be judged.

3.9 DEGREE OF REACTION

If we recall equation 3.42:

$$L = \frac{c_2^2 - c_1^2}{2} + \frac{w_1^2 - w_2^2}{2} + \frac{u_2^2 - u_1^2}{2} \tag{3.93}$$

we can define the first term as the work exchanged by action, while the other two terms represent the work exchanged by reaction.

Action work is the direct transfer of energy to/from the fluid by accelerating/decelerating it in the direction of the absolute flow (from rotor inlet 1 to rotor exit 2). In other words, it is the energy added to or removed from the fluid to change its overall motion relative to the stationary frame of reference:

$$L_a = \frac{c_2^2 - c_1^2}{2} \text{ (pump, compressor)} \tag{3.94}$$

$$L_a = \frac{c_1^2 - c_2^2}{2} \text{ (turbine)} \tag{3.95}$$

Reaction work is the energy exchange due to the relative motion between the fluid and the moving components of the machine. It encompasses the changes in both the relative flow velocity and the blade velocity. Using the first law of thermodynamics in rotating frames, it immediately follows that the reaction work corresponds to the static enthalpy rise (in compressors) or drop (in turbines) occurring in the rotor:

$$L_r = h_2 - h_1 = \frac{w_1^2 - w_2^2}{2} + \frac{u_2^2 - u_1^2}{2} \text{ (compressor)} \tag{3.96}$$

$$L_r = h_1 - h_2 = \frac{w_2^2 - w_1^2}{2} + \frac{u_1^2 - u_2^2}{2} \text{ (turbine)} \tag{3.97}$$

Let us define the *degree of reaction*, ε_r, as the fraction of reaction work done by the rotor compared to the total work:

$$\varepsilon_r = \frac{L_r}{L_r + L_a} \tag{3.98}$$

The degree of reaction helps understanding how the energy exchange is carried out.

- When $\varepsilon_r = 0$, the machine operates as an *impulse machine* (or *pure action machine*), in which all energy transfer occurs exclusively through changes in the absolute velocity c, with no enthalpy change across the rotor.
- When $\varepsilon_r = 1$, the machine is a *pure reaction machine*, where all energy transfer arises from changes in enthalpy (or pressure, in the case of a hydraulic machine) across the rotor. In this case, energy exchange is entirely due to changes in the relative velocity w and blade tangential speed u.
- For $0 < \varepsilon_r < 1$, the machine exhibits a *mixed impulse-reaction* character, where both action (impulse) and reaction effects contribute to the overall energy transfer.

The degree of reaction correlates strongly with the shape of the velocity triangles and, therefore, with efficiency.

3.9.1 PUMPS AND COMPRESSORS

In pumps and compressors, an impulse stage, characterized by a low degree of reaction ε_r, involves a large change in absolute velocity c and relatively small changes in both relative velocity w and tangential velocity u. The resulting velocity triangle in such a case exhibits a significant change in c between the inlet and the outlet. Since most of the pressure rise occurs in the stator, the rotor blades primarily serve to redirect the relative flow without contributing significantly to compression. Consequently, rotor design focuses mainly on flow guidance and minimizing losses due to turning, rather than on pressure recovery. This option is very rarely adopted in practice since it leads to relatively lower efficiencies. Exceptions are fans with forward-curved blades, where such a solution is implemented for compact installations.

In contrast, a reaction stage, where the degree of reaction is high ε_r, displays a relatively large change in the relative velocity w and tangential velocity u, with only a small change in the absolute velocity c. The corresponding velocity triangle for a high-ε_r pump/compressor stage reflects this by showing prominent variations in w and u, while c remains relatively steady. In this case, the stator vanes contribute significantly to the pressure rise. As a result, stator design becomes more critical in managing flow diffusion and maintaining aerodynamic efficiency, especially at higher flow coefficients.

Typical values of degree of reaction for pumps and compressors vary with the type of machine. In axial types, the degree of reaction typically ranges from 0.4 to 0.6, with 0.5 being ideal for symmetrical velocity triangles and balanced pressure/enthalpy rise between the rotor and stator. In centrifugal pumps and compressors, the degree of reaction is generally higher, often ranging from 0.7 to 0.9, as most of the pressure/enthalpy rise occurs in the impeller. In view of the above characteristics, reaction pumps and compressors are the standard choice.

3.9.2 TURBINES

Similar to pumps and compressors, in turbines, a low degree of reaction ε_r in an impulse stage results in a large change in absolute velocity c and small changes in both the relative velocity w and the tangential velocity u, with the velocity triangle clearly illustrating the significant difference in c from inlet to outlet. Here, the stator vanes (or nozzles) are responsible for most of the energy conversion from pressure to kinetic energy. They accelerate the flow significantly before it enters the rotor, which mainly redirects the high-speed flow to extract work. Therefore, stator design emphasizes converting pressure energy efficiently into kinetic energy with minimal losses. Such a solution is widely adopted for compact and high-expansion ratio installations.

In a reaction turbine stage, characterized by a high degree of reaction ε_r, the opposite behavior is observed: there is a large change in relative velocity w and tangential velocity u, while the change in absolute velocity c is relatively minor. The velocity triangle in this case highlights the dominant shifts in w and u. Since the pressure drop is more evenly split between the stator and rotor, the stator vanes in a high-reaction turbine also contribute to the expansion process. As a result, the stator must handle both the redirection and partial expansion of the flow, requiring more careful aerodynamic shaping to ensure smooth pressure gradients and avoid flow separation.

In axial turbines, the degree of reaction typically ranges from 0.4 to 0.6, with values around 0.5 being ideal for balanced expansion between rotor and stator, leading to high aerodynamic efficiency. However, impulse turbines, such as the classic Curtis stage, are widely adopted as high-pressure steam expanders and operate with a low reaction close to zero, where almost all the expansion occurs in the nozzles. Although the low reaction leads to efficiency penalties in the Curtis stages compared to the reaction ones, their ability to extract larger amounts of specific work per unit of blade speed u is appreciated.

3.9.3 EFFICIENCY AND DEGREE OF REACTION

At very low degrees of reaction (impulse-type), high deflections in the relative velocities can lead to greater frictional losses in the rotor. At the same time, significant differences in the absolute velocity can impair the efficiency of stator vanes. On the other hand, degrees of reaction that are too high can lead to excessive diffusion in the rotor and may result in boundary layer separation losses.

There exists an optimal range of ε_r, generally around 0.5 in axial machines and around 0.7 in radial ones, where the static energy exchange is optimally shared between the stator and rotor, minimizing losses and ensuring smoother velocity transitions. This typically leads to the most favorable efficiency. In axial machines, this balanced configuration is often deliberately targeted both for compressors and turbines.

3.10 PROBLEMS

1. In a centrifugal pump impeller, the inlet velocity triangle has a meridional velocity of $c_{m1} = 15\,\text{m/s}$ and a tangential velocity $c_{u1} = 5\,\text{m/s}$. At the outlet, the tangential velocity c_{u2} is corrected for slip, given by $\sigma_{\text{slip}} = 0.85$ and the meridional velocity is $c_{m2} = 20\,\text{m/s}$. The blade angle at the outlet is $\beta_2' = 40^\circ$ and the peripheral velocity is $u_2 = 40\,\text{m/s}$. Calculate the absolute velocity c_1 at the inlet, the tangential velocity c_{u2} at the outlet and the specific work spent to drive the pump.

   [15.8 m/s, 13.7 m/s, 500 J/kg]

2. A centrifugal pump impeller operates at a rotational speed of $N = 3000$ rpm. The impeller has an outer radius $r_2 = 0.2\,\text{m}$ and an inner radius $r_1 = 0.1\,\text{m}$. The blade angle at the inlet is $\beta_1' = 30^\circ$ and the blade angle at the outlet is $\beta_2' = 20^\circ$. The absolute meridional velocity c_m is constant and equal to $10\,\text{m/s}$. Determine the Eulerian work without corrections for slip. Next, correct for slip using Stodola's slip factor for $Z = 6$ blades.

   [1,779 J/kg, 1,072 J/kg]

3. A mixed flow turbine rotor operates with an inlet radius of $r_1 = 0.3\,\text{m}$ and an outlet radius of $r_2 = 0.1\,\text{m}$. The blade angles are $\beta_1' = 40^\circ$ at the inlet and $\beta_2' = 25^\circ$ at the outlet. The axial velocity, which is constant through the rotor, $c_m = 50\,\text{m/s}$ and the rotational speed is $N = 6000\,\text{rpm}$. The slip factor is 0.88. Determine the Eulerian work and correct for slip to calculate the actual work.

   [4,494.8 J/kg, 3,576.1 J/kg]

4. A centrifugal compressor impeller processes an air stream with the following parameters. The stagnation enthalpy at the inlet is $h_1^0 = 26{,}000\,\text{J/kg}$ and the pre-swirl tangential velocity at the inlet is $c_{u1} = 30\,\text{m/s}$. The relative velocity at the inlet is $w_1 = 221\,\text{m/s}$ and the peripheral velocity at the inlet is $u_1 = 184.31\,\text{m/s}$. At the outlet, the peripheral velocity is $u_2 = 293.22\,\text{m/s}$ and the relative velocity is $w_2 = 187.99\,\text{m/s}$. The number of blades is $Z = 16$ and the blade exit angle is

$\beta_2' = 90°$. The rotational speed of the compressor is $\omega = 80{,}000\,\text{rpm}$ and the isentropic efficiency is 0.85. Calculate the actual outlet stagnation enthalpy h_2^0 and the isentropic outlet stagnation enthalpy $h_{2,\text{isentropic}}^0$.

[94,100 J/kg, 83,900 J/kg]

5. An axially mounted horizontal pump impeller processes liquid oxygen ($\rho = 1135$ kg/m^3) achieving a hydraulic efficiency of 0.92. The pump is equipped with a vaned diffuser that ensures the absolute velocity at the pump outlet is equal to that at the pump inlet. However, the diffuser causes a reduction in hydraulic efficiency to $\eta_h = 0.87$ compared to the efficiency of the impeller alone. Under these conditions, the pump generates a pressure rise of 40 MPa. Calculate the pump head $H_{\text{effective}}$ in meters of liquid oxygen equivalent.

[3,125 m]

6. A gas turbine system includes a compressor and a turbine. The compressor operates with a total inlet temperature of $T_{c,i}^0 = 300\,\text{K}$, a total outlet temperature of $T_{c,e}^0 = 700\,\text{K}$, a polytropic efficiency of $\eta_{\text{pol,c}} = 0.88$ and a specific heat ratio of $\gamma = 1.4$. The turbine operates with a total inlet temperature of $T_{t,i}^0 = 1200\,\text{K}$, a total outlet temperature of $T_{t,e}^0 = 800\,\text{K}$ and an isentropic efficiency of $\eta_{\text{is,t}} = 0.90$. Calculate the compression ratio π_c for the compressor and the expansion ratio π_t for the turbine.

[13.6, 4.7]

7. A compressor operates with a total inlet temperature of $T_i^0 = 300\,\text{K}$, a total outlet pressure of $p_e^0 = 800\,\text{kPa}$ and a total inlet pressure of $p_i^0 = 100\,\text{kPa}$. The polytropic efficiency of the compressor is $\eta_{\text{pol,c}} = 0.87$, the specific heat at constant pressure is $c_p = 1005\,\text{J/kg·K}$ and the specific heat ratio is $\gamma = 1.4$. Calculate the total outlet temperature and the specific work input.

[593.9 K, 295 kJ/kg]

8. An axial turbine rotor operates under the following conditions: absolute inlet flow velocity $c_1 = 320\,\text{m/s}$, absolute outlet flow velocity $c_2 = 240\,\text{m/s}$, inlet flow angle $\alpha_1 = 35°$, outlet flow angle $\alpha_2 = 60°$, both flow angles measured from the tangential direction. At the rotor outlet, c_{u2} is oriented in the opposite direction to u. The blade speed u is initially 220 m/s. To optimize the turbine performance, the degree of reaction ε_r is required to be 0.5. Determine the optimal blade speed u_{opt} that achieves the desired degree of reaction.

[117.2 m/s]

NOTES

1. Figure 3.2 refers to a power absorber machine, therefore equations from 3.10 to 3.14 are valid for pumps and compressors.
2. Leonhard Euler (1707–1783) did not directly derive the equations. Rather, they are named in honor of Euler's fundamental contributions to fluid dynamics and mechanics, which

established the theoretical framework for understanding the laws governing turbomachinery.

3. Stodola, A., *Steam and Gas Turbines*, McGraw-Hill Book Company, New York, N. Y. (1927).
4. Wiesner, F. J., "A Review of Slip Factors for Centrifugal Impellers," *ASME Journal of Engineering for Power*, 89(4): 558–566 (1967).
5. Flow separation on the blade's suction side may result from excessive pressure loading. The exit flow angle is further distorted by secondary flow patterns (Chapter 8) that are produced by this separation. The interaction between the main flow and the boundary layers can also produce secondary flows, which add to the overall deviation.
6. See Appendix A.1.

4 Dimensional Analysis of Turbomachines

4.1 GENERAL FRAMEWORK

Dimensional analysis is a mathematical method used to explore the relationships between physical quantities by examining their fundamental dimensions. This method simplifies complex physical phenomena by reducing the number of variables involved, which is particularly useful in the fields of engineering and physics. By expressing variables in terms of fundamental dimensions such as mass [M], length [L], and time [T], we can derive dimensionless parameters that govern the behavior of physical systems.

Dimensional analysis is essential to scaling laws, similarity studies, and performance prediction in turbomachinery. This method allows engineers to create more effective and efficient machinery by using dimensionless characteristics and develop an in-depth comprehensive understanding of machine operational, or *functional*, characteristics.

One of the most powerful tools in dimensional analysis is the Buckingham theorem. This theorem states that if a physical phenomenon involves n variables and these variables are expressed in terms of m fundamental dimensions (3 in our case), the same phenomenon can be described in its completeness by using a reduced set of $n - m$ dimensionless parameters called π groups. These dimensionless parameters capture the essence of the physical problem, making it easier to study and analyze.

4.2 TURBOMACHINES FOR INCOMPRESSIBLE FLOWS

In incompressible flow turbomachines, the dynamic nature of energy exchange is governed by the transfer of mechanical energy through pressure forces acting on the fluid without significant changes in density. As a result, the primary performance indicators—namely the specific energy exchange gh, the efficiency η and the power involved P—are functions of a set of physical parameters that influence the momentum and viscous behavior of the flow. These *dependent variables* are determined by the interaction of the flow with the geometry of the machine, its rotational speed, and the fluid's properties.

The *independent variables* must therefore include: the volumetric flow rate Q, which sets the operating point on the machine's performance map; the rotational speed N, which governs the blade tip velocity and dynamic head; the characteristic length D, which scales the geometry; the fluid density ρ, which appears directly in the momentum balance and energy expressions; and the dynamic viscosity μ, which influences boundary layer development, skin friction losses and ultimately

 DOI: 10.1201/9781003619956-4

the efficiency. Additionally, geometric ratios l_i/D account for shape effects such as aspect ratios, blade angles, hub-to-tip ratios, etc.

These variables, the dependent and independent ones, form a complete set for dimensional analysis and are sufficient to capture the physical mechanisms influencing incompressible turbomachine performance.

To illustrate, consider a centrifugal water pump of fixed geometry, operating at a known inlet temperature and pressure. Given that the density and viscosity of water can be considered constant under these conditions, the fluid properties are fixed. Furthermore, for a specific impeller geometry (i.e. fixed D and $\frac{l_i}{D}$, the only remaining free parameters are the rotational speed N and the flow rate Q. Thus, the machine's head, efficiency and required power become functions solely of N and Q. This justifies the reduction of the performance dependency to two operational variables.

Let us formalize these concepts in a structured and general way. The entire set of variables of interest includes the following:

- Volumetric flow rate Q $[L^3T^{-1}]$
- Rotational speed N $[T^{-1}]$
- Energy per unit mass gh $[L^2T^{-2}]$[1]
- Power P $[ML^2T^{-3}]$
- Fluid density ρ $[ML^{-3}]$
- Dynamic viscosity μ $[ML^{-1}T^{-1}]$
- Characteristic length, e.g. diameter D $[L]$
- i-th Geometric ratio $\frac{l_i}{D}$ [-]
- Efficiency η [-]

where the dependent variables are:

$$(gh, \eta, P) \tag{4.1}$$

while the independent variables are:

$$\left(Q, N, D, \rho, \mu, \frac{l_i}{D}\right) \tag{4.2}$$

such that:

$$(gh, \eta, P) = \mathbf{f}\left(Q, N, D, \rho, \mu, \frac{l_i}{D}\right) \tag{4.3}$$

In this equation, the total number of variables is 9, which means that, after applying the Buckingham theorem, a set of $9 - 3 = 6$ dimensionless groups is expected.

The set of dimensionless groups that satisfy the Buckingham theorem is not unique; therefore, some constraints need to be specified. A common choice is to set ρ, N and D as the repeating variables (i.e. they must not form a dimensionless group by themselves). Using this hypothesis, one can easily derive the π_1 group for gh:

$$\pi_1 = gh \cdot \rho^a \cdot N^b \cdot D^c \tag{4.4}$$

which, after substituting the dimensions and equating the exponents of each dimension to zero:

$$[L^2T^{-2}]\cdot[ML^{-3}]^a\cdot[T^{-1}]^b\cdot[L]^c = L^{2-3a+c}T^{-2-b}M^a = M^0L^0T^0 \tag{4.5}$$

can be written as:

$$\begin{cases} a = 0 & \text{(mass)} \\ 2-3a+c = 0 & \text{(length)} \\ -2-b = 0 & \text{(time)} \end{cases} \tag{4.6}$$

and solving this gives: $a = 0, b = -2, c = -2$, which leads to:

$$\pi_1 = \frac{gh}{N^2D^2} \tag{4.7}$$

Similarly, for η:

$$\pi_2 = \eta\cdot\rho^a\cdot N^b\cdot D^c = \eta \text{ (already dimensionless)} \tag{4.8}$$

For P:

$$\pi_3 = P\cdot\rho^a\cdot N^b\cdot D^c \tag{4.9}$$

upon substituting the dimensions:

$$[ML^2T^{-3}]\cdot[ML^{-3}]^a\cdot[T^{-1}]^b\cdot[L]^c = M^{1+a}L^{2-3a+c}T^{-3-b} = M^0L^0T^0 \tag{4.10}$$

Once again, we equate the exponents of each dimension to zero:

$$\begin{cases} 1+a = 0 & \text{(mass)} \\ 2-3a+c = 0 & \text{(length)} \\ -3-b = 0 & \text{(time)} \end{cases} \tag{4.11}$$

Solving these equations gives: $a = -1$, $b = -3$, $c = -5$, therefore:

$$\pi_3 = \frac{P}{\rho N^3D^5} \tag{4.12}$$

Performing the same steps for Q gives:

$$\pi_4 = \frac{Q}{ND^3} \tag{4.13}$$

while, for μ:

$$\pi_5 = \frac{\rho ND^2}{\mu} \tag{4.14}$$

Geometric ratios are already dimensionless and can be grouped into a vector $\mathbf{N_s}$:

$$\mathbf{N_s} = \left(\frac{l_1}{D}, \frac{l_2}{D}, \ldots, \frac{l_i}{D}, \ldots, \frac{l_N}{D}\right) \tag{4.15}$$

As a result, the functional relationships established by equation 4.3 can be expressed as:

$$\psi = \frac{gh}{N^2D^2} = f_1\left(\frac{Q}{ND^3}, \frac{\rho ND^2}{\mu}, \mathbf{N_s}\right) = f_1\left(\phi, \mathrm{Re}, \mathbf{N_s}\right) \tag{4.16}$$

$$\eta = f_2\left(\frac{Q}{ND^3}, \frac{\rho ND^2}{\mu}, \mathbf{N_s}\right) = f_2\left(\phi, \mathrm{Re}, \mathbf{N_s}\right) \tag{4.17}$$

$$\Lambda = \frac{P}{\rho N^3D^5} = f_3\left(\frac{Q}{ND^3}, \frac{\rho ND^2}{\mu}, \mathbf{N_s}\right) = f_3\left(\phi, \mathrm{Re}, \mathbf{N_s}\right) \tag{4.18}$$

where:

- ψ is the *pressure*, or *head coefficient*, $\psi = \frac{gh}{N^2D^2}$ [2]
- ϕ is the *flow coefficient*, $\phi = \frac{Q}{ND^3}$ [3]
- Re is the *machine's Reynolds number*, $\mathrm{Re} = \frac{\rho ND^2}{\mu}$
- $\mathbf{N_s}$ is the *shape number*
- Λ is the *power coefficient*, $\Lambda = \frac{P}{\rho N^3D^5}$

4.2.1 CONDITIONS FOR DYNAMIC SIMILARITY

To ensure that performance predictions made from scaled models or different operating conditions remain valid, dynamic similarity must be established between the systems being compared. This requires that certain key similarity conditions are satisfied, allowing the flow and performance characteristics of a turbomachine to be extrapolated accurately across different sizes or operating regimes. These conditions include geometric similarity, kinematic similarity, and Reynolds similarity.

4.2.1.1 Geometric Similarity

Geometric similarity refers to the similarity in shape and relative proportions of geometric features between two or more machines. Specifically, it implies that two different geometries (such as models and prototypes) have corresponding dimensions that are proportional to each other, i.e. they have the same $\mathbf{N_s}$.

4.2.1.2 Kinematic Similarity

Kinematic similarity refers to the condition in which the flow patterns are dynamically similar between different machines satisfying geometric similarity. Kinematic similarity implies that, at corresponding locations within the flow domain, the directions and relative magnitudes of velocity vectors are preserved. As geometric similarity is also satisfied, this results in similar velocity triangles throughout the machine. Consequently, the flow coefficient ϕ remains invariant across the compared systems.

4.2.1.3 Reynolds Similarity

In experimental findings on flow machines, it has been observed that when the Reynolds number exceeds $5 \cdot 10^5$, its influence on turbomachine performance typically decreases significantly, although exact thresholds depend on geometry and flow conditions. This phenomenon occurs because at high Reynolds numbers, the boundary layers on the blades predominantly become turbulent and extremely thin. As a result, these layers have minimal impact on the overall flow characteristics. The parameter most significantly affected by the Reynolds number is efficiency η, which typically increases by 1%–2% as the Reynolds number increases by an order of magnitude.

For turbomachines that handle water, the kinematic viscosity $\nu = \frac{\mu}{\rho}$ is very low. Even when the product of rotational speed and characteristic diameter is relatively low, the corresponding Reynolds number remains high. Consequently, in initial approximations, the effects of the Reynolds number on performance can be disregarded.

4.2.1.4 Dynamic Similarity

Equations from 4.16 to 4.18 state that when geometric, kinematic, and Reynolds similarities are satisfied among different machines, *dynamic similarity* is guaranteed, i.e. ψ, η and Λ do not change as functions of ϕ (Figure 4.1).[4]

4.2.2 AFFINITY LAWS

The affinity laws help predict the performance changes of a turbomachine when the speed is altered. As flow rate Q is directly proportional to rotational speed N and the cube of impeller diameter D (equation 4.13):

$$Q \propto N \cdot D^3 \tag{4.19}$$

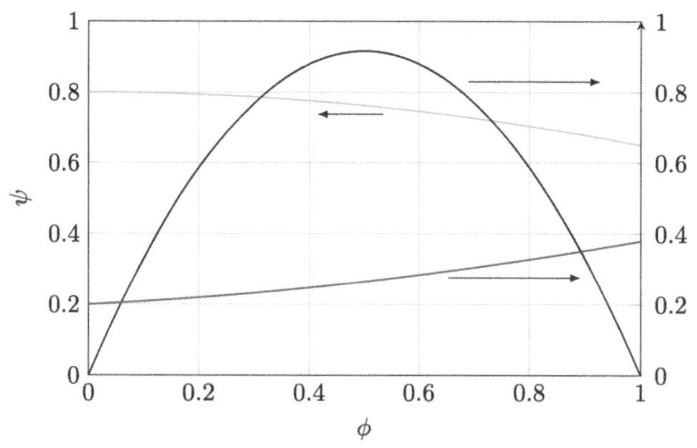

Figure 4.1 Exemplary $(\psi, \eta, \Lambda) - \phi$ curves of a pump.

and comparing two different operating conditions (subscripts 1 and 2) for a given machine:

$$\frac{Q_2}{Q_1} = \frac{N_2}{N_1} \tag{4.20}$$

Analogously, the head h developed by a given turbomachine is proportional to the square of the rotational speed and the square of the impeller diameter (equation 4.7):

$$h \propto (N \cdot D)^2 \implies \frac{h_2}{h_1} = \left(\frac{N_2}{N_1}\right)^2 \tag{4.21}$$

Finally, the power P of a turbomachine is proportional to the cube of the rotational speed and the fifth power of the impeller diameter (equation 4.12):

$$P \propto (N \cdot D)^3 \cdot D^2 = N^3 \cdot D^5 \implies \frac{P_2}{P_1} = \left(\frac{N_2}{N_1}\right)^3 \tag{4.22}$$

Since for a given diameter D, equations 4.19 and 4.21 are parametric curves in N, it follows that the locus of dynamically similar points in the $h-Q$ plane lies on a parabola described by:

$$h = k \cdot Q^2 \tag{4.23}$$

where k is a constant. As a result, we can also state that, if Q^* and h^* define the nominal operating point of a pump, all the points belonging to:

$$h = \frac{h^*}{(Q^*)^2} \cdot Q^2 \tag{4.24}$$

are dynamically similar and, therefore, also share the same value of the peak efficiency η^* (Figure 4.2).

It is worth noting that for the affinity laws and the dynamic similarity to hold, the Reynolds number must be sufficiently high, ensuring that the flow is fully turbulent.

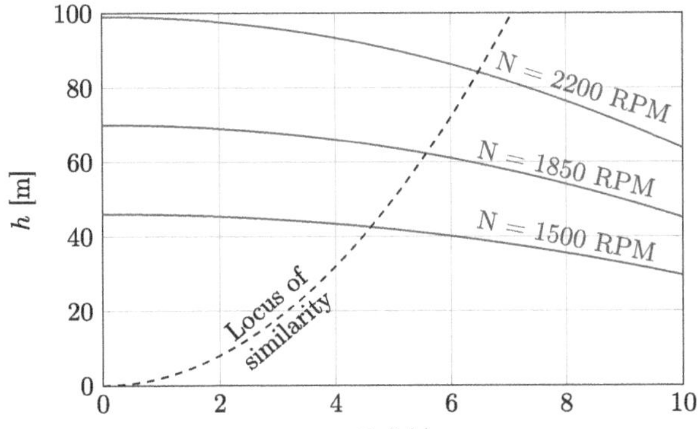

Figure 4.2 Exemplary $h-Q$ curves of a pump at different rpm. The loci of similar operating conditions are parabolas.

When Q becomes very low, the Reynolds number also decreases. If it drops below a critical value, the flow may transition from turbulent to laminar or transitional flow. In such cases, viscous effects become significant and the affinity laws no longer apply accurately.

4.3 TYPE NUMBER OR SPECIFIC SPEED

Engineers have sought to understand how different design parameters affected the performance metrics such as efficiency, flow rate, and head of turbomachines. To this purpose, performance data from numerous machines of different sizes, designs, and operating conditions have been collected and analyzed over time to identify patterns and correlations.

Using dimensional analysis, dimensionless parameters were derived, which are found to correlate strongly with performance characteristics. One of these parameters is the *type number* or *specific speed*:

$$K = \omega_s = \omega \frac{Q^{0.5}}{(gh)^{0.75}} \tag{4.25}$$

By statistically comparing the performance of a large number of machines, the ranges of K values that corresponded to optimal performance could be identified. Machines operating within these optimal ranges were found to be more efficient. These inks are still widely used to determine the dimensions of a machine during preliminary design.

As K is derived from dimensional analysis, it follows that dynamically similar machines do have the same type number. In view of these properties, engineers can use it predict the performance of a turbomachine under different conditions. For example, if a machine with a known K value operates efficiently at a certain flow rate and head, a similar machine with the same K value will exhibit comparable performance characteristics when scaled appropriately.

Concerning hydraulic pumps, as K varies, the most efficient shape of pumps changes as follows (Figure 4.3).

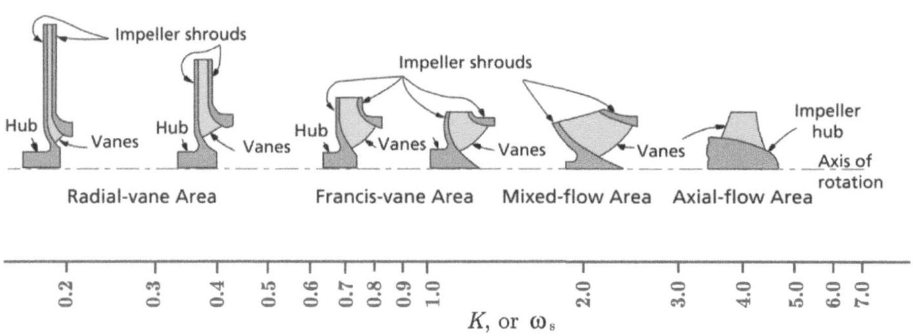

Figure 4.3 Type number for hydraulic pumps.

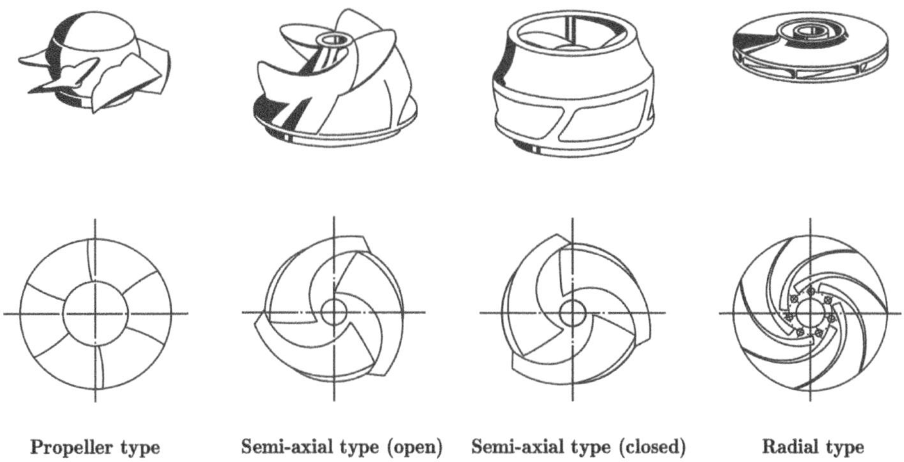

Figure 4.4 Pump impeller shapes.

As the specific speed increases, the impeller geometry transitions through distinct flow categories: starting with radial-vane impellers, moving through Francis-vane and mixed-flow types, and finally reaching axial-flow impellers. In the *radial-vane area*, impellers have vanes that direct the flow radially outward, with narrow, deep channels. These designs are suited for low-flow, high-head applications and correspond to low values of ω_s. As ω_s increases further, the impellers evolve into the *mixed flow area*, where the flow exits the impeller in both radial and axial directions, with wider and more open geometries. At the highest range of ω_s, the impellers enter the *axial-flow area*, where the vanes are almost parallel to the axis of rotation, and the flow is predominantly axial.

Figure 4.4 provides a graphic overview of pump impeller types.

As already mentioned, closed impellers have shrouds on both sides of the blades. For low-to-medium K numbers, these are preferred because they provide better handling of high-pressure conditions. Double shrouds also help minimize leakage and improve hydraulic performance, making them suitable for high-pressure applications.

Open impellers have no shrouds, while semi-axial impellers usually have one on the hub side. These designs handle larger volumes of fluid with lower drag and make them less prone to clogging by debris. They are also easier to clean and maintain. The reduced mechanical strength compared to closed impellers is less of a problem because operating pressures are lower.

4.3.1 DYNAMIC SIMILARITY AND CAVITATION

Cavitation is the formation of vapor-filled cavities (bubbles) in a liquid flow when local pressure falls below the liquid's vapor pressure. In pump systems, it typically occurs at the impeller inlet (eye) where the pressure is lowest. If the pressure in this region drops below the vapor pressure corresponding to the liquid temperature, the liquid begins to boil locally, forming vapor bubbles or cavities. As these convect to higher-pressure regions (for example, moving from the impeller inlet towards the impeller outlet), they suddenly collapse back into liquid. This sequence of bubble formation and violent collapse is the essence of cavitation.

Cavitation significantly impacts dynamic similarity because of its complex influence on fluid flow. For dynamic similarity, the flow patterns in the model and prototype should be identical. However, the presence of cavitation bubbles makes it challenging to replicate these patterns accurately, as the scale of bubble formation and collapse may not directly translate between the model and the prototype.

To achieve dynamic similarity, it is essential to match the cavitation number Ca between the model and the prototype. The cavitation number is given by:

$$\mathrm{Ca} = \frac{p - p_v}{\frac{1}{2}\rho c^2} \tag{4.26}$$

where p is the local pressure, p_v the vapor pressure of the fluid and c a characteristic velocity of the flow.

This number indicates the likelihood of the fluid to cavitate. If the cavitation number is not matched, the onset and severity of cavitation will differ between the model and the prototype, leading to discrepancies in flow characteristics.

4.4 TURBOMACHINES FOR COMPRESSIBLE FLOWS

Compressible flows involve variations in fluid density, pressure, and temperature, all of which significantly influence the flow field and energy exchange processes within the machine.

To accurately capture these effects, two additional variables must be introduced beyond those used in incompressible similarity analysis. First, the speed of sound evaluated at inlet stagnation conditions, a_1^0, is used to define a Mach-like number that reflects the ratio of blade speed to the acoustic propagation speed in the fluid. Second, the specific heat ratio γ is included to account for the thermodynamic behavior of the working fluid, as it governs the relationship between thermodynamic properties and affects the energy transformation through isentropic relations.

In this context, the choice of stagnation quantities such as the isentropic stagnation enthalpy change Δh_s^0 and the stagnation speed of sound a_1^0 is motivated by the need to represent the total energy exchange occurring in the machine which includes both static and dynamic contributions.

The relevant physical variables in compressible flow turbomachinery include:

- Mass flow rate $\dot{m}$ $[MT^{-1}]$
- Rotational speed N $[T^{-1}]$
- Isentropic stagnation enthalpy change[5] Δh_s^0 $[L^2T^{-2}]$
- Power P $[ML^2T^{-3}]$
- Density at inlet stagnation conditions ρ_1^0 $[ML^{-3}]$
- Speed of sound at inlet stagnation conditions a_1^0 $[LT^{-1}]$

- Specific heat ratio γ [-]
- Viscosity μ $[ML^{-1}T^{-1}]$
- Characteristic length, e.g. diameter D $[L]$
- i-th Geometric ratio $\left(\frac{l_i}{D}\right)$ [-]
- Efficiency η [-]

The dependent variables are:

$$\left(\Delta h_s^0, \eta, P\right) \tag{4.27}$$

while the independent ones are:

$$\left(\dot{m}, N, D, \rho_1^0, \mu, a_1^0, \gamma, \frac{l_i}{D}\right) \tag{4.28}$$

such that:

$$\left(\Delta h_s^0, \eta, P\right) = \mathbf{f}\left(\dot{m}, N, D, \rho_1^0, \mu, a_1^0, \gamma, \frac{l_i}{D}\right) \tag{4.29}$$

This gives us a total of 11 variables, which means that, after applying the Buckingham theorem, we expect a set of $11 - 3 = 8$ dimensionless groups.

Choosing ρ_1^0, N and D as repeating variables, we derive the dimensionless groups as follows.

For Δh_s^0:

$$\pi_1 = \Delta h_s^0 \cdot (\rho_1^0)^a \cdot N^b \cdot D^c \tag{4.30}$$

from which, upon substitution of the dimensional expressions, we have:

$$[L^2T^{-2}] \cdot [ML^{-3}]^a \cdot [T^{-1}]^b \cdot [L]^c = L^{2-3a+c}T^{-2-b}M^a = M^0L^0T^0 \tag{4.31}$$

and equating the exponents of each dimension to zero:

$$\begin{cases} a = 0 & \text{(mass)} \\ 2 - 3a + c = 0 & \text{(length)} \\ -2 - b = 0 & \text{(time)} \end{cases} \tag{4.32}$$

which returns: $a = 0, b = -2, c = -2$. Thus, we get:

$$\pi_1 = \frac{\Delta h_s^0}{N^2 D^2} \tag{4.33}$$

For η:

$$\pi_2 = \eta \cdot (\rho_1^0)^a \cdot N^b \cdot D^c = \eta \tag{4.34}$$

For P:

$$\pi_3 = P \cdot (\rho_1^0)^a \cdot N^b \cdot D^c \tag{4.35}$$

from which, upon substitution of the dimensional expressions, we obtain:

$$[ML^2T^{-3}] \cdot [ML^{-3}]^a \cdot [T^{-1}]^b \cdot [L]^c = M^{1+a}L^{2-3a+c}T^{-3-b} = M^0L^0T^0 \tag{4.36}$$

and equating the exponents of each dimension to zero:

$$\begin{cases} 1+a=0 & \text{(mass)} \\ 2-3a+c=0 & \text{(length)} \\ -3-b=0 & \text{(time)} \end{cases} \tag{4.37}$$

we obtain: $a=-1, b=-3, c=-5$. Thus:

$$\pi_3 = \frac{P}{\rho_1^0 N^3 D^5} \tag{4.38}$$

For $\dot{m}$:

$$\pi_4 = \dot{m} \cdot (\rho_1^0)^a \cdot N^b \cdot D^c \tag{4.39}$$

by substituting the corresponding dimensions, we obtain:

$$[MT^{-1}] \cdot [ML^{-3}]^a \cdot [T^{-1}]^b \cdot [L]^c = M^{1+a} L^{-3a+c} T^{-1-b} = M^0 L^0 T^0 \tag{4.40}$$

and equating the exponents of each dimension to zero:

$$\begin{cases} 1+a=0 & \text{(mass)} \\ -3a+c=0 & \text{(length)} \\ -1-b=0 & \text{(time)} \end{cases} \tag{4.41}$$

we obtain: $a=-1, b=-1, c=-3$. Thus:

$$\pi_4 = \frac{\dot{m}}{\rho_1^0 N D^3} \tag{4.42}$$

For μ, a_1^0 and γ:

$$\pi_5 = \frac{\rho_1^0 N D^2}{\mu}; \; \pi_6 = \frac{ND}{a_1^0}; \; \pi_7 = \gamma \tag{4.43}$$

For geometric ratios:

$$\mathbf{N_s} = \left(\frac{l_1}{D}, \frac{l_2}{D}, \ldots, \frac{l_i}{D}, \ldots, \frac{l_N}{D} \right) \tag{4.44}$$

4.4.1 FUNCTIONAL RELATIONSHIPS

Based on the results of the dimensional analysis, the functional relationships among the resulting π groups can be expressed as:

$$\psi_s = \frac{\Delta h_s^0}{N^2 D^2} = f_1 \left(\frac{\dot{m}}{\rho_1^0 N D^3}, \frac{\rho_1^0 N D^2}{\mu}, \frac{ND}{a_1^0}, \gamma, \mathbf{N_s} \right) = f_1 (\phi, \mathrm{Re}, \mathrm{Ma}, \gamma, \mathbf{N_s}) \tag{4.45}$$

$$\eta = f_2 \left(\frac{\dot{m}}{\rho_1^0 N D^3}, \frac{\rho_1^0 N D^2}{\mu}, \frac{ND}{a_1^0}, \gamma, \mathbf{N_s} \right) = f_2 (\phi, \mathrm{Re}, \mathrm{Ma}, \gamma, \mathbf{N_s}) \tag{4.46}$$

$$\Lambda = \frac{P}{\rho_1^0 N^3 D^5} = f_3\left(\frac{\dot{m}}{\rho_1^0 N D^3}, \frac{\rho_1^0 N D^2}{\mu}, \frac{ND}{a_1^0}, \gamma, \mathbf{N_s}\right) = f_3\left(\phi, \mathrm{Re}, \mathrm{Ma}, \gamma, \mathbf{N_s}\right) \quad (4.47)$$

where:

- ψ_s is the *isentropic work*, or *isentropic loading coefficient*,[6] $\psi_s = \frac{\Delta h_s^0}{N^2 D^2}$
- ϕ is the *flow coefficient*, $\phi = \frac{\dot{m}}{\rho_1^0 N D^3}$
- Re is the *machine's Reynolds number*, $\mathrm{Re} = \frac{\rho_1^0 N D^2}{\mu}$
- Ma is the *machine's Mach number*, $\mathrm{Ma} = \frac{ND}{a_1^0}$
- $\mathbf{N_s}$ is the *shape number*
- Λ is the *power coefficient*, $\Lambda = \frac{P}{\rho_1^0 N^3 D^5}$

This set of dimensionless groups, though physically sound, does not allow for a straightforward interpretation of the characteristic curves, which typically overlap as the Mach number changes (Figure 4.5).

To derive more useful charts, dimensionless groups can be changed. As ND is dimensionally equivalent to $\sqrt{\gamma R T_1^0}$ and $\rho_1^0 = p_1^0/(R T_1^0)$, equation 4.45 modifies into:

$$\frac{\Delta h_s^0}{\gamma R T_1^0} = \hat{f}_1\left(\frac{\dot{m}\sqrt{\gamma R T_1^0}}{p_1^0 \gamma D^2}, \frac{\rho_1^0 N D^2}{\mu}, \frac{ND}{a_1^0}, \gamma, \mathbf{N_s}\right)$$

which, for a machine of assigned diameter and shape operating with a known gas, D, γ and R are known and therefore can be omitted:

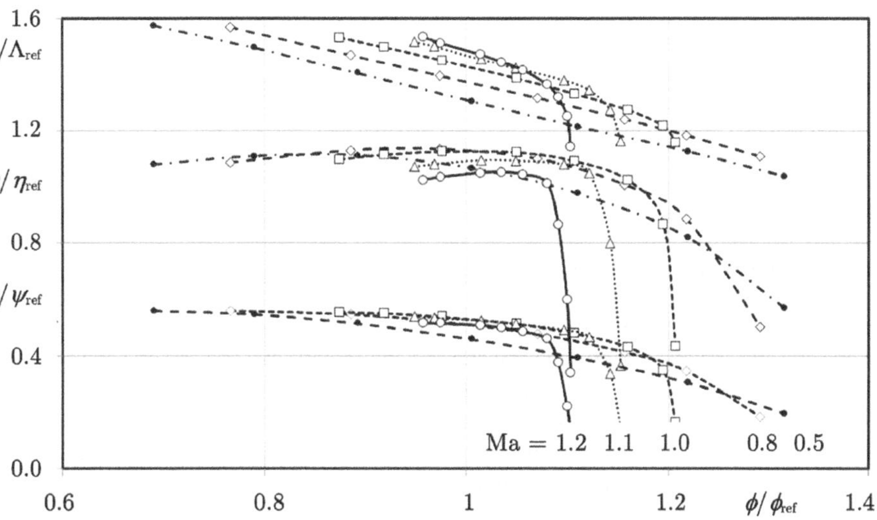

Figure 4.5 Dimensionless characteristic map for a compressor.

$$\frac{\Delta h_s^0}{\gamma R T_1^0} = \tilde{f}_1 \left(\frac{\dot{m}\sqrt{T_1^0}}{p_1^0}, \text{Re}, \frac{N}{\sqrt{T_1^0}} \right)$$

Furthermore, as (e.g. for a compressor) the following equation holds:

$$\frac{\Delta h_s^0}{\gamma R T_1^0} = \frac{1}{\gamma - 1} \left[\left(\frac{p_2^0}{p_1^0} \right)^{\frac{\gamma-1}{\gamma}} - 1 \right] = \tilde{\tilde{f}}(p_2^0/p_1^0)$$

we can account for the total pressure ratio instead of $\frac{\Delta h_s^0}{\gamma R T_1^0}$:

$$\frac{p_2^0}{p_1^0} = \bar{f}_1 \left(\frac{\dot{m}\sqrt{T_1^0}}{p_1^0}, \text{Re}, \frac{N}{\sqrt{T_1^0}} \right) \tag{4.48}$$

and, including the remaining groups:

$$\left(\frac{p_2^0}{p_1^0}, \eta, \Lambda \right) = \bar{\mathbf{f}} \left(\frac{\dot{m}\sqrt{T_1^0}}{p_1^0}, \text{Re}, \frac{N}{\sqrt{T_1^0}} \right) \tag{4.49}$$

A typical compressor map is shown in Figure 4.6 which plots the *total pressure ratio* p_2^0/p_1^0 against the *corrected mass flow rate* $\frac{\dot{m}\sqrt{T_1^0}}{p_1^0}$. It includes multiple speed lines representing different *corrected rotational speeds* $\frac{N}{\sqrt{T_1^0}}$ (the machine's Mach numbers). Additionally, iso-efficiency lines are incorporated to indicate regions of constant compressor efficiency η. The *surge line* indicates the boundary where the compressor begins to experience surge, a form of flow instability that can cause significant damage. Operating near the surge line is generally avoided. The *choke line* represents the maximum mass flow rate the compressor can handle at each speed. Beyond this line, the compressor cannot increase the mass flow rate regardless of the pressure ratio.

4.4.2 DYNAMIC SIMILARITY

When geometric, kinematic, and Reynolds similarity[7] are ensured, in compressible flow turbomachines, dynamic similarity can be guaranteed only when Mach similarity is achieved, i.e. when Ma is invariant.

In such conditions, in both compressor and turbine maps, performance data are independent of the thermodynamic conditions of the flow at the inlet, thanks to corrections to mass flow and rotational speed.

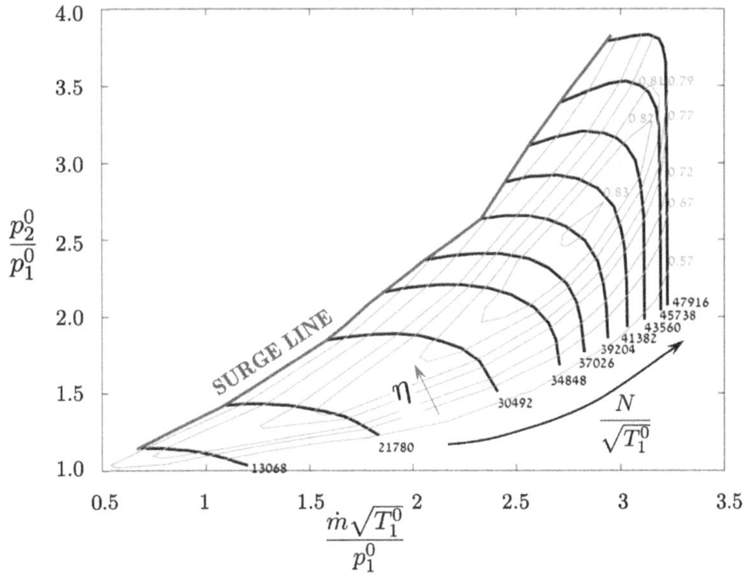

Figure 4.6 Example of a compressor map.

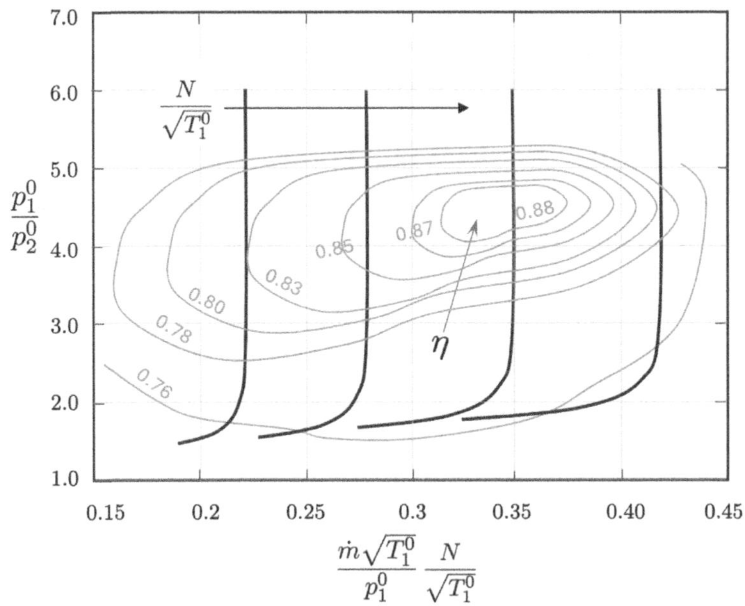

Figure 4.7 Example of a turbine map.

A typical turbine map is given in Figure 4.7 showing the total pressure ratio (also called the *expansion ratio*) $\frac{p_1^0}{p_2^0}$ against the corrected mass flow rate $\frac{\dot{m}\sqrt{T_1^0}}{p_1^0}$. In order to better distinguish the speed lines, in turbines, on the x-axis corrected mass flow rate is multiplied by the corrected rotational speeds $\frac{N}{\sqrt{T_1^0}}$. This adjustment allows for a clearer representation of performance across different operational conditions and makes it possible to visualize iso-efficiency curves.

From the figure, it is evident that compressors and turbines exhibit opposite behaviors with respect to the dependency of energy exchange versus corrected mass flows. The dependency of the speed line trend on corrected flow is as follows: while compressors exhibit a decreasing pressure ratio with increasing corrected flow, turbines display the opposite behavior. This characteristic is also supported by predictions from one-dimensional theory of turbomachinery illustrated in the previous chapter.

A further difference exists in their choking behavior: in compressors, the choking corrected mass flow varies substantially with the rotational speed, whereas in turbines, this dependency is modest. The corrected choking mass flow in turbines is almost independent of the rotational speed. This difference is often attributed to the design characteristics: in compressors, the rotors are prone to choking, while in turbines, the first nozzle is designed to choke regardless of the rotor speed. Another notable difference pertains to efficiency. In compressors, efficiency varies significantly across the map, reaching its maximum values relatively far from the choking lines. In contrast, the efficiency of a turbine remains almost constant and near its maximum values, even when the turbine operates in choke mode.

4.5 THE CORDIER DIAGRAM FOR PUMPS AND FANS

Cordier (1953)[8] gathered extensive performance data from a variety of turbomachines, mainly pumps and fans, operating under different conditions. These data included a derivation of the dimensionless specific diameter, defined as:

$$D_s = D\frac{(gh)^{0.25}}{Q^{0.5}} \tag{4.50}$$

where D is the impeller diameter.

Through regression analysis, Cordier determined a mathematical function $D_s = f(K)$ which describes how the specific diameter corresponding to the best-efficiency configurations varies with specific speed (Figure 4.8). Similar plots relating specific diameter to specific speed have been presented by other authors for various types of turbomachines. For example, Stepannof's chart is widely used for pumps, Balje's diagrams are commonly referenced for compressors and turbines and Pfleiderer's plots are frequently used for fans and blowers.[9]

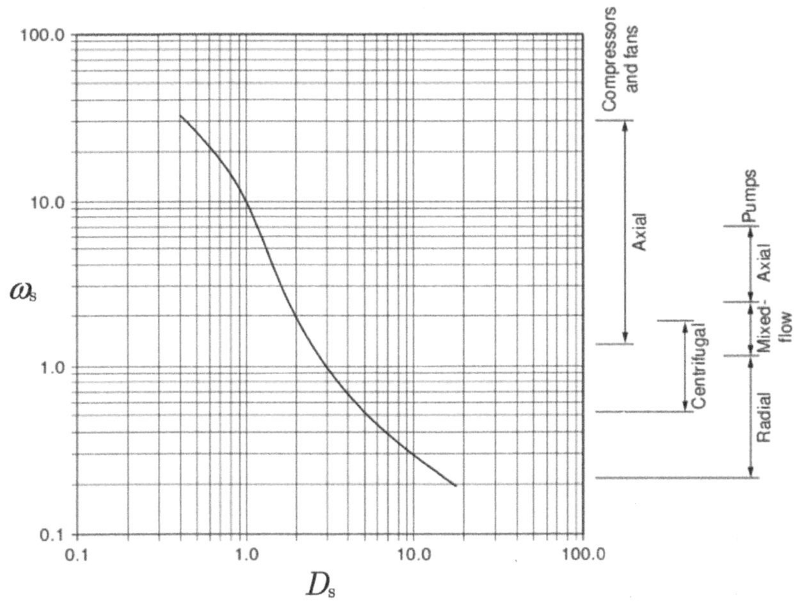

Figure 4.8 The Cordier diagram.

4.6 PROBLEMS

1. Under maximum efficiency conditions, a hydraulic pump handles a flow rate $Q = 26.00\,\text{m}^3/\text{min}$ and generates a head $h = 21\,\text{m}$. The pump efficiency is $\eta = 0.87$. Determine the power absorbed by the pump when its rotational speed is halved, keeping its efficiency constant.

   [12.8 kW]

2. Prove that two pumps operating under maximum efficiency conditions have equal flow coefficient ϕ, pressure coefficient ψ, and specific speed K. The pumps handle respective flow rates $Q_1 = 0.2\,\text{m}^3/\text{s}$ and $Q_2 = 0.35\,\text{m}^3/\text{s}$ and heads $h_1 = 47\,\text{m}$ and $h_2 = 88\,\text{m}$. Their rotational speeds are $N_1 = 1500.00\,\text{rpm}$ and $N_2 = 1814.93\,\text{rpm}$, respectively. The pumps are assumed to be geometrically similar, with a diameter ratio $k = 1.13$.
3. A pump with an impeller diameter $D = 0.5$ m and speed $N = 1200$ rpm delivers a flow rate $Q = 10\ \text{m}^3/\text{min}$ and a head of $h = 30$ m. The efficiency curve is nonlinear and given by:

$$\eta(Q) = \eta_{\text{max}}\left[1 - \left(\frac{Q - Q_{\text{opt}}}{Q_{\text{opt}}}\right)^2\right],$$

   where $\eta_{\text{max}} = 0.85$ and $Q_{\text{opt}} = 12\ \text{m}^3/\text{min}$. Predict the flow rate, head, and efficiency if the speed is reduced to $N = 900$ rpm. Then, account for head losses proportional to Q^2 with a loss coefficient $k_h = 0.1\ \text{m}/(\text{m}^3/\text{min})^2$ and recalculate the pump's head.

[7.5 m^3/min, 16.9 m, 6.3 kW, 11.25 m]

4. A centrifugal pump operates with a total flow rate $Q = 15\ \text{m}^3/\text{min}$, a total head $h = 200$ m and a rotational speed $N = 1200$ rpm. Calculate the specific speed K Next, analyze a design that achieves the same head and flow rate using two identical parallel pumps. Recalculate K for each pump.

[0.21, 0.15]

5. A centrifugal pump must handle $50,000\ \text{m}^3/\text{day}$ at a head of 40 m while maintaining efficiency above 85%. The pump operates at 1200 rpm. Compute the specific speed $\omega_s = K$, and suggest the appropriate pump type and impeller diameter D.

[1.08, 0.5 m]

6. A centrifugal compressor operates at maximum efficiency under sea-level conditions where the ambient pressure is $p_{\text{sea}} = 1\,\text{bar} = 100,000\,\text{Pa}$ and the ambient temperature is $T_{\text{sea}} = 20°\text{C}$. The compressor has a mass flow rate of $\dot{m}_{\text{sea}} = 5.6\,\text{kg/s}$ and operates at a rotational speed of $N_{\text{sea}} = 3600\,\text{rpm}$. Later, the same compressor is installed at a high-altitude location where the ambient pressure decreases to $p_{\text{alt}} = 74,000\,\text{Pa}$ and the ambient temperature drops to $T_{\text{alt}} = -4°\text{C} = 268.15\,\text{K}$. Assess whether, under high-altitude conditions, the working point of the compressor moves toward stall or choking.

[towards choking]

7. Assume the impeller diameter of the compressor in the previous problem is $D_1 = 1.2\,\text{m}$. At a higher altitude, where the ambient conditions are $p_{\text{alt}} = 74,000\,\text{Pa}$ and $T_{\text{alt}} = -4°\text{C}$, a second compressor is designed with a new impeller diameter D_2 for $\dot{m}_2 = \dot{m}_{\text{sea}} = 5.6\,\text{kg/s}$. The design goal is to maintain the same maximum efficiency as the compressor in the previous problem. Determine the impeller diameter D_2 of the second compressor.

[1.37 m]

8. Two geometrically similar gas turbines, Turbine A and Turbine B, operate both under choking conditions. Turbine A has a diameter $D_A = 1.2\,\text{m}$, an inlet stagnation temperature $T_{1A}^0 = 300\,\text{K}$, an inlet stagnation pressure $p_{1A}^0 = 100\,\text{kPa}$ and operates at a rotational speed $N_A = 12,000\,\text{rpm}$. Turbine B has a reduced size with a diameter $D_B = 0.8\,\text{m}$, an inlet stagnation temperature $T_{1B}^0 = 290\,\text{K}$ and an inlet stagnation pressure $p_{1B}^0 = 120\,\text{kPa}$. Determine the rotational speed N_B of Turbine B that ensures dynamic similarity. Verify that the corrected mass flow rates are the same for both turbines.

[17,640 rpm]

NOTES

1. Dimension of head h is [L].
2. As defined in the previous Chapter, a non-dimensional group $\psi = \frac{L}{u_2^2}$ was introduced which was termed *work coefficient*, or *theoretical head coefficient*. This coefficient is linked to the pressure coefficient via the internal, or hydraulic, efficiency, i.e.:

$$\text{Work coefficient} = \text{Pressure coefficient} \cdot \eta_{\text{hyd}}^{\pm 1}$$

where + and − in the exponent are applicable to turbines and pumps, respectively. The same symbol ψ is kept for consistency with the findings, which will be derived in the case of compressible flow machines, where the work coefficient is actually equal to the pressure coefficient.
3. In some cases, the velocity coefficient Φ is used as an alternative to the expression $\frac{Q}{ND^3}$. The velocity coefficient Φ is defined as the ratio of the meridional velocity component c_m to the mean blade speed u, $\Phi = c_m/u$. In pumps, these velocities are measured at the impeller outlet, in turbines at runner's inlet
4. When dimensions vary greatly, such as in large machines compared to small ones, the relative roughness of the surfaces becomes lower. This phenomenon implies reduced friction loss coefficients. Consequently, larger machines often exhibit higher efficiencies.
5. Some authors prefer to use the actual stagnation enthalpy change Δh^0 as this is representative of the specific work exchange.
6. The *work coefficient* $\psi = \frac{\Delta h^0}{N^2 D^2}$ is of popular use also.
7. For moderately high machine Reynolds numbers $\text{Re} > 5 \cdot 10^5$, the influence of Re on performance tends to vanish. However, it is still essential to consider Re for accurate predictions, especially at lower values.
8. Cordier, O., "Ähnlichkeitsbedingungen für strömungsmaschinen," *BWK Bd*, 6(Nr):3 37–340 (1953).
9. Stepanoff, A. J., *Centrifugal and Axial Flow Pumps*, John Wiley & Sons Inc., New York (1957); Balje, O. E., "A Study on Design Criteria and Matching of Turbomachines: Part B—Compressor and Pump Performance and Matching of Turbocomponents," *ASME Journal of Engineering for Power*, 84(1), 103-114 (1962); Pfleiderer, C., *Die Kreiselpumpen*, Springer-Verlag, Berlin Gottingen Heidelberg (1961).

5 Pump Systems

Pump systems are integral to a wide range of civil and industrial applications and serve the essential function of transporting fluids from one location to another. Among the various types of pumps, centrifugal pumps are the most commonly used due to their versatility, efficiency, and simplicity of design. This chapter aims to provide a broad view of centrifugal pumps and focuses on the fundamental fluid dynamics within the pump components, as well as important considerations in the design, operation, and regulation of these systems. Special attention is given to the various forms of energy losses that can occur within the pump, including frictional and mechanical losses. Additionally, this chapter addresses the critical issue of cavitation—a damaging phenomenon that arises when local pressure within the pump drops below the vapor pressure of the fluid—leading to noise, vibration, and potential material erosion.

5.1 GENERAL CHARACTERISTICS OF CENTRIFUGAL PUMPS

A centrifugal pump consists of two essential components: an impeller and a casing (Figure 5.1). The impeller drives the liquid into rotary motion, while the pump casing directs the fluid toward the impeller (entry section) and then discharges it at a higher pressure (delivery section).

In some applications (Figure 5.2), the impeller can have an inducer upstream of it, a high-solidity axial-flow bladed element whose role is to pre-pressurize the liquid, thus reducing the chances of cavitation at the impeller eye. The inducer can be

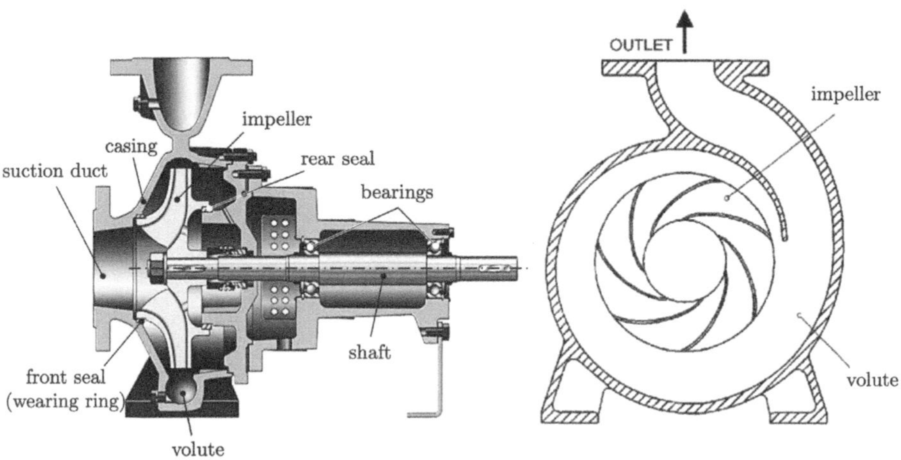

Figure 5.1 Sectional view and a schematic diagram of a centrifugal pump.

DOI: 10.1201/9781003619956-5

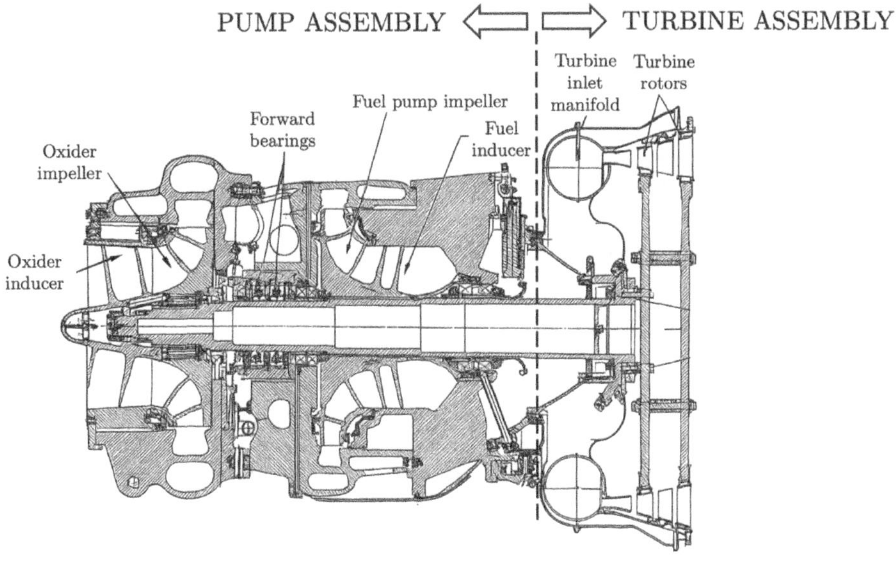

Figure 5.2 The F-1 turbopump assembly for the Saturn V rocket's first stage features a single-shaft, dual-pump unit driven by an axial turbine producing about 40.5 MW. It pressurizes liquid oxygen and RP-1 fuel to feed the main combustion chamber at roughly 103 and 83 bar, respectively. The oxidizer and fuel pumps use axial inducers and centrifugal impellers.

followed by inlet guide vanes (IGV), which can help further deflect the flow before it enters the impeller and provide the mechanical connection between the hub and casing.

The impeller is mounted on a shaft supported by bearings and is driven by a motor through a coupling. Motors can be electric or endothermic depending on the application. The pump casing includes suction and discharge ducts, houses the rotor assembly, and supports the bearings while also being packed around the shaft to prevent external leakage.

To minimize high-pressure liquid leaking back to the suction side, wear rings are fitted between the impeller and the casing. The liquid enters the impeller through the suction duct and is forced into a spiral-like motion by the impeller blades. Radial impellers typically feature shrouds perpendicular to the shaft, while mixed-flow impellers often have double-curvature blades.

As liquid exits the impeller, both its absolute velocity and pressure typically increase, while the relative velocity is ideally reduced to enhance the pressurization effect. The casing then partially converts the kinetic energy of the fluid into pressure, either in a volute or a bladed diffuser.

Volute casings gradually increase in area, while diffusion casings employ guide vanes for more effective conversion of velocity into pressure. The volute may not be radially aligned with the impeller (*overhung*) in order to allow for a more compact design and avoid the need for complex piping that would be required if the fluid exited radially. The pump discharge nozzle can be positioned tangentially on the volute casing or radially in line with the shaft plane by using a gooseneck configuration (Figure 5.1).

In some cases, where maximum efficiency is required at a given operating flow rate, after exiting the impeller, the liquid enters a bladed diffuser, which typically consists of stationary vanes that improve the deceleration of the fluid compared to a vaneless configuration or a simple volute. By reducing the velocity efficiently, the diffuser promotes rapid conversion of the kinetic energy into pressure energy, thus improving the pump's peak efficiency.

In applications requiring higher pressures than a single impeller can provide, multistage pumps are used, with multiple impellers arranged in series (bottom of Figure 5.3). Each stage comprises an impeller, a diffuser, and return guide vanes (typically integrated with the diffuser), all housed within a single stage casing. Parallel, back-to-back impellers are found as well (top of Figure 5.3), which feature a double-suction arrangement and a single casing. Such a solution is adopted when a combination of high-pressure and relatively high-flow rates is needed in a compact installation.

5.2 FUNDAMENTAL FLUID DYNAMICS OF CENTRIFUGAL PUMPS

Fluid dynamics within centrifugal pumps is inherently viscous, three-dimensional, and exhibits complex flow patterns. A simplified yet effective approach to understanding pump behavior involves analyzing the flow as it was inviscid in both the meridional plane and the blade-to-blade set of planes.

This method of modeling is fundamental to understanding the primary phenomena occurring within the pump channels. While it does not account for three-dimensional and viscous effects, it successfully captures the majority of relevant hydrodynamic phenomena that influence the pump's performance, including head generation and cavitation risk. Furthermore, this two-dimensional approach serves as a useful tool in pump design, allowing engineers to predict key performance metrics and optimize the geometry of the impeller and other components.

5.2.1 THE INVISCID MERIDIONAL FLOW

A first-step approach to the analysis of the flow in the meridional plane consists in ignoring the presence of the impeller blades. This allows for an assessment of how the hub and shroud contours influence the flow and enables an understanding of the distribution of velocity and pressure in such a plane. Although simplistic, this model yields realistic results when the pump operates near its design point. Under these conditions, the influence of the blades—responsible for introducing additional source terms in the governing equations—only slightly modifies the flow pattern without fundamentally altering its nature.

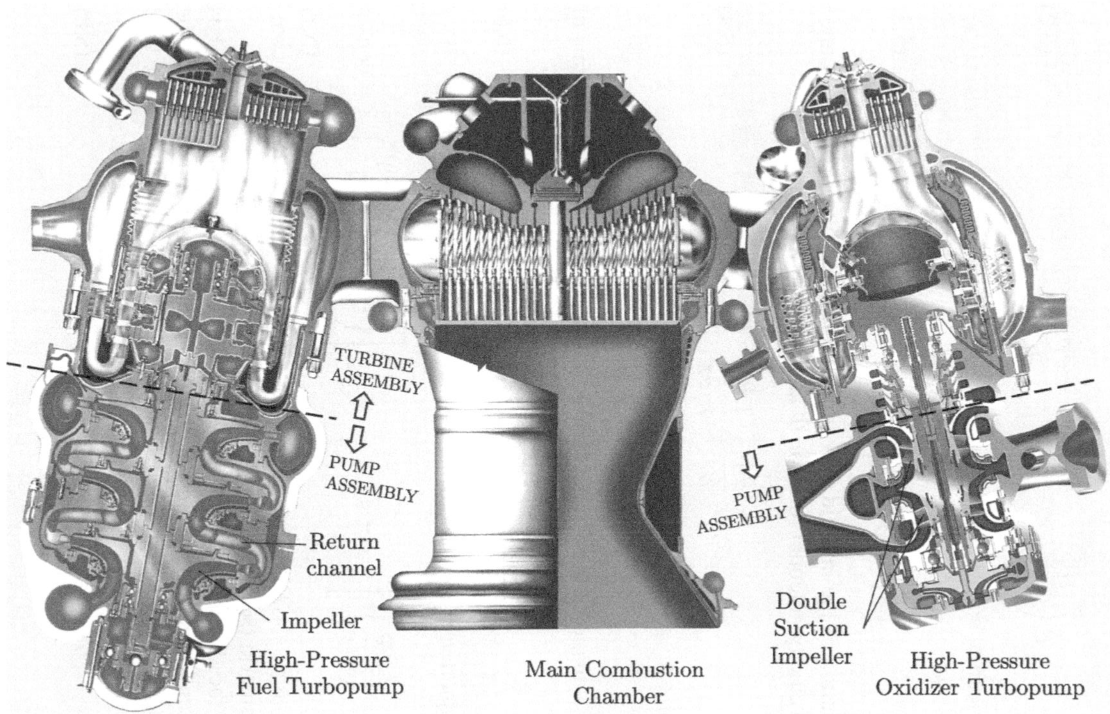

Figure 5.3 The powerhead of the Space Shuttle Main Engine (SSME), featuring two high-pressure turbopumps, each driven by a close-coupled preburner. The high-pressure fuel turbopump is a three-stage centrifugal pump, fully insulated to minimize chilldown time, maintain liquid hydrogen at −253°C, and prevent liquid air formation (−194°C) on the pump. The high-pressure oxidizer turbopump is a double-entry, twin-impeller centrifugal pump with a small boost pump integrated into the shaft.

In the meridional plane, the continuity equation is as follows:

$$\frac{1}{r}\frac{\partial}{\partial r}(rc_r)+\frac{\partial c_z}{\partial z}=0 \tag{5.1}$$

while, assuming inviscid flow, the radial momentum equation is:

$$\rho\left(c_r\frac{\partial c_r}{\partial r}+c_z\frac{\partial c_r}{\partial z}\right)=-\frac{\partial p}{\partial r}+\rho\frac{c_u^2}{r} \tag{5.2}$$

and the axial momentum equation is:

$$\rho\left(c_r\frac{\partial c_z}{\partial r}+c_z\frac{\partial c_z}{\partial z}\right)=-\frac{\partial p}{\partial z} \tag{5.3}$$

An elegant solution of the inviscid flow in the meridional plane can be obtained by introducing a velocity potential function Φ and a stream function Ψ:

$$c_r=\frac{\partial\Phi}{\partial r}=\frac{1}{r}\frac{\partial\Psi}{\partial z} \tag{5.4}$$

$$c_z=\frac{\partial\Phi}{\partial z}=-\frac{1}{r}\frac{\partial\Psi}{\partial r} \tag{5.5}$$

The continuity equation for potential flow is automatically satisfied in terms of stream function:

$$\frac{\partial^2\Psi}{\partial z^2}+\frac{\partial^2\Psi}{\partial r^2}-\frac{1}{r}\frac{\partial\Psi}{\partial r}=0 \tag{5.6}$$

while Bernoulli equation makes it possible to solve for the pressure field:

$$\frac{p}{\rho}+\frac{1}{2}\left[\left(\frac{\partial\Phi}{\partial r}\right)^2+\left(\frac{\partial\Phi}{\partial z}\right)^2\right]=\text{constant},\ \frac{p}{\rho}+\frac{1}{2r^2}\left[\left(\frac{\partial\Psi}{\partial r}\right)^2+\left(\frac{\partial\Psi}{\partial z}\right)^2\right]=\text{constant}. \tag{5.7}$$

Equation 5.6 can be solved using the finite-difference method (FDM) on a computational cartesian grid (see Figure 5.4).[1]

In Figure 5.4, the distribution of the stream function from $\Psi=0$ to $\Psi=1$ is displayed for a sample geometry representative of a centrifugal pump impeller. The range $\Delta\Psi$ between two consecutive values is set such that the i-th fraction of the total flow rate is:

$$Q_i=\frac{Q}{\text{Number of streamtubes}}=\text{constant} \tag{5.8}$$

where

$$Q_i=\pi(r^2_{\Psi=i}-r^2_{\Psi=i-1})c_{m,i} \tag{5.9}$$

being $c_{m,i}=\sqrt{c^2_{z,i}+c^2_{r,i}}$. In particular, when $i=1$, Q_1 corresponds to the fraction of total flow rate Q between $\Psi=0$ (hub contour) and $\Psi=1/(\text{Number of streamtubes})$.

The portion of streamtube between two consecutive values of the stream function is often referred to as an *elementary machine*.

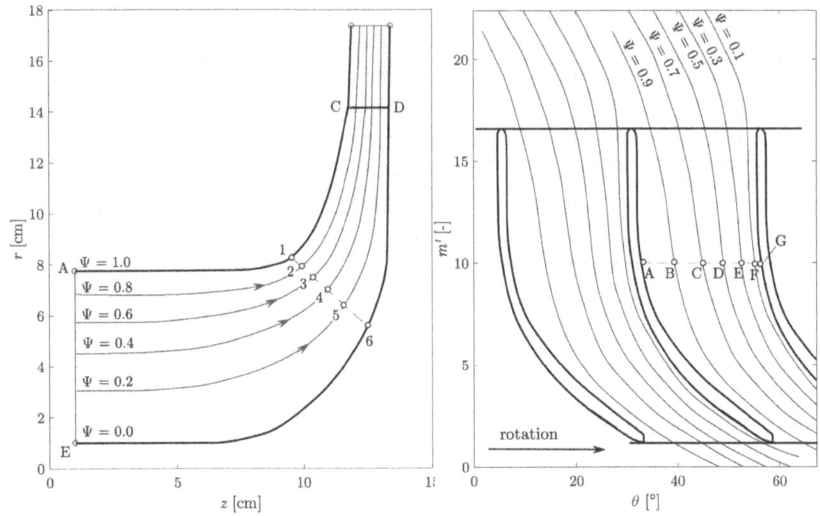

Figure 5.4 Streamlines in a centrifugal pump impeller as a result of potential flow solution.

From the figure, it is immediately apparent that the shape of the meridional channel dictates the distribution of the stream function and consequently the configuration of the streamlines. These tend to crowd near regions where the shroud has the smallest radius of curvature while, to satisfy the continuity equation, they must spread out as they approach the hub contour. This distribution indicates that the maximum meridional velocity occurs near the zone of greatest streamline curvature, which corresponds to minimum static pressure. Moreover, it suggests that a pressure gradient, directed inward from the point of minimum pressure, is established. This velocity and pressure distribution is essential when drawing velocity triangles for each elementary section of the machine and must be carefully considered when assessing the pump's susceptibility to cavitation.

5.2.2 THE INVISCID BLADE-TO-BLADE FLOW

To simplify the analysis, we assume that the flow occurs in a purely radial plane at $z =$ constant, meaning the z -components of the flow quantities are zero. Given that the blade-to-blade relative flow is rotational, Euler's equations cannot be directly solved using a potential flow approach. However, a solution can still be obtained by superimposing two solutions derived from velocity potentials: (i) Φ_1, associated with zero flow through the rotating impeller and (ii) Φ_2, associated with finite flow through a stationary impeller. Since both the boundary conditions and the governing differential equation for continuity:

$$\nabla^2\Phi = \frac{\partial^2\Phi}{\partial r^2} + \frac{1}{r}\frac{\partial\Phi}{\partial r} + \frac{1}{r^2}\frac{\partial^2\Phi}{\partial\theta^2} = 0 \ (z = \text{constant}) \tag{5.10}$$

are linear, the velocity potential for the entire flow can be determined by summing the solutions from the individual cases, i.e. $\Phi = \Phi_1 + K\Phi_2$, where K is a function of flow rate and impeller tip speed. The *eddy-flow solution* Φ_1 represents the inviscid flow in a rotating impeller with the throttle closed and no axial flow. In contrast, the *through-flow solution* Φ_2 describes the flow through an annular space that matches the impeller's hub-to-shroud profile but lacks blades. An example of such a solution is shown in Figure 5.4, where rotating blades are plotted in a conformal plane along with relevant streamlines. The figure illustrates how the flow deviates from the Eulerian one-dimensional pattern as it enters the impeller and exhibits a net slip as it exits.

5.3 SLIP FACTOR IN CENTRIFUGAL PUMPS

As already mentioned in Chapter 3 and, more in detail, in the previous subsection, predicting the slip factor:

$$\sigma_{slip} = \frac{c_{u2}}{c_{u2,\mathrm{E}}} = \frac{c_{u2,\mathrm{E}} - c_s}{c_{u2,\mathrm{E}}} = 1 - \frac{c_s}{c_{u2,\mathrm{E}}} = 1 - \frac{c_s/u_2}{1 - \phi \cot \beta_2'}$$

in centrifugal flow machines involves the superposition of throughflow and rotational flow fields under the hypothesis of inviscid flow. Other factors contributing to the slip include the effects of blade shape and number, as well as, even though to a minor extent, the differences in boundary layer thicknesses on the pressure and suction sides of the blades (especially for long and thick profiles at off-design). Furthermore, thicker boundary layers, which occur at lower Reynolds numbers, enhance these effects, as do the turbulent wakes downstream of rotor blades and disks. Overall, it is very important to note that viscous effects play a much less significant role in determining the slip compared to inviscid effects.

One of the most popular and rigorous approaches to predict the slip factor in centrifugal pumps is that developed by Busemann[2] making use of a sophisticated potential flow theory (Figure 5.5): this is a general method for impeller blades having logarithmic spiral camber lines $\beta' = \mathrm{const}$. Although the details of his method are out of the scope here, it is worth mentioning that the results he obtained were interpolated by Wiesner (Figure 5.6), leading to:

$$\sigma_W = 1 - \frac{\sqrt{\sin \beta_{2'}}}{Z^{0.70}} \tag{5.11}$$

if $\varepsilon_{\mathrm{limit}} = r_1/r_2 < \frac{1}{\ln^{-1}(8.16 \sin \beta_{2'}/Z)}$, or:

$$\sigma_B = \left(1 - \frac{\sqrt{\sin \beta_{2'}}}{Z^{0.70}}\right) \cdot \left[1 - \left(\frac{r_1/r_2 - \varepsilon_{\mathrm{limit}}}{1 - \varepsilon_{\mathrm{limit}}}\right)^3\right] \tag{5.12}$$

otherwise.

Other correlations have been developed over the years, mostly relying on semi-empirical approaches. Among these, one of the most popular is the one proposed by

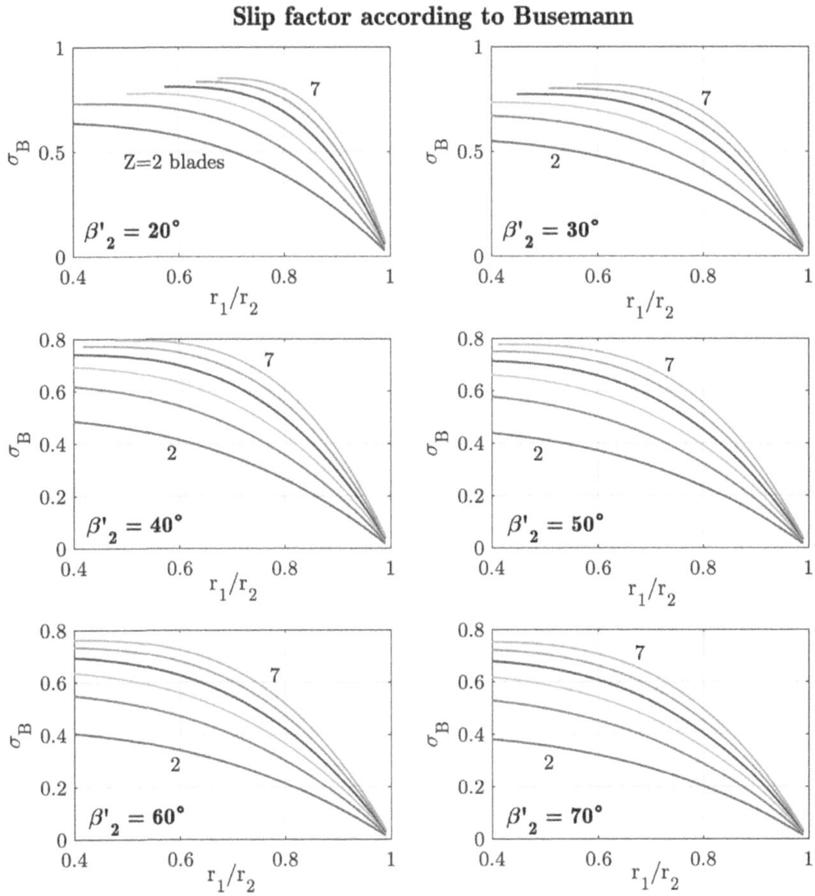

Figure 5.5 Slip factor σ_B as a function of the radius ratio r_1/r_2 for different blade numbers Z and blade exit angles β_2' according to Busemann.

Pfleiderer[3]:

$$\sigma_P = \frac{c_{u2}}{c_{u2,\mathrm{E}}} = \frac{1}{1 + \frac{2Y}{Z}\frac{1}{1-\left(\frac{d_1}{d_2}\right)^2}} \tag{5.13}$$

where $Y = a(1 + \beta_{2'}[°]/60)$ and $a = 0.6$ for impellers with a vaned diffuser, while $a = 0.65 - 0.84$ in the case where only a volute is present.

Once the slip factor has been estimated, the theoretical head h_t of the pump is derived from:

$$gh_t = \frac{gh}{\eta_{\mathrm{hyd}}} = u_2 c_{u2} = u_2^2 \sigma_{\mathrm{slip}} \left(1 - \phi \cot \beta_2'\right) \tag{5.14}$$

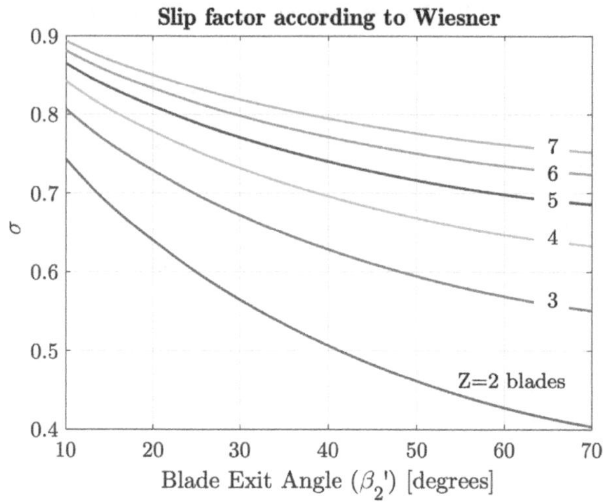

Figure 5.6 Slip factor σ_{slip} as a function of the blade exit angle β_2' for different blade numbers Z according to Wiesner's model.

5.4 VOLUTES AND DIFFUSERS

A volute is responsible for: (i) collecting fluid discharged from the impeller and directing it to the pump outlet, and (ii) converting the velocity head imparted to the fluid by the impeller into pressure, thus allowing efficient pump operation.

In smaller, single-stage pumps, volutes are typically used to connect the impeller to the discharge pipe. This is contrary to medium to high-power single-stage pumps and all multistage pumps, where vaned diffusers are used downstream of the impellers for the same purpose.

5.4.1 VOLUTES

The volute begins at the tongue, where the fluid enters and proceeds along a spiral path until reaching the volute throat, the final section across which the entire discharge flow rate passes.

The governing equations for the current line in a volute build on principles of conservation of momentum and continuity. As shown in Figure 5.7, the angular distribution of the velocity components c_u (from momentum) and c_m (from continuity) defines the trajectory. In particular, the local angle α of a generic fluid particle can be defined as:

$$\tan\alpha = \frac{c_m}{c_u} \tag{5.15}$$

where c_m is the local meridional component and c_u is the local circumferential component of velocity. The meridional velocity is given by:

$$c_m = \frac{Q}{2\pi rb} \tag{5.16}$$

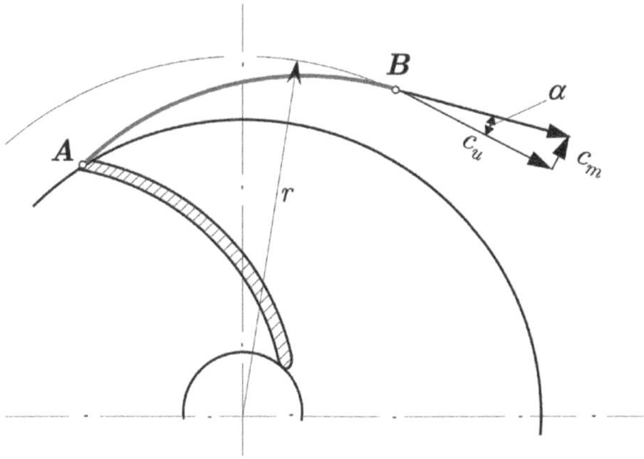

Figure 5.7 Sample trajectory of a particle exiting the volute of a centrifugal pump.

and the circumferential velocity, assuming rc_u is constant along the streamline owing to a free-vortex distribution,[4] is expressed as:

$$rc_u = r_2 c_{u2} = \text{const.} \tag{5.17}$$

When b is constant, the volute cross section is rectangular, and this leads to the following equation for $\tan\alpha$:

$$\tan\alpha = \frac{c_m}{c_u} = \frac{Q}{2\pi b(r_2 c_{u2})} = \text{const.} \tag{5.18}$$

i.e. the equation of a logarithmic spiral.

When b is proportional to r, i.e. the volute cross section is trapezoidal, $b(r) = k \cdot r$, where k is a proportionality constant:

$$\tan\alpha = \frac{Q}{2\pi kr(r_2 c_{u2})} = \frac{\text{const.}}{r} \tag{5.19}$$

i.e. the equation of an Archimedes spiral.

The cross section is often chosen to be circular. In this case, b is a more complex piecewise function of r. The determination of area distribution follows a simple yet effective criterion where area is simply proportional to the angular coordinate:

$$A(\theta) = A_v \frac{\theta(^\circ)}{360} \tag{5.20}$$

where A_v is the throat area. In the case of circular cross sections, at the volute throat the mean velocity is approximately 0.5 to 0.65 times the impeller's outlet circumferential velocity c_{u2}. This relationship can be written as:

$$c_{th} = (0.5 \div 0.65)c_{u2} \tag{5.21}$$

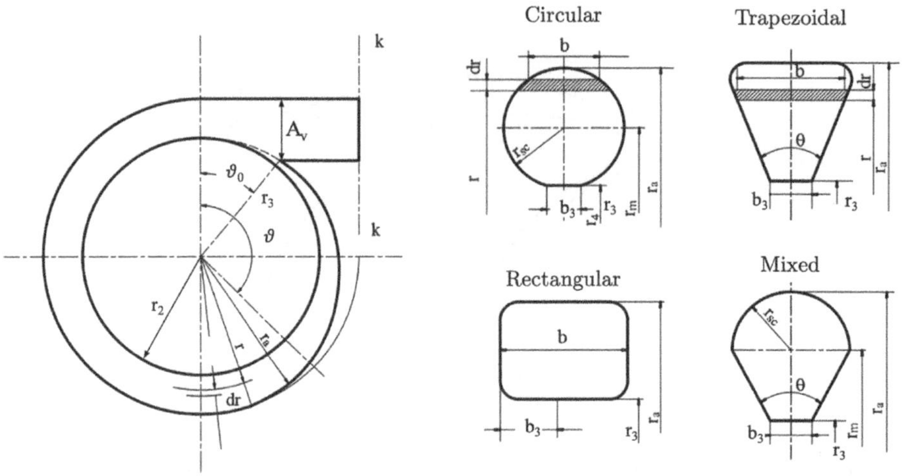

Figure 5.8 Volute geometry and common cross-sectional shapes.

Interestingly, the criterion given in equation 5.21 allows for a constant average velocity across all sections of the volute.

A more detailed sizing of the volute sections, as shown in Figure 5.8, can be achieved through the following procedure. Various shapes can be adopted for these sections, and the outer contour of the volute can be analytically determined. We can make the following assumptions.

i) The flow freely expands in the volute without friction, following a free vortex distribution.
ii) The local flow rate in the circumferential direction, denoted as $Q(\theta)$, is proportional to the local circumferential coordinate θ.

This relationship is described by:

$$q(\theta) = \frac{Q}{2\pi}\theta \tag{5.22}$$

It follows that:

$$q(\theta) = \int c_u dA = \int_{r_3}^{r_a(\theta)} c_u b(r)\,dr = \int_{r_3}^{r_a(\theta)} r c_u b(r)\frac{dr}{r} = c_{u3} r_3 \int_{r_3}^{r_a(\theta)} b(r)\frac{dr}{r} \tag{5.23}$$

By combining the above equations, we obtain:

$$\theta = \frac{2\pi c_{u3} r_3}{Q}\int_{r_3}^{r_a(\theta)} b(r)\frac{dr}{r} = \frac{c_{u3}}{c_{r3}}\int_{r_3}^{r_a(\theta)} \frac{b(r)}{b_3}\frac{dr}{r} = \frac{1}{\tan\alpha_3}\int_{r_3}^{r_a(\theta)} \frac{b(r)}{b_3}\frac{dr}{r} \tag{5.24}$$

from which the following cases can be discussed:

- $b(r) = \text{constant} = b_3$

$$\theta - \theta_0 = \frac{1}{\tan \alpha_3} \ln \frac{r_a(\theta)}{r_3} \implies r_a(\theta) = r_3 \cdot e^{(\theta-\theta_0)\cdot \tan \alpha_3} \tag{5.25}$$

- $b(r) = b_3 + kr$

$$\theta - \theta_0 = \frac{1}{\tan \alpha_3} \left[\ln \frac{r_a(\theta)}{r_3} + \frac{k}{b_3} [r_a(\theta) - r_3] \right] \tag{5.26}$$

- $b(r)$ is the chord of a circle (see Figure 5.8). This case is complicated, as $b(r)$ can be made explicit only when both the radius R and the r -coordinate of the circle's center r_0 of the circle are defined. In this case,

$$b(r) = 2\sqrt{R^2 - (r - r_0)^2} \tag{5.27}$$

However, since R and r_0 themselves are a function of θ, obtaining the solution is not straightforward. In this case, equations 5.20 and 5.21 can be used instead.

The choice of a circular cross section is motivated by advantages related to fluid dynamics, ease of manufacturing, and operational efficiency. In some specific circumstances, rectangular and trapezoidal cross-sections also have specific advantages over circular ones (see Table 5.1).

Independent of the shape of the cross section, the flow velocity across the volute is not uniform (Figure 5.9). Along any given streamline, the flow diffuses, resulting in a mean velocity ranging between 0.78 and 0.92 of the maximum velocity at the center of the flow path. Losses at the volute tongue, particularly *ram losses* and *separation losses*, are minimized by ensuring that the volute angle α_3 at the tongue corresponds to the absolute velocity vector at the impeller outlet. To account for the *free vortex diffusion* between the impeller exit and the volute tongue, a gap is provided between them. The diameter of the tongue, D_3, is usually positioned as follows:

$$D_3 = (1.08 \div 1.2)D_2 \tag{5.28}$$

This allows the flow to adjust with minimum losses. Additionally, the volute width at the tongue, b_3, is set to accommodate a range of impeller dimensions, typically:

$$b_3 = (1.5 \div 2) \cdot b_2 \tag{5.29}$$

This flexibility in design ensures that pumps can operate efficiently across a wide range of operating conditions.

5.4.2 DIFFUSERS

A vaned diffuser is designed to improve hydraulic efficiency by converting the kinetic energy of the fluid exiting the impeller into pressure energy. This is achieved through a series of stationary vanes, which guide the flow as it leaves the impeller.

Table 5.1
Comparison of Rectangular, Trapezoidal, and Circular Cross-Sections in a Pump Volute

Design Aspect	Rectangular	Trapezoidal	Circular
Use of Space	Excellent for compact designs; efficient use of tight spaces, fitting along flat surfaces.	Efficient in wide-area flow systems like channels.	Best suited for circular or semi-circular piping.
Flow Characteristics	Can cause turbulence or flow separation at corners, but good for controlled, confined flows.	Better for open flow systems, reduces turbulence in wide flow areas.	Smooth flow, minimal turbulence or separation.
Manufacturing	Easy to fabricate using flat plates; cheaper for non-complex designs.	Moderately easy to fabricate; common in civil and hydraulic engineering.	Slightly more expensive due to curved geometries.
Structural Stability	Performs well in constrained environments but can create stress at corners.	Stable in open-channel flows and hydraulic structures, distributes forces well.	Superior for pressure distribution, reduces stress concentration.
Application Example	HVAC ducts, electric machine cooling channels, low-specific-speed pumps.	Open channels, spillways, irrigation systems, some volute designs.	Pump volutes, pipelines, pressure vessels, closed-channel flows.

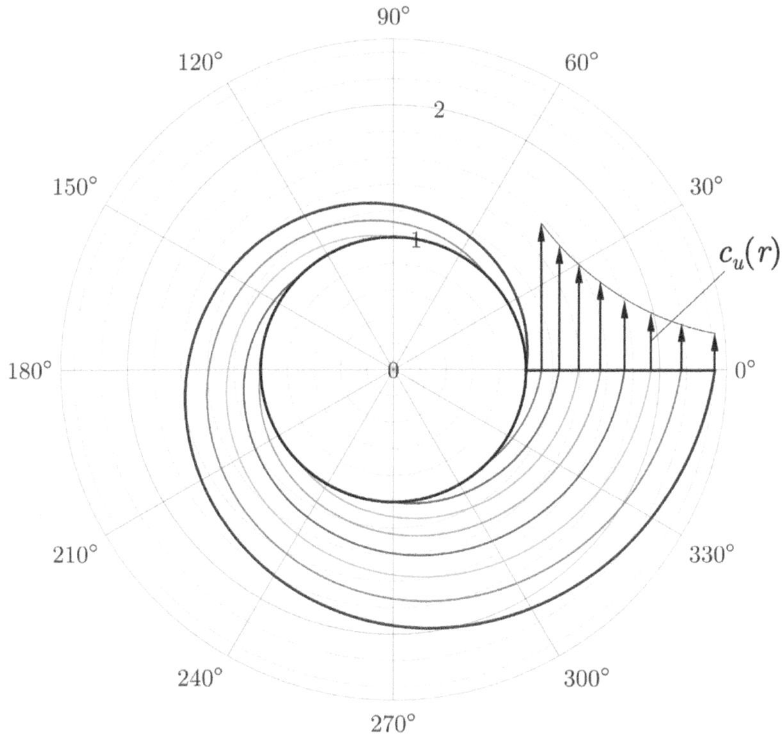

Figure 5.9 Logarithmic spiral-shaped streamlines in a volute with b_3 = constant, released from the impeller exit with an angular pitch of 45°.

The primary role of the vaned diffuser is to decelerate the fluid in a controlled manner, promoting the conversion from kinetic energy to pressure energy in a more effective manner than using just a volute. This can be realized in vaned diffusers by shortening the length of the path the flow would need to reach the same pressure rise as a vaneless diffuser. To avoid excessive viscous losses, this process must be carried out within a limited radial distance from the impeller exit (3) to gain appreciable efficiency gain at the Best Efficiency Point (BEP). A pressure recovery coefficient for a diffuser can be defined as follows:

$$C_{p,r} = \frac{p_4 - p_3}{\frac{1}{2}\rho c_3^2} \tag{5.30}$$

In formal terms, the role of the diffuser is to obtain the maximum pressure recovery with minimum hydraulic losses. To demonstrate the usefulness of a vaned diffuser in pursuing this objective, let us consider the following situations where a free diffusion, i.e. with no vanes, is compared to a forced diffusion achieved by vanes. In case of free diffusion, a fluid particle leaving the impeller follows a logarithmic

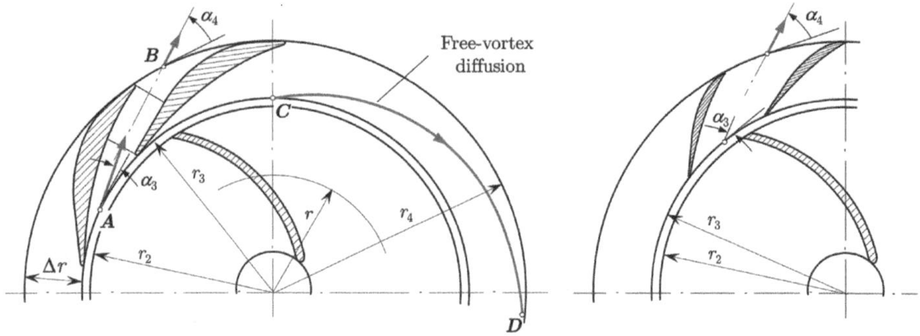

Figure 5.10 Geometry of bladed and bladeless diffusers.

spiral trajectory (from C to D in Figure 5.10) in the polar coordinates r, θ:

$$r = r_3 e^{k\theta} \tag{5.31}$$

such that $k = \tan\alpha_3$, the tangent to the trajectory, is constant. Suppose the particle has to move along this trajectory from r_3 to r_4, with $r_4 > r_3$ $\Delta r = r_4 - r_3$. Let us compute the length of the curve from r_3 to r_4:

$$S_{C-D} = \int \sqrt{dr^2 + (rd\theta)^2} \tag{5.32}$$

Using the relation for the logarithmic spiral $d\theta = \frac{1}{\tan\alpha_3}\frac{dr}{r}$, we get:

$$S_{C-D} = \int_{r_3}^{r_4} \sqrt{1 + \left(\frac{1}{\tan\alpha_3}\right)^2}\, dr = \frac{1}{\sin\alpha_3}(r_4 - r_3) \tag{5.33}$$

On the other hand, if we assume that by using vanes the flow follows almost a straight line in the diffuser, we find that the curve length can be well approximated by the length of segment A-B (see Figure 5.10):

$$S_{\text{A-B}} = \frac{\Delta r}{\sin\alpha_4} \tag{5.34}$$

From the geometric analysis of a straight line diffuser whose inlet angle is α_3, the outlet angle α_4 is a function of Δr, or, as an alternative, of $\frac{r_4}{r_3}$,[5] as displayed in Figure 5.11.

Figure 5.12 shows a quantitative comparison between vaned and vaneless (free-vortex) diffuser configurations. The effect on the ideal pressure recovery coefficient $C_{p,r}$ of increasing the radius ratio r_4/r_3 for any given α_3 is much more significant in vaned diffusers than in vane-less ones where free vortex diffusion is obtained. At the same time, we clearly see that for a given radius ratio, the length of diffusing flow paths A-B and C-D is comparable only for radius ratios tending to one, which,

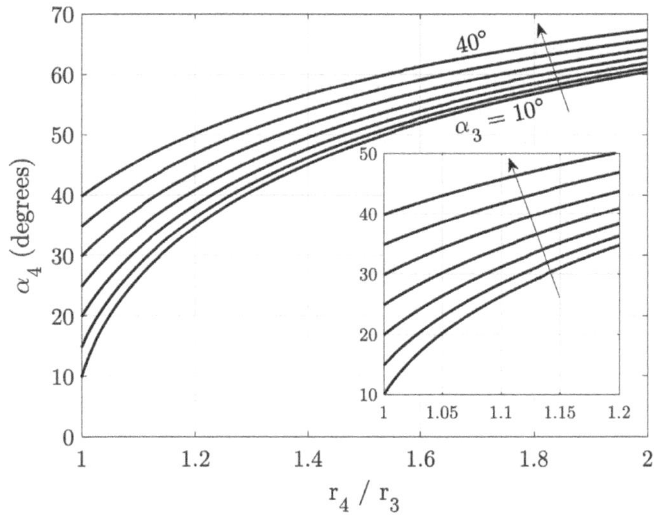

Figure 5.11 Geometrical characteristics of a straight line bladed diffuser.

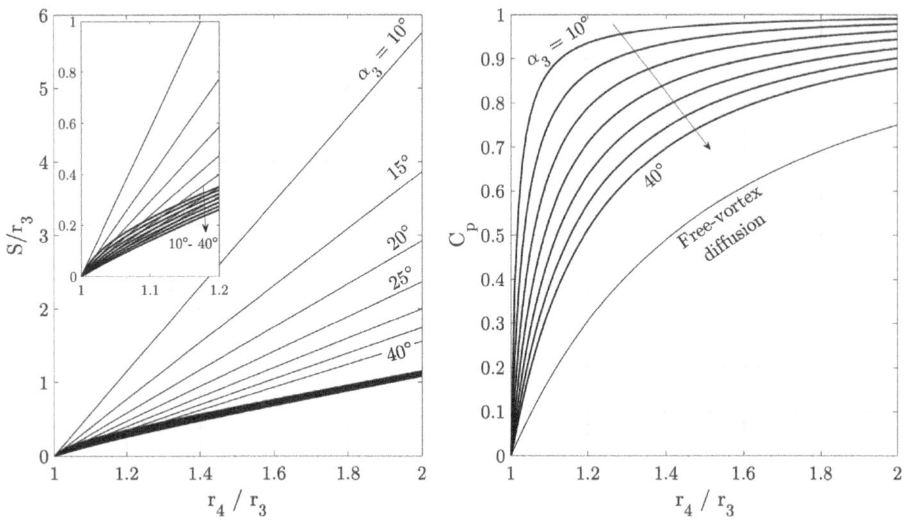

Figure 5.12 Comparison between vaned and vaneless (free-vortex) diffuser configurations.

in fact, means no diffusion at all. As r_4/r_3 increases, the logarithmic spiral exhibits long arc lengths, which suggests large friction losses. This justifies the use of vaned diffusers to increase peak hydraulic efficiency. These vanes not only enhance diffuser efficiency but also allow for a reduction in compressor size by decreasing the diffuser radius ratio.

Vaned diffusers show inferior performance far from the BEP. At off-design conditions, the fluid flow is less uniform, leading to increased flow separation, turbulence, and hydraulic losses within the diffuser. This results in reduced pressure recovery and lower efficiency. The fixed geometry of the vanes becomes less suited for the distorted flow patterns experienced away from the BEP, causing a performance drop. In off-design, incidence effects at the vane leading edge also limited the operating range. Consequently, pumps and blowers that require a wide operating range typically avoid using vaned diffusers.

The hydraulic design of vaned diffusers is guided by principles similar to those used in the design of volute casings. The throat area of the diffuser inlet is designed to match the throat area of a volute casing for similar operating conditions. Despite the vaned diffuser being more restrictive to flow, it achieves similar performance to a volute casing by providing more orderly flow through its channels.

The base diameter D_3 and the vane entrance angle α_3 are determined using the same methods as for volute design. These parameters must ensure smooth flow entry into the diffuser channels while minimizing energy losses. Typically, the diffuser vane entrance angle α_3 is aligned with the direction of the absolute velocity of the fluid leaving the impeller.

The width of the vaned diffuser b_3 is slightly greater than that of the impeller exit width b_2, with $b_3 \approx 1.1\,b_2$. This allows for a smoother transition of fluid from the impeller to the diffuser while accommodating for slight variations in the impeller's exit flow.

The number of vanes in the diffuser must be carefully chosen to ensure the required throat area is maintained and the channel shape between two vanes supports smooth flow. Minimizing the number of vanes reduces the likelihood of flow separation and turbulence within the channels.

The outer diameter of the vaned diffuser is generally constrained by the overall space limitations of the pump design. This diameter typically ranges from 1.35 D_2 to 1.6 D_2, where D_2 is the diameter at the impeller exit.

The selection of the diffuser vane shape and number of vanes is informed by the theory of cascades, which describes the pressure recovery characteristics of a blade row. When selecting the number and shape of vanes, it is important to ensure that the angle of divergence of the diffuser channel is appropriate for the expected flow conditions. If the divergence angle is too large, flow separation will occur, leading to increased losses and reduced pressure recovery. The optimal divergence angles depend on the type of diffuser cross-section: for circular conical diffusers, the divergence angle is around 8°; for square-section diffusers, the optimum divergence angle is approximately 6°; for rectangular-section diffusers, such as those formed between two parallel walls, the divergence angle is approximately 11°.

Low-solidity diffusers (Figure 5.10, right) are also common choices: in these cases,. a distinct throat is not identifiable. Experimental tests have demonstrated that low-solidity cascades, in single-row and double-row tandem configurations, offer improved pressure recovery compared to vaneless diffusers and maintain a broad operating range.

5.5 ELEMENTARY QUANTIFICATION OF HYDRAULIC LOSSES IN PUMPS

The energy transfer is almost perfectly isothermal in hydraulic machines. Consequently, rather than using the local rate of entropy generation per unit volume, σ_v 1.57, it is more convenient to express the losses in terms of pressure, or head, drop.

In an elementary view, head losses can be categorized into two types: distributed wall losses and concentrated losses. Distributed wall losses $h_{r,c}$ arise from friction between the fluid and surfaces of machine components. Concentrated losses $h_{r,\text{in}}$ occur in regions where the flow experiences abrupt changes, such as flow separation close to the blades' leading edges, thus creating localized disturbances. Therefore:

$$h_r = h_{r,c} + h_{r,\text{in}} \tag{5.35}$$

The generalized Bernoulli equation is used to incorporate the pressure drops associated with these phenomena from inlet i to delivery e of the machine:

$$\frac{p_e}{\rho g} + \frac{c_e^2}{2g} + z_e = \frac{p_i}{\rho g} + \frac{c_i^2}{2g} + z_i + h_t - \sum h_r \text{ (pump)}$$

$$\frac{p_i}{\rho g} + \frac{c_i^2}{2g} + z_i = \frac{p_e}{\rho g} + \frac{c_e^2}{2g} + z_e + h_t - \sum h_r \text{ (turbine)}$$

Specifically, for a hydraulic pump, the distributed head loss can be expressed as:

$$h_{r,c} = f_w \frac{w_2^2}{2g} + f_c \frac{c_2^2}{2g} \tag{5.36}$$

where f_w is the friction factor for wall frictional losses in the impeller; f_c is the friction factor for wall frictional losses in the stationary components.[6]

The coefficients f_w and f_c depend on several factors, including the flow Reynolds number, surface roughness, and the geometric characteristics of the flow passages.

Concentrated head losses, denoted as $h_{r,\text{in}}$, can be expressed as:

$$h_{r,\text{in}} = \lambda_r \frac{\Delta c_1^2}{2g} + \lambda_s \frac{\Delta c_2^2}{2g} \tag{5.37}$$

where Δc_1 and Δc_2 are the losses of kinetic energy due to mismatched incidence between the rotor-flow at the rotor inlet and the stator-flow at the stator inlet, respectively (see Figure 5.13):

$$\Delta c_1 = u_1 - w_{u1-} = u_1 - \frac{c_{1-}}{\tan \beta_1'} \tag{5.38}$$

$$\Delta c_2 = c_{u2} - \frac{c_{m2}}{\tan \alpha_2'} \tag{5.39}$$

Coefficients λ_r and λ_s can be determined empirically and depend mainly on blade shape, blade number, and blade angles. It is important to note that $h_{r,\text{in}}$ approaches

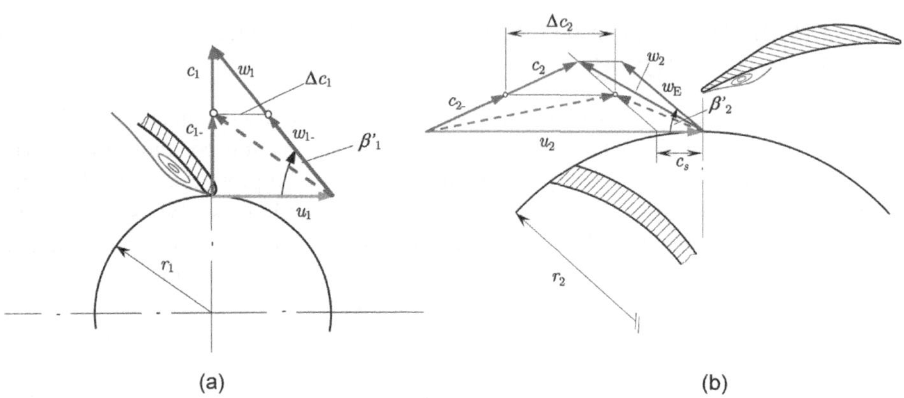

Figure 5.13 Velocity triangles and flow representation for pump off-design conditions. (a) Inlet velocity triangle. (b) Exit velocity triangle.

zero when the flow is properly aligned with respect to the rotor and stator blades, that is, under design conditions.

When all hydraulic losses $\sum h_r$ have been estimated, the pump's head h and pertinent hydraulic efficiency can be finally computed as:

$$h = h_t - \sum h_r \tag{5.40}$$

$$\eta_{\text{hyd}} = \frac{h}{h_t} \tag{5.41}$$

An example of characteristic curves $(h, \eta_{\text{hyd}}) = f(Q)$ for a centrifugal pump at a given rotational speed is provided in Figure 5.14. The nominal, or design, operating condition is set at maximum η_{hyd}, when $\sum h_r$ reaches a minimum. The influence of

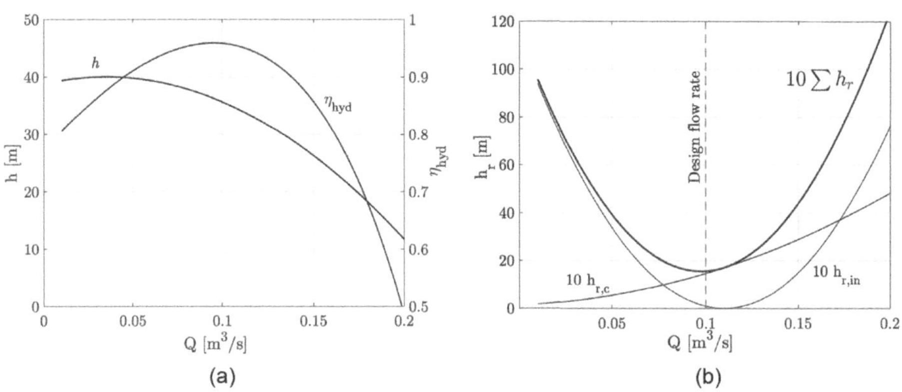

Figure 5.14 Characteristic curves of a pump. (a) Head h and hydraulic efficiency η_{hyd} as functions of the flow rate Q. (b) Distributed head losses $10h_{r,c}$, concentrated head losses $10h_{r,\text{in}}$ and their sum $10\Sigma h_r$. The dashed red line marks the design flow rate.

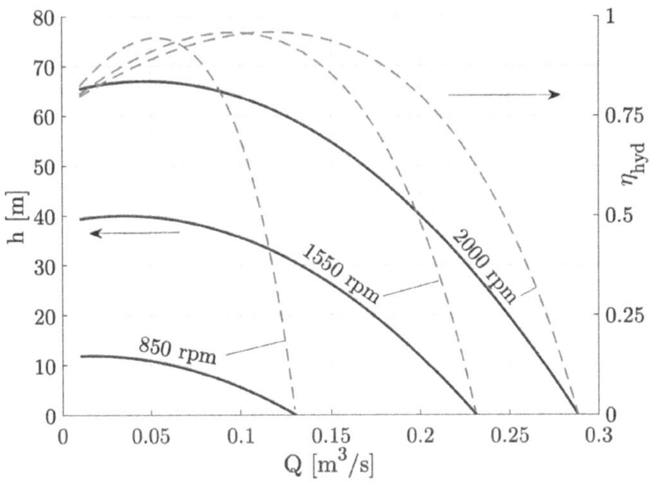

Figure 5.15 Pump performance curves for different rotational speeds $N = 850$ rpm, $N = 1550$ rpm and $N = 2000$ rpm.

rotational speed on pump characteristics is shown in Figure 5.15 under the assumption that Reynolds number does not affect loss coefficients.

5.6 ELEMENTARY QUANTIFICATION OF OTHER LOSSES IN PUMPS

5.6.1 VOLUMETRIC LOSSES

Volumetric efficiency η_v quantifies the effectiveness of the pump in delivering the intended flow rate, accounting for internal leakage losses. It is defined as the ratio of the actual flow rate Q discharged by the pump to the theoretical flow rate Q_t that would be achieved without any leakage:

$$\eta_v = \frac{Q}{Q_t} \tag{5.42}$$

Internal leakage occurs through the clearances between the rotating impeller and the stationary casing, such as front and rear seals (Figure 5.1) and balance holes. These leakages cause a portion of the fluid to recirculate within the pump rather than be discharged, reducing the pump's volumetric efficiency.

Volumetric efficiency is a function of the specific speed: in fact, in pumps with low specific speeds (radial flow designs), the impeller geometry typically results in higher internal leakage losses due to larger pressure differentials across the impeller. Its values range from 97% for $\omega_s = 0.75$ and reach >99% for $\omega_s > 1.2$.

5.6.2 WINDAGE LOSSES

Windage losses, also known as disk friction losses, occur due to friction between the rotating impeller surfaces and the surrounding fluid outside (i.e. the fluid contained between external surfaces of the impeller and the casing). These losses convert mechanical energy into heat, reducing the pump's overall efficiency. The power loss due to disk friction can be computed taking into account the friction force exerted on the shrouded disk at radius r:

$$dF = Ku^2\,dA \tag{5.43}$$

where K is a constant related to fluid properties and surface conditions, $dA = 2\pi r\,dr$ is the elementary area of the disk element and u the peripheral velocity of the disk element. It follows that:

$$dF = K2\pi u^2 r\,dr = K2\pi(\omega r)^2 r\,dr \tag{5.44}$$

The power required to overcome this resisting force is obtained by multiplying dF by the velocity u:

$$P_d = K2\pi \int_0^r u^3 r\,dr = K2\pi\omega^3 \int_0^r r^4\,dr = \frac{K2\pi\omega^3 r^5}{5} \tag{5.45}$$

These losses typically vary as a function of the specific speed. Indicative values show that losses account for approximately 10% of the shaft power at $\omega_s = 0.25$, but decrease significantly to 5% at $\omega_s = 0.75$ and drop to less than 2% for $\omega_s > 1.2$.

5.6.3 MECHANICAL LOSSES

Mechanical losses are the result of energy dissipation due to friction and wear in the pump moving parts, such as bearings and ring-wear. They typically consume between 0.5% and 2% of the input power, with the percentage generally decreasing as the pump's power increases. It is important to note that the exact percentage of mechanical losses can vary based on factors such as bearing type (e.g. ball bearings vs. journal bearings), load and alignment conditions, and maintenance practices. The designer should refer to specific literature or technical catalogs for more accurate estimates.

5.6.4 OVERALL PUMP EFFICIENCY AND CHARACTERISTICS

The overall efficiency can be defined as:

$$\eta_{\text{overall}} = \eta = \frac{\text{Useful hydraulic power output}}{\text{Input shaft power}} = \eta_{\text{hyd}} \cdot \eta_{\text{v}} \cdot \eta_{\text{mech}} \tag{5.46}$$

such that the estimated shaft power absorbed by the pump is:

$$P = \frac{\rho Q g h}{\eta_{\text{overall}}} \tag{5.47}$$

Figure 5.16 provides an estimate of the overall efficiency as a function of specific speed, accompanied by a representative breakdown of its individual components. Figure 5.17 illustrates the performance characteristics at various rotational speeds.

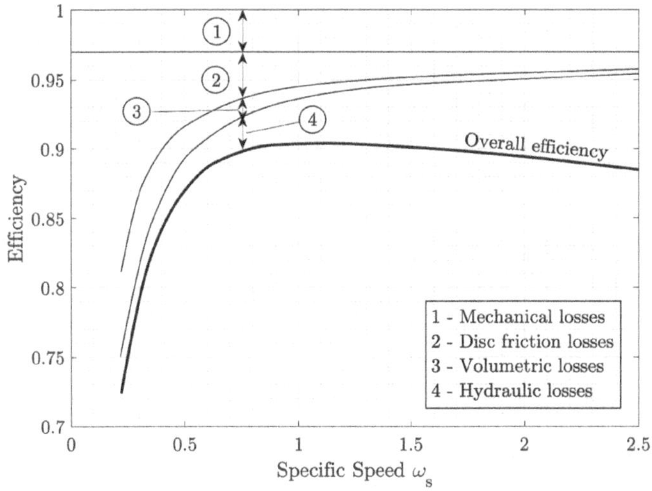

Figure 5.16 Variation of overall efficiency and loss components with specific speed ω_s.

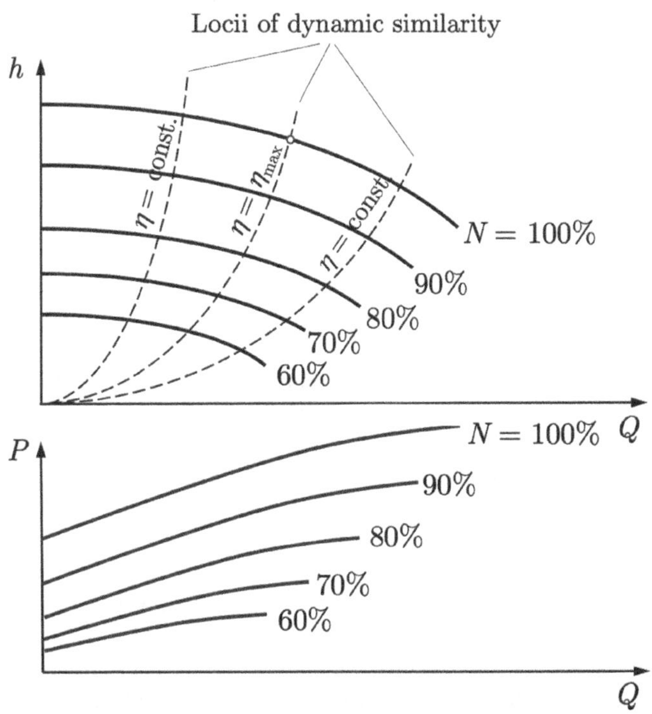

Figure 5.17 Head and power curves vs. flow rate and rotational speed.

In the top plot, the curves of constant efficiency are also provided. In the bottom plot, shaft power P curves are shown. In particular, at zero flow $Q = 0$, the shaft power is not zero due to internal fluid recirculation and mechanical losses. The spinning impeller causes hydraulic losses as fluid recirculates within the machine, and additional power is required to overcome mechanical losses like friction in bearings and seals.

5.7 PUMP CAVITATION

As already mentioned, cavitation occurs when the local pressure in the fluid drops below the vapor pressure, leading to the formation of vapor bubbles. These bubbles can violently implode as they move to regions of higher pressure, causing significant damage to the impeller and other pump components.

Every liquid has a characteristic vapor pressure $p_v(T)$ that depends on temperature, i.e. the pressure at which the liquid is in equilibrium with its vapor (i.e. it boils at that pressure for a given temperature). In a flowing system, if the local static pressure p drops to $p_v(T)$ or below, the liquid will vaporize and form bubbles of its own vapor. Thus, a key cavitation criterion is $p_{\text{local}} \leq p_v(T)$. For water at room temperature, p_v is low (approximately 2–3 kPa absolute at 20°C), but for hotter liquids or cryogenic fluids, p_v can be substantial. Maintaining pressure above p_v at all points in the pump's flow path is essential to prevent cavitation inception.

However, even if the absolute pressure just equals p_v, cavitation may not immediately occur without a nucleation site. Real liquids often contain microscopic gas bubbles or impurities that serve as nuclei. If no suitable nuclei are present, a liquid can sustain tension below p_v (metastable state) up to a point. In practice, though, most engineering fluids have enough microbubbles or dissolved gas such that once p falls to (or slightly below) p_v, cavitation begins at those nucleation sites.

In summary, the fundamental processes involved in cavitation are:

- **Inception**. A tiny nucleus (gas or vapor pocket, often 10–100 μm in size) enters a low-pressure region and grows explosively as local pressure $p \rightarrow p_v$. The bubble can expand rapidly.
- **Advection**. The vapor-filled bubble is carried along with the flow into higher-pressure regions of the pump (for example, towards the impeller blades' downstream side).
- **Collapse**. Once the bubble encounters pressure significantly above p_v, it implodes violently. The implosion can generate local pressures on the order of hundreds of bar, creating shock waves.
- **Damage/noise**. Repeated implosions near solid boundaries generate surface fatigue and pit the material. They also produce audible noise and vibrations.

From a fluid mechanics perspective, cavitation is intimately linked to Bernoulli's principle and energy conservation. As liquid accelerates into the impeller eye, kinetic energy increases at the expense of pressure energy, causing a pressure drop. If the design or operating conditions cause this pressure drop to reach vapor pressure, cavitation results. Thus, cavitation avoidance is largely about maintaining adequate pressure energy on the suction side of pumps.

In centrifugal pumps with shrouded impellers, cavitation typically occurs near the leading edge of a blade at the suction station. More specifically, the section side of profiles located near the shroud is the most common place where cavitation begins and develops due to the localized minimum pressure caused by both the shape of the meridional contour and the hydrodynamic action of the rotating blades. The curvature of the profile can further amplify the phenomenon.

To estimate the conditions under which a pump is subject to cavitation, we can start by reformulating the generalized Bernoulli equation into the following form:

$$\frac{p_1 - p_v}{\rho g} - \sum_i h_{r,i} - z_B = \frac{p_B - p_v}{\rho g} + \frac{c_B^2}{2g} \tag{5.48}$$

where p_1 is the pressure acting on the free surface of the suction reservoir, p_v the vapor pressure of the fluid, z_B the elevation of station B with respect to the level of the free surface in the suction reservoir, p_B the pressure at the pump inlet, c_B the fluid velocity at station B and $\sum_i h_{r,i}$ the overall head losses in the suction pipe.

It is critical to emphasize that the specific characteristics of the pumping circuit establish the boundary conditions necessary for cavitation to occur. These boundary conditions must then be compared against the internal characteristics of the pump to evaluate the actual onset of cavitation. Under acceptable operating conditions, the pressure p_B calculated using the Bernoulli equation at the pump inlet must exceed the vapor pressure of the fluid to avoid cavitation.

When the valve in the suction pipe is throttled, the equation remains valid; however, increased pressure losses due to throttling lead to a decrease in pressure at the pump inlet. If throttling continues to intensify, a threshold will be reached where the inlet pressure falls to the vapor pressure, marking the onset of cavitation, or cavitation inception. This condition is quantified as the available net positive suction head ($NPSH_a$) on the left side of the equation, which must exceed the required net positive suction head ($NPSH_r$) on the right side to prevent cavitation. Thus, the cavitation condition can be expressed as:

$$NPSH_a = NPSH_r + x$$

where x represents the anti-cavitation margin.

To maximize $NPSH_a$ and minimize $NPSH_r$, various strategies can be employed. These include controlling the installation height of the pump to ensure sufficient submergence in the suction reservoir, optimizing the dimensions and layout of the suction pipe (including reducing bends and using larger diameters to decrease friction losses) and ensuring that the free surface pressure of the suction reservoir is adequately maintained, such as through pressurization or reduction of vapor pressure. Furthermore, minimizing any obstacles in the flow path and ensuring smooth transitions in the piping can enhance flow characteristics, thereby increasing $NPSH_a$ and reducing the risk of cavitation.

5.7.1 DETERMINATION OF $NPSH_r$

An accurate assessment of $NPSH_r$ is carried out only experimentally. The procedure for its determination is described in International Standards such as ISO 9906:2012. Measurements can be done either in an open sump or in a closed-loop circuit. In Figure 5.18, an example of closed configuration is given. In such a case, the pump draws liquid through an inlet pipe from a pressurized reservoir, in which the level of the free liquid surface is kept fixed, and the pressure of the liquid entering the pump is adjusted by means of a foot valve installed in the inlet pipe at the lowest practicable level. To prevent loss of priming, the foot valve acts as a non-return valve and also facilitates pump priming when it is in suction. A further throttle valve is placed far downstream of the pump to allow for effective regulation of the flow rate within the circuit.

The test usually starts with the downstream throttle valve fully open in order to save power and preserve the electric motor from overloads. Once the pump is started and the required rotational speed is stably achieved, both the foot and the throttle valves are adjusted to obtain the target flow rate at which the $NPSH_r$ has to be measured.

Pressure is measured upstream and downstream of the pump using absolute or differential pressure transducers (Bourdon dial gauge, electronic pressure transducers, etc.). Vapor pressure is usually derived from standard data and the measurement of the temperature of the liquid entering the pump. The inlet measuring section should generally be positioned two diameters upstream from the pump inlet flange, while the outlet measuring section must be arranged in a straight pipe section that is coaxial with the pump outlet flange and of the same diameter.

The speed of rotation is measured typically using a direct indicating tachometer, an optical or magnetic counter or a stroboscope. Flow rate is measured using a flow meter (orifice plates, nozzles, Venturi tubes, or others). The piping upstream of the

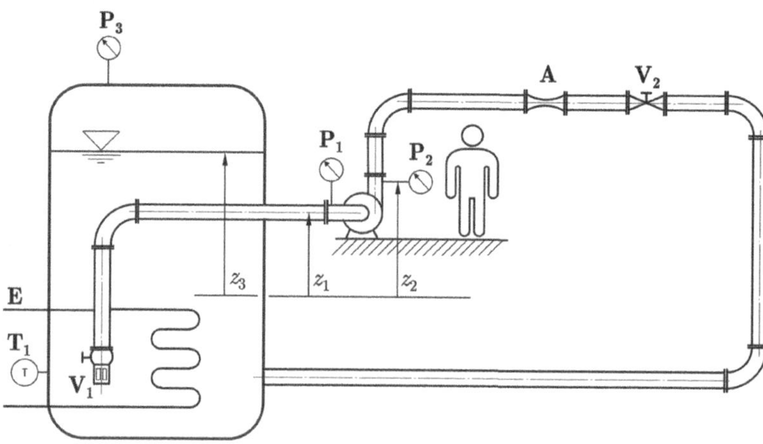

Figure 5.18 Schematic of a test bench for cavitation experiments in pumps.

flow meter must be straight, match the diameter of the flow meter and extend at least 10 times the pipe diameter. The downstream piping should also meet these requirements, with a minimum length of 5 times the pipe diameter measured from flange to flange.

Measurement of torque is performed using a suitable dynamometer or torque meter.

From the measurement data, the actual working point of the pump can be determined in terms of both flow rate Q and head h. Furthermore, the right-hand side of equation 5.49 can be calculated. By lowering the suction pressure through synchronous adjustment of both the foot and throttle valves, the point of cavitation inception is eventually reached. For the flow rate under investigation, the $NPSH_r$ is identified as the value of the right-hand side of the equation 5.49 at which a slight drop (conventionally a 3% drop) in head is observed. The test is repeated at different fixed flow rates, and once again, the corresponding suction pressure and flow rate are recorded at the onset of cavitation. After completing the series of tests, a smooth curve is drawn through the recorded data points and plotted. Figure 5.19 shows a typical set of test results and the resulting $NPSH_r$ curve.

In addition to experiments, a simplistic way to estimate $NPSH_r$ for a pump of a given size is to use the following, empirically based, equation:

$$NPSH_r = \frac{1}{2g}\left[\frac{Q^2 k_m^2}{\pi^2 N^2}(1+\lambda_m+\lambda_w)+\lambda_w \omega^2 r_1^2\right] \tag{5.49}$$

where the empirical values of the coefficients are $k_m = 1.12$, $\lambda_m = 0.05$ and $\lambda_w = 0.35$.

5.7.2 SIMILARITY IN PUMP CAVITATION

In hydraulic pumps, the Thoma number Th is frequently used instead of the cavitation number because it is easier to measure and more directly correlated to the

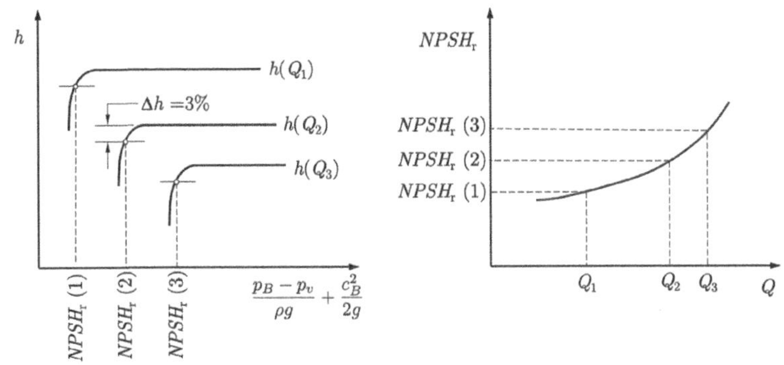

Figure 5.19 Determination of the required Net Positive Suction Head $NPSH_r$ based on a 3% head drop criterion.

pump's characteristics. The Thoma number is defined as:

$$\text{Th} = \frac{NPSH}{h} \tag{5.50}$$

The fluid will cavitate if the Thoma number is smaller than the critical cavitation parameter, also referred to as the critical Thoma number:

$$\text{Th}_c = \frac{NPSH_r}{h} \tag{5.51}$$

Many pump manufacturers provide specific empirical correlations or charts based on extensive testing of their pump models.

5.7.3 CAVITATION SUPPRESSION AND MITIGATION STRATEGIES

Preventing or mitigating cavitation in pumps involves both design choices and operational measures. Key strategies include:

- **Providing adequate NPSH margin**. Ensure that the system $NPSH_a$ is comfortably above the pump's $NPSH_r$. This may involve placing the pump below the source tank, pressurizing the suction vessel, using shorter or larger-diameter suction pipes to cut losses and avoiding filters or fittings that rob suction pressure. According to pump standards, a common guideline is a margin of 1–3 m above $NPSH_r$ to account for uncertainties.
- **Reducing pump speed or impeller diameter**. Since cavitation risk increases with tip speed and suction specific speed, slowing the pump (especially during off-design low-flow operation) can alleviate cavitation. Trimming the impeller or running at a lower rpm reduces the required h for a given flow, hence raises σ_{Th}. This, of course, also reduces pump performance, so it is a trade-off.
- **Using an inducer (for high-speed pumps)**. An inducer (Figure 5.2) is a small axial-flow impeller stage added in front of the main impeller. Its purpose is to slightly increase the pressure of the fluid before it enters the main impeller, thus increasing $NPSH_a$ as seen by the main impeller. Inducers are designed with very low necessary $NPSH_r$ themselves and usually operate in a partially cavitating mode (but in a controlled way). By the time the fluid exits the inducer, its pressure is higher, so the main impeller sees no cavitation. Inducers are common in rocket turbopumps and other applications with low $NPSH_a$ (such as pumping near-boiling liquids). The design of inducers involves helically twisted blades that are very good at handling two-phase mixtures and are less sensitive to cavitation.
- **Improving suction geometry**. Flow disturbances or swirl entering the pump can exacerbate cavitation by causing uneven pressure distribution. Using straightening vanes, suction bells, or sufficiently long straight runs can ensure uniform flow into the impeller eye. Avoiding vortices in the suction sump (by providing anti-vortex baffles or sufficient submergence) prevents a strong vortex from feeding into the pump and potentially cavitating (vortex cavitation at pump intake).

- **Material and coatings**. While they do not prevent cavitation, using cavitation-resistant materials (e.g. stainless steels, coatings like hard facings or elastomeric coatings) can mitigate damage. For instance, softer materials can absorb the impact without cracking (rubber linings in some pumps). Hardened surfaces last longer against pitting.

5.8 PRELIMINARY SIZING AND MECHANICAL DESIGN

5.8.1 PRELIMINARY HYDRODYNAMIC DESIGN

Preliminary sizing relies on experience from previous pump configurations that have demonstrated satisfactory performance under similar operating conditions.

In this respect, the laws of similarity and affinity help reduce the design space by providing an initial estimate of the key design parameters, such as impeller diameter and major shape factors. Designers can select initial parameters that can later be fine-tuned and optimized using more advanced computational analyses, such as Computational Fluid Dynamics (CFD) and validated through experimental tests.

Here below guidelines for the preliminary design are provided. Let us assume knowledge of the following quantities: Q, h, and $NPSH_r$.

5.8.1.1 Selection of Rotational Speed

The rotational speed is selected on the basis of several factors:

- Type of driver used.
- Efficiency requirements.
- Total head and total flow rate.

The two most common types of drivers are electric motors (particularly direct drive electric motors) and endothermic prime movers, such as steam turbines or gas turbines. Direct drive electric motors have rotational speed determined by the frequency of the electrical power supply and the number of poles in the motor. This relationship is governed by the following equation for synchronous speed:

$$n_s = \frac{120 \cdot f}{\text{poles}} \tag{5.52}$$

where n_s is synchronous speed in rpm, f is the network frequency in Hz and $poles$ the number of motor poles. In practical applications, most electric motors are induction motors, which run slightly more slowly than the calculated synchronous speed due to an electromagnetic slip $s > 0$. The actual speed N of an induction motor can be calculated as:

$$N = N_s(1 - s) \tag{5.53}$$

where s is generally in the range of 1%–3% for modern industrial motors.

Turbines are typically used as prime movers in large-scale industrial processes, such as power plants, petrochemical facilities, and refineries, as well as in aerospace

Table 5.2
Summary of Driver Characteristics

Driver Type	Typical Speed Range (rpm)	Mechanical Coupling	Power Range	Applications
Direct Drive Electric Motor	≤3,000 (50 Hz); ≤3,600 (60 Hz)	Direct or with minor adjustments	Small to medium (kW to MW range)	Industrial pumps (Oil and Gas, Chemical Plants, Food and Beverage, Pharmaceutical), water supply systems, sewage and sludge pumping, HVAC systems, mining and mineral processing, renewable energy systems
Steam/Gas Turbines	5,000–20,000	Requires reduction gearbox	High power (MW range)	Power plants, refineries, large-scale industrial applications, aerospace and marine propulsion

applications (e.g. rockets), where high mechanical power is needed. The rotational speed of turbines is generally much higher than what centrifugal pumps typically require, which leads to using reduction gearboxes to match the pump's operational speed range. A turbine may operate at several thousand rpm (e.g. 6000–20,000 rpm), while the pump may only need to run at a speed in the range of 1000–3000 rpm.

Table 5.2 summarizes the typical driver types, speed and power ranges, as well as popular applications of centrifugal pumps.

5.8.1.2 Selection of Head and Flow Coefficients

Based on the specifications and after selecting the rotational speed, the specific speed can be calculated:

$$\omega_s = \frac{\pi N}{30} \frac{Q^{0.5}}{(gh)^{0.75}}$$

which defines the type of pump. On the other hand, fixing ω_s and deriving ω allows for the selection of high-performance machines. Pumps with $0.6 < \omega_s < 1.5$ offer maximum efficiency and robust off-design performance, while those with $\omega_s > 1.5$ have a narrower operating range.

Regardless of the criterion used to determine ω_s, the head coefficient ψ and flow coefficient ϕ can be derived from the statistical chart in Figure 5.20. This chart is based on data regression from well-performing pump realizations.

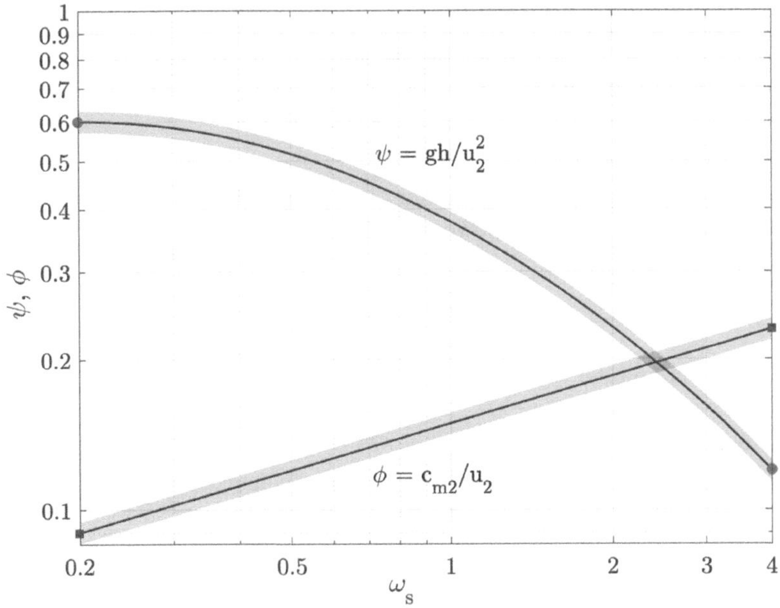

Figure 5.20 Head coefficient $\psi = gh/u_2^2$ and flow coefficient $\phi = c_{m2}/u_2$ as functions of specific speed ω_s in pumps. The shaded bands represent the variance of feasible values.

The peripheral speed u_2 of the impeller is linked to the head coefficient ψ:

$$u_2 = \sqrt{\frac{gh}{\psi}} \tag{1.6}$$

General guidelines to the maximum allowable peripheral speed based on common impeller materials are provided in Table 5.3.

Table 5.3
Maximum Allowable Peripheral Velocity $u_{2,\max}$ for Various Materials and Their Examples

Material	**Example**	$u_{2,\max}$ (m/s)
Grey cast iron	EN-JL1040, EN-GJL-250	40
Spheroidal graphite cast iron	EN-JS1030, EN-GJS-400-15	50
Bronze and brass	2.1050, G-CuSn 10	50
Stainless steels (normal)	1.4008, GX7CrNiMo12-1	95
Stainless steels (special)	1.4317, GX4CrNi13-4	110

5.8.1.3 Dimensional Characteristics

Pump dimensions are derived from the operating conditions and specific speed. The impeller diameter is given by:

$$D_2 = \frac{60u_2}{\pi N} \tag{5.54}$$

Efficiencies η, η_{hyd}, η_{v}, and η_{mech} must be postulated, for example, upon consultation of Figure 5.16. The shaft power is computed as:

$$P = \frac{\rho Q g h}{\eta} \tag{5.55}$$

while the shaft torque is:

$$\tau = \frac{30P}{\pi N} \tag{5.56}$$

from which the maximum shear shaft stress can be calculated and used to get a preliminary dimension of shaft diameter given the allowable shear stress of the material τ_{adm}:

$$d_s = \sqrt[3]{\frac{16\tau}{\pi \tau_{\text{adm}}}} \tag{5.57}$$

Outer impeller width b_2 can be derived from:

$$b_2 = \frac{Q/\eta_{\text{v}}}{\pi D_2 \xi_2 \phi u_2} \tag{5.58}$$

which must be ≥ 4 mm to avoid excessive flow blockage at the impeller outlet. The blockage factor ξ_2 is initially guessed and then iteratively determined (as a function of the blade thickness t_2 and blade number Z) based on the following equation:

$$\xi_2 = 1 - \frac{Z \cdot t_2}{\pi D_2 \sin \beta_2'} \tag{5.59}$$

At the impeller eye, diameters are chosen to maintain appropriate inlet velocity c_{m0} to avoid cavitation:

$$Q = c_{m0}\frac{\pi}{4}\left(D_{0,s}^2 - D_{0,h}^2\right) = c_{m0}\frac{\pi}{4}D_{0,s}^2\left(1 - \frac{D_{0,h}^2}{D_{0,s}^2}\right) \tag{5.60}$$

Inlet velocity c_{m0} should typically not exceed (10–12 m/s) for water under standard conditions. The ratio of hub to shroud diameter, $\frac{D_{0,h}}{D_{0,s}}$, usually ranges from 0.3 to 0.5, depending on the pump type and application.

Impeller inlet sizing depends mainly on specific speed ω_s. In radial pumps, $c_{m1} \approx c_{m2}$, $1.2 \leq \frac{D_0}{D_{1,s}} \leq 1.5$. For high-flow, low-head pumps, a larger suction diameter ratio is ideal to maintain low inlet velocities $1.6 \leq \frac{D_0}{D_{1,s}} \leq 2.0$.

Table 5.4
Typical Ranges for $\frac{D_{1,s}}{D_{1,h}}$ in Radial Impellers

Pump Type	Specific Speed ω_s	Typical Range for $\frac{D_{1,s}}{D_{1,h}}$
Low-flow, high-head	$\omega_s < 0.5$	$1.6 - 2.0$
Medium-flow, medium-head	$0.5 \leq \omega_s \leq 1.5$	$1.8 - 2.2$
High-flow, low-head	$\omega_s > 1.5$	$2.0 - 2.5$

The theoretical head is given by:

$$gh_t = \frac{gh}{\eta_{\text{hyd}}} = u_2\, c_{u2} = u_2^2\, \sigma_{\text{slip}} \left(1 - \phi \cot \beta_2'\right) \tag{5.61}$$

Assuming a number of impeller blades (see next item 4.) and based on impeller dimensions, the slip factor σ can be calculated (see Section 5.3). Next, the blade exit angle is computed from 5.61 using an iterative approach on β_2'. Coming back to the inlet section, assuming for $\frac{D_{1,s}}{D_{1,h}}$ a value as in Table 5.4, the mean inlet diameter is:

$$D_1 = \left(D_{1,s} + D_{1,h}\right)/2 \tag{5.62}$$

from which:

$$u_1 = \omega D_1/2 \tag{5.63}$$

Blade width b_1 and inlet blade angle β_1 are computed as:

$$b_1 \simeq \frac{Q}{\pi D_1\, c_{m1}} \tag{5.64}$$

$$\beta_1 = \arctan \frac{c_{m1}}{u_1} \tag{5.65}$$

Finally, it is customary to ensure that the deceleration of the relative flow is moderate, which is verified by checking the condition:

$$\frac{w_2}{w_1} \geq 0.7 \tag{5.66}$$

where:

$$w_1 = \sqrt{u_1^2 + c_{m1}^2}, \quad w_2 = \sqrt{(u_2 - c_{u2})^2 + c_{m2}^2} \tag{5.67}$$

For the impeller blades shown in Figure 5.21, profiles with a constant thickness and two circular arcs were used.

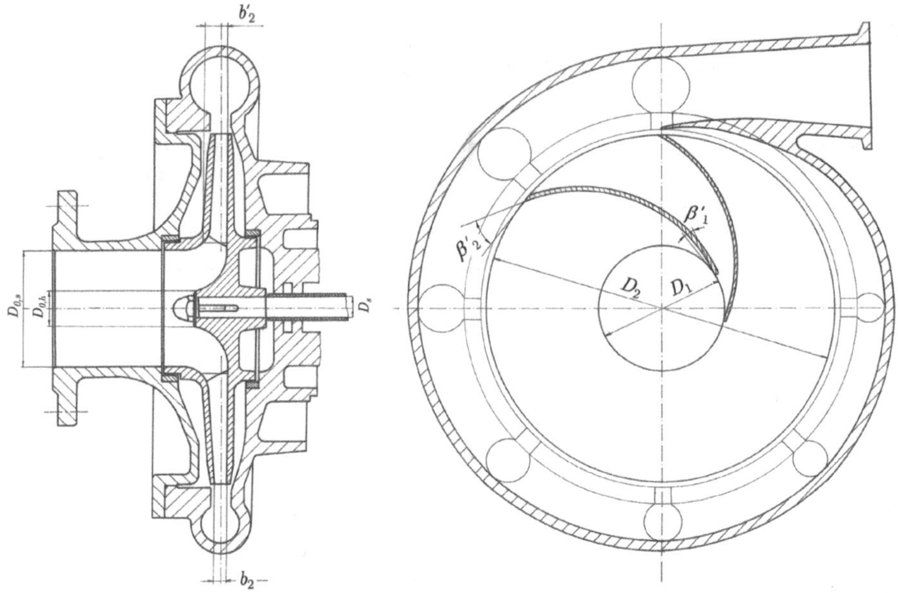

Figure 5.21 Meridional and blade-to-blade views of a centrifugal pump.

5.8.1.4 Number of Blades

The number of impeller blades depends on the specific speed, pump size, and total head. Smaller and lower-head pumps typically have fewer blades, while larger, high-head pumps require more blades. For special services, such as pumping paper stock or sewage, non-clogging impellers may have as few as two blades. From purely hydrodynamic considerations, the selection of the number of blades is such as to ensure that the relative velocity of the fluid in the $r-\theta$ plane does not approach zero along the blade-to-blade surface, in particular at the impeller exit on the pressure side. If the relative velocity drops to zero, it can result in flow separation on the pressure side of the blade, leading to efficiency losses, increased vibration and potential mechanical damage due to unsteady flow patterns. To prevent this, the number of blades must be chosen so that the relative flow is well-guided between blades, with minimal risk of separation.

A simplified yet effective method to analyze and satisfy this constraint involves assuming an incompressible flow with relative streamlines that follow a consistent path at an angle, denoted β (see Appendix A). By modeling the relative flow such that the tangential component of the velocity, w_θ, remains proportional to the radial component, w_r, through the relation $w_\theta = \frac{w_r}{\tan\beta}$, designers can gain insight into the blade spacing required to maintain steady flow. In practice, the appropriate number of blades balances adequate flow guidance with minimal blockage, as increasing the blade count can raise frictional losses and reduce the impeller's hydraulic efficiency. The chosen blade number should therefore ensure optimal fluid acceleration and con-

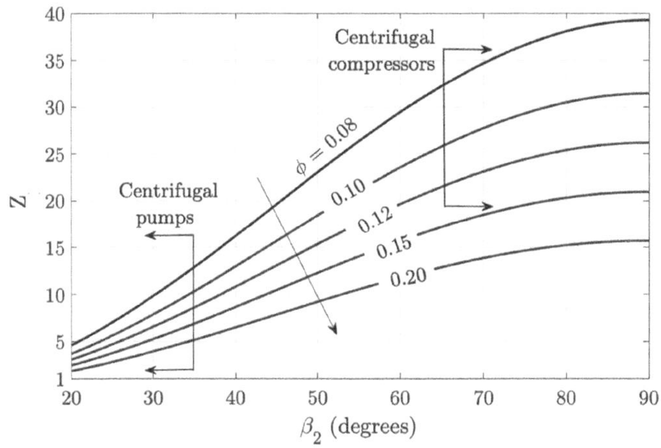

Figure 5.22 Chart for estimating the required number of impeller blades.

trol while minimizing the likelihood of flow separation, enhancing overall impeller performance. An indication of the minimum number of impeller blades can be given from (see Appendix A.3):

$$Z > \frac{\pi}{\phi\left(\frac{1}{\tan^2\beta_2}+1\right)} \tag{5.68}$$

which is plotted in Figure 5.22.

5.8.1.5 Reduction Factors for Similar Pumps

In some cases, it is straightforward to use reduction factors or multipliers to obtain a pump's sizing on the basis of simple affinity criteria. If an impeller with diameter D_1 produces flow Q_1 at head h_1 and speed N_1 and a new impeller with diameter D_2 is required to produce flow Q_2 at head h_2 and speed N_2, the specific speed of both impellers must remain the same. The following relationship holds:

$$N_1\frac{Q_1^{1/2}}{h_1^{3/4}} = N_2\frac{Q_2^{1/2}}{h_2^{3/4}} \tag{5.69}$$

The following affinity laws for flow and head can be used:

$$Q_2 = Q_1 f^3\left(\frac{N_2}{N_1}\right) \tag{5.70}$$

$$h_2 = h_1 f^2\left(\frac{N_2}{N_1}\right)^2 \tag{5.71}$$

where $f = \frac{D_2}{D_1}$ is the reduction factor or multiplier.

An alternative expression for the reduction factor can be derived from equation 5.71:

$$f = \frac{N_1\sqrt{h_2}}{N_2\sqrt{h_1}} \tag{5.72}$$

5.8.1.6 Cavitation Check

$$NPSH_a = NPSH_r + \text{margin} \tag{5.73}$$

5.8.1.7 Volute Sizing

See Section 5.4.

5.9 PUMP-CIRCUIT MATCHING

Pump-circuit matching involves aligning the pump's performance curve $h(Q)$ with the system characteristic curve $h_r(Q)$ at the duty, or working, point. The duty point is where the total head generated by the pump equals the head required by the system or, in other words, when the pump's head characteristic intersects the circuit resistance characteristic:

$$h(Q) = h_r(Q) \tag{5.74}$$

At this point, the pump delivers the required flow rate Q while achieving specific values of absorbed power P, efficiency η, and $NPSH_r$.

Deviations in the actual system curve $h_r(Q) = h_s + \Delta h_f(Q)$ can shift the working point. Steep $h_r(Q)$ curves result in minimal flow rate variation ΔQ with system changes, while flat curves cause larger ΔQ, making them more sensitive. Flat curves allow flexibility for throttling, suitable for variable flow, whereas steep curves ensure consistent flow.

5.9.1 STABLE VS. UNSTABLE OPERATION

Head decreases as the flow rate increases. This relationship is quantified by a measure known as the steepness ratio, which is calculated as:

$$\frac{h_0 - h^*}{h^*} \tag{5.75}$$

where h_0 is the head at zero flow rate and h^* denotes the head at the BEP. The value of the steepness ratio depends on the type of pump and its specific speed and cannot be selected arbitrarily. For instance, radial flow pumps have ratios between 0.10 and 0.25. Mixed flow pumps generally fall within the range 0.25 to 0.80, and axial flow pumps exhibit steepness ratios greater than 0.80.

Stable $h(Q)$ curves are characterized by a consistent decrease in head as the flow rate increases. For such curves, there is only one corresponding flow rate value for any given head value.

Unstable $h(Q)$ curves are characterized by local positive $h(Q)$ slopes, $\frac{dh}{dQ}>0$. In fact, a decrease in discharge Q leads to a decrease in head h. This reduction further reduces the driving energy for flow causing Q to decrease even more. This feedback amplifies the initial reduction and potentially leads to oscillatory behavior in the pumping system.

The characteristic shape of an unstable $h(Q)$ curve is often seen in axial or mixed flow pumps, especially those designed for high specific-speed applications.

5.9.2 PUMP STARTING AND REGULATION

Centrifugal pumps must be filled with liquid before starting, that is, they must be *primed*. In fact, standard pumps cannot evacuate air from the suction line and require external methods for priming. Self-priming centrifugal pumps make use of built-in ejectors to progressively evacuate gases out from the suction pipe.

Proper starting procedures are necessary to avoid transient conditions that can cause mechanical stress and damage. Improper starting typically occurs when instantaneous torque required by the pump is not adequately supported by the torque delivered by the motor. Typically, three starting options are available:

- **Starting against a closed valve**. In centrifugal pumps shaft power at $Q=0$ is minimum, while torque increases with the square of the rotational speed. Such torque must overcome static friction in bearings and shaft seals, which is particularly high at low speeds. Shaft is accelerated at $Q=0$ until desired rotational speed is reached, and then the throttle is opened gradually at a constant rotational speed. For centrifugal pumps, this procedure is followed, while in axial flow pumps, where the power input at $Q=0$ exceeds that at the duty point, this starting method should be avoided.
- **Starting against an open non-return valve**. The starting torque curve follows the same shape as when starting against a closed valve, up to a point where the non-return valve opens due to pump pressure, initiating flow. Beyond this point, the torque curve is determined by constructing intermediate $h(Q)$ curves and calculating torque from power input. Once again, this procedure is suited for centrifugal pumps and should be avoided for axial flow ones.
- **Starting with an open valve and empty delivery pipeline**. When the delivery pipeline is empty, the pump provides nearly zero head, and torque increases with the square of the rotational speed along the $h \approx 0$ line. This method suits pumps where power input decreases with an increasing flow rate, such as axial flow pumps.

Other hybrid procedures can be adopted on a case-by-case basis.

Effective *regulation* can be achieved through various methods such as throttling, adjusting impeller speed, or through pre-rotation carried out by inlet guide vanes. Each method has implications for the efficiency and wear of the pump.

Figure 5.23 illustrates the effects of throttling and speed regulation on the efficiency η and shaft power P of a pump when the required flow rate decreases from

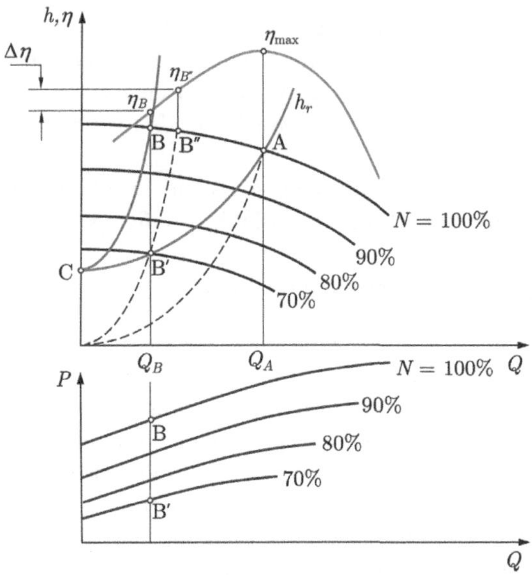

Figure 5.23 Comparison of throttling and speed regulation for flow rate reduction from Q_A to Q_B.

Q_A to Q_B. When throttling is applied, the operating point shifts from A to B as seen in the top plot. This transition causes the pump efficiency to decrease from its maximum value $\eta_{\max}$ at A to a lower efficiency, η_B, at B. The reduction in flow rate is achieved by increasing system resistance, which results in a reduction in shaft power, as shown in the bottom plot, corresponding to the move from point A to B.

In contrast, when speed regulation is used instead of throttling, the required flow rate Q_B is reached at point B' by reducing the pump's rotational speed to 70% of its nominal value (Figure 5.23). At B', dynamic similarity ensures that the pump efficiency $\eta_{B'}$ equals the efficiency at B'', as both points lie on the same locus of dynamic similarity. This results in an efficiency improvement of $\Delta\eta$ compared to throttling. Additionally, speed regulation significantly decreases the shaft power required, as indicated by the lower power consumption at B' in the bottom plot.

Pre-rotation control adjusts the flow direction entering the pump. Positive pre-rotation (aligned with impeller rotation) reduces $h(Q)$, while negative pre-rotation (opposite to impeller rotation) increases it. This is achieved by using inlet guide vanes and by altering their setting angle. The effectiveness of pre-rotation depends on the pump's specific speed. It has minimal impact on radial flow pumps but becomes significant for mixed-flow and axial-flow pumps, where it enables adjustments to the operating point with minimal efficiency loss. Using a variable geometry flap diffuser instead of standard guide vanes maintains nearly constant efficiency over a wider range.

5.10 PROBLEMS

1. The meridional view of the pump in Figure 5.4 contains points of known coordinates (z, r): [A(1,7.75), E(1,1), 1(9.5,8.5), 2(9.85,8), 3(10.3,7.5), 4(10.9,7), 5(11.6,6.4), 6(12.5, 5.6)] cm. The pump handles a flow rate $Q = 70$ l/s of water entering with uniform axial velocity along $\bar{AE}$. Assuming segment $\bar{16}$ to be orthogonal to local meridional velocity c_m, calculate the discrete distribution of c_m. Assuming the inlet pump pressure is 1 bar, compute pertinent discrete distribution of pressure along $\bar{16}$.

   [15.81 m/s, 3.77 m/s, 500.8 J/kg]

2. The blade-to-blade plane $m' - \theta$ of the radial impeller with radial blades $\beta_2' = 90°$ depicted in right of Figure 5.4 is such that $m' = 10$ corresponds to a radius $r = 50$ mm and a blade height $b = 8$ mm. The impeller features 16 blades and handles 550 l/min of incompressible fluid. Assuming the following θ, r coordinates for points A to G, A(33,50), B(39.9,50), C(45.4,50), D(49,50), E(51.9,50), F(54.4,50), G(55.5,50) and constant slopes of streamlines $\beta = 80°$ in points A to G, find the difference in the relative velocity from A to G Δw_{A-G} and the corresponding difference in pressure energy $\Delta p_{A-G}/\rho$.

   [6.4 m/s, $-28\,\text{m}^2/\text{s}^2$]

3. A set of centrifugal pump impellers with 7 blades operate with the following parameters: blade outlet angle $\beta_{2'} = 20°$, impeller radius $r_2 = 0.5$ m, inlet radius ratio r_1/r_2 varies from 0.3 to 0.6, rotational speed $N = 1450$rpm, flow rate Q varies from 0.01 to 0.06 m^3/s for each pump. Assume $c_{u2,\text{E}} = 18 - 100 \cdot Q$ for each pump. Compute the Eulerian head h_{Euler} and the theoretical head $h_{\text{corrected}}$ according to the Wiesner equatiom for all combinations of Q and r_1/r_2. Prove that the ratio $h_{\text{Euler}}/h_{\text{corrected}}$ is independent of Q.
4. Derive and plot the trajectory $r_a(\theta)$ of a fluid particle in a 360°-degree volute when b_3 = constant and the cross-section is rectangular. Assume $Q = 0.0377\,\text{m}^3/\text{s}$, $b_3 = 0.02$ m, $r_3 = 0.1$ m, $c_{u3} = 25$ m/s. Next, assuming $p_3 = 300$ kPa, $\rho = 1000\,\text{kg/m}^3$ find the pressure at the end of the trajectory.

   [546.8 kPa]

5. Solve the previous exercise when the cross-section is trapezoidal, $b(r) = b_3 + 0.05r$.

   [515.2 kPa]

6. A trapezoidal section volute has $b(r) = b_3 + kr$. Derive the governing equation for $r_a(\theta)$ incorporating k. Assume $b_3 = 0.02$ m and analyze the trajectory for $k = 0.01, 0.02, 0.03$. Create plots of $r_a(\theta)$ and $\tan\alpha$ versus θ for each k.
7. A straight line vaned diffuser of a centrifugal pump impeller has an inlet radius $r_3 = 0.25$ m, a constant depth $b = 0.012$ m and operates with $Q = 0.1\ m^3/s$ and $\alpha_3 = 20°$ to the tangential direction. The outlet radius is $r_4 = 0.3$ m. No flow losses occur between the inlet and outlet. Determine the pressure recovery coefficient of such a diffuser.

   [0.789]

8. A centrifugal pump operates at its nominal conditions with the following parameters: $Q^* = 0.07\,\mathrm{m}^3/\mathrm{s}$, impeller exit diameter $D_2 = 0.22\,\mathrm{m}$, impeller inlet and exit blade angles, $\beta_1' = 9.9^\circ$ and $\beta_2' = 40^\circ$, respectively, and number of blades $Z = 8$. The inlet mean radius is $r_1 = 0.07$ the impeller inlet and exit widths are $b_1 = 0.08$, $b_2 = 0.02\,\mathrm{m}$. The rotational speed is $N = 1550\,\mathrm{rpm}$, the friction factor of impeller $f_w = 0.15$. The volute parameters are: $r_3 = 0.17\,\mathrm{m}$, $r_4 = 0.25\,\mathrm{m}$, $b_3 = 0.022$ m, $f = 0.11$, $k_c = 0.4$. Determine the pump's head h and its hydraulic efficiency η_{hyd}.
   [16.2 m, 0.93]

9. The pump in the previous problem operates at off-design $Q = 0.11\,\mathrm{m}^3/\mathrm{s}$ under unchanged rotational speed. Incidence loss coefficients are $\lambda_r = 0.15$ and $\lambda_s = 0.25$. Determine the pump's head h and its hydraulic efficiency η_{hyd} at the updated working point.
   [8.42 m, 0.68]

10. Determine the characteristic curves of the pump in the previous exercise when Q is varied from 0.01 to 0.2 m^3/s.

NOTES

1. The differential equation can be solved using the following finite difference approximations. We compute the central difference for second derivatives:

$$\frac{\partial^2 \Psi}{\partial z^2} \approx \frac{\Psi_{i,j+1} - 2\Psi_{i,j} + \Psi_{i,j-1}}{(\Delta z)^2}; \quad \frac{\partial^2 \Psi}{\partial r^2} \approx \frac{\Psi_{i+1,j} - 2\Psi_{i,j} + \Psi_{i-1,j}}{(\Delta r)^2}$$

then use a central difference for the radial derivative:

$$\frac{\partial \Psi}{\partial r} \approx \frac{1}{r_i}\frac{\Psi_{i+1,j} - \Psi_{i-1,j}}{2\Delta r}$$

and the Jacobi method update formula:

$$\Psi_{i,j}^{(k+1)} = \frac{(\Delta z)^2\left(\Psi_{i+1,j}^{(k)} + \Psi_{i-1,j}^{(k)}\right) + (\Delta r)^2\left(\Psi_{i,j+1}^{(k)} + \Psi_{i,j-1}^{(k)}\right) - (\Delta r)^2 \frac{1}{r_j} \frac{\Psi_{i,j+1}^{(k)} - \Psi_{i,j-1}^{(k)}}{2}}{2\left[(\Delta z)^2 + (\Delta r)^2\right]}$$

where $\Psi_{i,j}^{k+1}$ is the updated value of Ψ at grid point (i, j) at iteration k, Δr is the spacing between grid points in the r -direction, Δz is the spacing between grid points in the z-direction, r_i is the radial coordinate at grid point i, $\Psi_{i+1,j}$, $\Psi_{i-1,j}$, $\Psi_{j+1,i}$ and $\Psi_{j-1,i}$ are the values of Ψ at neighboring grid points.
2. Busemann, A., "Das Förderhöhenverhältnis radialer Kreiselpumpen mit logarithmisch-spiraligen Schaufeln," *ZAMM - Zeitschrift für Angewandte Mathematik und Mechanik*, 8(5): 372–384 (1928).
3. Wiesner, F. J., "A Review of Slip Factors for Centrifugal Impellers," *Journal of Engineering for Power*, 89(4): 558–566 (1967).
4. A free-vortex distribution allows for the flow to evolve without angular momentum exchange along its path, which implies that no net momentum is exchanged between the flow and the pump volute casing.

5. Finding $\alpha_4 = f\left(\alpha_3, \frac{r_4}{r_3}\right)$ can be solved using analytic geometry. The solution of this simple problem is not algebraically written in closed form and requires numerics.
6. A slightly more accurate model for the losses in a volute at on-design operation is as follows. In a simple volute where a free-vortex flow takes place, $rc_u =$ constant and the flow is confined between two parallel walls at a distance b, the head loss due to wall friction can be estimated using the Darcy-Weisbach equation:

$$h_f = f\frac{S}{D}\frac{c_m^2}{2g} \tag{5.76}$$

where f is a modified Darcy friction factor, S the flow path length along the logarithmic spiral, D the hydraulic diameter and c_m the mean velocity in the spiral direction. An approximate expression for D is $D \approx 4\pi \cdot r_3 \cdot \tan\alpha_3$. It is often found that the friction factor f is two to three times greater than the value for fully developed pipe flow. As already proven, the length is;

$$S = \frac{1}{\sin\alpha_3}(r_4 - r_3) \tag{5.77}$$

Thus, the friction loss is:

$$h_f = f\frac{1}{4\pi \cdot r_3 \cdot \tan\alpha_3}\frac{1}{\sin\alpha_3}(r_4 - r_3)\frac{c_m^2}{2g} \tag{5.78}$$

The head loss due to flow bending (curvature-induced losses) is given by:

$$h_c = k_c\frac{c_m^2}{2g} \tag{5.79}$$

where k_c is an empirical constant function of $\frac{R_c}{D}$, R_c is the mean radius of curvature of the volute and D its hydraulic diameter), typically around 0.4 when $\frac{R_c}{D} < 3$ or 0.2 when $3 \leq \frac{R_c}{D} \leq 5$ for turbulent flows. An approximate expression is $R_c \approx r_3(1 + \pi\tan\alpha_3)$. The total head loss is the sum of friction and curvature losses:

$$h_{\text{total}} = h_f + h_c \tag{5.80}$$

The mean velocity c_m in the spiral direction can be approximated as the average azimuthal velocity between r_3 and r_4:

$$c_m \approx \frac{1}{r_4 - r_3}\int_{r_3}^{r_4} c_u(r)\,dr = \frac{K}{r_4 - r_3}\ln\left(\frac{r_4}{r_3}\right) \tag{5.81}$$

where $c_u(r) = \frac{K}{r}$ is the azimuthal velocity in a free-vortex flow.

6 Two-Dimensional Cascades

6.1 GENERAL CHARACTERISTICS

A two-dimensional cascade is a series of side-by-side airfoil-like blade sections (or *blade elements*) arranged in a row. Each blade section in the cascade can be thought of as an individual airfoil interacting with its neighboring ones, such that the flow is two-dimensional and periodic.

In a two-dimensional flow, the variations in the flow properties are confined to the plane defined by the blades, thus disregarding effects that occur in the third dimension.

A periodic flow is characterized by a flow pattern repeating itself from one blade passage to the next: essentially, the flow characteristics (velocity, pressure, etc.) at a point in one passage are the same as at the corresponding point in the adjacent passage. The distance between these two corresponding points is called spacing.

Two-dimensional cascades serve as an effective model for studying axial flow machines by analyzing the flow on cylindrical stream surfaces (flow surfaces of revolution), found at a prescribed value of the machine radius, unwrapped on a 2-D plane (Figure 6.1).

6.1.1 A COMMON CLASSIFICATION

Turbomachinery cascades can be classified into decelerating and accelerating, based on the behavior of the flow through the blade passages.

In decelerating cascades, the fluid velocity decreases as it passes through the blade passages. This occurs when the blades are designed to decelerate the flow by increasing the cross-sectional area available to the flow.

In accelerating cascades, the opposite occurs, since the flow velocity increases as it passes through the blade passages. The blade sections are staggered to reduce the cross-sectional area of the passage, thus causing flow acceleration.

Decelerating and accelerating cascades must be understood in relation to the frame of reference in which they operate. In a relative frame, where relative velocities are analyzed, decelerating cascades are typically employed in compressor rotors. Here, the pressure in the flow increases as a result of the deceleration of the relative velocity.

Conversely, in stationary blading, such as in compressor stators, we examine absolute velocities. In this context, the absolute velocity decelerates, which results in a pressure increase through the conversion of the kinetic energy into pressure energy.

In turbines, accelerating cascades are particularly important in both stators (nozzles) and rotors. In nozzles, the flow's pressure is converted into kinetic energy by

DOI: 10.1201/9781003619956-6

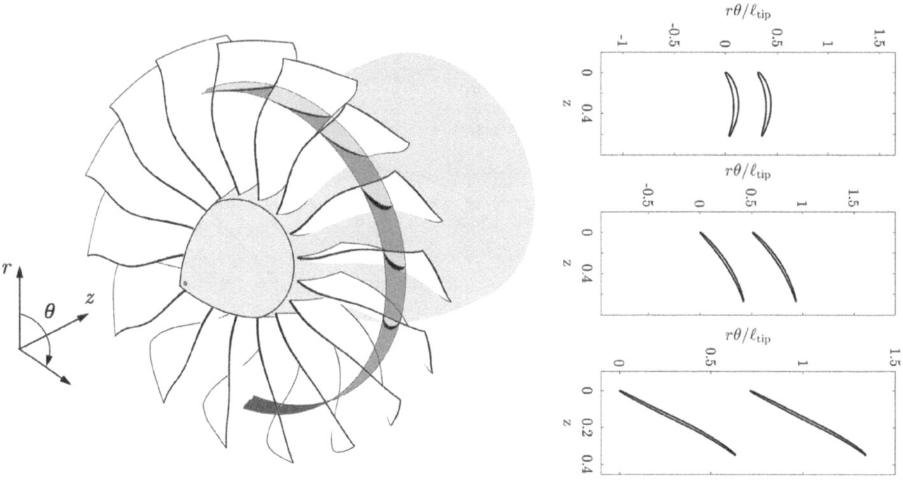

(a) (b)

Figure 6.1 Schematic of a turbomachinery rotor stage intersected by a cylindrical surface (a); blade-to-blade views at three radial positions obtained by unwrapping three cylindrical sections onto planes (b).

accelerating the absolute flow. In turbine rotors, the acceleration of the relative flow allows for further expansion of the flow.

An exception to this is found in pure impulse rotor cascades, which are infrequently used in compressors but are commonly found in turbines. In these rotors, the relative flow is neither accelerated nor decelerated; instead, it is merely deviated in its direction by the blades within an almost constant pressure environment. This deviation changes the angular momentum of the flow causing the turbine rotor to spin.

6.2 DECELERATING CASCADES

In approaching the study of decelerating cascades, incompressible flow with friction can be considered a valid assumption when $\mathrm{Ma} < 0.3$. Additionally, considering frictional forces in these analyses makes it possible to draw preliminary information on efficiency.

6.2.1 CASCADE GEOMETRY AND VELOCITY DIAGRAMS

Figure 6.2 shows a schematic of a decelerating cascade of airfoils, where the key elements and parameters are illustrated. The top-left part of the figure shows the profile of each blade element, highlighting key geometric parameters.

Flow angles and velocities are defined as follows. $\mathbf{c}_1$ and $\mathbf{w}_1$ are the absolute and relative velocity vectors, respectively, at the inlet of the cascade. These are assumed to be uniformly distributed and are referenced to an imaginary station 1 located well

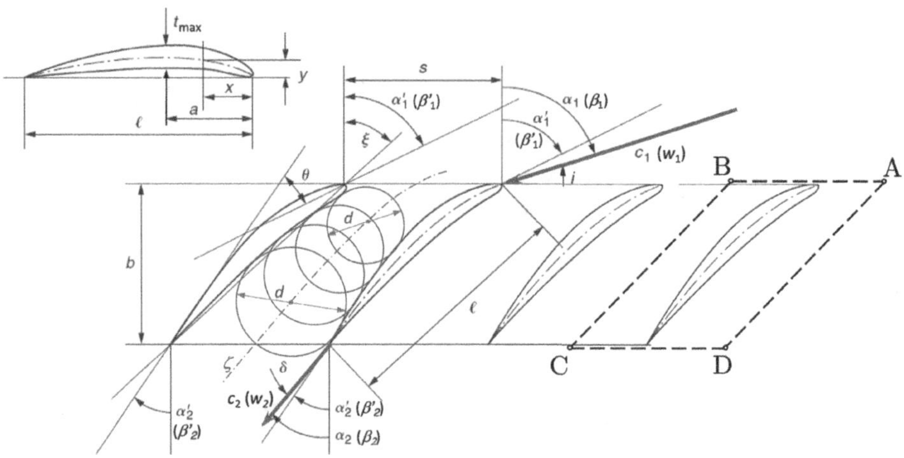

Figure 6.2 Geometry of a decelerating cascade. In brackets are quantities in the relative frame.

upstream of the cascade. $\mathbf{c}_2$ and $\mathbf{w}_2$ denote the absolute and relative velocity vectors, respectively, at exit station 2, assumed to be located well downstream of the cascade. Figure 6.2 shows all these vectors with their pertinent magnitude, indicated as c_1, w_1, c_2, and w_2. α_1 and β_1 are the inlet flow angles of the absolute and relative flow, respectively, measured with respect to the axial (or vertical) direction, while α_2 and β_2 are the corresponding exit flow angles. α_1', β_1', α_2' and β_2' are the *metal* (or *blade*) angles at inlet and exit respectively.

Cascade parameters include s, the spacing between adjacent airfoils, b the axial chord length, i the incidence angle, as well as θ, the camber angle, ξ the stagger angle and δ, which denotes the exit deviation angle, i.e. the angular slip between the flow and the blade angle.

Under the assumption that compressibility effects are negligible, the vertical projection of the velocities is constant across the cascade, which leads to $|\mathbf{c}_2| < |\mathbf{c}_1|$, (or $|\mathbf{w}_2| < |\mathbf{w}_1|$), indicating a decrease in absolute (or relative) velocity across the cascade. This is also confirmed by analyzing the diverging distribution of the diameter of the blue circles inscribed into the blade passage in Figure 6.2, which indicates that the cascade behaves as a diffuser in its frame of reference.

Figure 6.3 shows the velocity diagram in a decelerating rotor cascade. As discussed earlier, rotating and stationary cascades can be analyzed using the same methods, provided that relative and absolute quantities (namely, flow velocity components and stagnation properties) are considered, respectively.

6.2.2 ANALYSIS OF CASCADE FORCES IN LOW-SPEED FLOWS

The performance and mechanical loading of turbomachinery blades—both rotating and stationary—can be rigorously analyzed through the application of linear momentum conservation to a control surface enclosing an arbitrary blade in the row.

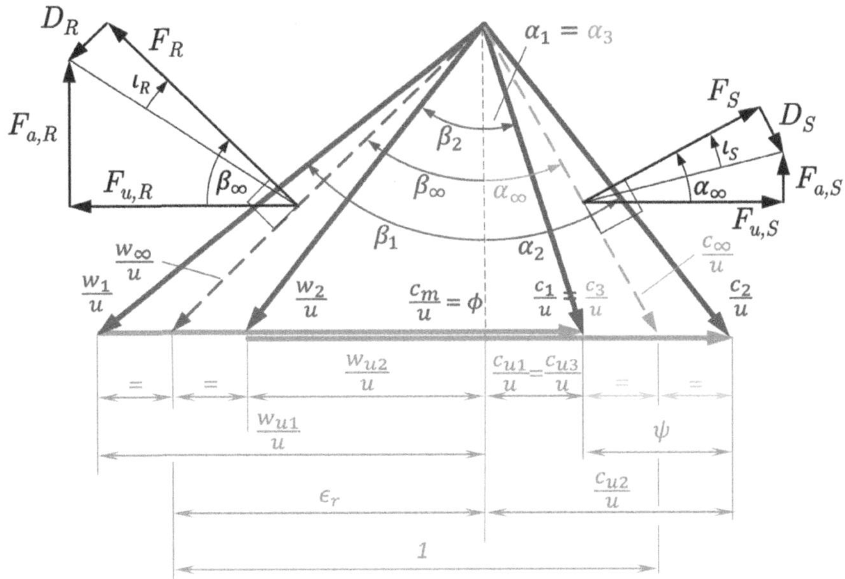

Figure 6.3 Non-dimensional velocity diagram in a decelerating rotor cascade $|\mathbf{w}_2| < |\mathbf{w}_1|$. When $\mathbf{c}_3 = \mathbf{c}_1$ (repeating stage), the right-hand side of the diagram shows a non-dimensional velocity diagram for a decelerating stator cascade downstream of the rotor one, such as $|\mathbf{c}_3| < |\mathbf{c}_2|$.

Such a closed surface is defined by two parallel segments, each of length s, drawn at the inlet and outlet stations (1 and 2), and connected by two additional parallel lines joining the ends of these segments, overall forming the closed path ABCD in Figure 6.2.

For rotating blades, the change in the relative velocity of the fluid (viewed in the rotating frame) leads to a net force on the blade, which, by Newton's third law, corresponds to a reaction torque on the shaft. This torque is directly responsible for work extraction (in turbines) or work input (in compressors). The dominant force component is tangential and governed by the change in angular momentum across the blade. The axial and radial components, while often secondary in energy considerations, are essential for structural design and shaft thrust calculations.

For stationary blades, the force arises from changes in absolute velocity as the flow is turned to align with downstream conditions. Although no shaft work extracted or supplied by these blades, they experience aerodynamic loading due to changes in both magnitude and direction of the absolute velocity vectors.

6.2.2.1 Forces Acting on Rotating Blades

The momentum equation in axial and tangential directions gives:

$$F_{a,R} = (p_2 - p_1)s \tag{6.1}$$

$$F_{u,R} = \rho s c_m (w_{u1} - w_{u2}) = \rho s c_m^2 (\tan\beta_1 - \tan\beta_2) \tag{6.2}$$

Introducing the following definition of mean velocity and mean flow angle in the cascade:

$$w_\infty = \frac{c_m}{\cos\beta_\infty}, \tan\beta_\infty = \frac{1}{2}(\tan\beta_1 + \tan\beta_2) \tag{6.3}$$

the lift and drag forces can be computed:

$$\begin{pmatrix} F_R \\ D_R \end{pmatrix} = \begin{bmatrix} \sin\beta_\infty & \cos\beta_\infty \\ -\cos\beta_\infty & \sin\beta_\infty \end{bmatrix} \begin{pmatrix} F_{a,R} \\ F_{u,R} \end{pmatrix} = \begin{pmatrix} F_{a,R}\sin\beta_\infty + F_{u,R}\cos\beta_\infty \\ F_{u,R}\sin\beta_\infty - F_{a,R}\cos\beta_\infty \end{pmatrix} \tag{6.4}$$

i.e.:

$$F_R = (p_2 - p_1) s \sin\beta_\infty + \rho c_m s (w_{u1} - w_{u2}) \cos\beta_\infty$$

If the generalized Bernoulli equation holds (Ma < 0.3), the lift force can also be written as:

$$\begin{aligned} F_R &= \left(p_{w2}^0 - \rho\frac{w_2^2}{2} - p_{w1}^0 + \rho\frac{w_1^2}{2}\right) s \left(\frac{w_{u,\infty}}{w_\infty}\right) + \rho c_m s (w_{u1} - w_{u2}) \left(\frac{c_m}{w_\infty}\right) \\ &= \rho s (w_{u1} - w_{u2}) \left(\frac{c_m^2}{w_\infty}\right) + \frac{\rho s}{2}\left(\frac{w_{u,\infty}}{w_\infty}\right)(c_m^2 + w_{u1}^2 - c_m^2 - w_{u2}^2) - \Delta p_w^0 s \left(\frac{w_{u,\infty}}{w_\infty}\right) \\ &= \rho s (w_{u1} - w_{u2}) \left(\frac{c_m^2}{w_\infty}\right) + \rho s \left(\frac{w_{u,\infty}}{w_\infty}\right)\left(\frac{w_{u1} + w_{u2}}{2}\right)(w_{u1} - w_{u2}) - s\Delta p_w^0 \left(\frac{w_{u,\infty}}{w_\infty}\right) \\ &= \rho s c_m (w_{u1} - w_{u2}) - s\Delta p_w^0 \left(\frac{w_{u,\infty}}{w_\infty}\right) \end{aligned}$$

while drag force becomes:

$$\begin{aligned} D_R &= \rho s c_m (w_{u1} - w_{u2}) \sin\beta_\infty - (p_2 - p_1) s \cos\beta_\infty \\ &= \rho s c_m (w_{u1} - w_{u2}) \frac{w_{u,\infty}}{w_\infty} - \left(p_{w2}^0 - \rho\frac{w_2^2}{2} - p_{w1}^0 + \rho\frac{w_1^2}{2}\right) s \frac{c_m}{w_\infty} \\ &= \rho s c_m (w_{u1} - w_{u2}) \frac{w_{u,\infty}}{w_\infty} - \rho s c_m (w_{u1} - w_{u2}) \frac{w_{u,\infty}}{w_\infty} + s\Delta p_w^0 \left(\frac{c_m}{w_\infty}\right) \\ &= s\Delta p_w^0 \left(\frac{c_m}{w_\infty}\right) \end{aligned}$$

from which the following expressions can be derived for the lift and drag coefficients of the rotor cascade:

$$C_{L,R} = \frac{F_R}{(\rho/2) w_\infty^2 \ell}, C_{D,R} = \frac{D_R}{(\rho/2) w_\infty^2 \ell} \tag{6.5}$$

and finally:

$$C_{L,R} = \frac{2(w_{u1} - w_{u2})}{w_\infty} \cdot \frac{s}{\ell} - C_{D,R} \tan\beta_\infty = \frac{2\Delta w_u}{w_\infty \sigma_R} - C_{D,R}\tan\beta_\infty \tag{6.6}$$

where σ_R is the rotor cascade solidity.

If $C_{D,R}$ is small compared to $C_{L,R}$, as in high efficiency airfoils near their design conditions:

$$C_{L,R} \approx \frac{2\Delta w_u}{w_\infty \sigma_R} \tag{6.7}$$

This relationship is particularly useful in preliminary blade design as it provides a direct link between aerodynamic performance and geometric parameters. For instance, by prescribing a target lift coefficient—often based on efficiency, loading, or flow turning requirements—the corresponding cascade solidity can be estimated to achieve the desired aerodynamic behavior.

6.2.2.2 Forces Acting on Stationary Blades

An almost perfect dualism exists between stationary and rotating cascade which is obtained by substituting absolute velocities with relative ones (Figure 6.2). The following are the forces computed in the absolute frame from the inlet to the outlet of a generic stationary cascade.

$$F_{a,S} = (p_3 - p_2)s_S \tag{6.8}$$

$$F_{u,S} = \rho s_S c_m (c_{u2} - c_{u3}) = \rho s_S c_m^2 (\tan\alpha_2 - \tan\alpha_3) \tag{6.9}$$

$$c_\infty = \frac{c_m}{\cos\alpha_\infty}, \; \tan\alpha_\infty = \frac{1}{2}(\tan\alpha_2 + \tan\alpha_3) \tag{6.10}$$

Lift and drag forces are given by:

$$F_S = F_{a,S}\sin\alpha_\infty + F_{u,S}\cos\alpha_\infty, \; D_S = F_{u,S}\sin\alpha_\infty - F_{a,S}\cos\alpha_\infty \tag{6.11}$$

i.e.:

$$F_S = \rho s_S c_\infty (c_{u2} - c_{u3}) - s_S \Delta p^0 \left(\frac{c_{u,\infty}}{c_\infty}\right), D_S = s_S \Delta p^0 \left(\frac{c_m}{c_\infty}\right) \tag{6.12}$$

Lift and drag coefficients of the stator cascade:

$$C_{L,S} = \frac{F_S}{(\rho/2)c_\infty^2 \ell_S} = \frac{2\Delta c_u}{c_\infty \sigma_S} - C_{D,S}\tan\alpha_\infty \tag{6.13}$$

$$C_{D,S} = \frac{D_S}{(\rho/2)c_\infty^2 \ell_S} \tag{6.14}$$

and, finally, when $C_{D,S}$ is small compared to $C_{L,S}$, as in high efficiency airfoils near their design conditions:

$$C_{L,S} \approx \frac{2\Delta c_u}{c_\infty \sigma_S} \tag{6.15}$$

6.2.2.3 Performance Figures of the Stage

6.2.2.3.1 Total Pressure Loss Coefficient

When a generic cascade experiences a non-null drag force, a corresponding total pressure loss occurs, which can be quantified by a total pressure loss coefficient:

$$\zeta_R = \frac{\Delta p_w^0}{(\rho/2)w_2^2} = \frac{p_{w1}^0 - p_{w2}^0}{(\rho/2)w_2^2} \tag{6.16}$$

$$\zeta_S = \frac{\Delta p^0}{(\rho/2)c_3^2} = \frac{p_2^0 - p_3^0}{(\rho/2)c_3^2} \tag{6.17}$$

with which the aerodynamic efficiency can be computed as:

$$\frac{1}{E_R} = \frac{C_{D,R}}{C_{L,R}} = \tan(\iota_R) = \frac{\zeta_R}{2}\frac{\cos^2\beta_\infty}{\cos^2\beta_2}\left[\frac{1}{(\tan\beta_1 - \tan\beta_2)}\right] \tag{6.18}$$

$$\frac{1}{E_S} = \frac{C_{D,S}}{C_{L,S}} = \tan(\iota_S) = \frac{\zeta_S}{2}\frac{\cos^2\alpha_\infty}{\cos^2\alpha_3}\left[\frac{1}{(\tan\alpha_2 - \tan\alpha_3)}\right] \tag{6.19}$$

6.2.2.3.2 Stage Efficiency

Recalling the following expressions for fundamental performance figures:

$$\phi = \frac{c_m}{u} \text{ (flow coefficient)} \tag{6.20}$$

$$\psi = \frac{c_{u2} - c_{u1}}{u} = \frac{w_{u1} - w_{u2}}{u} \text{ (work coefficient)} \tag{6.21}$$

$$\varepsilon_r = 1 - \frac{c_{u2} + c_{u1}}{2u} = \frac{w_{u2} + w_{u1}}{2u} \text{ (degree of reaction)} \tag{6.22}$$

we introduce a stage efficiency by comparing the actual specific work needed to keep rotor rotation u, i.e. $L = \psi u^2 = \frac{1}{\rho}\Delta p_{1-3}^0$, and the ideal work that would be spent in case of inviscid flow, $L_{id} = \psi_{id} u^2 = \frac{1}{\rho}\Delta p_{1-3}^{0*} = \frac{1}{\rho}\Delta p_{1-3}$ (as in the case of a frictionless flow, Bernoulli would hold in its conservative form, i.e. $\Delta p_{1-3}^{0*} = \Delta p_{1-3}$:

$$\eta = \frac{L_{id}}{L} = \frac{\psi_{id}}{\psi} = \frac{\Delta p_{1-3}}{\rho\psi u^2} = \frac{\Delta p_R + \Delta p_S}{\rho\psi u^2} \tag{6.23}$$

Now, let us recall that:

$$\Delta p_R = \frac{F_{a,R}}{s_R}, \Delta p_S = \frac{F_{a,S}}{s_S} \tag{6.24}$$

where Δp_R and Δp_S are the static pressure increases in rotor and stator, respectively. Force F_a can be determined from the trigonometric tangent difference identity for small angles ι:

$$F_{a,R} = F_{u,R}\tan(\beta_\infty - \iota_R) = F_{u,R}\frac{\tan\beta_\infty - \iota_R}{1+\iota_R\tan\beta_\infty} \tag{6.25}$$

$$F_{a,S} = F_{u,S}\tan(\alpha_\infty - \iota_S) = F_{u,S}\frac{\tan\alpha_\infty - \iota_S}{1+\iota_S\tan\alpha_\infty} \tag{6.26}$$

From Figure 6.2, the following relationships hold:

$$\tan\beta_\infty = \frac{\varepsilon_r}{\phi},\ \tan\alpha_\infty = \frac{(1-\varepsilon_r)}{\phi} \tag{6.27}$$

Thus, using equation 6.24:

$$\Delta p_R = \rho\psi\phi u^2\left(\frac{\varepsilon_r - \phi\iota_R}{\phi+\iota_R\varepsilon_r}\right),\ \Delta p_S = \rho\psi\phi u^2\left(\frac{1-\varepsilon_r-\phi\iota_S}{\phi+\iota_S(1-\varepsilon_r)}\right) \tag{6.28}$$

and the overall static pressure rise in the stage is therefore:

$$\Delta p = \Delta p_R + \Delta p_S = \rho\psi\phi u^2\left[\frac{\varepsilon_r-\phi\iota_R}{\phi+\iota_R\varepsilon_r} + \frac{1-\varepsilon_r-\phi\iota_S}{\phi+\iota_S(1-\varepsilon_r)}\right] \tag{6.29}$$

By rearranging previous equations, from equation 6.23 it is easy to arrive at:

$$\eta = \phi\left[\frac{\varepsilon_r-\phi\iota_R}{\phi+\iota_R\varepsilon_r} + \frac{1-\varepsilon_r-\phi\iota_S}{\phi+\iota_S(1-\varepsilon_r)}\right] \tag{6.30}$$

which is plotted in Figure 6.4 assuming $1/\iota_R = 1/\iota_S = 25$ and 60.

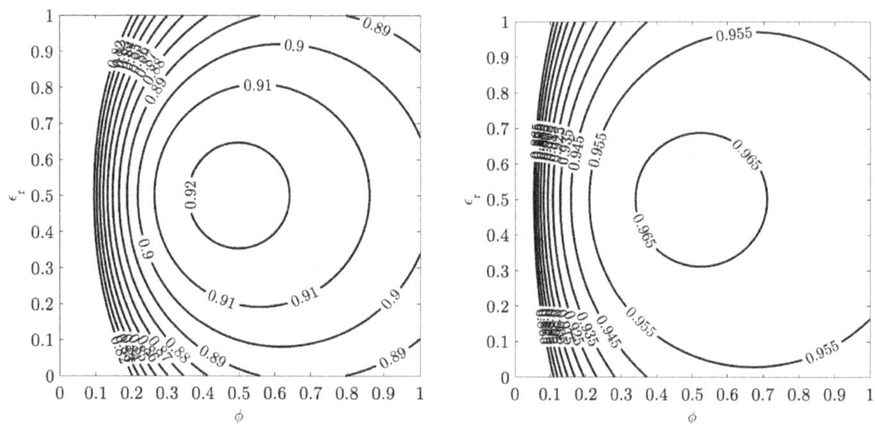

(a) (b)

Figure 6.4 Contour plots of η for two values of E. Left: $E = 25$; right: $E = 60$.

6.2.2.4 Optimal Stage Degree of Reaction $(\varepsilon_r)_{opt}$ and Related $(\phi)_{opt}$ and $(\psi)_{opt}$

To find the optimal degree of reaction of the stage, let us compute $\frac{\partial \eta}{\partial \varepsilon_r} = 0$:

$$\frac{\partial \eta}{\partial \varepsilon_r} = \phi \left[\frac{\phi(1+\iota_R^2)}{(\phi+\iota_R\varepsilon_r)^2} - \frac{\phi(1+\iota_S^2)}{(\phi+\iota_S(1-\varepsilon_r))^2} \right] = 0 \tag{6.31}$$

Solving this for ε_r gives a rather complicated equation, which is not reported here for the sake of brevity. However, if ι_R is assumed equal to ι_S, the maximum efficiency is reached when:

$$(\varepsilon_r)_{\text{opt}} = \frac{1}{2} \tag{6.32}$$

and the corresponding maximum value of stage efficiency is:

$$\eta_{\max} = \phi \left[\frac{0.5 - \frac{\phi}{E}}{\phi + \frac{\phi}{E}} + \frac{0.5 - \frac{0.5}{E}}{\phi + \frac{0.5}{E}} \right] = 2\phi E \frac{1 - \frac{2\phi}{E}}{1 + 2\phi E} \tag{6.33}$$

The contour plots in Figure 6.4 confirm that a stage degree of reaction $\varepsilon_r = 0.5$ corresponds to the optimal stage efficiency independently of E. The optimal flow coefficient ϕ_{opt} exhibits only a mild dependence on the loading parameter E, decreasing slightly from $\phi_{\text{opt}} \approx 0.5$ at $E = 25$ to $\phi_{\text{opt}} \approx 0.47$ at $E = 60$.

For instance, if $E = 20$ and $\phi = 0.5$

$$\eta_{\max} = \frac{19}{21} \tag{6.34}$$

As a further example, in a low-speed cascade of NACA 65-(12)10 blade sections with $\sigma = 1.0$, experimental tests have recorded E as high as 70. Therefore, the expected maximum value of cascade efficiency in this case is $\phi = 0.5$:

$$\eta_{\text{max, NACA 65-(12)10}} = 0.967 \tag{6.35}$$

Figure 6.5 illustrates how the maximum cascade efficiency $\eta_{\max}$ varies with the flow coefficient ϕ for different values of profile efficiency E (15, 20, 30 and 50). Optimal values of the flow coefficient ϕ can be found by solving the following equation:

$$\frac{\partial \eta}{\partial \phi} = 0 \Rightarrow \frac{(2E - 8\phi)(1 + 2\phi E) - (2E\phi - 4\phi^2)}{(1 + 2\phi E)^2} = 0 \tag{6.36}$$

which, neglecting the negative solution, leads to:

$$\phi_{\text{opt}} = -\frac{1}{2E} + \sqrt{\frac{4}{E^2} + \frac{1}{4}} \tag{6.37}$$

Once again, for instance if $E = 20$, $\phi_{\text{opt}} = 0.485$.

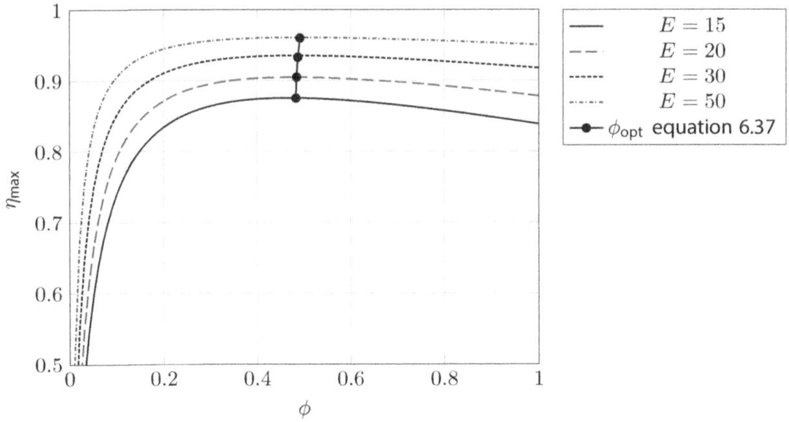

Figure 6.5 Optimal values of cascade efficiency as a function of the flow coefficient for several values of profile efficiency $\varepsilon_r = 0.5$.

The corresponding optimal values of the work coefficient are obtained from the velocity triangles:

$$\psi = \frac{c_{u2} - c_{u1}}{u}; \; (\varepsilon_r)_{opt} = \frac{1}{2} = 1 - \frac{c_{u1} + c_{u2}}{2u} \tag{6.38}$$

from which, upon substitution of $(\varepsilon_r)_{opt}$, the following expression is obtained:

$$\psi_{opt} = 1 - 2\tan\alpha_1 \left(-\frac{1}{2E} + \sqrt{\frac{4}{E^2} + \frac{1}{4}} \right) \tag{6.39}$$

From experiments carried out on decelerating cascades, it is found that tolerable values for ψ should not exceed 0.45 to avoid cascade stall. As a consequence, under the assumption that $E = 20$, it is found that α_1 should be greater than $\approx 30°$.

Since α_1 is linked to β_1 via the velocity triangles:

$$\frac{u - c_m \tan\alpha_1}{u} = \frac{c_m \tan\beta_1}{u} \implies \tan\beta_1 = \frac{1}{\phi} - \tan\alpha_1 \tag{6.40}$$

it is possible to determine optimal values for β_1:

$$\tan[(\beta_1)_{opt}] = \frac{1}{-\frac{1}{2E} + \sqrt{\frac{4}{E^2} + \frac{1}{4}}} - \tan\alpha_1 \tag{6.41}$$

Optimal values of the work coefficient ψ_{opt} as a function of the flow angle α_1 are shown in Figure 6.6. The plots illustrate that ψ_{opt} and related parameters such as ϕ_{opt} (not shown) are scarcely influenced by the airfoil efficiency E. The values of $\beta_{1,opt}$ in degrees, shown on the secondary y-axis, also demonstrate minimal dependence on E.

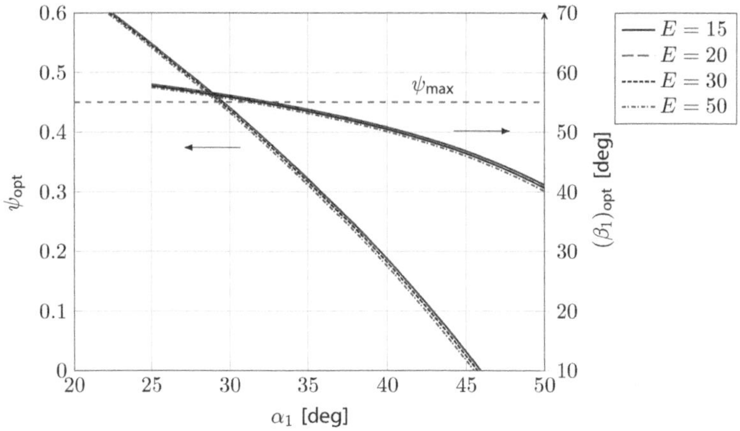

Figure 6.6 Optimal values of the work coefficient as a function of the flow angle α_1 $\varepsilon_r = 0.5$, ϕ_{opt}.

6.2.2.5 Degree of Reaction of Frontal Stages

In some specific cases, such as in frontal compressor stages without IGV, higher than 1/2 degree of reaction is often found. For instance, in inlet stages with $\psi = 0.4$, the degree of reaction is $\varepsilon_r = 1 - \psi/2 = 1 - 0.4/2 = 0.8$. In such cases, the non-dimensional velocity diagram in the rotor blade looks like the one shown in Figure 6.7. The corresponding values for cascade efficiency can be computed using the preceding formulas.

6.2.2.6 Low-Reaction Cascades

In practical applications, the optimal choice for ψ is often constrained by aerodynamic and structural limitations. For instance, in modern large fans with low hub-to-tip diameter ratios, cascades are designed subject to no velocity swirl at the inlet $c_{u1} = 0$. For airfoils located near the hub sections, this results in β_1 values lower than 40° and $\phi < 0.5$. Consequently, the degree of reaction is significantly less than 0.5 and frequently approaches zero. In such scenarios, selecting the appropriate ψ must be guided by alternative criteria. One effective approach is to use loading criteria based on the diffusion factor.

6.2.3 DIFFUSION FACTOR AND BLADE LOADING CRITERIA

The diffusion factor for a blade is a measure of aerodynamic loading and indicates the extent to which the velocity is reduced through the blade passage.

For rotor blades, it can be defined as:

$$DF_R = 1 - \frac{w_2}{w_1} + \frac{|\Delta w_u|}{2\sigma_R w_1} \tag{6.42}$$

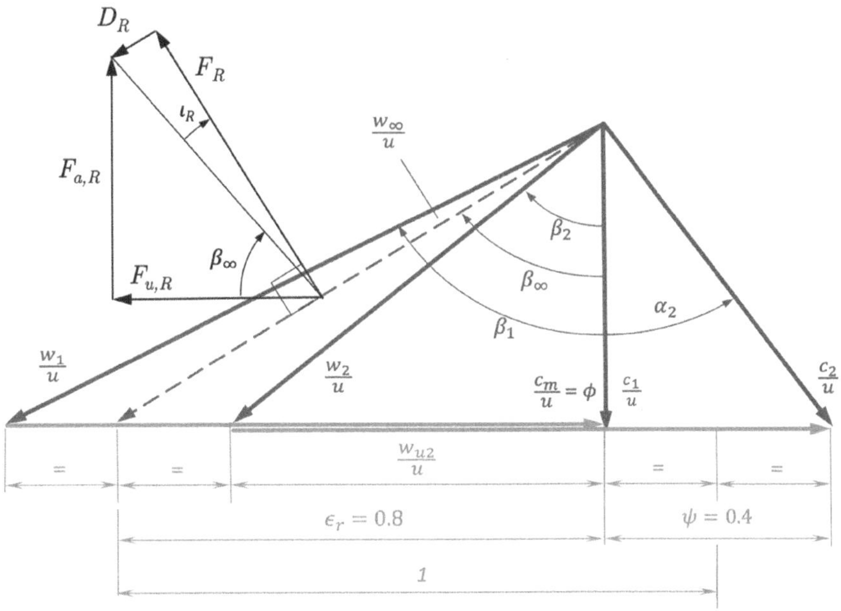

Figure 6.7 Non-dimensional velocity diagram in a decelerating rotor cascade $c_{u1}/u = 0; \varepsilon_r = 0.8; \psi = 0.4$.

where $\Delta w_u = w_{u2} - w_{u1}$ is the change in the tangential component of the relative velocity through the rotor.

For stator blades, it can be defined as:

$$DF_S = 1 - \frac{c_3}{c_2} + \frac{\Delta c_u}{2\sigma_S c_2} \tag{6.43}$$

where $\Delta c_u = c_{u2} - c_{u3}$ is the change in the tangential component of the absolute velocity through the stator.

High diffusion factors indicate high aerodynamic loading, which can lead to boundary layer separation and loss of efficiency. From experiments in low-speed cascades, it has been found that values of DF higher than 0.65 are directly correlated with blade stall for a large set of airfoil shapes.

To link the load coefficient $\psi = \frac{c_{u2}-c_{u1}}{u}$ to DF, we can express c_{u2} in terms of ψ and c_{u1}:

$$c_{u2} = \psi u + c_{u1} \tag{6.44}$$

such that:

$$|\Delta w_u| = |c_{u1} - (\psi u + c_{u1})| = \psi u \tag{6.45}$$

and, under the assumption that $c_{m1} = c_{m2}$:

$$DF_R = 1 - \frac{\cos\beta_1}{\cos\beta_2} + \frac{\psi u}{2\sigma_R w_1} = 1 - \frac{\cos\beta_2}{\cos\beta_1} + \frac{\psi}{2\sigma_R \cos\beta_1} \tag{6.46}$$

from which:

$$\psi = 2\sigma_R \cos\beta_1 \left(DF_R - 1 + \frac{\cos\beta_1}{\cos\beta_2} \right) \tag{6.47}$$

Furthermore, by looking at the velocity triangles, $\psi = (w_1 \sin\beta_1 - w_2 \sin\beta_2)/u$ and $\varepsilon_r = (w_1 \sin\beta_1 + w_2 \sin\beta_2)/(2u)$, thus the following relationship can be obtained:

$$\frac{\psi}{\varepsilon_r} = \frac{2(\cos\beta_2 \tan\beta_1 - \sin\beta_2)}{\cos\beta_2 \tan\beta_1 + \sin\beta_2} \implies \varepsilon_r = \frac{\psi(\cos\beta_2 \tan\beta_1 + \sin\beta_2)}{2(\cos\beta_2 \tan\beta_1 - \sin\beta_2)} \tag{6.48}$$

Carter developed a correlation for the maximum allowed deflection:

$$\tan\beta_1 - \tan\beta_2 = 0.1125\sigma_R(6 - \sigma_R)\frac{\cos\beta_\infty}{\cos^2\beta_2} \tag{6.49}$$

Figure 6.8 illustrates the relationship between $\Delta\beta = (\beta_1 - \beta_2)$, ψ and σ as functions of β_1 for given diffusion factors $DF_R = 0.65$ and 0.45. The shaded region below each line indicates the permissible values of $\Delta\beta_{\max}$ and ψ for a given σ, ensuring that the diffusion factor remains at or below the limits. As σ increases, the lines shift upward, indicating that a higher solidity enables greater allowable flow deflections and work coefficients while maintaining the same diffusion factor.

A similar procedure makes it possible to compute the diffusion factor in stator cascades.

6.2.4 COMPRESSOR AIRFOIL SHAPES

Over the years, various profile shapes have been developed and employed, each with its own unique characteristics and advantages.

6.2.4.1 NACA 65-Series Profiles

The NACA 65-series profiles are derived from NACA aircraft wing airfoils designed for approximately uniform loading. The NACA 65-series blades use a camber line derived from the base airfoil NACA 65-(10)10. The camber line coordinates are scaled by the lift coefficient C_{L_0} to generate other airfoil configurations. The equation for the camber line can be written as:

$$y_c(x) = C_{L_0} \cdot y_{\text{base}}(x) \tag{6.50}$$

where C_{L_0} is the lift coefficient of the isolated airfoil and $y_{\text{base}}(x)$ is the base camber line coordinate for $C_{L_0} = 1.0$.

The thickness distribution t_b/ℓ defines the blade's shape along the chord x/ℓ. Both camber line and thickness data for the NACA 65-(10)10 profile are shown in Table 6.1.

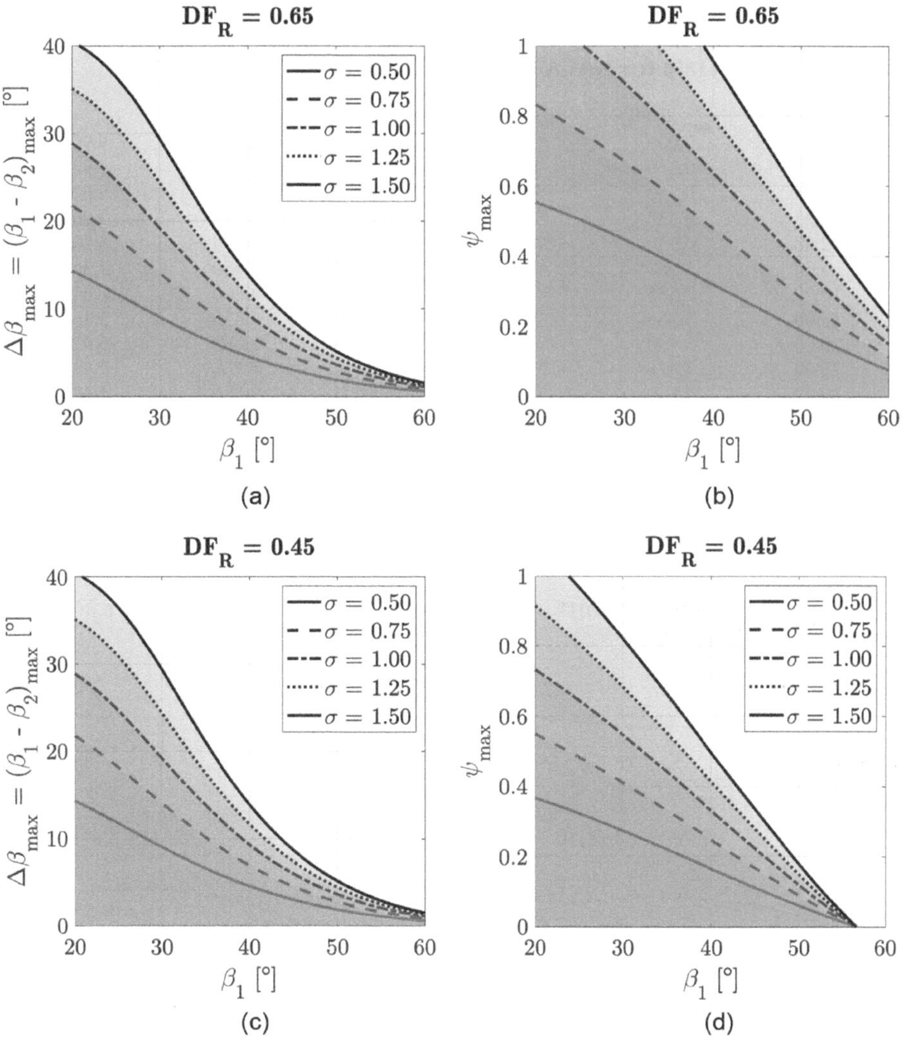

Figure 6.8 Rotor cascade characteristics having $DF_R = 0.65$ (a,b) and 0.45 (c,d).

The final blade coordinates are calculated by combining the camber line and thickness distribution (Figure 6.9):

$$y_{\text{upper}}(x) = y(x) + \frac{t_b(x)}{2},\; y_{\text{lower}}(x) = y(x) - \frac{t_b(x)}{2} \tag{6.51}$$

To standardize experimental data, the NACA 65-series camber line is often approximated using an equivalent circular-arc camber line. The radius of curvature for

Table 6.1
Dimensionless Data for NACA 65-(10)10 Profile

x/ℓ (%)	y_{base}/ℓ (%)	t_b/ℓ (%)	x/ℓ (%)	y_{base}/ℓ (%)	t_b/ℓ (%)
0.00	0.000	0.000	40.00	5.355	9.992
0.50	0.250	1.544	45.00	5.475	9.926
0.75	0.350	1.864	50.00	5.515	9.624
1.25	0.535	2.338	55.00	5.475	9.060
2.50	0.930	3.148	60.00	5.355	8.292
5.00	1.580	4.354	65.00	5.150	7.364
10.00	2.585	6.080	70.00	4.860	6.312
15.00	3.365	7.332	75.00	4.475	5.168
20.00	3.980	8.286	80.00	3.980	3.974
25.00	4.475	9.006	85.00	3.365	2.770
30.00	4.860	9.520	90.00	2.585	1.620
35.00	5.150	9.848	95.00	1.580	0.612
			100.00	0.000	0.000

The leading edge radius is 0.687% of ℓ.

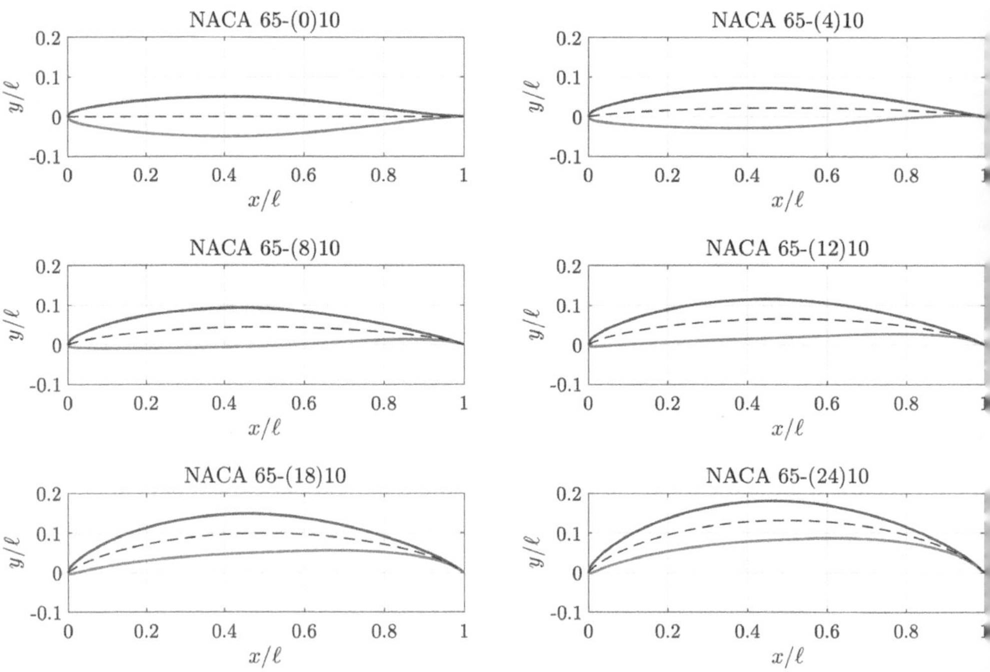

Figure 6.9 NACA 65-series airfoils with different camber line shapes defined by their camber parameter (0, 4, 8, 12, 18, and 24).

this camber line is given by:

$$R = \frac{\ell}{2\sin(\theta/2)} \tag{6.52}$$

where θ is the camber angle. The coordinates of the circular-arc camber line are:

$$y_c(x) = R - \sqrt{R^2 - x^2}. \tag{6.53}$$

NACA 65-series airfoils (Figure 6.9) are designated as NACA 65-$(C_{L_0})t_b$, where C_{L_0} is the lift coefficient in tenths (e.g. $C_{L_0} = 1.5$ becomes 15) and t_b is the maximum thickness-to-chord ratio as a percentage. For example, a 10% thick airfoil with a lift coefficient of 1.5 is designated as NACA 65-(15)10.

6.2.4.2 Parabolic-Arc Camber Line Profiles

The parabolic-arc camber line allows for more flexibility in blade loading by varying the location of the point of maximum camber (x_A, y_A) (Figure 6.10a). The camber line equation $y_c(x)$ is defined as:

$$y_c(x) = \begin{cases} \frac{\frac{y_A}{\ell}}{\left(\frac{x_A}{\ell}\right)^2}\left(2\frac{x_A}{\ell}x - x^2\right), & 0 \le x \le \frac{x_A}{\ell} \\ \frac{\frac{y_A}{\ell}}{\left(1-\frac{x_A}{\ell}\right)^2}\left[\left(1 - 2\frac{x_A}{\ell}\right) + 2\frac{x_A}{\ell}x - x^2\right], & \frac{x_A}{\ell} < x \le 1 \end{cases} \tag{6.54}$$

The slope of the camber line $\frac{dy_c}{dx}$ at $\frac{x}{\ell} = 0.05$ conventionally defines the blade inlet angle (or the leading edge angle) (Top of Figure 6.10):

$$\tan\theta_i = \frac{\frac{y_A}{\ell}}{\left(\frac{x_A}{\ell}\right)^2}\left(2\frac{x_A}{\ell} - 2\cdot 0.05\right) \tag{6.55}$$

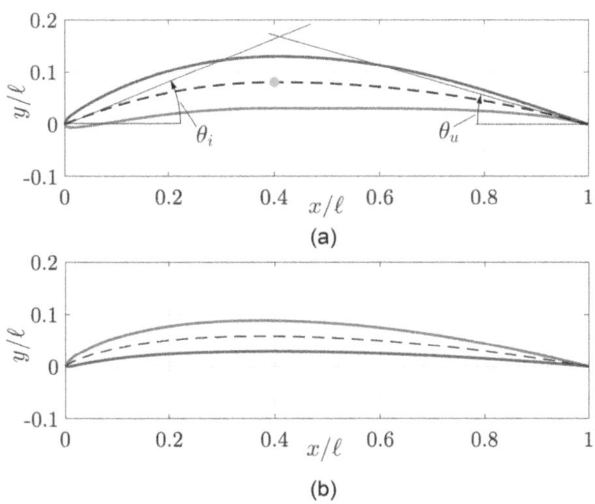

Figure 6.10 Top: NACA 65 airfoil with parabolic-arc camber line. Bottom: NACA A4K6 63-Series guide vane.

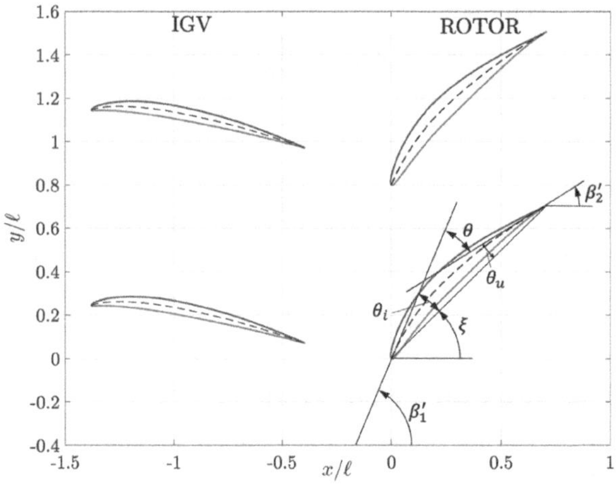

Figure 6.11 Staggered configuration of IGV and rotor cascades.

while at $\frac{x}{\ell} = 0.95$ the slope:

$$|\tan\theta_u| = \left| \frac{\frac{y_A}{\ell}}{\left(1 - \frac{x_A}{\ell}\right)^2} \left[\left(1 - 2\frac{x_A}{\ell}\right) + 2\frac{x_A}{\ell} \cdot 0.95 - 0.95^2 \right] \right| \tag{6.56}$$

conventionally defines the blade outlet angle (or the trailing edge angle).

6.2.4.3 Circular-Arc Camber Line

As already seen, the circular-arc camber line can be used along with various thickness distributions to generate several airfoil shapes. The circular-arc camber line is defined by angle θ and chord length ℓ. The radius of curvature and camber line coordinates are:

$$R = \frac{\ell}{2\sin(\theta/2)}, \; y_c(x) = R - \sqrt{R^2 - x^2}, \tag{6.57}$$

6.2.4.4 British C.4 Series Profiles

The C.4 profile is thicker toward the leading edge compared to the NACA 65 series, with a maximum thickness of 10% at 30% chord (Figure 6.12, Table 6.2). It typically uses either a circular-arc or parabolic-arc camber line.

The blade shape is defined by combining the camber line and the thickness distribution:

$$y_{\text{upper}}(x) = y(x) + \frac{t_b(x)}{2}, \; y_{\text{lower}}(x) = y(x) - \frac{t_b(x)}{2} \tag{6.58}$$

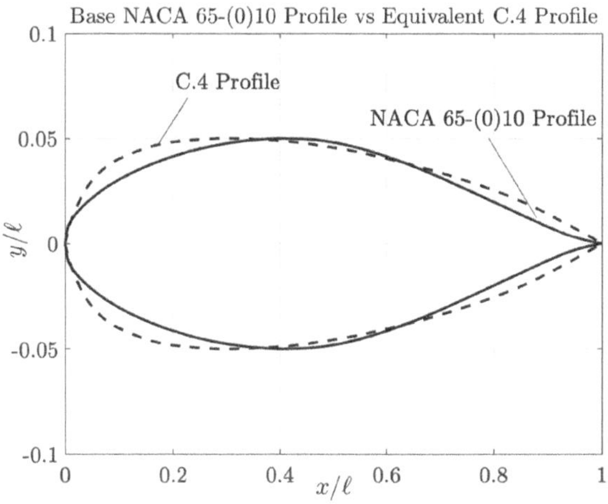

Figure 6.12 Comparison between NACA 65-(0)10 and an equivalent C.4 profile.

Table 6.2
Thickness Distribution for British C.4 Series and NACA A4K6 63-Series

(a) British C.4 Series Profiles		(b) NACA A4K6 63-Series Guide Vane		
x/ℓ (%)	t_b/ℓ (%)	x/ℓ (%)	y/ℓ (%)	t_b/ℓ (%)
0	0.000	0	0.000	0.000
10	8.04	1.25	0.792	1.542
20	9.66	2.5	1.357	2.114
30	10.00	5.0	2.248	2.924
40	9.78	10.0	3.531	4.020
50	9.14	20.0	5.040	5.312
60	8.10	40.0	5.820	5.942
80	5.08	80.0	2.922	2.876
100	0.00	100.0	0.000	0.000

C.4 profiles are designated using the format tb-C4/$\theta_c P(a/\ell)$, where tb is the maximum thickness-to-chord ratio (%), θ_c the camber angle (in degrees), and $P(a/\ell)$ denotes a parabolic camber line with maximum camber at a/ℓ, or $C(a/\ell)$ for a circular camber line. For example, a C.4-based blade section with 10% thickness, 20° camber angle and a parabolic camber line with $a/\ell = 0.4$ would be designated as 10C4/20P40.

6.2.4.5 Double-Circular-Arc Profiles

A double-circular-arc (DCA) profile is designed with both the upper and lower surfaces consisting of circular arcs that transition smoothly into either a circular or elliptical shape at the leading and trailing edges (Figure 6.13). The construction follows the same principles used for defining circular-arc camber lines.

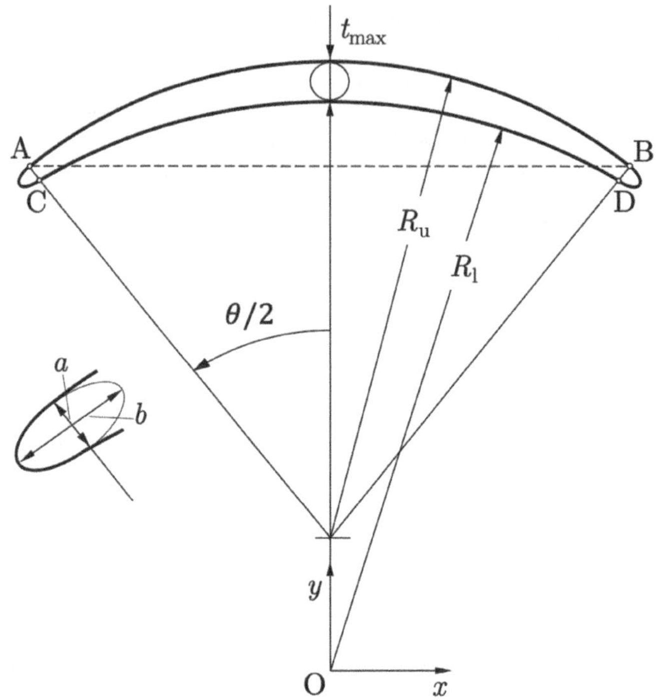

Figure 6.13 A DCA profile. The upper and lower surfaces are defined by circular arcs with radii R_u and R_l.

6.2.4.6 NACA A4K6 63-Series Guide Vane Profiles

NACA A4K6 63-Series guide vane profile The NACA A4K6 63-series guide vane uses a parabolic-arc camber line, while the thickness distribution is provided in Table 6.2.

The upper and lower surface coordinates of the blade are computed in the usual way:

$$y_{\text{upper}}(x) = y_c(x) + \frac{t_b(x)}{2}, \; y_{\text{lower}}(x) = y_c(x) - \frac{t_b(x)}{2} \tag{6.59}$$

6.2.4.7 Staggered Profiles

Figure 6.11 provides a sample of a staggered arrangement of blade sections, including inlet guide vanes followed by a rotor cascade. The IGV blades are designed to guide the incoming flow with minimal deviation while ensuring an optimal flow direction for the rotor blades downstream. The following relationships exist among the angles shown:

$$\theta = |\theta_u| + \theta_i,\ \beta_1' = \beta_2' + \theta,\ \xi = \beta_1' - \theta_i \tag{6.60}$$

where these can be combined to obtain the desired interdependency.

6.2.5 SURFACE PRESSURE DISTRIBUTION

The pressure coefficient, defined as:

$$C_p = \frac{p_1^0 - p}{\frac{1}{2}\rho c_1^2} \tag{6.61}$$

represents the normalized pressure difference between the inlet total pressure and the local static pressure on the airfoil surface.

Figure 6.14 illustrates the distribution of C_p along the normalized chord x/ℓ for a stationary cascade of NACA65-(12)10 blades at low speed. As the stagger angle decreases, the flow incidence $i = \beta_1 - \beta_1'$ for a rotor, (or $\alpha_1 - \alpha_1'$ for a stator), increases, causing larger pressure differences between the suction and pressure sides, which enhances the lift coefficient, thus indicating stronger local acceleration/deceleration and blade loading.

At the cascade exit, $C_p < 1$ highlights that the static pressure is higher than at the inlet, signifying flow deceleration and pressure recovery.

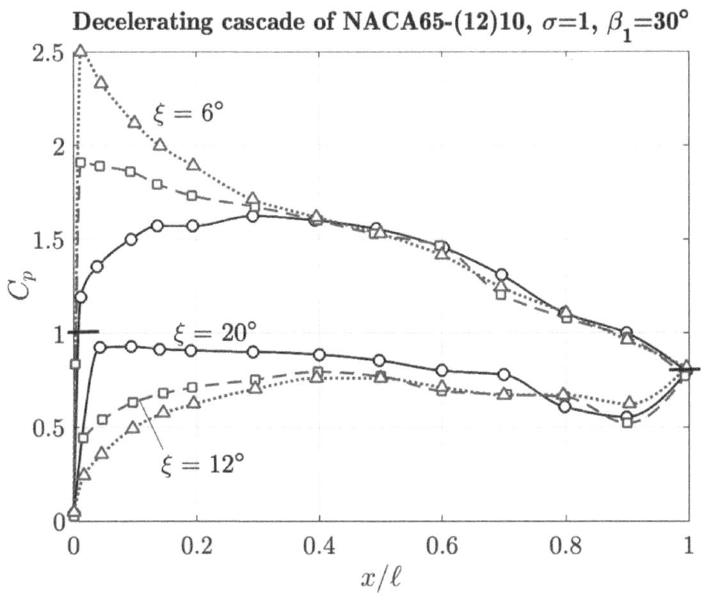

Figure 6.14 Pressure coefficient C_p along the chord x/ℓ of a decelerating cascade of NACA65-(12)10 blades, solidity σ of 1 and inlet blade angle β_1 of 30°. Results are presented for three stagger angles $\xi = 6°, 12°, 20°$.

6.3 ACCELERATING CASCADES

Accelerating cascades, where the working fluid—such as steam or gas—is continuously accelerated as it passes through the passages, are essential in axial turbines. Accelerating the flow is key for maintaining its stability and minimizing losses: in fact, flow acceleration helps stabilize the boundary layer, reducing the risk of flow separation and turbulence, which in turn minimizes energy losses and enhances efficiency.

An example of a velocity diagram in this cascade arrangement is shown in Figure 6.15. The blue circles superimposed on the cascade indicate the cross-sectional areas of the flow passages. These areas typically form a converging passage in the respective frame of reference, consistent with the accelerating nature of the flow. Due to this acceleration, turbine cascades can achieve significantly higher flow deflections compared to decelerating cascades. These larger deflections result in higher profile loading and increased tangential forces, necessitating the use of thicker airfoils.

The magnitude of flow acceleration is a function of the degree of reaction:

- in reaction cascades, the flow is always accelerated in its frame of reference. The blades are designed so that both fixed and moving blades contribute to the pressure drop, continuously accelerating the fluid. This results in a continuous decrease in static enthalpy and pressure through the cascades. Configurations of this type are often referred to as *Parsons cascades* after Sir Charles Parsons invented them (1884), for which a degree of reaction around 0.5 provides the best results (left of Figure 6.16);
- in action-or De Laval (1882)-stages, significantly higher acceleration, and corresponding pressure drop, occur in the nozzles compared to Parsons' architecture. In the rotors, the fluid maintains a constant velocity magnitude from inlet to outlet, with the primary change being in the direction of the flow. This leads to a high

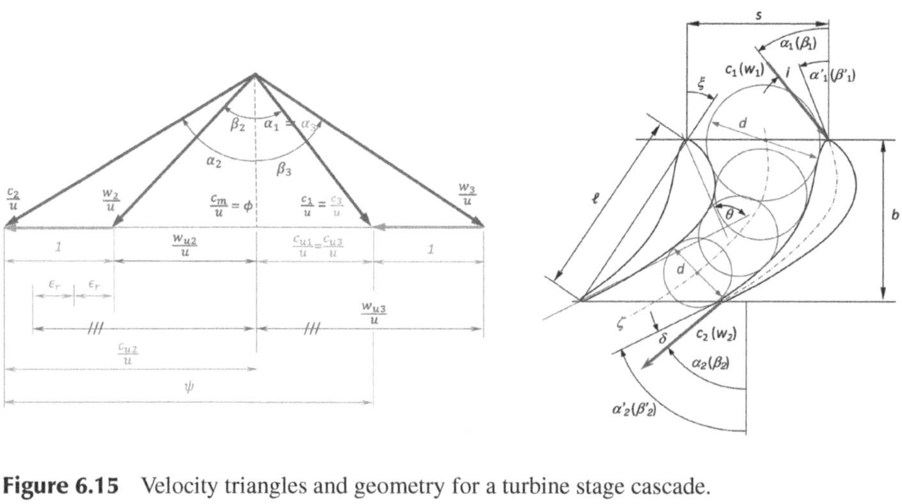

Figure 6.15 Velocity triangles and geometry for a turbine stage cascade.

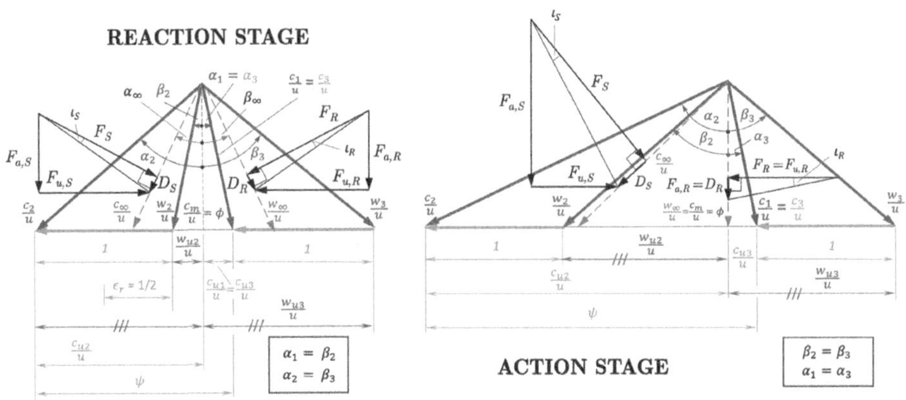

Figure 6.16 Velocity triangles in reaction and action stages.

tangential momentum change across the rotor, which generates a comparatively high amount of specific work. The static enthalpy drops significantly in the stators but remains constant in the rotors. As a consequence, the pressure is nearly constant across the rotor passages (right of Figure 6.16).

Figure 6.16 illustrates several key differences between reaction and action stages. For the same tangential speed u and flow coefficient $\phi = c_m/u$, reaction stages involve smaller flow deflections $\Delta\alpha_{1-2} = \alpha_2 - \alpha_1$ (or $\Delta\beta_{2-3} = \beta_3 - \beta_2$) and lower kinetic energy loads $\frac{1}{2}\Delta c_{1-2}^2$ (or $\frac{1}{2}\Delta w_{2-3}^2$) in both the nozzle and the rotor compared to action stages. This observation provides a preliminary justification for the higher efficiency and lower work coefficients observed in reaction stages.

In reaction stages, the forces exerted on the blades include axial and tangential components of similar magnitude. The tangential force component on the rotor blades drives rotor motion and generates shaft torque, while the axial component must be counterbalanced by appropriate components such as thrust bearings.

In contrast, in action stages, the force pattern differs significantly between the nozzle and the rotor. In the nozzle, large forces are generated due to substantial flow deflections and high kinetic energy loads. However, in the rotor, the force is primarily tangential with a negligible axial component. This requires a robust and rigid nozzle arrangement, often requiring nozzle vanes fixed to both the hub and the shroud. Conversely, the rotor assembly in action stages does not demand special thrust bearings.

Rotor blades in action stages experience lower pressure forces, allowing the use of thinner blades capable of operating at higher rotational speeds without significant mechanical stress. This makes action stages particularly suitable for high-speed applications, such as in aircraft engines and certain power generation turbines.

Regardless of the type of turbine cascade, the procedure outlined in Section 6.2.2 can be easily adapted to compute forces in an accelerating cascade for low-speed flows. Attention must be paid to changes in the relevant subscripts (1 and 2 refer to the nozzle inlet and outlet, respectively, while 2 and 3 refer to the rotor inlet and outlet) and to the direction of the force components. In this regard, it is important to note that the axial component of the force on turbine cascade airfoils is consistent with the sign of the pressure difference $p_1 - p_2$ (or $p_2 - p_3$), while the direction of the tangential force is opposite to the net change in the tangential flow momentum.

In the following subsections, we will derive the relevant equations to describe the performance of turbine cascades. It is crucial to note that while deriving these equations, all absolute flow velocities have positive peripheral components c_u when they point in the direction of the peripheral speed u. However, in most turbine stages, c_u at the rotor outlet is often oriented in the opposite direction to u to maximize work extraction and therefore must be regarded as negative. Finally, for the sake of completeness, it is important to emphasize that, to ensure full consistency in the derived equations, w_u, the generic peripheral component of the relative flow velocity, has an opposite sign with respect to c_u.

6.3.1 GENERAL PERFORMANCE COEFFICIENTS

6.3.1.1 Enthalpy Loss Coefficients

For generic nozzle and rotor cascades, we define a performance index based on how the actual transformation is close to the ideal, or isentropic, one:

$$\xi_N = \frac{h_2 - h_{2s}}{\frac{1}{2}c_2^2}, \ \xi_R = \frac{h_3 - h_{3s}}{\frac{1}{2}w_3^2} \tag{6.62}$$

for which it is immediate to recognize that the coefficients are 0 when the transformation is isentropic.

6.3.1.2 Zero Reaction Stage

As the degree of reaction is defined as:

$$\varepsilon_r = \frac{h_2 - h_3}{h_3^0 - h_1^0} = \frac{w_{u3} - |w_{u2}|}{2u} = \phi\left(\frac{\tan\beta_3 - \tan\beta_2}{2}\right) \tag{6.63}$$

when $\varepsilon_r = 0$:

$$w_{u3} = |w_{u2}| \rightarrow \tan\beta_3 = \tan\beta_2 \rightarrow \beta_3 = \beta_2 \tag{6.64}$$

On the other hand:

$$\psi = \phi(\tan\alpha_2 + \tan\alpha_3) = \phi\tan\alpha_2 + \frac{c_3\sin\alpha_3}{u} = 2\phi\tan\alpha_2 - 2 \tag{6.65}$$

as, from the velocity triangles, $\frac{c_3\sin\alpha_3}{u} = \frac{c_2\sin\alpha_2}{u} - 2$.

If the flow exiting the rotor is in axial direction $\alpha_3 = 0$ (Figure 6.17 right):

$$\psi = \phi\tan\alpha_2 \tag{6.66}$$

and, since in this case $\tan\beta_2 = 1/\phi$:

$$\psi = 2 \text{ (zero reaction, axial exit)} \tag{6.67}$$

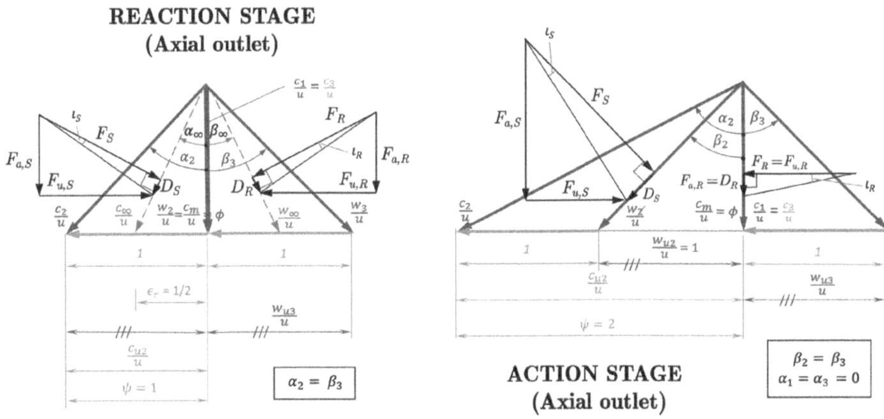

Figure 6.17 Velocity triangles in reaction and action stages with axial outlet.

Under the assumption that the flow is isentropic, $h_3 - h_2 = h_3 - h_{2s}$ and no pressure change occurs in the rotor, i.e. $p_3 = p_2$. In this case, we refer to a *perfect impulse* rotor cascade.

6.3.1.3 0.5 Reaction Stage

When $\varepsilon_r = 0.5$, $\Delta h_{2-3} = \Delta h_{1-2}$ and the velocity triangles become symmetric, that is, $\alpha_2 = \beta_3$ and $\alpha_3 = \beta_2$. In this case, the work coefficient becomes:

$$\psi = \frac{c_{u2} + c_{u3}}{u} = \phi(\tan\alpha_2 + \tan\beta_3) - 1 = \phi(\tan\alpha_2 + \tan\alpha_3) \tag{6.68}$$

or equivalently:

$$\psi = \frac{w_{u2} + w_{u3}}{u} = 2\phi(\tan\beta_2) \tag{6.69}$$

Since $c_{u2}/u + w_{u3}/u$ is the overall basis of the velocity triangle (see left of Figure 6.16) we derive:

$$2(c_{u3}/u) = 2\phi\tan\alpha_2 - 2 \tag{6.70}$$

which finally leads to:

$$\psi = \phi\tan\alpha_2 + (c_{u3}/u) \implies \psi = \phi\tan\alpha_2 + \phi\tan\alpha_2 - 1 = 2\phi\tan\alpha_2 - 1 \tag{6.71}$$

In case of $\alpha_3 = 0$ (Figure 6.17 left):

$$\psi = 1 \text{ (0.5 reaction, axial exit)} \tag{6.72}$$

6.3.2 OPTIMUM STAGE PERFORMANCE

We are now interested in analyzing $\psi = f(\phi)$ for both the cases of zero and 0.5 reaction to derive optimal stage performance and, therefore, to obtain effective design criteria. Let us introduce and elaborate on the concept of stage isentropic efficiency (also referred to as *total-to-total* isentropic efficiency):

$$\begin{aligned}
\eta_{\text{stage}} = \eta_{TT} &= \frac{h_1^0 - h_3^0}{h_1^0 - h_{3ss}^0} \\
&= \frac{h_1^0 - h_3^0}{h_1^0 - h_3^0 + (h_3^0 - h_{3ss}^0)} \quad \text{(add and subtract } h_3^0 \text{ at the denominator)} \\
&= \frac{1}{1 + (h_3^0 - h_{3ss}^0)/(h_1^0 - h_3^0)} \\
&= \frac{1}{1 + (h_3 - h_{3s} + (h_2 - h_{2s}))/(h_1^0 - h_3^0)}
\end{aligned} \tag{6.73}$$

[1]

i.e.,

$$\begin{aligned}
\eta_{\text{stage}} &= \frac{1}{1 + (\xi_R \frac{w_3^2}{2} + \xi_N \frac{c_2^2}{2})/(h_1^0 - h_3^0)} = \frac{1}{1 + (\xi_R \frac{w_3^2}{2} + \xi_N \frac{c_2^2}{2})/(u(c_{u2} + c_{u3}))} \\
&= \frac{1}{1 + \frac{\xi_R \frac{w_3^2}{u^2} + \xi_N \frac{c_2^2}{u^2}}{2\left(\frac{c_{u2}}{u} + \frac{c_{u3}}{u}\right)}}
\end{aligned} \tag{6.74}$$

and finally:

$$\eta_{TT} = \frac{1}{1 + \left[\xi_R \left(\frac{w_3}{u}\right)^2 + \xi_N \left(\frac{c_2}{u}\right)^2\right] \frac{1}{2\psi}} \tag{6.75}$$

Under the following assumptions:

- the enthalpy loss coefficients are functions of the flow deflection according to the empirical correlation by Soderberg[2]:

$$\xi_N = 0.025 \left[1 + \left(\frac{\Delta\alpha^\circ}{90}\right)^2\right] \left(1 + 3.2\frac{b_N}{H}\right), \quad \xi_R = 0.025 \left[1 + \left(\frac{\Delta\beta^\circ}{90}\right)^2\right] \left(1 + 3.2\frac{b_R}{H}\right) \tag{6.76}$$

  where H is the blade height, constant for both nozzle and rotor, and $\frac{b_N}{H} = \frac{b_R}{H} = 1/3$, $\Delta\alpha = \alpha_2 + \alpha_3$ and $\Delta\beta = \beta_2 + \beta_3$ are the flow deflections for nozzle and rotor in degrees, respectively;
- the Reynolds number based on each cascade throat diameter is 10^5;
- each axial cascade solidity $\ell/b = 2.5/\phi$;

the two cases (zero and 0.5 reaction) can be discussed.

6.3.2.1 Zero Reaction

Below are the relevant equations.

$$\psi = 2\phi \tan\alpha_2 - 2 \tag{6.77}$$

$$\psi = 2\phi \tan\beta_2 \implies \beta_2 = \arctan\frac{\psi}{2\phi} \implies \Delta\beta = 2\arctan\frac{\psi}{2\phi} \tag{6.78}$$

$$\eta_{TT} = \frac{1}{1 + \left[\xi_R\left(\frac{1}{\cos\beta_2}\right)^2 + \xi_N\left(\frac{1}{\cos\alpha_2}\right)^2\right]\frac{\phi^2}{2\psi}} \tag{6.79}$$

6.3.2.2 0.5 Reaction

Here below are the relevant equations.

$$\psi = 2\phi \tan\alpha_2 - 1 \tag{6.80}$$

from which:

$$\Delta\alpha = \Delta\beta = \arctan\left(\frac{\psi + 1}{2\phi}\right) + \arctan\left(\tan\alpha_2 - \frac{1}{\phi}\right) \tag{6.81}$$

$$\eta_{TT} = \frac{1}{1 + (\xi_R + \xi_N)\frac{\phi^2}{2\psi\cos^2\alpha_2}} \tag{6.82}$$

Figures 6.18 and 6.19 illustrate the preceding relationships in the form of $\psi = f(\phi)$, with parametric values of α_1, $\Delta\beta$, and η_{TT}. These charts, known as Hawthorne's diagrams, are essential for the insights they provide into the influence of stage parameters on stage performance.

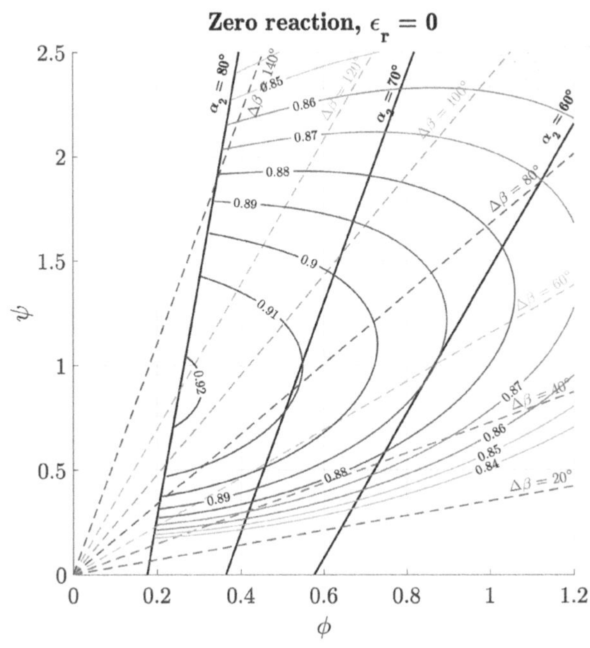

Figure 6.18 Hawthorne's diagram for 0 reaction turbine stage.

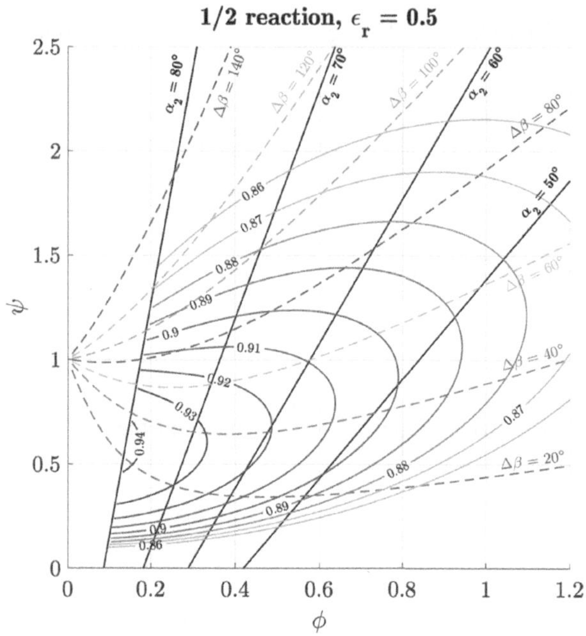

Figure 6.19 Hawthorne's diagram for 0.5 reaction turbine stage.

6.3.2.3 Optimal Cascade Solidity for Reaction Stages, the Zweifel Criterion

The empirical Zweifel criterion is often employed to determine optimal blade loading and solidity for turbine cascades, ensuring efficient performance while minimizing losses. The criterion is derived from the balance between the tangential load on the blades and the ideal load capacity without risking fluid separation.

With reference to Figure 6.20, the ideal tangential load is postulated by Zweifel[3] as:

$$(F_u)_{id} = (p_1^0 - p_2)\ell \tag{6.83}$$

while the actual tangential load is given by a momentum balance:

$$F_u = \rho s c_m (c_{u1} + c_{u2}) \tag{6.84}$$

assuming the axial velocity is constant through the cascade.

Let us define the Zweifel number Zw as:

$$\mathrm{Zw} = \frac{F_u}{(F_u)_{id}} \tag{6.85}$$

Substituting 6.84 and the ideal tangential load 6.83 into Z we get:

$$\mathrm{Zw} = \frac{\rho s c_m (c_{u1} + c_{u2})}{(p_1^0 - p_2) b} = \frac{\rho s c_m (c_{u1} + c_{u2})}{\frac{1}{2}\rho c_2^2 \ell} = \frac{2\left(\frac{s}{\ell}\right)\cos^2\alpha_2(\tan\alpha_1 + \tan\alpha_2)}{\frac{1}{2}c_m^2 \sec^2\alpha_2} \tag{6.86}$$

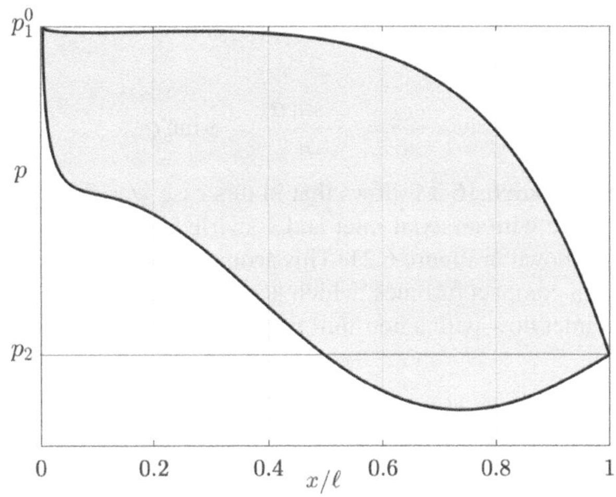

Figure 6.20 Pressure distribution along the blade surface of a reaction turbine blade section.

and finally:

$$\text{Zw} = \frac{F_u}{(F_u)_{id}} = 2\left(\frac{s}{\ell}\right)\cos^2\alpha_2(\tan\alpha_1 + \tan\alpha_2) \tag{6.87}$$

According to the Zweifel criterion, the optimal blade loading coefficient lies in the range 0.8–1.2. This range ensures that the blades are not too lightly loaded (leading to high losses due to increased secondary flows and mixing) nor too heavily loaded (causing boundary layer separation and increased profile losses). Owing to this assumption, the optimal solidity can be found as:

$$\sigma_{\text{opt}} = \left(\frac{\ell}{s}\right)_{\text{opt}} = \frac{0.4 - 0.6}{\cos\xi\cos^2\alpha_2(\tan\alpha_1 + \tan\alpha_2)} \tag{6.88}$$

6.3.3 ACTION STAGES

6.3.3.1 De Laval Cascades

A De Laval cascade configuration with axial inlet and axial outlet is characterized by the velocity diagram shown in Figure 6.16 right.

Typically, the angle α_2 of the absolute velocity c_2 downstream of the nozzle vanes is selected to maximize work coefficient $\psi = \Delta c_{u_{2-3}}/u$. However, practical limitations arise because angles greater than approximately 78° cannot be achieved in a cascade due to the finite thickness of the blades and to the lower bounds on the admissible outlet cross-section area. As a result, both α_2 and β_2 should be kept below roughly 75°. In turbines that handle large flow rates, the flow coefficient $\phi = c_m/u$ is relatively large (>1) such that angles smaller than 75° are commonly found.

In a De Laval cascade with null rotor swirl velocity at the outlet ($c_{u3} = 0$), the work coefficient becomes:

$$\psi = \frac{c_{u2}}{u} = \frac{c_2 \sin\alpha_2}{u} = \phi \tan\alpha_2 \tag{6.89}$$

By examining Figure 6.16 it follows that in this case $\psi = 2$.

A De Laval stage with an axial inlet and a swirled outlet is characterized by the velocity diagram shown in Figure 6.21. This arrangement is particularly appreciated as frontal stages in compact turbines, which need reaction stages downstream, as the latter require an inlet flow with a non-null tangential flow component.

In such case,

$$\psi = \frac{c_{u2} - c_{u3}}{u} = \frac{c_{u2} + |c_{u3}|}{u} \tag{6.90}$$

or:

$$\psi = \frac{c_2 \sin\alpha_2 + c_3 \sin\alpha_3}{u} = \phi\,(\tan\alpha_2 + \tan\alpha_3) \tag{6.91}$$

By examining Figure 6.21 it follows that, when α_2 reaches its maximum values, then $\psi >> 1$.

A Curtis turbine (Figure 6.22), frequently referred to as a *velocity-compound* turbine, is a development of the De Laval stage, in which the fluid moves through the

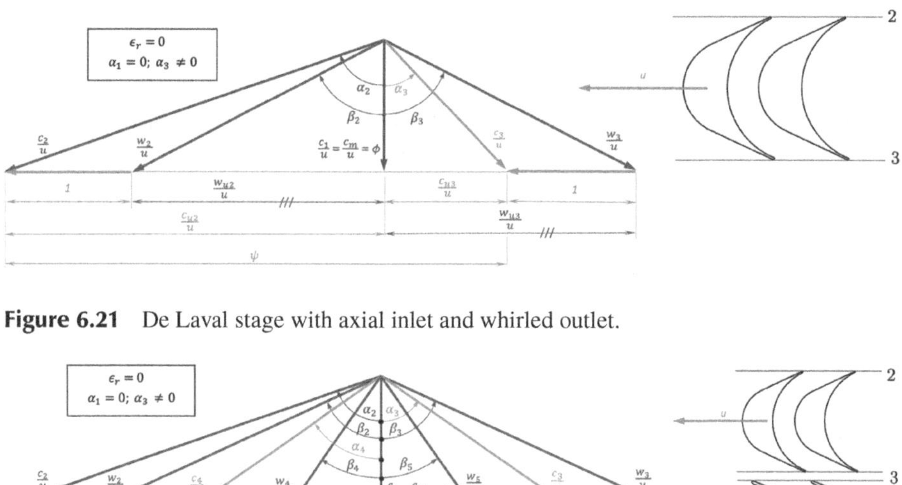

Figure 6.21 De Laval stage with axial inlet and whirled outlet.

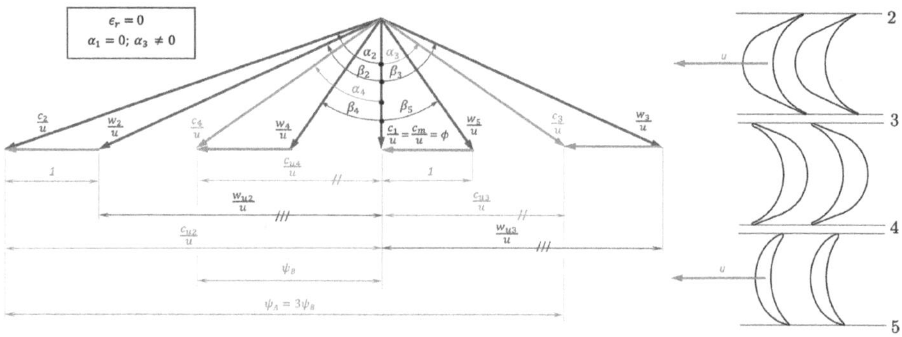

Figure 6.22 The Curtis stage.

first set of rotor blades after being fully accelerated at the first nozzle. After leaving the nozzle, the fluid is deflected so that it enters the next set of rotor blades without expanding further. Therefore, in a Curtis stage with zero exit angle:

$$\psi = \frac{c_{u2} + |c_{u3}| + c_{u4}}{u} \tag{6.92}$$

which, assuming $c_{u2}/u = 4$, $|c_{u3}|/u = c_{u4}/u = 2$, as in Figure 6.22, brings:

$$\psi_{\text{Curtis}} = 8 \tag{6.93}$$

6.3.4 TURBINE PROFILES

Turbine blades are typically classified according to the type of flow within the cascades, specifically whether the blades are designed for impulse or reaction cascades. Commonly adopted geometries are those studied by NACA. For a reaction turbine, NACA proposes the primary series, known as A_3K_7 (Figure 6.23). In contrast, an impulse turbine has nearly equal entering and leaving velocities and blades designed for this condition are classified by NACA as the secondary series, labeled $B_1E_1I_1$.

Although modern airfoils are usually studied based on specific flow conditions and desired pressure distribution, the above-mentioned NACA airfoil designs continue to be of fundamental interest since they serve as the basic shapes from which later developments evolved. Table 6.3 provides the coordinates of the mean line and thickness distribution for the two series of airfoils, both referring to a nominal lift coefficient for an isolated airfoil $C_{lo} = 1$ and a maximum thickness of 20% of the chord. The coordinates for different profiles characterized by an arbitrary C_{lo} are obtained as:

$$y_c(x) = C_{lo} \cdot (y_c)_{C_{lo}=1} \tag{6.94}$$

while conversion to other maximum thicknesses are made by multiplying the ordinates of the uncambered airfoil section by the ratio of the desired thickness to the tabulated thickness.

The following equations govern the computation of the upper and lower surface coordinates of the turbine blade.

First, the arc length s along the camber line is computed as:

$$S(x) = \int_0^x \sqrt{1 + \left(\frac{dy_c}{dx}\right)^2}\, dx \tag{6.95}$$

For a discrete set of points, this can be approximated as:

$$S_i = \sum_{j=1}^{i} \sqrt{\Delta x_j^2 + \Delta y_{c,j}^2} \tag{6.96}$$

where

$$\Delta x_j = x_{j+1} - x_j, \quad \Delta y_{c,j} = y_{c,j+1} - y_{c,j} \tag{6.97}$$

Table 6.3
Mean-Line and Thickness Distributions for the A_3K_7 Primary Turbine Blade Series and $B_1E_1I_1$ Secondary Turbine Blade

A_3K_7 Blade Series				$B_1E_1I_1$ Blade Series			
Mean-line Coordinates		Thickness Distribution		Mean-line Coordinates		Thickness Distribution	
x/ℓ (%)	y_c/ℓ (%)	x_t/ℓ (%)	t_t/ℓ (%)	x/ℓ (%)	y_c/ℓ (%)	x_t/ℓ (%)	t_t/ℓ (%)
0	0.000	0	0.000	0	0.000	0	0.000
0.5	0.397	1.25	3.469	0.5	0.336	1.25	2.583
1.25	0.836	2.5	4.972	1.25	0.718	2.5	3.282
2.5	1.428	5.0	6.918	2.5	1.253	5.0	4.041
5.0	2.359	10	9.007	5.0	2.128	10	5.007
10	3.689	15	9.827	10	3.480	15	5.952
15	4.597	20	10.000	15	4.506	20	6.995
20	5.271	25	9.899	20	5.291	25	8.063
25	5.623	30	9.613	25	5.862	30	9.025
30	5.852	35	9.106	30	6.261	35	9.727
35	5.936	40	8.594	35	6.514	40	10.000
40	5.897	45	7.913	40	6.642	45	9.725
45	5.753	50	7.152	45	6.652	50	9.009
50	5.516	55	6.339	50	6.551	55	8.016
55	5.200	60	5.500	55	6.342	60	6.908
60	4.814	65	4.661	60	6.016	65	5.848
65	4.367	70	3.848	65	5.551	70	5.000
70	3.870	75	3.087	70	4.983	75	4.312
75	3.328	80	2.406	75	4.332	80	3.624
80	2.746	85	1.830	80	3.680	85	2.936
85	2.133	90	1.387	85	2.824	90	2.248
90	1.485	95	1.101	90	1.953	95	1.560
95	0.801	100	0.000	95	0.948	100	0.000
100	0.000			100	0.000		

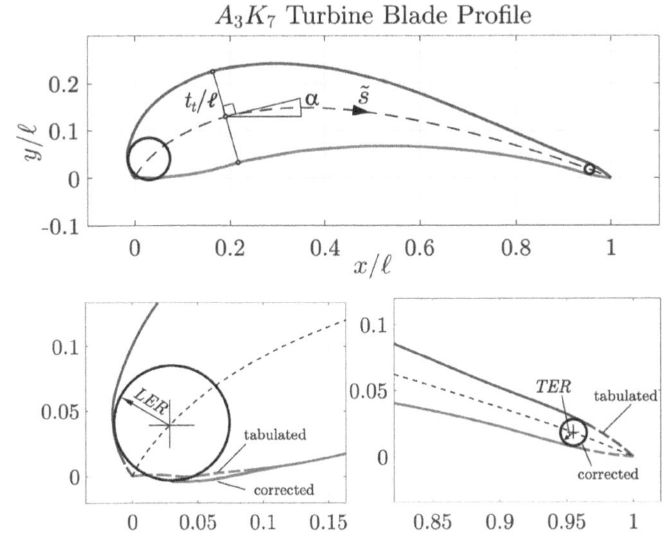

Figure 6.23 Blade profile of an A_3K_7 turbine airfoil, $C_{lo} = 2.5, t/\ell = 0.2$.

The arc length is then normalized:

$$\tilde{S}_i = \frac{S_i}{S_{\max}} \tag{6.98}$$

The thickness distribution t_t is interpolated along the normalized arc length:

$$t_t(\tilde{s}) = \text{SplineInterp}(\tilde{S}, t_t) \tag{6.99}$$

and the local slope of the camber line, along with the pertinent camber line angle, are calculated:

$$\frac{dy_c}{dx}, \quad \alpha = \arctan\left(\frac{dy_c}{dx}\right) \tag{6.100}$$

Finally, the coordinates of the upper and lower surfaces are calculated as:

$$x_{\text{upper}} = x - t_t \sin\alpha, \quad x_{\text{lower}} = x + t_t \sin\alpha \tag{6.101}$$

$$y_{\text{upper}} = y_c + t_t \cos\alpha, \quad y_{\text{lower}} = y_c - t_t \cos\alpha \tag{6.102}$$

While the A_3K_7 geometry, with certain modifications, remains in use today, the $B_1E_1I_1$ configuration is incompatible with the requirements of modern impulse rotor airfoils. This limitation arises primarily from its large leading-edge radii, which induce significant flow losses at high Mach numbers. To address these challenges, alternative shapes have been proposed, such as the one illustrated in Figure 6.24. These designs aim to maintain a nearly constant flow passage area between adjacent sections (i.e. between the upper surface of one blade and the lower surface of the next). This approach ensures a nearly uniform magnitude of the relative velocity, while also incorporating a slight acceleration of the flow to mitigate the risk of undesired flow separations.

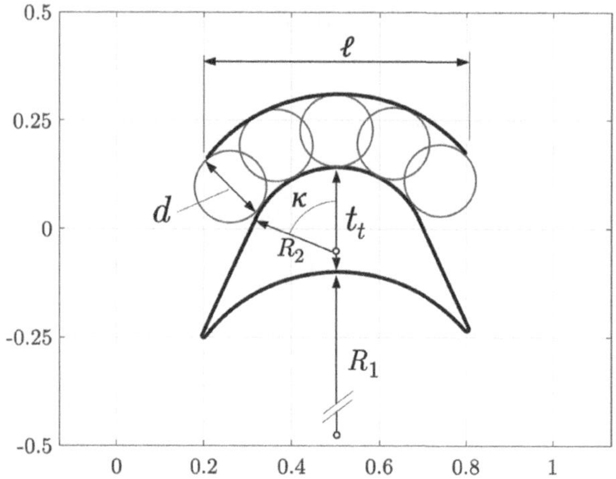

Figure 6.24 Schematic of an impulse blade geometry.

6.3.5 SURFACE PRESSURE DISTRIBUTION

In turbine cascades, the surface pressure distribution depends on the degree of reaction.

The left side of Figure 6.25 shows the distribution of C_p over a reaction cascade A_3K_7, where exit pressure p_2 is much lower than inlet pressure p_1, especially at higher stagger angles $\xi = 14°, 26°$. This indicates strong flow acceleration, which stabilizes the boundary layer by reducing adverse pressure gradients and minimizing the risks of separation.

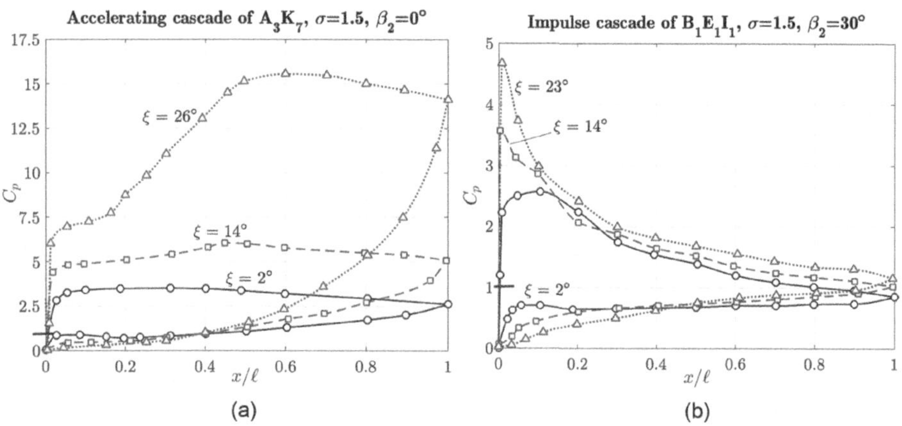

Figure 6.25 Pressure coefficient C_p distribution along the normalized chord x/ℓ for turbine cascades. (a) Reaction cascade A_3K_7 with a solidity of $\sigma = 1.5$ and inlet flow angle $\beta_2 = 0°$. (b) Impulse cascade $B_1E_1I_1$ with $\sigma = 1.5$ and $\beta_2 = 30°$.

The right side of Figure 6.25 represents C_p in a nearly zero reaction impulse cascade $B_1E_1I_1$, where the exit pressure is nearly conserved across all incidences $\xi = 2°, 14°, 23°$. This results in large fractions x/ℓ of flow deceleration, leading to adverse pressure gradients and potential boundary layer instability.

6.4 QUANTIFICATION OF FLOW LOSSES IN 2D CASCADES

6.4.1 GENERAL CHARACTERISTICS

The process of energy conversion in turbomachines is affected by various irreversibilities that occur within the flow of the working fluid. Even in an adiabatic flow, these irreversibleities manifest themselves as energy losses due to friction and viscous dissipation, both of which can be referred to as *viscous effects*.

Viscous effects are inherently related to frictional forces that convert mechanical energy into thermal energy, a form of energy that, owing to the second law of thermodynamics, is degraded compared to mechanical work. This process is inherently irreversible, leading to an increase in entropy, which is a measure of the system's disorder or the unavailability of energy for work. Every form of frictional force leads to entropy generation: relevant examples include, but are not limited to, flows in attached boundary layers and separation flows.

Viscous losses in a 2D cascade are primarily caused by: (i) boundary layer development along the blade surfaces and (ii) the mixing of the wake after the fluid passes through the blade row.

Regarding (i), as the fluid flows over the blade, the viscosity of the fluid causes shear stress and leads to a velocity gradient within the boundary layer. The dissipation of kinetic energy due to this shear stress manifests itself as frictional losses. These losses are more pronounced in regions with adverse pressure gradients, such as the suction side (SS) of an airfoil, which can lead to boundary layer thickening and, in severe cases, particularly relevant to compressors, to flow separation (stall).

Concerning (ii), after the fluid passes through the blade row, it produces a wake region characterized by a deficit in kinetic energy and increased turbulence. The wake is a direct consequence of the thickening and partial separation of the boundary layer at the trailing edge and of the mixing downstream. The mixing losses occur when the high-momentum fluid outside the wake mixes with the low-momentum fluid within the wake, leading to a redistribution of energy and an irreversible loss of total pressure. At a station far downstream of the cascade, the turbulent mixing will be complete, and the mean steady flow will again be uniform.

A graphic representation of boundary layer formation and wake development throughout a compressor cascade is provided in Figure 6.26.

The flow losses in compressor and turbine cascades differ significantly due to the contrasting pressure gradients in each device. In compressors, the adverse pressure gradient causes flow deceleration, increasing the boundary layer thickening and likelihood of its separation and wake formation, which leads to higher aerodynamic losses (Figure 6.14). In contrast, turbines operate on average under a favorable pres-

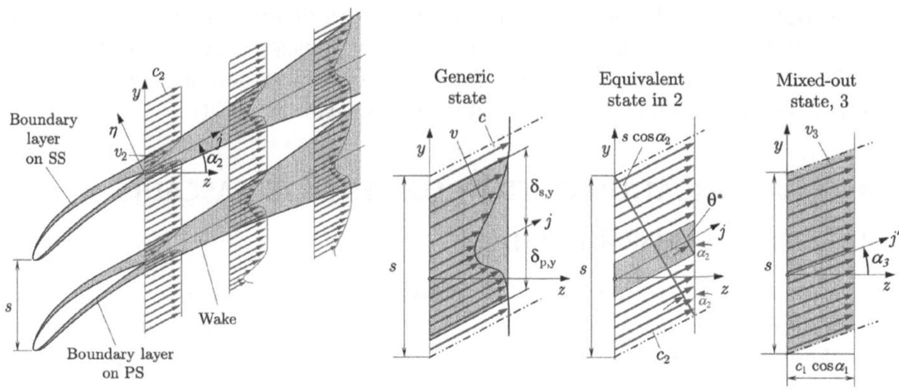

Figure 6.26 Schematic representation of boundary layer and wake development through a cascade blade passage.

sure gradient (in particular reaction blades, Figure 6.25), which accelerates the flow, reducing the tendency for separation and generally resulting in lower losses.

6.4.2 TWO-DIMENSIONAL CASCADE VISCOUS LOSSES

Accurately determining laminar and turbulent boundary layer behavior over arbitrary profile shapes and within wakes remains challenging and limits the accuracy of theoretical methods for estimating viscous losses in cascades. This issue is particularly significant in compressors, where losses, in addition to aerodynamic efficiency, are closely related to flow stability. Conversely, flow losses in turbines are generally less critical provided the cascade operates near its design point.

Despite significant challenges, Lieblein (1965)[4] developed a comprehensive theoretical methodology for incompressible flows in compressor cascades to determine: (i) the boundary layer characteristics on blade surfaces and (ii) the total pressure defect induced by these characteristics in the downstream wake. A simplified version of Lieblein's method is presented here which captures the core insights of his findings. Starting from the definition of a total pressure loss coefficient past a cascade of airfoils:

$$\zeta = \frac{p_1^0 - \bar{p}_2^0}{\rho_1 c_1^2/2} \tag{6.103}$$

the contribution of (i) can be evaluated using the wake momentum thickness θ^* at the cascade exit station (see Figure 6.26):

$$\theta^* = \int_{-s/2}^{s/2} \left(\frac{v}{c}\right)\left(1 - \frac{v}{c}\right) dy \tag{6.104}$$

where $v = v(y)$ is the local flow velocity at the cascade exit station and c is the flow velocity measured at the same station outside the wake in the y direction.

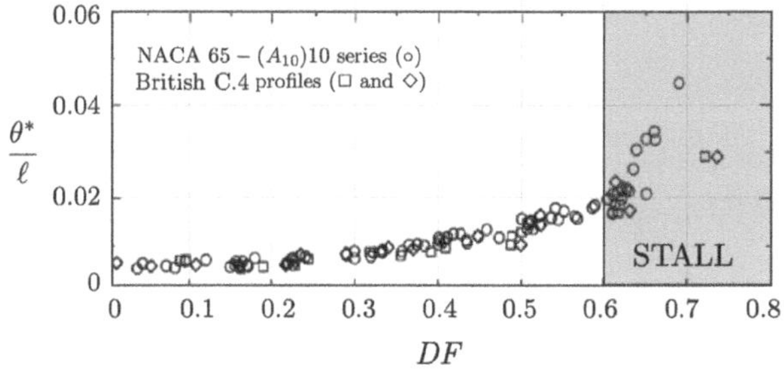

Figure 6.27 Profile total pressure loss parameter as a function of the diffusion factor.

Lieblein discovered a correlation between design, or minimum, profile loss (established by equation 6.103) and stall inception, resembled by the diffusion factor DF, which is virtually independent of the profile type.[5] This correlation is shown in Figure 6.27 in the form:

$$\frac{\theta^*}{\ell} = f(DF) \tag{6.105}$$

Lieblein found that the loss in a blade row increases sharply as the flow starts to separate from the suction side, and this happens when the diffusion factor exceeds 0.6. Typically, a well-designed blade with moderate loading operates with a diffusion factor around 0.45. Although this principle was originally derived from a narrow set of compressor blade designs under minimum loss conditions, the diffusion factor is now widely used in preliminary design across different compressor types, including both compressible and incompressible flows.

Regarding total pressure loss in the wake (i.e. contribution ii)), a more complex model is required, which accounts for the shape factor of the velocity profile at the trailing edge. A total cascade loss coefficient is defined as:

$$\zeta_c = \frac{p_1^0 - \bar{p}_2^0}{\frac{1}{2}\rho v_3^2} + \frac{\bar{p}_2^0 - \bar{p}_3^0}{\frac{1}{2}\rho v_3^2} = \frac{1}{v_3^2}\left[c_2^2 - \frac{\int_0^1 v_2^3(\xi)\,d\xi}{\int_0^1 v_2(\xi)\,d\xi}\right] + \frac{\lambda}{v_3^2}\left[\int_0^1 v_2(\xi)\,d\xi\right]^2 \tag{6.106}$$

v being the local velocity in the wake and $\lambda = \frac{\bar{p}_2^0 - \bar{p}_3^0}{\frac{\rho}{2}\left[\int_0^1 v_2(\xi)\,d\xi\right]^2}$ the mixing loss coefficient, can be studied as a function of $x = \frac{\theta^*}{\ell}\frac{\sigma_S}{\cos\alpha_2}$, where θ^* is the total momentum thickness of the boundary layer at cascade exit station 2 (see Appendix A.2).

In Figure 6.28, the loss coefficient ζ_c is presented for three values of H, i.e. $H =$ 1.4, 2 and 2.6. Experience has demonstrated that the shape parameters of unstalled turbulent boundary layers typically do not exceed a value of 2, with an average shape parameter H ranging from 1.2 to 1.6, while an H value of 2 corresponds to highly loaded blades where the flow remains attached. H value of 2.6 is very rarely found as it indicates heavily loaded profiles prone to stall.

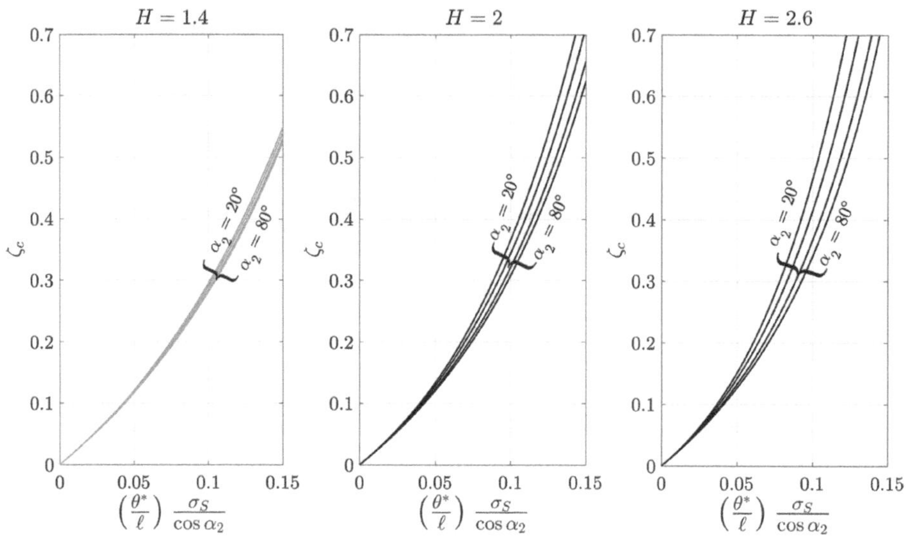

Figure 6.28 Total cascade loss coefficient as function of $x = \frac{\theta^*}{\ell}\frac{\sigma_S}{\cos\alpha_2}$ for three values of $H =$: moderately loaded $H = 1.4$, highly loaded $H = 2$, near-stalled $H = 2.6$ blade sections.

The results shown in Figure 6.28 are of paramount importance to understand the role of cascade solidity in determining cascade performance. In well-designed cascades, it is generally assumed that less than 10% of the cascade's mixed-out kinetic energy, $\frac{v_3^2}{2}$, is lost as total pressure drop, $h_3^0 - h_1^0$. Under these conditions, the cascade loss coefficient ζ_c shows a negligible dependence on the outlet flow angle α_2 and only a minor dependence on the wake shape factor H. For our analysis, we will assume $H \approx 1.4$, which is characteristic of turbulent attached flows in moderately loaded cascade airfoils.

6.4.2.1 Case 1

Consider the case where both $\cos\alpha_2$ and $\frac{\theta^*}{\ell}$ are constant. This situation corresponds to a cascade flow with a fixed outlet inclination and a fixed ratio of the equivalent wake thickness to the chord length. For an airfoil with a given chord length ℓ, this implies a constant boundary layer momentum deficit. To illustrate this, we can use the flat plate analogy to analyze the velocity profiles of turbulent boundary layers developing over the same length. Schlichting proposed the following relation for turbulent flow over a flat plate:

$$\theta^* = 0.035\frac{\delta}{Re_\delta^{1/5}}$$

where $Re_\delta = \frac{\ell\delta}{\nu}$ is the Reynolds number based on boundary layer thickness δ developed over the flat plate, chord length ℓ and kinematic viscosity ν. This implies that if two cascades have the same c_2 and Reynolds number Re_δ, they will have the same boundary layer thickness δ and hence the same momentum deficit θ^*. Consequently, they will exhibit the same wake shape factor H. Under these assumptions, Figure 6.28 demonstrates that ζ_c is an increasing function of the cascade solidity σ. This result is representative of a cascade operating at a fixed outlet flow angle, where the viscous losses are influenced by the cascade wake profile and boundary layer characteristics. As the momentum deficit increases due to excessive solidity, the cascade loss coefficient ζ_c also increases, therefore reflecting the impact of boundary layer behavior on aerodynamic performance.

6.4.2.2 Case 2

Consider a cascade with constant chord length ℓ operating at varying outlet flow angles α_2. When α_2 decreases, several aerodynamic parameters are affected, including loss coefficient ζ_c, deflection $\varepsilon = \alpha_1 - \alpha_2$ (or $\beta_1 - \beta_2$ in the case of rotor cascades) and aerodynamic loading. When α_2 decreases, the boundary layer thickness δ typically grows because the flow undergoes more significant turning, causing increased momentum deficits and adverse pressure gradients. Increased boundary layer thickness leads to:

- **Higher total losses.** A thicker boundary layer results in a more pronounced wake and increased aerodynamic losses, including higher frictional and wake losses.
- **Increased separation losses.** A thicker boundary layer is more susceptible to separation, especially with sharp flow turning, leading to additional drag and reduced efficiency.

To describe this phenomenon quantitatively, we can rely on turbulent boundary layer theory over a cascade of flat plates subject to adverse pressure gradient (or deceleration). Assuming incompressible flow, Spence (1956)[6] developed a momentum integral equation with the approximation $H = 1.5$:

$$\left(\frac{\theta^*}{\ell}\right)^{1.2}\left(\frac{c_e}{c_1}\right)^{4.2} = \frac{0.016}{\mathrm{Re}_\ell^{0.2}}\int_0^1 \left(\frac{c_e}{c_1}\right)^4 d\left(\frac{x}{\ell}\right) \tag{6.107}$$

where c_e is the local velocity outside the boundary layer, x is along the chord direction and Re_ℓ is the Reynolds number based on chord length ℓ. Spence's equation can be integrated along x provided that a law for c_e is specified. Figure 6.29 shows the solution obtained for both linear and parabolic deceleration from cascade inlet to outlet and for several deceleration factors c_2/c_1. As is apparent, a more pronounced momentum thickness is registered at the trailing edge as deceleration intensifies and a parabolic deceleration, which has far more intense adverse pressure gradient toward the exit, is established.

To counteract the effects of increased boundary layer thickness and mitigate overall losses, adjusting the solidity of the cascade is essential. In particular, increasing solidity for a given velocity deceleration factor c_2/c_1 reduces the intensity of the pressure gradient dp/dx all along the chord length, with significant thinning of the

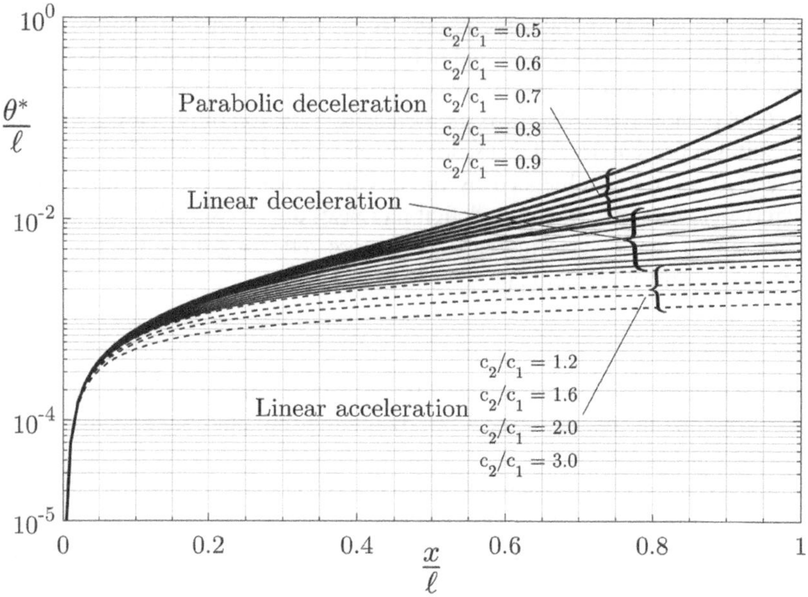

Figure 6.29 Non-dimensional chord-wise momentum thickness distribution along a cascade of flat plates in turbulent flow conditions and under adverse and favorable pressure gradients.

boundary layer and associated losses (Figure 6.29). In other words, higher solidity means closer-spaced blades, which better manage the wake and reduce its thickness, thereby lowering associated losses. In addition, increased solidity improves flow guidance through the cascade, helping to maintain attached flow and reduce separation losses even with increased deflection and boundary layer thickness. An optimal solidity can be determined, using the previously shown equations, which minimizes overall losses.

As expected, a favorable pressure gradient, illustrated in Figure 6.29 under acceleration conditions, results in a significantly reduced boundary layer momentum thickness at the cascade exit, a characteristic commonly observed in turbines.

6.4.3 HOWELL'S CRITERION FOR COMPRESSOR CASCADES

Howell's criterion provides a framework for selecting the optimal solidity in compressor/pump cascades operating in the incompressible flow regime.

Howell introduces the concept of a *nominal deflection*, denoted as ε^*. This is a design parameter set at 80% of the low-speed stalling deflection. The nominal deflection acts as a conservative threshold that provides operational margin before the onset of stall, ensuring that the compressor operates efficiently without hitting the critical stall condition. The corresponding fluid outlet angle, α_2^*, associated with this nominal deflection, plays a crucial role in determining the performance characteristics of the axial compressor.

The nominal deflection is influenced by specific geometric and aerodynamic factors, i.e. fluid outlet angle $\alpha_{ex} = \alpha_2$, cascade solidity σ, and profile camber angle γ.

Howell observed that ε^* does not change significantly within the range of incidence of flows $+5°$ to $-5°$ $i = \alpha_1 - \alpha_1'$. This implies that under optimal design conditions, the nominal deflection is relatively insensitive to small changes in camber, simplifying the design process. More in detail, he noted that the nominal deflection ε^* and the corresponding nominal outlet angle, α_2^*, are only functions of cascade solidity provided that the camber angle is within the range 20°–40°, a common range for decelerating cascades, and that the flow is fully turbulent $Re_\ell \geq 3 \cdot 10^5$:

$$(\varepsilon^*, \alpha_{ex}^*) = f(\sigma) \tag{6.108}$$

The correlation set by this formula is shown in Figure 6.30a, which can be used by any designer to obtain the optimal cascade solidity for prescribed (design) values of ε^* and α_{ex}^*. A good approximation of the nominal values in Figure 6.30 can be obtained using the following empirical relation:

$$\tan\alpha_1 - \tan\alpha_2 = \frac{1.55}{1 + 1.5\frac{1}{\sigma}} \tag{6.109}$$

To obtain the blade exit angle, Howell suggests using the already mentioned Carter's rule (see Section 3.5.1.3) for a selected value of profile camber θ_c:

$$\delta_{rot} = \beta_{2,rot} - \beta_{2,rot}' = \frac{m\theta_{c,rot}}{\sigma_{rot}^n} \tag{6.110}$$

$$\delta_{stat} = \alpha_{2,stat} - \alpha_{2,stat}' = \frac{m\theta_{c,stat}}{\sigma_{stat}^n} \tag{6.111}$$

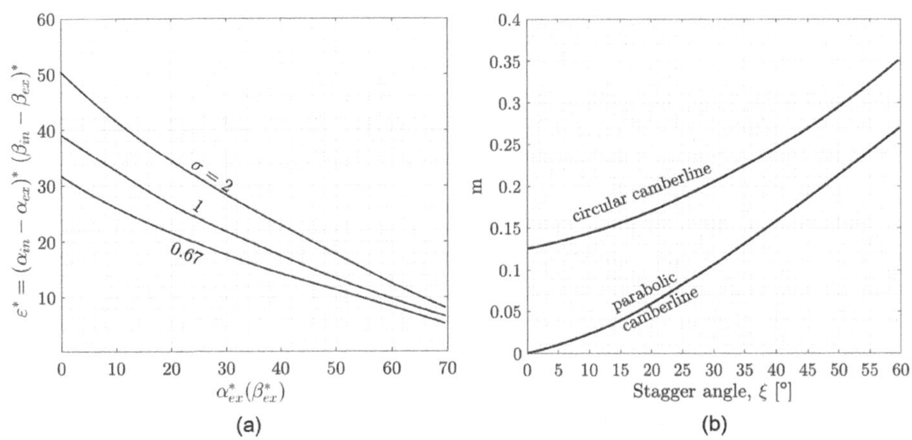

Figure 6.30 (a) Howell's correlation. (b) Variation of the empirical factor m in Carter's deviation angle rule.

where m can be evaluated from Figure 6.30b. For circular-arc camber lines, the following analytic expression can be used:

$$m = 0.216 + 0.000875\xi + 0.00002625\xi^2 \tag{6.112}$$

In parabolic-arc camber lines, the following expression is also widely used to estimate m in rotors and stators, respectively:

$$m_r = 0.23\left(\frac{2a_r}{\ell_r}\right)^2 + \frac{\beta_2[^\circ]}{500}, \quad m_s = 0.23\left(\frac{2a_s}{\ell_s}\right)^2 + \frac{\alpha_2[^\circ]}{500} \tag{6.113}$$

where a/ℓ denotes the non-dimensional chord-wise position of maximum camber.

6.5 ON/OFF-DESIGN PERFORMANCE OF 2-D CASCADES UNDER COMPRESSIBLE FLOW

In the previous sections, we explored the nominal working conditions of 2-D cascades, focusing on how these conditions are optimized to achieve peak performance under standard, design-intended scenarios. Furthermore, we assumed that the flow was incompressible, a condition that strictly applies to hydraulic and aeraulic machines but does not accurately represent the physical behavior of modern compressors and turbines operating with gases.

Design conditions represent the ideal operating point where flow angles, deflections, and aerodynamic loads are optimized for maximum efficiency and energy exchange. However, in practice, turbomachines often face off-design conditions, where actual operating parameters deviate from the design intent.

The set of on/off operating conditions forms the *characteristic curves*, or the *map*, of the turbomachine, an intrinsic feature that illustrates how performance metrics—especially energy exchange, efficiency, and power—depend on rotational speed and mass flow rate. As a dynamic system, a turbomachine's actual operating point on the map (its working point) is determined by the equilibrium conditions defined by the continuity, momentum, and energy equations applied to the system in which the machine operates.

Indeed, modeling involves understanding the behavior of the entire machine, not only its cascades, in its complex three-dimensional configuration. However, the role played by two-dimensional blade sections, as key fundamental components of axial machines, is relevant in determining some of the most critical characteristics of the map.

In both decelerating and accelerating cascades operating under off-design conditions, experimental evidence shows that the downstream flow angle remains nearly constant, regardless of variations in the inlet stage velocity direction over a wide range of inlet flow angles. We will adopt this hypothesis in the development of the design characteristics.

6.5.1 OFF-DESIGN PERFORMANCE OF DECELERATING CASCADES

Figure 6.31 illustrates an arbitrary velocity diagram for a stage, depicting both design and off-design conditions. Under off-design conditions, the flow coefficient ϕ deviates from its design value. The figure specifically shows a scenario where the flow coefficient is reduced. This can result either from a reduction in through-flow velocity c_m, which represents a decrease in volumetric flow rate, or from an increase in peripheral speed u, which indicates a rise in rotational speed.

Under reduced ϕ, the increase in flow incidence angles causes the boundary layer on the suction side of the cascade profiles to thicken with respect to design operation. This is accompanied by a stronger adverse pressure gradient acting on the boundary layer at the front of the profile. When the incidence angle remains within certain limits, it can lead to increased work exchange and only a slight reduction in cascade efficiency. However, if the incidence angle rises too sharply, the boundary layer may become unstable and eventually separate from the surface, leading to a stalling condition that significantly impairs both work transfer and efficiency.

Denoting with an asterisk $*$ the design quantities, the updated values of the work coefficient and degree of reaction can be found as follows.

At design conditions:

$$\psi^* = \frac{c_{u2}^* - c_{u1}^*}{u}, \quad \phi^* = \frac{c_m^*}{u}, \quad \tan\beta_1^* = \frac{w_{u1}^*}{c_m^*}, \quad \tan\alpha_2^* = \frac{c_{u2}^*}{c_m^*} \tag{6.114}$$

Since $\frac{c_{u1}}{c_m} = \frac{c_{u1}^*}{c_m^*}$ and $\frac{w_{u2}}{c_m} = \frac{w_{u2}^*}{c_m^*}$ remain constant at off-design conditions, and given that $c_{u1} = u - w_{u1}$ and $c_{u2} = u - w_{u2}$ always holds true, we can write:

$$\psi = \frac{u - w_{u2}^* \frac{\phi}{\phi^*} - c_{u1}^* \frac{\phi}{\phi^*}}{u} = 1 - \frac{\phi}{\phi^*}\frac{w_{u2}^*}{u} - \frac{\phi}{\phi^*}\frac{c_{u1}^*}{u} \tag{6.115}$$

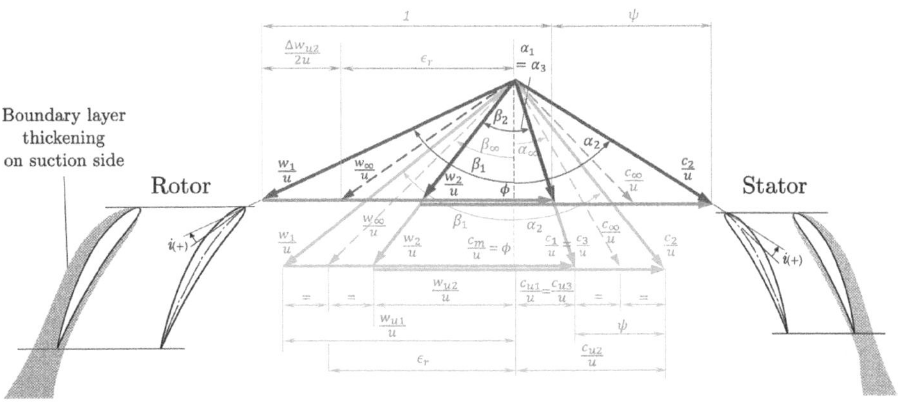

Figure 6.31 Decelerating cascade at off-design, reduced ϕ operation.

from which:

$$1-\psi=\frac{\phi}{\phi^{*}}\left(\frac{w_{u2}^{*}+c_{u1}^{*}}{u}\right) \tag{6.116}$$

and, with similar considerations:

$$1-\psi^{*}=\left(\frac{w_{u2}^{*}+c_{u1}^{*}}{u}\right) \tag{6.117}$$

Now, dividing both sides of 6.116 and 6.117:

$$\psi=1-(1-\psi^{*})\frac{\phi}{\phi^{*}} \tag{6.118}$$

is the equation of a line in the $\psi-\phi$ plane with slope $-(1-\psi^{*})\frac{1}{\phi^{*}}$. Since in decelerating cascades $\psi^{*}<1$, the slope is always negative, which means that in off-design conditions the work coefficient always decreases with the flow coefficient. In other words, for a given rotational speed, the energy exchange always decreases with the volumetric flow rate.

As far as the degree of reaction is concerned, we can start from its definitions:

$$\varepsilon_r=\frac{w_{u2}+w_{u1}}{2u}, \quad \varepsilon_r^{*}=\frac{w_{u2}^{*}+w_{u1}^{*}}{2u} \tag{6.119}$$

and by adding and subtracting w_{u2} and w_{u2}^{*} in the pertinent numerators:

$$\varepsilon_r=\frac{w_{u2}+w_{u1}-w_{u1}+w_{u1}}{2u}=\frac{\psi}{2}+\frac{w_{u2}}{u}, \quad \varepsilon_r^{*}=\frac{\psi^{*}}{2}+\frac{w_{u2}^{*}}{u} \tag{6.120}$$

from which:

$$\frac{\varepsilon_r-\frac{\psi}{2}}{\varepsilon_r^{*}-\frac{\psi^{*}}{2}}=\frac{\phi}{\phi^{*}} \tag{6.121}$$

Recalling and manipulating equation 6.118:

$$\frac{\psi}{2}-\frac{\psi^{*}}{2}\frac{\phi}{\phi^{*}}=\frac{1}{2}-\frac{1}{2}\frac{\phi}{\phi^{*}} \tag{6.122}$$

from equation 6.121 we finally get:

$$\varepsilon_r=\frac{1}{2}+\frac{\phi}{\phi^{*}}\left(\varepsilon_r^{*}-\frac{1}{2}\right) \tag{6.123}$$

which is the equation of a line in the $\varepsilon_r-\phi$ plane with slope $\frac{1}{\phi^{*}}\left(\varepsilon_r^{*}-\frac{1}{2}\right)$. It is worth noting that, if $\varepsilon_r^{*}=0.5$ the degree of reaction is no longer a function of ϕ and assumes a constant value of 0.5.

To find how rotor and stator inlet flow angles β_1 and α_2 change in off-design conditions, let us develop expressions of ε_r and ψ.

Given:

$$\varepsilon_r = \frac{w_{u2} + w_{u1}}{2u} = 1 - \frac{c_{u2} + c_{u1}}{2u} \tag{6.124}$$

substitute $w_{u2} = c_m \tan\beta_2^*$, $w_{u1} = c_m \tan\beta_2$, $\frac{c_m}{u} = \phi$, $c_{u2} = c_m \tan\alpha_2$, $c_{u1} = c_m \tan\alpha_1^*$ to get:

$$\varepsilon_r = \frac{\phi(\tan\beta_2^* + \tan\beta_1)}{2} = 1 - \frac{\phi(\tan\alpha_2 + \tan\alpha_1^*)}{2} \tag{6.125}$$

On the other hand, given:

$$\psi = \frac{w_{u1} - w_{u2}}{u} = \frac{c_{u2} - c_{u1}}{u} \tag{6.126}$$

upon the same substitution, we obtain:

$$\psi = \phi(\tan\beta_1 - \tan\beta_2^*) = \phi(\tan\alpha_2 - \tan\alpha_1^*) \tag{6.127}$$

Casting equations 7.35 and 6.127 to eliminate $\tan\beta_2^*$ and $\tan\alpha_1^*$, and isolate $\tan\beta_1$ and $\tan\alpha_2$ gives:

$$\tan\beta_1 = \frac{2\varepsilon_r + \psi}{2\phi}, \quad \tan\alpha_2 = \frac{1 - \varepsilon_r + \frac{\psi}{2}}{\phi} \tag{6.128}$$

which, after substituting equations 6.118 and 6.123 reads:

$$\tan\beta_1 = \frac{1}{\phi} + \frac{\varepsilon_r^* + \frac{\psi^*}{2} - 1}{\phi^*}, \quad \tan\alpha_2 = \frac{1 - \frac{\phi}{\phi^*}\varepsilon_r^* + \frac{\psi^*\phi}{2\phi^*}}{\phi} \tag{6.129}$$

6.5.1.1 Characteristic Curves for Incompressible Flows

Provided that $\tan\beta_1$ and $\tan\alpha_2$ remain constant under off-design conditions, the equations derived thus far are generally valid and apply regardless of whether the flow is incompressible or compressible. However, the values and distribution of cascade efficiency $\eta = f(\phi, \varepsilon_r^*)$ under off-design conditions do depend on the flow's compressibility. When flow can be considered incompressible, the stage efficiency can be calculated using equation 6.30, which is provided here as a reference:

$$\eta = \phi\left[\frac{\varepsilon_r - \phi\iota_R}{\phi + \iota_R\varepsilon_r} + \frac{1 - \varepsilon_r - \phi\iota_S}{\phi + \iota_S(1 - \varepsilon_r)}\right]$$

In this formula, the aerodynamic efficiencies of the profile $1/\iota_R$ and $1/\iota_S$ can be maintained nearly constant over a relatively narrow range around ϕ^*. This is supported by Howell's findings on flow deflection (refer to the previous chapter), which demonstrate that flow deflection is virtually unaffected by variations in the outlet flow angle (denoted as α_2 for stators and β_2 for rotors) when cascade solidity is set. However, for more accurate computations, a distribution for ι becomes indispensable. Based on low-speed tests on cascades of NACA-65 profiles (see Figure 6.33), the aerodynamic efficiency exhibits a typical parabolic distribution as a function of

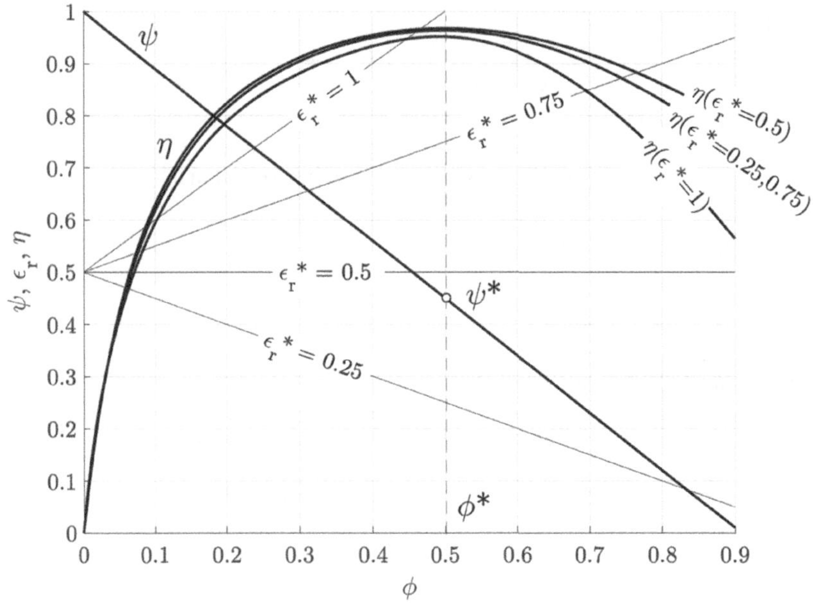

Figure 6.32 Example of characteristic curves of a decelerating cascade. $\phi^* = 0.5$, $\iota_{\min} = \frac{1}{70}$, $\psi^* = 0.45$. $\varepsilon_r^* = [0.25, 0.5, 0.75, 1]$.

inlet flow velocity, with a maximum under the design condition. Below is an example:

$$\iota_R = \frac{\iota_{\min}}{12^2}\left(\beta_1^\circ - \beta_1^\circ(\phi^*)\right)^2 + \iota_{\min},\ \iota_S = \frac{\iota_{\min}}{12^2}\left(\alpha_2^\circ - \alpha_2^\circ(\phi^*)\right)^2 + \iota_{\min} \tag{6.130}$$

Introducing equation 6.123 into the expression of η, after some manipulation, we obtain:

$$\eta = \phi\left[\frac{\frac{1}{2} + \frac{\phi}{\phi^*}\left(\varepsilon_r^* - \frac{1}{2}\right) - \phi\,\iota_R(\beta_1)}{\phi + \frac{\iota_R(\beta_1)}{2} + \frac{\phi}{\phi^*}\left(\iota_R(\beta_1)\varepsilon_r^* - \frac{\iota_R(\beta_1)}{2}\right)} + \frac{\frac{1}{2} + \frac{\phi}{\phi^*}\left(-\varepsilon_r^* + \frac{1}{2}\right) - \phi\,\iota_S(\alpha_2)}{\phi + \frac{\iota_S(\alpha_2)}{2} + \frac{\phi}{\phi^*}\left(-\iota_S(\alpha_2)\varepsilon_r^* + \frac{\iota_S(\alpha_2)}{2}\right)}\right] \tag{6.131}$$

a plot of which is given in Figure 6.32.

6.5.1.2 Characteristic Curves for Compressible Flows

In this case, the characteristic curves are influenced not only by the variables introduced in the previous section but also by the Mach numbers that characterize the inlet flow conditions. Specifically, the inlet flow Mach number $\mathrm{Ma}_1 = \frac{c_1}{a_1}$ and the tangential total Mach number $\mathrm{Ma}_{u1,0} = \frac{u}{a_1^0}$ must be considered when analyzing the performance of the stage. In other words, for a given thermodynamic state of the flow at the stage inlet, the characteristic curves will exhibit a dependence on the peripheral cascade velocity.

Let us start with defining the total pressure loss coefficients for both the rotating and stationary cascades:

$$\zeta_{\text{rotor}} = \zeta_r \equiv \frac{p_{w1}^0 - p_{w2}^0}{\rho_1 w_1^2/2} = \frac{2p_{w1}^0}{\rho_1 w_1^2}\left(1 - \frac{p_{w2}^0}{p_{w1}^0}\right) = \frac{2p_1\left(1+\frac{\gamma-1}{2}\text{Ma}_{w1}^2\right)^{\frac{\gamma}{\gamma-1}}}{\rho_1 w_1^2}\left(1 - \frac{p_{w2}^0}{p_{w1}^0}\right) \tag{6.132}$$

$$\zeta_{\text{stator}} = \zeta_s \equiv \frac{p_2^0 - p_3^0}{\rho_2 c_2^2/2} = \frac{2p_2^0}{\rho_2 c_2^2}\left(1 - \frac{p_3^0}{p_2^0}\right) = \frac{2p_2\left(1+\frac{\gamma-1}{2}\text{Ma}_2^2\right)^{\frac{\gamma}{\gamma-1}}}{\rho_2 c_2^2}\left(1 - \frac{p_3^0}{p_2^0}\right) \tag{6.133}$$

which exhibit a strong dependency on incidence and Mach (Figure 6.33). It follows that:

$$\begin{aligned}\frac{p_{w2}^0}{p_{w1}^0} &= 1 - \frac{\rho_1 w_1^2 \zeta_r}{2p_1\left(1+\frac{\gamma-1}{2}\text{Ma}_{w1}^2\right)^{\frac{\gamma}{\gamma-1}}} \\ &= 1 - \frac{\gamma w_1^2 \zeta_r}{2\gamma R T_1\left(1+\frac{\gamma-1}{2}\text{Ma}_{w1}^2\right)^{\frac{\gamma}{\gamma-1}}} = 1 - \frac{\zeta_r \gamma \text{Ma}_{w1}^2}{2\left(1+\frac{\gamma-1}{2}\text{Ma}_{w1}^2\right)^{\frac{\gamma}{\gamma-1}}}\end{aligned} \tag{6.134}$$

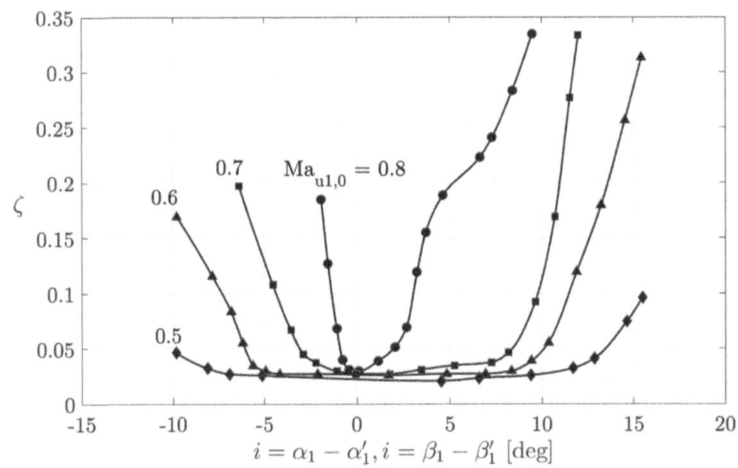

Figure 6.33 Sample plot of cascade total pressure loss coefficient as a function of the incidence angle and blade Mach number $\text{Ma}_{u1,0} = \frac{u}{a_1^0}$.

and

$$\frac{p_3^0}{p_2^0} = 1 - \frac{\zeta_s \gamma \mathrm{Ma}_2^2}{2\left(1+\frac{\gamma-1}{2}\mathrm{Ma}_2^2\right)^{\frac{\gamma}{\gamma-1}}} \tag{6.135}$$

The stage isentropic efficiency can be computed as

$$\eta_{TT} = \frac{T_{3ss}^0 - T_1^0}{T_3^0 - T_1^0} = \frac{T_1^0\left(\frac{T_{3ss}^0}{T_1^0} - 1\right)}{T_1^0\left(\frac{T_3^0}{T_1^0} - 1\right)} = \frac{\pi_s^{\frac{\gamma-1}{\gamma}} - 1}{\tau_s - 1} \tag{6.136}$$

where:

$$\pi_s = \frac{p_3^0}{p_1^0} = \left(\frac{p_3^0}{p_2^0}\right)\left(\frac{p_2^0}{p_{w2}^0}\right)\left(\frac{p_{w2}^0}{p_{w1}^0}\right)\left(\frac{p_{w1}^0}{p_1^0}\right) \tag{6.137}$$

$$\tau_s - 1 = \frac{(\gamma-1)\psi u^2}{\gamma R T_1^0} = (\gamma-1)\psi \mathrm{Ma}_{u1,0}^2 \tag{6.138}$$

In addition, we see that:

$$\frac{p_{w1}^0}{p_1^0} = \left(\frac{1+\frac{\gamma-1}{2}\mathrm{Ma}_{w1}^2}{1+\frac{\gamma-1}{2}\mathrm{Ma}_1^2}\right)^{\frac{\gamma}{\gamma-1}}, \quad \frac{p_2^0}{p_{w2}^0} = \left(\frac{1+\frac{\gamma-1}{2}\mathrm{Ma}_2^2}{1+\frac{\gamma-1}{2}\mathrm{Ma}_{w2}^2}\right)^{\frac{\gamma}{\gamma-1}} \tag{6.139}$$

To express Mach numbers as functions of stage parameters, consider that:

$$a_2^2 = (\gamma-1)\left[h_2^0 - \frac{c_2^2}{2}\right] = (\gamma-1)\left[h_1^0 + \omega\left(r_2 c_{u2} - r_1 c_{u1}\right) - \left(\frac{c_{z2}^2 + c_{u2}^2}{2}\right)\right] \tag{6.140}$$

thus:

$$\mathrm{Ma}_1^2 = \frac{c_1^2}{a_1^2} = \frac{c_m^2 + c_{u1}^2}{\gamma R T_1^0 - (\gamma-1)\left[\frac{c_m^2 + c_{u1}^2}{2}\right]} = \frac{\phi^2\left(1+\tan^2\alpha_1^*\right)}{\frac{1}{\mathrm{Ma}_{u1,0}^2} - (\gamma-1)\left[\frac{\phi^2\left(1+\tan^2\alpha_1^*\right)}{2}\right]} \tag{6.141}$$

$$\mathrm{Ma}_{w1}^2 = \frac{w_1^2}{a_1^2} = \frac{c_m^2 + (u - c_{u1})^2}{\gamma R T_1^0 - (\gamma-1)\left[\frac{c_m^2 + c_{u1}^2}{2}\right]} = \frac{1 - 2\phi\tan\alpha_1^* + \phi^2\left(1+\tan^2\alpha_1^*\right)}{\frac{1}{\mathrm{Ma}_{u1,0}^2} - (\gamma-1)\left[\frac{\phi^2\left(1+\tan^2\alpha_1^*\right)}{2}\right]} \tag{6.142}$$

$$\begin{aligned}\mathrm{Ma}_2^2 = \frac{c_2^2}{a_2^2} &= \frac{c_m^2 + c_{u2}^2}{\gamma R T_1^0 + (\gamma-1)\left[\omega r\left(c_{u2} - c_{u1}\right) - \frac{c_m^2 + c_{u2}^2}{2}\right]} \\ &= \frac{\phi^2\left(1+\tan^2\alpha_2\right)}{\frac{1}{\mathrm{Ma}_{u1,0}^2} + (\gamma-1)\left[\psi - \frac{\phi^2\left(1+\tan^2\alpha_2\right)}{2}\right]}\end{aligned} \tag{6.143}$$

$$\begin{aligned} \mathrm{Ma}_{w2}^2 = \frac{w_2^2}{a_2^2} &= \frac{c_m^2 + (u - c_{u2})^2}{\gamma R T_1^0 + (\gamma - 1)\left[\omega r (c_{u2} - c_{u1}) - \frac{c_m^2 + c_{u2}^2}{2}\right]} \\ &= \frac{1 - 2\phi \tan\alpha_2 + \phi^2 \left(1 + \tan^2\alpha_2\right)}{\frac{1}{\mathrm{Ma}_{u1,0}^2} + (\gamma - 1)\left[\psi - \frac{\phi^2\left(1 + \tan^2\alpha_2\right)}{2}\right]} \end{aligned} \tag{6.144}$$

A sample plot of a cascade map is shown in Figure 6.34. To improve readability, the values of the flow coefficient on the x axis have been scaled by a factor of $\mathrm{Ma}_{u1,0}/\mathrm{Ma}_{u1,0}^*$, where $\mathrm{Ma}_{u1,0}^* = 0.5$. The plot illustrates that increasing the Mach number associated with the peripheral speed results in higher stage compression ratios, with only minor effects on peak efficiency. However, a higher Mach number reduces the range for efficient operation. The dashed lines represent zones with positive characteristic slope, i.e. where pressure ratios increase with flow rate, suggesting that the cascade is likely approaching stall conditions. The minimum stable flow rates are marked by small circles connected by thin continuous lines, referred to as *stall lines.* At flow rates exceeding the nominal values, inefficiencies due to improper incidence become evident, further decreasing the pressure ratio as the flow rate increases. Additionally, varying the degree of reaction from 0.5 noticeably affects the operating range for stable operation. While only the case of a unitary degree of reaction is plotted, similar trends are expected for other values.

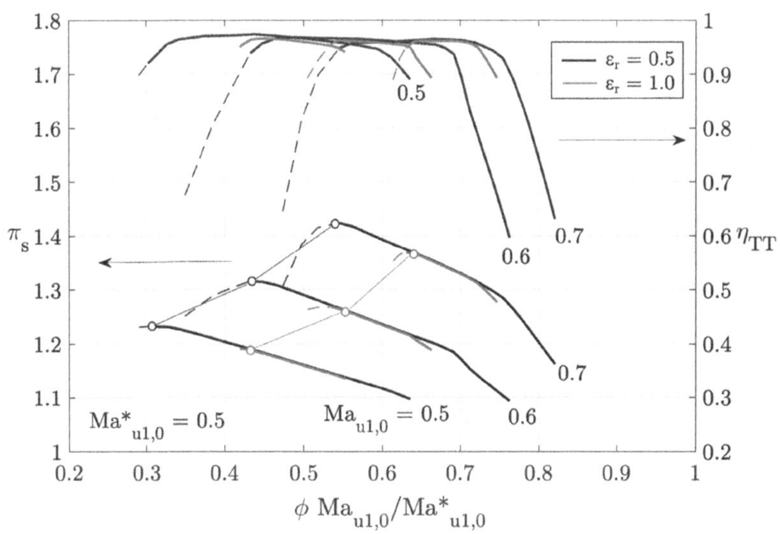

Figure 6.34 Compressor cascade map. ζ values of Figure 6.33. $\phi^* = 0.5$; $\gamma = 1.4$; $\psi^* = 0.45$; $\mathrm{Ma}_{u1,0} = [0.5\ 0.6\ 0.7]$; $\varepsilon_r^* = [0.5, 1]$.

At flow rates exceeding the maximum values depicted in the graph, the stage is choked. Choking (not shown in the figure) occurs when the flow through the rotor reaches a critical condition (Ma = 1), typically near the leading edge, preventing further increases in mass flow rate. This results in a significant drop in performance and efficiency.

6.5.2 OFF-DESIGN PERFORMANCE OF ACCELERATING CASCADES

Following the approach used for the compressor cascade, we will now derive the characteristic curves for a generic turbine stage. As is customary in turbines, 1 denotes the nozzle entry, 2 the rotor inlet, and 3 the rotor outlet (Figure 6.35). Consider the following:

$$\psi^* = \frac{c^*_{u2} + c^*_{u3}}{u}, \quad \phi^* = \frac{c^*_m}{u}, \quad \tan\alpha^*_2 = \frac{c^*_{u2}}{c^*_m}, \quad \tan\beta^*_3 = \frac{w^*_{u3}}{c^*_m}$$

$$\frac{c_{u2}}{c_m} = \frac{c^*_{u2}}{c^*_m}, \quad \frac{w_{u3}}{c_m} = \frac{w^*_{u3}}{c^*_m}$$

$$c_{u2} = u + w_{u2}, \quad c_{u3} = w_{u3} - u$$

thus:

$$\psi^* = \phi^*(\tan\alpha^*_2 + \tan\beta^*_3) - 1 \tag{6.145}$$

$$\psi = \phi(\tan\alpha_2 + \tan\beta_3) - 1 \tag{6.146}$$

From these equations:

$$\psi = \phi\left(\frac{\psi^*}{\phi^*} + \frac{1}{\phi^*}\right) - 1 \implies \frac{\psi}{\psi^*} = \frac{\phi}{\phi^*}\left(1 + \frac{1}{\psi^*}\right) - \frac{1}{\psi^*} \tag{6.147}$$

The graph of ψ against ϕ is therefore a straight line with a positive slope $\frac{\psi^*}{\phi^*} + \frac{1}{\phi^*}$, dependent only on the design value of the work coefficient ψ^* (see Figure 6.36).

Interestingly, the degree of reaction ε_r in off-design has the same dependency on ϕ found for compressor cascades. In fact:

$$\varepsilon_r = \frac{w_{u3} - |w_{u2}|}{2u} = \frac{\phi\tan\beta^*_3 - (c_{u2}/u - 1)}{2} = \frac{\phi\tan\beta^*_3 - \phi\tan\alpha^*_2 + 1}{2} \tag{6.148}$$

which, compared to the same expression referred to the design condition:

$$\varepsilon^*_r = \frac{\phi^*\tan\beta^*_3 - \phi^*\tan\alpha^*_2 + 1}{2} \tag{6.149}$$

gives:

$$\varepsilon_r = \frac{\phi}{\phi^*}\left(\varepsilon^*_r - \frac{1}{2}\right) + \frac{1}{2} \tag{6.150}$$

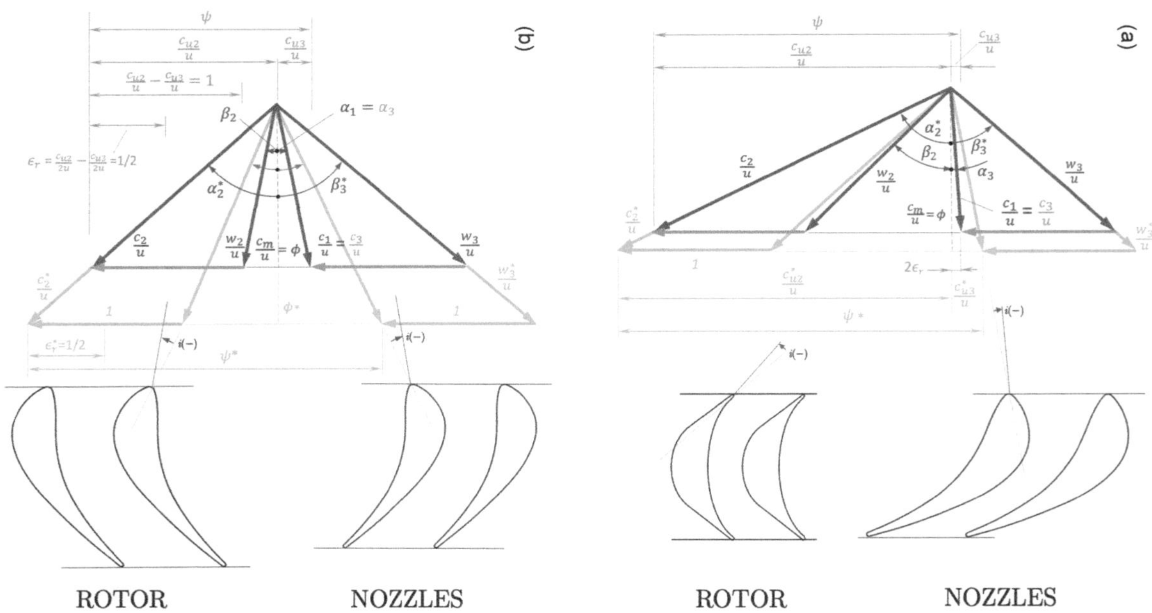

Figure 6.35 Velocity triangles and blade geometries for two off-design axial turbine stages with different reaction levels. **A** shows a zero-reaction ($\varepsilon_r = 0$) impulse cascade. **B** shows a 50% reaction ($\varepsilon_r = 0.5$) stage.

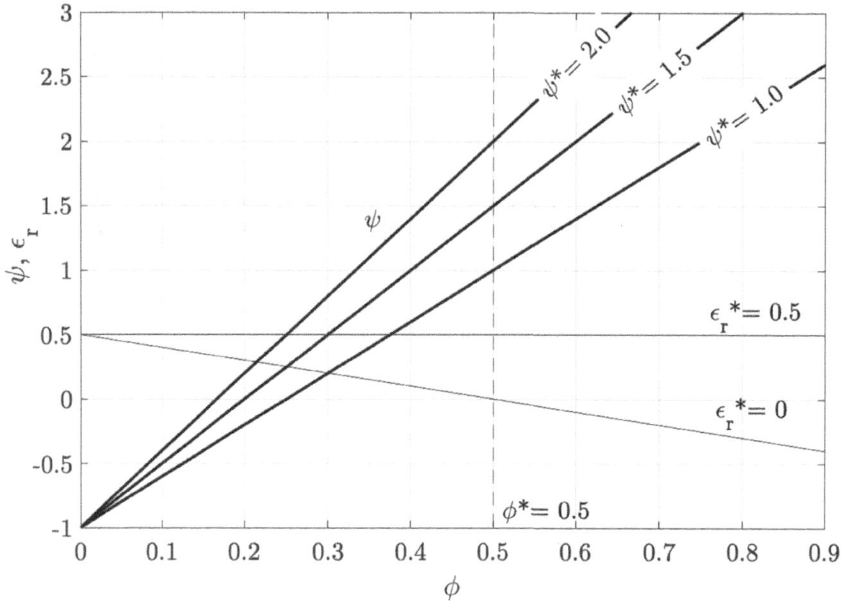

Figure 6.36 Cascade of turbine stage: work coefficient as function of flow coefficient for $\varepsilon_r^* = 0$ and $\varepsilon_r^* = 0.5$, $\phi^* = 0.5$, $\psi^* = [1, 1.5, 2]$.

Flow angles can be computed recognizing that from the expressions of ψ and ε_r:

$$\frac{w_{u2}}{u} = \psi - \frac{w_{u3}}{u}, \quad \frac{w_{u3}}{u} = \varepsilon_r + \frac{\psi}{2} \tag{6.151}$$

Thus:

$$\tan\beta_2 = \frac{w_{u2}}{c_m} = \frac{1}{\phi}\left(\frac{\psi}{2} - \varepsilon_r\right) \tag{6.152}$$

$$\tan\alpha_1 = \tan\alpha_3 = \frac{w_{u3} - u}{c_m} = \frac{1}{\phi}\left(\frac{\psi}{2} + \varepsilon_r - 1\right) \tag{6.153}$$

6.5.3 CHARACTERISTIC CURVES FOR TURBINE CASCADES

In turbine cascades, consider that the effect of compressibility on performance is essential for realistic predictions in off-design conditions. Consequently, it is customary to define total pressure loss coefficients that inherently account for variations in Mach number:

$$Y_{\text{nozzle}} = Y_n \equiv \frac{p_1^0 - p_2^0}{p_2^0 - p_2} = \frac{p_2^0\left(\frac{p_1^0}{p_2^0} - 1\right)}{p_2^0\left(1 - \frac{p_2}{p_2^0}\right)} = \frac{\left(\frac{p_1^0}{p_2^0} - 1\right)}{1 - \left(1 + \frac{\gamma-1}{2}\text{Ma}_2^2\right)^{\frac{\gamma}{1-\gamma}}} \tag{6.154}$$

$$Y_{\text{rotor}} = Y_r \equiv \frac{p_{w2}^0 - p_{w3}^0}{p_{w3}^0 - p_3} = \frac{p_{w3}^0\left(\frac{p_{w2}^0}{p_{w3}^0} - 1\right)}{p_{w3}^0\left(1 - \frac{p_3}{p_{w3}^0}\right)} = \frac{\left(\frac{p_{w2}^0}{p_{w3}^0} - 1\right)}{1 - \left(1 + \frac{\gamma-1}{2}\text{Ma}_{w3}^2\right)^{\frac{\gamma}{1-\gamma}}} \tag{6.155}$$

from which:

$$\frac{p_1^0}{p_2^0} = 1 + Y_n\left[1 - \left(1 + \frac{\gamma-1}{2}\text{Ma}_2^2\right)^{\frac{\gamma}{1-\gamma}}\right] \tag{6.156}$$

$$\frac{p_{w2}^0}{p_{w3}^0} = 1 + Y_r\left[1 - \left(1 + \frac{\gamma-1}{2}\text{Ma}_{w3}^2\right)^{\frac{\gamma}{1-\gamma}}\right] \tag{6.157}$$

On the other hand:

$$\frac{p_2^0}{p_{w2}^0} = \left(\frac{1 + \frac{\gamma-1}{2}\text{Ma}_2^2}{1 + \frac{\gamma-1}{2}\text{Ma}_{w2}^2}\right)^{\frac{\gamma}{\gamma-1}}, \quad \frac{p_{w3}^0}{p_3^0} = \left(\frac{1 + \frac{\gamma-1}{2}\text{Ma}_{w3}^2}{1 + \frac{\gamma-1}{2}\text{Ma}_3^2}\right)^{\frac{\gamma}{\gamma-1}} \tag{6.158}$$

All Mach numbers can be computed as functions of stage parameters and $\text{Ma}_{u2,0}^2$ (since now, differently from compressors, the rotor inlet is denoted by 2 and the stagnation enthalpy does not change across the nozzle):

$$\text{Ma}_2^2 = \frac{\phi^2\left(1 + \tan^2\alpha_2^*\right)}{\frac{1}{\text{Ma}_{u2,0}^2} - (\gamma-1)\left[\frac{\phi^2\left(1+\tan^2\alpha_2^*\right)}{2}\right]} \tag{6.159}$$

$$\text{Ma}_{w2}^2 = \frac{1 - 2\phi\tan\alpha_2^* + \phi^2\left(1 + \tan^2\alpha_2^*\right)}{\frac{1}{\text{Ma}_{u2,0}^2} - (\gamma-1)\left[\frac{\phi^2\left(1+\tan^2\alpha_2^*\right)}{2}\right]} \tag{6.160}$$

$$\text{Ma}_3^2 = \frac{\phi^2\left(1 + \tan^2\alpha_3\right)}{\frac{1}{\text{Ma}_{u2,0}^2} + (\gamma-1)\left[\psi - \frac{\phi^2\left(1+\tan^2\alpha_3\right)}{2}\right]} \tag{6.161}$$

$$\text{Ma}_{w3}^2 = \frac{\phi^2\left(1 + \tan^2\beta_3^*\right)}{\frac{1}{\text{Ma}_{u2,0}^2} + (\gamma-1)\left[\psi - \frac{\phi^2\left(1+\tan^2\alpha_3\right)}{2}\right]} \tag{6.162}$$

Finally, the isentropic turbine stage efficiency based on two-dimensional cascade analysis can be derived as:

$$\eta_{TT} = \frac{T_1^0 - T_3^0}{T_1^0 - T_{3ss}^0} = \frac{1 - \frac{1}{\tau_s}}{1 - \left(\frac{1}{\pi_s}\right)^{\frac{\gamma-1}{\gamma}}} \tag{6.163}$$

where:

$$\pi_s = \frac{p_1^0}{p_3^0} = \left(\frac{p_1^0}{p_2^0}\right)\left(\frac{p_2^0}{p_{w2}^0}\right)\left(\frac{p_{w2}^0}{p_{w3}^0}\right)\left(\frac{p_{w3}^0}{p_3^0}\right) \tag{6.164}$$

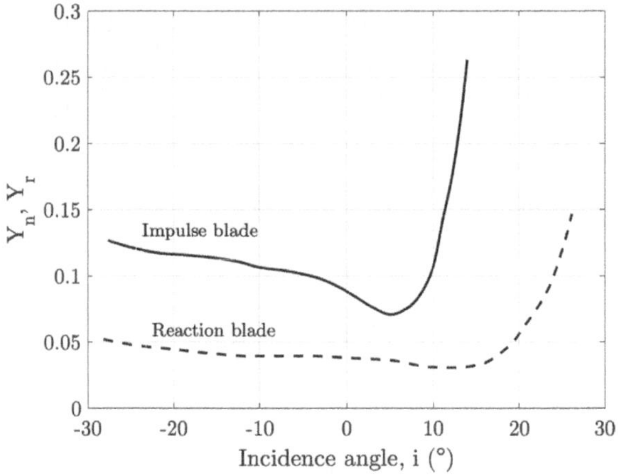

Figure 6.37 Typical total pressure loss coefficients for impulse (solid line) and reaction (dashed line) turbine blades versus incidence angle i.

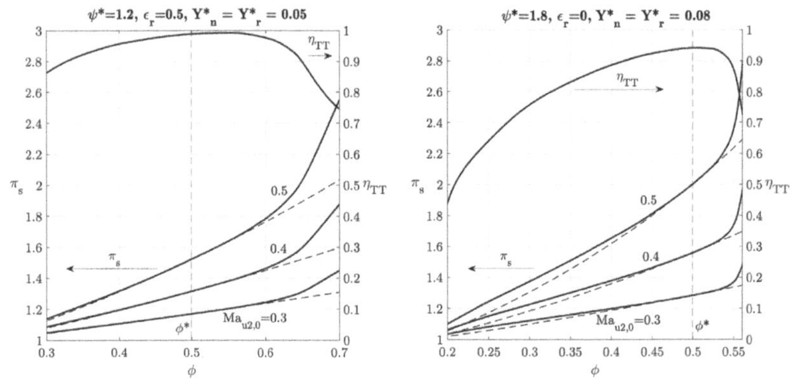

Figure 6.38 Characteristic curves of turbine stages based on cascade analysis: expansion ratio and isentropic efficiency as functions of flow coefficient for $\varepsilon_r^* = 0.5$ (a) and $\varepsilon_r^* = 0$ (b). Peripheral total Mach number 0.3–0.5. No choking displayed. Dashed curves correspond to the model described in Section 6.5.2.

and

$$\tau_s = \frac{T_1^0}{T_3^0} \Longrightarrow 1 - \frac{1}{\tau_s} = (\gamma - 1)\mathrm{Ma}_{u2,0}^2 \psi \tag{6.165}$$

are the stage expansion ratio and stage temperature ratio, respectively.

As a first approximation, the total pressure loss coefficients Y_n and Y_r are functions of the incidence angle and the degree of reaction (Figure 6.37). The data shown in the figure can be used to estimate the characteristics of the turbine stage (Figure 6.38). In the plots, the dashed lines represent the expansion ratio that could be achieved without accounting for profile losses.

On average, reaction stages exhibit higher peak efficiency compared to impulse stages. Moreover, in reaction stages, efficiency remains close to its maximum over a wide range of flow rates. This is due to the lower total pressure losses under both design and off-design conditions (Figure 6.37).

In contrast, at nominal flow rate ϕ^*, impulse stages are characterized by higher expansion ratios. As expected, in both reaction and impulse stages, the expansion ratio increases with the peripheral Mach number.

Additionally, the turbine map is largely unaffected by profile stall due to the asymmetric nature of total pressure loss as a function of the incidence angle (Figure 6.37). At flow rates lower than the nominal, negative incidence angles help protect the stage from stalling. However, at flow rates higher than nominal, positive incidence angles lead to increased losses, causing a sharp drop in efficiency and creating a natural tendency toward stall.

In practical cases, however, choking occurs before the stage reaches stall (not shown in the figures). Choking limits the maximum mass flow rate through the blades, causing the expansion ratio to increase exponentially. This typically occurs at relatively small positive incidence angles when the turbine operates close to the nominal flow rate, where stage efficiency is near its peak.

6.6 PROBLEMS

1. A rotor cascade operates with an incompressible fluid that has axial velocity $c_m = 80\,\text{m/s}$, flow angles $\beta_1 = 65°$, $\beta_2 = 45°$. A target lift coefficient $C_{L,R} = 1.5$ is desired and a drag coefficient $C_{D,R} = 0.06$ is expected. Determine the needed cascade solidity σ_R.

   [0.77]

2. A rotor cascade operates with air $\rho = 1.22\,\text{kg/m}^3$ at inlet temperature $T_1 = 288\,\text{K}$ and under the following conditions: axial velocity $c_m = 40\,\text{m/s}$, inlet angle $\beta_1 = 65°$, outlet angle $\beta_2 = 45°$. A target lift coefficient $C_{L,R} = 1.5$ is desired and a drag coefficient $C_{D,R} = 0.06$ is expected. The blade chord length is $\ell = 35\,\text{mm}$. Determine the static pressure rise and the relative total pressure drop across the cascade.

   [3234 Pa, 293 Pa]

3. An air compressor stage features a repeated rotor-stator cascade arrangement $\mathbf{c_3} = \mathbf{c_1}$. The rotor parameters are: inlet air density: $\rho = 1.225\,\text{kg/m}^3$, inlet temperature: $T_1 = 288\,\text{K}$, axial velocity $c_m = 35\,\text{m/s}$, inlet angle: $\beta_1 = 65°$, outlet angle: $\beta_2 = 45°$, lift coefficient: $C_{L,R} = 1.2$, drag coefficient: $C_{D,R} = 0.06$, chord length: $\ell = 35\,\text{mm}$, solidity: $\sigma_R = 0.95$. The stator parameters are: inlet angle: $\alpha_2 = 38°$, outlet angle: $\alpha_3 = 15°$, lift coefficient: $C_{L,S} = 1.2$, drag coefficient: $C_{D,S} = 0.05$, chord length: $\ell_S = 30\,\text{mm}$. Compute pressure rise $p_3 - p_1$ and stagnation pressure rise $p_3^0 - p_1^0$ across the stage. Also, compute the stage degree of reaction.

   [3017 Pa, 3421 Pa, 0.65]

4. An air compressor stage is formed by a rotor and a stator cascade featuring aerodynamic efficiency $E_R = 65$ and $E_S = 50$, respectively. The rotor operates with air $\rho = 1.225\,\mathrm{kg/m^3}$ at $T_1 = 320\,\mathrm{K}$, $\beta_1 = 65°$ and deflects the relative flow by 10°. The stator has inlet flow angle $\alpha_2 = 30°$ and deflects the relative flow by 20°. The meridional velocity throughout the stage is $c_m = 45\,\mathrm{m/s}$. Compute total pressure loss coefficients for both rotor and stator cascades, as well as the flow and work coefficients of the stage.

   [0.016, 0.014, 0.43, 0.31]

5. A low-speed, repeated 0.5 reaction, compressor stage is designed with rotor and stator cascades of NACA 65-(12)10 blade sections. Both cascades have an aerodynamic efficiency $E_R = E_S = E$ of 65. The stage flow coefficient ϕ is 0.65 and the work coefficient $\psi = 0.33$. Compute flow angles β_1, β_2, α_2, α_3, draw non-dimensional velocity triangles, and estimate stage efficiency η_s.

   [$\beta_1 = 45.65°$, $\beta_2 = 27.27°$, $\alpha_2 = 45.65°$, $\alpha_3 = 27.27°$, $w_1/u = 0.930$, $w_2/u = 0.731$, $c_2/u = 0.930$, $c_3/u = 0.731$, $\eta_s = 0.959$]

6. Determine the values of rotor and stator solidity of the stage in the previous problem such that the diffusion factors $DF_R < 0.45$ and $DF_S < 0.4$.

   [$\sigma_R > 0.75$, $\sigma_S > 0.95$]

7. A blade section is to be designed using a parabolic-arc camber line, where the turning angles are specified as $\beta_1' = 65°$ (inlet angle) and $\beta_2' = 45°$ (outlet angle). Determine the camber line parameters, $\frac{x_A}{\ell}$ and $\frac{y_A}{\ell}$, along with the stagger angle, ξ, to satisfy these requirements.

   [0.5, 0.0955, 46°]

8. A double circular-arc profile features a chord length $\ell = 45$ mm and a camber angle $\Theta = 2\theta = 88°$. Determine the radius of curvature R of the circle that describes the suction side. When staggered at $\xi = 45°$, calculate the inlet and outlet angles β_1' and β_2'.

   [0.0487 m, 72.5°, 17.5°]

9. A turbine stage operates with isentropic efficiency $\eta = 0.85$ and axial outlet. The given parameters are: flow coefficient $\phi = 0.8$, absolute flow angle at nozzle outlet $\alpha_2 = 70°$, blade speed $u = 250\,\mathrm{m/s}$. Compute absolute and relative velocities at the rotor inlet and outlet, as well as the specific work generated in the stage.

   [$c_2 = 584.8$ m/s, $c_3 = 200$ m/s, $w_2 = 551.8$ m/s, $w_3 = 551.8$ m/s, $\Delta h^0 = 137.4$ kJ/kg]

10. A 0.5-reaction turbine stage operates with inlet gases at $T_1 = 900$ K, blade speed $u = 350\,\mathrm{m/s}$, absolute axial velocity $c_x = 180\,\mathrm{m/s}$ and absolute inlet flow angle $\alpha_2 = 72°$. The turbine nozzle operates with an efficiency $\eta_N = \frac{\Delta h_{1-2}}{\Delta h_{s,1-2}} = 0.95$. Determine the work coefficient ψ. In addition, determine the relative velocities across the rotor w_2 and w_3 and the the nozzle expansion ratio $\pi_s = p_1/p_2$. The fluid behaves as an ideal gas having $c_p = 1250\,\mathrm{J/(kg \cdot K)}$ and $\gamma = 1.33$.

    [$\psi = 2.17$, $w_2 = 272\,\mathrm{m/s}$, $w_3 = 582.5\,\mathrm{m/s}$, $\pi_s = 1.7$]

11. A 0-reaction turbine stage is to be designed for a total to total isentropic efficiency $\eta_{TT} = 0.87$. The design blade speed is 200 m/s, and inlet flow velocity in the nozzle is also 200 m/s. Determine the flow angle at the nozzle exit α_2 and the rotor flow turning $\Delta\beta$.

[63.4°, 90°]

12. A 0.5-reaction turbine stage is designed with a blade speed rotor $u = 350\,\text{m/s}$ for a stagnation enthalpy drop of $\Delta h^0 = 147\,\text{kJ/kg}$. Determine the maximum total-to-total efficiency, η_{TT}, of the stage and the pertinent flow coefficient, ϕ and nozzle exit flow angle α_2. Assume geometric ratios $b_N/H = b_R/H = 0.33$.

[0.90, 0.51, 64.9°]

13. Determine the rotor blade speed, u, flow coefficient ϕ and flow turning $\Delta\beta$ of a 0.5-reaction turbine stage that makes it possible to achieve a total-to-total efficiency, η_{TT}, of 0.92. The given parameters for the turbine stage are: stagnation enthalpy drop $\Delta h^0 = 120\,\text{kJ/kg}$, blade-to-height ratios $\frac{b_N}{H} = 0.33$, $\frac{b_R}{H} = 0.33$.

[$u = 358$ m/s, $\phi = 0.27$, $\Delta\beta = 68°$]

14. A 0.5-reaction, repeated compressor stage is to be designed for moderately-loaded operational conditions (shape factor H =1.4 for both the rotor and stator) with: inlet relative velocity $w_1 = 220$ m/s, $\beta_1 = 60°$, $w_2 = 154$ m/s and $\beta_2 = 40°$. To avoid rotor and stator stall, the diffusion factor for both components is set at 0.4. The design rotor's total pressure loss coefficient is expected to be $\zeta = 0.015$, while the total cascade loss coefficient ζ_c must be kept below 0.2. Determine the solidity of both the rotor and stator cascades that fulfill the requirements.

[1.35, 2]

15. A low-speed compressor cascade is to be designed for nominal conditions with $\alpha_1 = 50°$ and α_2 =33°. Determine optimal cascade solidity and blade angles α_1' and α_2' if a circular-arc camber line is selected and optimal incidence is expected at $i = \alpha_1 - \alpha_1'$ =1°.

[0.81, 48.9°, 25.6°]

16. Under design conditions, an axial stage consisting of decelerating cascades (rotor and stator) operates with $\psi^* = 0.8$, $\phi^* = 0.3$, and $\varepsilon_r = 0.3$. Determine the values of the incidence angles $i_R = \beta_1 - \beta_1^*$ (rotor) and $i_S = \alpha_2 - \alpha_2^*$ (stator) as functions of the flow coefficient ϕ. If both profiles exhibit stall at an incidence angle of 13°, identify which cascade (rotor or stator) stalls first and determine the corresponding value of ϕ.

[stator, 0.21]

17. An axial compressor rotor cascade in a repeated stage operates by ingesting air $\gamma = 1.4$, $R = 287\,\text{J/(kg·K)}$ at a stagnation temperature at the inlet $T_1^0 = 15°\text{C}$. At the design point, the parameters are as follows: relative total pressure loss coefficient $\zeta_r^* = 0.025$, axial velocity $c_m^* = 120\,\text{m/s}$, tangential component of velocity at the inlet $c_{u1}^* = 70\,\text{m/s}$, tangential component of velocity at the outlet

$c_{u2}^* = 160\,\text{m/s}$ and blade speed $u^* = 267\,\text{m/s}$. Determine the design rotor pressure ratio $\pi_R^* = p_2^0/p_1^0$.

[1.31]

18. The rotor cascade of the previous exercise operates off-design with increased u_2 compared to design conditions, such that off-design quantities are given as $\phi = 0.85\phi^*$ and $\zeta_R = 0.031$. Determine the off-design rotor pressure ratio.

[1.42]

19. A repeated axial compressor stage for air consists of a rotor and a stator cascade operating under the following conditions: the stagnation temperature at the stage inlet is $T_1^0 = 25°\text{C}$, the blade speed $u^* = 320\,\text{m/s}$, and the axial velocity $c_m^* = 166\,\text{m/s}$. The tangential components of the absolute velocity at the rotor inlet and outlet are $c_{u1}^* = 85\,\text{m/s}$ and $c_{u2}^* = 210\,\text{m/s}$, respectively. The total pressure loss coefficients for the rotor and stator are $\zeta_r = 0.025$ and $\zeta_s = 0.028$, respectively. Determine the stage pressure ratio and the total-to-total stage efficiency.

[1.52, 0.96]

20. The stage analyzed in the previous exercise is now operating under off-design conditions, where the incidence angles for both the rotor and stator blade sections increase by $+5°$. Under these updated conditions, the rotor loss coefficient is $\zeta_r = 0.21$, and the stator loss coefficient is $\zeta_s = 0.18$. Determine the off-design values of the flow coefficient, stage pressure ratio and total-to-total stage efficiency.

[0.45, 1.37 (likely to stall), 0.71]

21. An axial reaction turbine stage operates under the following conditions: $\psi^* = 0.9$, $\phi^* = 0.6$, $\alpha_2^* = 68°$, and $\beta_3^* = 50°$. In off-design operations, the flow coefficient becomes $\phi = 0.5$. Compute the off-design loading coefficient, the relative change of reaction $\Delta\varepsilon_{r\%}$ and the change in the rotor incidence angle.

[0.58, +56%, −16.5°]

22. A zero-reaction axial turbine stage operates with the following parameters: $\psi^* = 1.25$, $\phi^* = 0.7$, $\alpha_2^* = 75°$. Analyze the off-design sensitivity of the parameters ψ, ε_r, β_2 (relative flow angle at rotor inlet) and i_r (flow incidence angle at rotor inlet) as functions of ϕ, for a range of ϕ values from 0.4 to 0.8. As an example, derive the values of ψ, ε_r, β_2 and i_r at off-design $\phi = 0.75$.

[1.41, −0.04, 44.7°, 2.9°]

23. A nozzle of an axial turbine operates with an inlet axial velocity $c_1 = 140\,\text{m/s}$ at stagnation temperature $T_1^0 = 850\,\text{K}$. The specific heat ratio of the gas is $\gamma = 1.33$ and the gas constant $R = 287\,\text{J/(kg}\cdot\text{K)}$. The nozzle vane exit angle is $\alpha_2' = 76°$. The vane features a parabolic-arc camber line with camber angle $\theta = 73°$, the stagger angle is $\xi = 45°$ and the cascade solidity $\sigma_N = 1.2$. The nozzle total pressure loss coefficient is $Y_{\text{nozzle}} = 0.12$. Determine the total pressure ratio across the nozzle vanes $\frac{p_2^0}{p_1^0}$.

[0.98]

24. An axial turbine stage with zero reaction is designed with a work coefficient $\psi^* = 1.8$ and a flow coefficient $\phi^* = 0.35$. The tangential stagnation Mach number is $Ma_{u2,0} = 0.4$. The nozzle and rotor total pressure loss coefficients are expressed as functions of the flow coefficient ϕ by the following relations: $Y_n = \frac{0.17}{(\phi^*)^2}(\phi - \phi^*)^2 + 0.08$, $Y_r = \frac{0.15}{(\phi^*)^2}(\phi - \phi^*)^2 + 0.12$. Determine the stage pressure ratio $\pi_s = p_1^0/p_3^0$ and the total-to-total efficiency η_{TT} at both design operating condition $\phi = \phi^* = 0.35$ and off-design condition $\phi = 0.45$.
[Design: 1.53, 95%; off-design: 1.87, 66%]

NOTES

1. By analyzing a generic expansion transformation with reasonable efficiency, it follows that:

$$h_2 - h_{3s}^0 \approxeq h_{2s} - h_{3ss}^0$$

while, under the same assumption:

$$h_3^0 - h_3 \approxeq h_{3s}^0 - h_{3s} \approxeq h_{3ss}^0 - h_{3ss}$$

Combining the two yields:

$$h_3^0 - h_3 \approxeq h_2 - h_{3s} + h_{3ss}^0 - h_{3s} \implies h_3^0 - h_{3ss}^0 \approxeq (h_2 - h_{2s}) + (h_3 - h_{3s})$$

2. In Horlock, J. H., *Axial Flow Turbines: Fluid Mechanics and Thermodynamics*, Krieger Pub Co., Malabar, FL (1973).
3. Zweifel, O., "The Spacing of Turbomachine Blading, Especially with Large Angular Deflection," *Brown Boveri Review (BBR)*, 32(1): 436–444 (1945).
4. Lieblein, S., *Experimental Flow in Two-Dimensional Cascades*, Chapter V, pp. 183–226, NASA SP-36, Washington, D.C., USA (1965).
5. Lieblein's findings are based on test results on NACA65-(A10)10 series and two British C4 parabolic-arc profiles having maximum thickness-to-chord ratio of 0.1 and a Reynolds number $2.5 \cdot 10^5$.
6. Spence, D. A., "The Development of Turbulent Boundary Layers," *Journal of the Aeronautical Sciences*, 23(1): 3–15 (1956).

7 Three-Dimensional Flow in Axial Flow Machines

In this chapter, we will explore the design criteria for establishing the most desirable tangential velocity distribution downstream of blades in axial flow machines, including axial flow pumps, axial fans, and axial compressor stages. This discussion is closely related to the concepts introduced in Chapter 6, where we examined two-dimensional cascade flows. However, the focus here is to complement that discussion with the specifics of three-dimensional blade design. By integrating the knowledge from Chapter 6 with the new concepts presented in this chapter, students will gain a deeper understanding of the variations in velocity triangles from the hub to the tip of a real blade and how these variations lead to blade twist. It is important to note that these principles are not directly transferable to centrifugal or mixed flow machines due to their inherently three-dimensional flow characteristics.

7.1 SIMPLE RADIAL EQUILIBRIUM

To determine the behavior of the velocity triangles as a function of radius, it is necessary to study the radial equilibrium of a generic fluid element moving along a spiral trajectory along both axial, z, and tangential directions, θ. Let us assume an inviscid fluid, a flow without radial velocity components, $c_r = 0$, and subject to radial force equilibrium, where the pressure forces acting on a generic fluid particle are entirely counterbalanced by the centrifugal forces. Under these assumptions, the helical trajectory of the fluid particle must be the result of radial flow equilibrium governed by the following equation (see Section 1.4.7.2):

$$\frac{dp}{dr} = \rho\omega^2 r = \rho\frac{c_u^2}{r} \tag{7.1}$$

To obtain the equation of radial equilibrium, the definition of total enthalpy and the first law of thermodynamics are elaborated by differentiating both with respect to the radius and neglecting second-order differentials:

$$h^0 = h + \frac{1}{2}c_z^2 + \frac{1}{2}c_u^2 \Longrightarrow \frac{dh^0}{dr} = c_z\frac{dc_z}{dr} + c_u\frac{dc_u}{dr} \tag{7.2}$$

$$Tds = dh - \frac{dp}{\rho} \Longrightarrow \frac{d}{dr}(Tds) = \frac{dh}{dr} - \frac{d}{dr}\left(\frac{dp}{\rho}\right) \Longrightarrow T\frac{ds}{dr} = \frac{dh}{dr} - \frac{1}{\rho}\frac{dp}{dr} \tag{7.3}$$

The equation of radial equilibrium is then obtained by combining the preceding three equations, where $T\frac{ds}{dr} = 0$ to satisfy the assumption of isentropic flow:

DOI: 10.1201/9781003619956-7

$$\frac{dh^0}{dr} = c_z \frac{dc_z}{dr} + \frac{c_u}{r} \frac{d}{dr}(rc_u) \tag{7.4}$$

7.1.0.1 Free Vortex Flow with Uniform Total Enthalpy

At a generic station where flow has uniform total energy along the radius, we have:

$$\frac{dh^0}{dr} = 0 \tag{7.5}$$

Substituting equation 7.5 into equation 7.4, we get:

$$0 = c_z \frac{dc_z}{dr} + \frac{c_u}{r} \frac{d}{dr}(rc_u) \tag{7.6}$$

The free vortex is realized when:

$$rc_u = \text{const.} = b_1 \tag{7.7}$$

where b_1 is a constant. As a result:

$$\frac{d(rc_u)}{dr} = \frac{d}{dr}(b_1) = 0 \implies 0 = c_z \frac{dc_z}{dr} \tag{7.8}$$

which implies that either $c_z = 0$, which is non-physical as it implies no axial velocity, or:

$$c_z = \text{const.} \tag{7.9}$$

7.2 VORTEX LAWS FOR AXIAL MACHINES

7.2.1 FREE VORTEX FLOW ACROSS A ROTOR WITH UNIFORM INLET TOTAL ENTHALPY

Assume a rotor imparting specific work L as the flow crosses the rotor from inlet 1 to outlet 2. If we assume a free vortex flow both at 1 and at 2:

$$c_{u1} = \frac{b_1}{r} \tag{7.10}$$

$$c_{u2} = \frac{b_2}{r} \tag{7.11}$$

First of all, it is easy to calculate the specific work under these assumptions:

$$L(r) = \omega r \left(\frac{b_2}{r} - \frac{b_1}{r} \right) = \omega (b_2 - b_1) \tag{7.12}$$

which proves that $L(r) = L$ is constant with respect to r and consequently:

$$\frac{dh^0}{dr} = c_z \frac{dc_z}{dr} + \frac{c_{u2}}{r} \frac{d}{dr}(rc_{u2}) = 0 \tag{7.13}$$

For the specific work to be constant along the radius, c_z must also be constant along the radius, implying $\frac{dc_{z2}}{dr} = 0$. Therefore, the total enthalpy (and therefore the specific work) is constant along the radius at station 2 as well. A representation of such flow is given in Figures 7.1 and 7.2.

The degree of reaction ε_r is also calculated under the assumption that $c_{z1} = c_{z2} =$ const.:

$$\varepsilon_r(r) = 1 - \frac{c_{u2} + c_{u1}}{2\omega r} = 1 - \frac{\frac{b_2}{r} + \frac{b_1}{r}}{2\omega r} = 1 - \frac{b_2 + b_1}{2\omega r^2} \tag{7.14}$$

which means that the static enthalpy rise across the rotor, $h_2 - h_1$, decreases with radius. For an incompressible flow machine, this means that the static pressure increase, $p_2 - p_1$, is variable with r. In both cases, the static energy rise (i.e. the reaction work) is not under the control of the designer.

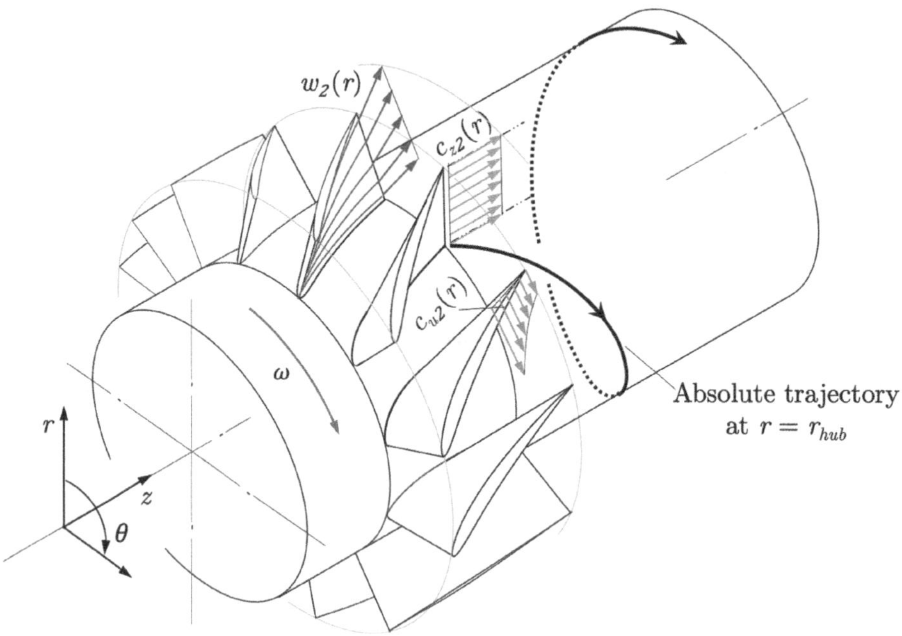

Figure 7.1 Free vortex flow across a rotor with uniform inlet total enthalpy.

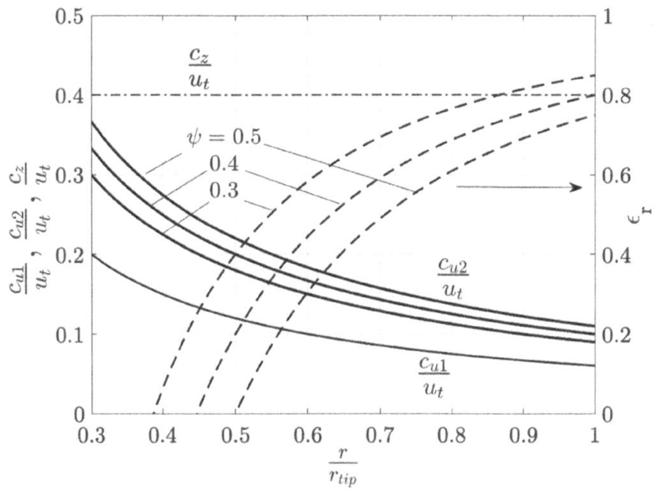

Figure 7.2 Characteristics of free vortex flow across a rotor with uniform inlet total enthalpy. Velocity components are normalized with respect to tip blade velocity, $u_t = \omega r_{\text{tip}}$.

7.2.2 FORCED VORTEX FLOW ACROSS A ROTOR WITH UNIFORM INLET TOTAL ENTHALPY AND INLET FREE VORTEX

The so-called *forced vortex flow* is identified with a solid body rotational motion, where the tangential velocity is directly proportional to its distance from the center of rotation:

$$c_{u2} = a_2 r \tag{7.15}$$

where $a_2 = \text{const}$. If $c_{u1} = \frac{b_1}{r}$ the specific work becomes:

$$L(r) = \omega r \left(a_2 r - \frac{b_1}{r} \right) = \omega a_2 r^2 - \omega b_1 \tag{7.16}$$

Therefore:

$$\frac{dL}{dr} = \frac{dh_2^0}{dr} - \frac{dh_1^0}{dr} = \frac{dh_2^0}{dr} - 0 = 2\omega a_2 r \tag{7.17}$$

The radial equilibrium equation reads:

$$\frac{dh^0}{dr} = c_z \frac{dc_z}{dr} + 2a_2^2 r \tag{7.18}$$

from which:

$$c_z \frac{dc_z}{dr} = 2\omega a_2 r - 2a_2^2 r \tag{7.19}$$

Integrating with respect to r and taking the square root:

$$c_{z2}(r) = \sqrt{2(\omega a_2 - a_2^2)r^2 + C} \tag{7.20}$$

The nominal degree of reaction $\varepsilon_r(r)^1$ is:

$$\varepsilon_r(r) = 1 - \frac{a_2 r + \frac{b_1}{r}}{2\omega r} \tag{7.21}$$

$$= 1 - \frac{a_2 r^2 + b_1}{2\omega r^2} \tag{7.22}$$

$$= 1 - \left(\frac{a_2}{2\omega} + \frac{b_1}{2\omega r^2}\right) = 1 - \frac{a_2}{2\omega} - \frac{b_1}{2\omega r^2} \tag{7.23}$$

The present case leads to a highly non-uniform distribution of axial velocity, work, and reaction distribution along the radius, meaning that its applicability to turbomachinery is unsuitable.

7.2.3 FORCED VORTEX FLOW ACROSS A ROTOR FOR AXIAL INFLOW

Let us assume that $c_{u2} = b_2 r$ and $c_{u1} = 0$ cause a forced vortex at the rotor outlet. The specific work becomes:

$$L(r) = \omega b_2 r^2 \tag{7.24}$$

such that:

$$\frac{dL}{dr} = 2\omega b_2 r \tag{7.25}$$

from which we reasonably argue that $\frac{dh_2^0}{dr} = 2\omega b_2$ unless dh_1^0/dr is known from other considerations (e.g. from the knowledge of total enthalpy distribution of a preceding stage).

While c_{z1} can be assumed constant with radius, from the radial equilibrium equation 7.4 it follows that:

$$c_{z2} = \sqrt{4\omega b_2 r - 2b_2^2 r^2 + C_2} \tag{7.26}$$

The nominal degree of reaction is:

$$\varepsilon_r = 1 - \frac{b_2}{2\omega} \tag{7.27}$$

which is independent of r. Characteristics of this vortex are provided in Figure 7.3.

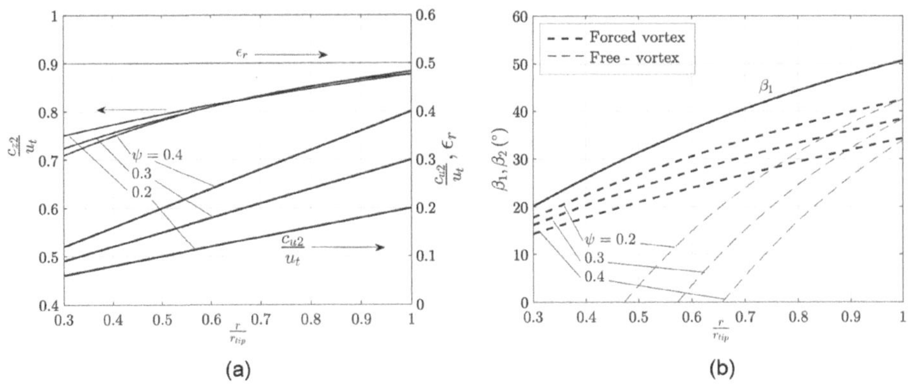

Figure 7.3 Variation in flow angles and velocity components along the normalized radial position $\frac{r}{r_{\text{tip}}}$ for different values of tip work coefficient $\psi = \psi_t$. (a) Normalized meridional velocity c_{z2}/u_t, tangential velocity c_{u2}/u_t and degree of reaction ε_r as functions of $\frac{r}{r_{\text{tip}}}$. (b) Inlet and outlet relative flow angles β_1, β_2 for forced and free vortex designs.

7.2.4 MIXED VORTEX FLOW ACROSS A ROTOR

The previous discussion suggests that by properly blending free and forced vortex distributions, a desirable radial distribution of both the specific work and the degree of reaction can be achieved.

For a mixed vortex distribution:

$$c_{u1} = ar - \frac{b}{r} \tag{7.28}$$

$$c_{u2} = ar + \frac{b}{r} \tag{7.29}$$

The specific work of the rotor is:

$$L(r) = \omega r(c_{u2} - c_{u1}) = 2\omega b = \text{const.} \tag{7.30}$$

The degree of reaction ε_r is:

$$\varepsilon_r = 1 - \frac{a}{\omega} = \text{const.} \tag{7.31}$$

Ensuring that both the specific work $L(r)$ and the degree of reaction ε_r remain constant and under the designer's control is the key to optimizing rotor performance. A constant $L(r)$ guarantees consistent energy transfer to the fluid, while a controllable ε_r allows the designer to fine-tune the balance between energy conversion and efficiency. This flexibility enables the designer to select the most appropriate value of ε_r based on the optimization criteria detailed in the previous chapter.

7.2.5 GENERALIZED VORTEX FLOW ACROSS A ROTOR

Let us generalize the vortex law by assuming:

$$c_{u2} = ar^n + \frac{b}{r}, \; c_{u1} = ar^n - \frac{b}{r} \tag{7.32}$$

The specific work L is:

$$L = 2\omega b = \text{const.} \tag{7.33}$$

thus:

$$\frac{dL}{dr} = \frac{dh_2^0}{dr} - \frac{dh_1^0}{dr} = \frac{d}{dr}[2\omega b] = 0 \Longrightarrow \frac{dh_2^0}{dr} = \frac{dh_1^0}{dr} \tag{7.34}$$

while the degree of reaction becomes:

$$\varepsilon_r = 1 - \frac{ar^{n-1}}{\omega} \tag{7.35}$$

from which it follows that the reaction is constant along the radius only if $n = 1$ (first power law). In non-dimensional terms, if ε_{r_m} is the reaction at the mean radius r_m:

$$\frac{\varepsilon_r(r)}{\varepsilon_{r_m}} = \frac{1 - \mathcal{K}\left(\frac{r}{r_m}\right)^{n-1}}{1 - \mathcal{K}}, \quad \mathcal{K} = \frac{ar_m^{n-1}}{\omega} \tag{7.36}$$

Once again, assuming $dh_1^0/dr = 0$, we derive the axial velocity distributions in 1 and 2 from the radial equilibrium equation:

For station 1:

$$0 = c_{z1}\frac{dc_{z1}}{dr} + a^2(n+1)\,r^{2n-1} - ab(n+1)\,r^{n-2} \tag{7.37}$$

For station 2:

$$0 = c_{z2}\frac{dc_{z2}}{dr} + a^2(n+1)\,r^{2n-1} + ab(n+1)\,r^{n-2} \tag{7.38}$$

or:

$$c_{z(1,2)}\frac{dc_{z1,2}}{dr} = -a^2(n+1)\,r^{2n-1} \pm \frac{ab(n+1)}{r} \tag{7.39}$$

which, upon integration ($n \neq 0$), yields to:

$$c_{z(1,2)}(r) = \sqrt{\pm 2ab(n+1)\ln r - \frac{a^2(n+1)}{n} r^{2n} + C_{(1,2)}} \tag{7.40}$$

For $n = 1$, the equation simplifies to:

$$c_{z1,2}\frac{dc_{z1,2}}{dr} = -2a^2 r \pm \frac{2ab}{r} \tag{7.41}$$

which leads to:

$$c_{z(1,2)} = \sqrt{\pm 4ab\ln r - 2a^2r^2 + C_{(1,2)}} \tag{7.42}$$

In non-dimensional fashion, equation 7.40 becomes:

$$\frac{c_{z(1,2)}}{u_t} = \sqrt{K_{(1,2)} + \left(\frac{2a}{\omega}\right)^2 \left[-\frac{1}{2}\left(\frac{r}{r_t}\right)^2 \pm \frac{b}{ar_t^2}\ln\left(\frac{r}{r_t}\right)\right]} \tag{7.43}$$

where

$$K_{(1,2)} = \frac{1}{\omega^2 r_t^2}(\pm 4ab\ln r_t + C_{(1,2)}) \tag{7.44}$$

7.2.5.1 Calculation of Constants

Constants a, b, and C are derived from the following boundary conditions. Introducing the expression for the area-averaged specific work done by the rotor:

$$\mathfrak{L} = \frac{2}{r_{\text{tip}}^2 - r_{\text{hub}}^2}\int_{r_{\text{hub}}}^{r_{\text{tip}}} L(r)\,r\,dr \tag{7.45}$$

and since $L(r) = \text{const.}$:

$$\mathfrak{L} = 2\omega b \implies b = \frac{\mathfrak{L}}{2\omega} \tag{7.46}$$

The value of a can be determined in two ways, depending on the preference of the designer.

- **1st choice: specification of ε_r.** From equation 7.35 the designer can specify a desired value of reaction ε_r^* at a given radius r^*:

$$a = \omega\frac{1-\varepsilon_r^*}{r^{*\,n-1}} \tag{7.47}$$

- **2nd choice: specification of $c_{u2,\text{tip}}$.** On the other hand, a suitable value of $c_{u2,t} = \frac{b}{r_{\text{tip}}} + ar_{\text{tip}}^n$ can be stipulated based on the desired tip profile loading ψ_{tip}, that is, $c_{u2,\text{tip}} = u_t\psi_{\text{tip}}$, where ψ_{tip} is under the control of the designer (see Chapter 6). The latter condition can be paired with equation 7.45 to obtain:

$$a = \frac{1}{r_{\text{tip}}^n}\left(c_{u2,\text{tip}} - \frac{b}{r_{\text{tip}}}\right) \tag{7.48}$$

The value of C_1 and C_2 (or, equivalently of $K_{(1,2)}$, as per equation 7.44) can be obtained by ensuring that the radial distribution of the axial velocity fulfills the continuity equation upstream and downstream of the rotor:

$$\dot{m} = 2\pi\int_{r_{\text{hub}}}^{r_{\text{tip}}} \rho_1 c_{z1}\,r\,dr = 2\pi\int_{r_{\text{hub}}}^{r_{\text{tip}}} \rho_2 c_{z2}\,r\,dr \tag{7.49}$$

where ρ is constant for an incompressible flow, while:

$$\dot{m} \approx 2\pi \bar{\rho}_2 \int_{r_{\text{hub}}}^{r_{\text{tip}}} c_{z2}\, r\, dr \tag{7.50}$$

and the mean value of fluid density can be obtained from:

$$\bar{\rho}_2 = \frac{\rho_{01}\pi_c^{1/m}}{\left(1+\frac{\gamma-1}{2}\bar{M}_2^2\right)^{\frac{1}{\gamma-1}}}, \quad m = \frac{\gamma\eta_p}{\gamma(\eta_p-1)+1} \tag{7.51}$$

by postulating a reasonable value for $\bar{M}_2$, say < 0.5.

A sample spanwise distribution of flow angles and flow turning in an axial fan rotor designed using a first power law and $\psi_{\text{tip}} = 0.15$ is provided on the left-hand side of Figure 7.4. The corresponding spanwise distribution of velocity components downstream of the rotor is given in the same figure. As a comparison, a free-vortex design for a stage handling the same flow rate and equal overall work is provided on the right-hand side. In the latter case, it is apparent that flow turning is more pronounced at the hub, thus leading to a higher degree of blade twist.

Tuning the values of n and ψ_{tip} makes it possible for the designer to establish a proper blade twist while setting the bases for satisfactory aerodynamic and structural performance. Also, the spanwise distribution of ε_r can be adjusted by acting on n and ψ_{tip}. The spanwise distribution of the reaction for the above designs is also shown, confirming that a constant value can be achieved when a first-power law is adopted.

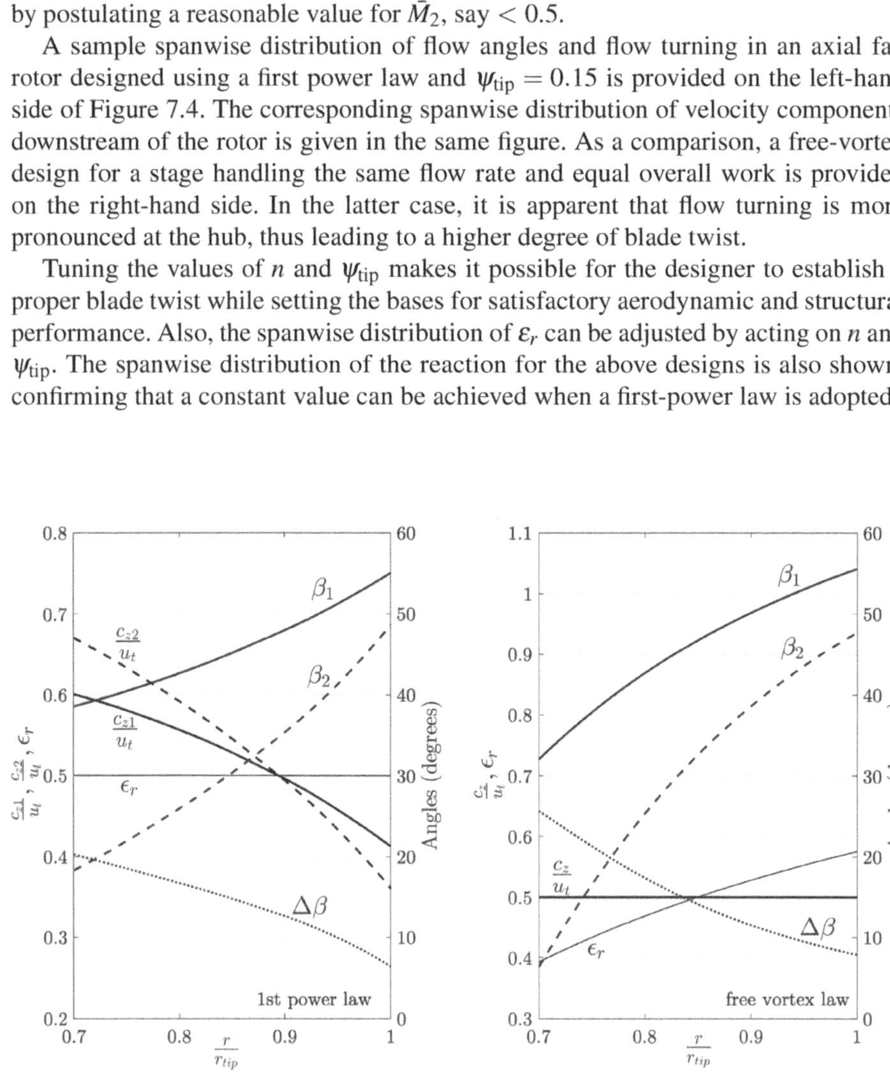

Figure 7.4 Spanwise distribution of velocity components in a first power law and a free-vortex design for a stage handling the same flow rate and equal overall work.

7.2.6 A SPECIAL CASE: GENERALIZED VORTEX FLOW ACROSS A ROTOR WITH AXIAL INLET

This scenario is a common occurrence in single-stage axial machines where the flow at station 1 has no prewhirl, $c_{u1} = 0$. Such cases are frequently observed in axial flow pumps, axial fans, and the first stage of axial compressors that lack inlet guide vanes (IGVs). Under these conditions:

$$c_{u2} = ar^n + \frac{b}{r}, \; c_{u1} = 0 \tag{7.52}$$

The specific work $L(r)$ of the rotor is given by:

$$L(r) = \omega \left(ar^{n+1} + b\right) \tag{7.53}$$

from which:

$$\frac{dL}{dr} = \frac{dh_2^0}{dr} - \frac{dh_1^0}{dr} = \frac{dh_2^0}{dr} - 0 \Longrightarrow \frac{dh_2^0}{dr} = \omega(n+1)ar^n \tag{7.54}$$

and the degree of reaction ε_r is:

$$\varepsilon_r(r) = 1 - \frac{\left(ar^n + \frac{b}{r}\right) + 0}{2\omega r} \tag{7.55}$$

By casting the two preceding equations, it is straightforward to prove that L and the degree of reaction are linked by:

$$L = 2\omega^2 r^2 (1 - \varepsilon_r) \tag{7.56}$$

This indicates that once the specific work is defined as a function of radius, the degree of reaction is inherently determined and cannot be adjusted by the designer.

The axial velocity component downstream of the rotor is derived in the usual way upon integration of the radial equilibrium equations:

$$c_z = \sqrt{-\frac{a^2(n+1)r^{2n}}{n} - \frac{a(n+1)br^{n-1}}{n-1} + C} \;\; (n \neq 1) \tag{7.57}$$

$$c_z = \sqrt{-2a^2r^2 - 4ab\ln(r) + C} \;\; (n = 1) \tag{7.58}$$

In the case where integral rotor work can be specified by the designer:

$$\mathfrak{L} = \frac{2}{r_{\text{tip}}^2 - r_{\text{hub}}^2} \int_{r_{\text{hub}}}^{r_{\text{tip}}} L(r)\, r\, dr \tag{7.59}$$

we derive:

$$\mathfrak{L} = \frac{2\omega a}{(n+3)(r_{\text{tip}}^2 - r_{\text{hub}}^2)} \left(r_{\text{tip}}^{n+3} - r_{\text{hub}}^{n+3}\right) + \omega b \tag{7.60}$$

On the other hand, the specification of the integral mass flow rate allows for an additional relationship between a and b:

$$\dot{m} = 2\pi \int_{r_{\text{hub}}}^{r_{\text{tip}}} \rho_2 c_{z2}\, r\, dr \tag{7.61}$$

where a, b and C can then be derived using a numerical procedure.

7.2.7 PRACTICAL ISSUES ON VORTEX LAWS IN COMPRESSORS

Free-vortex designs are classically associated with high on-design efficiency because they achieve a uniform radial work distribution. Recent studies have shown that free-vortex rotors tend to align well with the ideal Cordier optimum and yield higher efficiencies at small flow coefficients and hub/tip ratios. However, radial non-uniformity in the degree of reaction drives large radial pressure gradients and secondary flows (see next Chapter). In physical terms, excessive tangential velocity at the hub promotes cross-flows and end-wall losses. Thus, designers often moderate the free-vortex twist, for example, using a mild outward increase in swirl that can reduce hub loading.

Stall margin is also sensitive to the vortex distribution. Experiments and CFD studies show that giving the designer extra freedom in the swirl law can significantly extend the stall-free range. In one rotor-only case, a general vortex distribution with optimized exponent n in equation 7.57 markedly increases the stall margin compared to a strict free-vortex baseline. In fact, a properly selected non-uniform radial swirl distribution, which alters the axial velocity profile near the tips or hub, delays flow separation.

7.3 VORTEX LAWS FOR AXIAL TURBINES

Two specific laws are most commonly applied due to their effectiveness in achieving the desired turbine characteristics. The first is again the free vortex law, which assumes that the product of radius and tangential velocity remains constant along r. This approach is beneficial because it naturally minimizes secondary flows and reduces losses. The second commonly used method is a special case where the exit flow angle from the nozzle, α_2 is kept constant across the blade span. This approach simplifies the design process and ensures consistent performance, particularly in action stages.

7.3.1 THE FREE VORTEX TURBINE

Let us recall the specific work generated by a turbine rotor at radius r:

$$L = h_2^0 - h_3^0 = u(c_{u2} + c_{u3}) \tag{7.62}$$

and, according to the working hypothesis of a free vortex, assume $rc_{u2} = C_2$ and $rc_{u3} = C_3$, where subscript 2 denotes the nozzle exit and 3 the rotor exit. As a result:

$$L = \omega(C_2 + C_3) \tag{7.63}$$

is independent of r. Assume the flow entering the nozzle has a uniform stagnation enthalpy, h_2^0, then, according to equation 7.63, the stagnation enthalpy at the rotor outlet, h_3^0, is also constant. Consequently, from the simple radial equilibrium equation 7.13:

$$c_{z2} = \text{constant}, \quad c_{z3} = \text{constant} \tag{7.64}$$

The degree of reaction is expected to vary along r in the following manner:

$$\varepsilon_r(r) = 1 - \frac{C_2 - C_3}{2\omega r^2} \tag{7.65}$$

Let us define $\phi(r) = c_z/(\omega r)$, so that:

$$\tan\alpha_2 = \frac{c_{u2}}{c_z} = \frac{C_2}{\omega r^2 \cdot \phi(r)} \tag{7.66}$$

$$\tan\alpha_3 = \frac{c_{u3}}{c_z} = \frac{C_3}{\omega r^2 \cdot \phi(r)} \tag{7.67}$$

$$\tan\beta_2 = \frac{c_{u2} - \omega r}{c_z} = \frac{C_2 - \omega r^2}{\omega r^2 \cdot \phi(r)} \tag{7.68}$$

$$\tan\beta_3 = \frac{\omega r + c_{u3}}{c_z} = \frac{\omega r^2 + C_3}{\omega r^2 \cdot \phi(r)} \tag{7.69}$$

It is customary to impose $\varepsilon_r \geq 0$ at $r = r_{\text{hub}}$ such that the radial distribution of the reaction obeys (Figure 7.5):

$$\varepsilon_r(r) = 1 - \frac{2\omega r_{\text{hub}}^2}{2\omega r^2} = 1 - \frac{r_{\text{hub}}^2}{r^2} \tag{7.70}$$

Therefore, we solve the system of equations:

1. $L = \omega(C_2 + C_3)$,
2. $\varepsilon_r(r) = 1 - \frac{C_2 - C_3}{2\omega r^2}$,

with the constraint that $\varepsilon_r(r) = 0$ at $r = r_{\text{hub}}$. The solutions are:

$$C_2 = \frac{L}{2\omega} + \omega r_{\text{hub}}^2 \tag{7.71}$$

$$C_3 = \frac{L}{2\omega} - \omega r_{\text{hub}}^2 \tag{7.72}$$

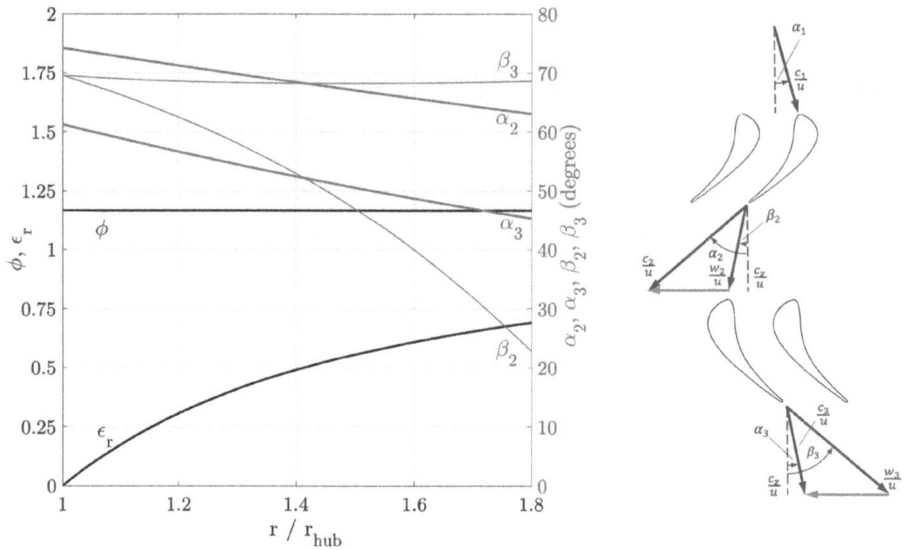

Figure 7.5 A free vortex turbine.

7.3.2 VORTEX FLOW ACROSS A ROTOR WITH UNIFORM INLET α_2

To deal with this case, it is useful to write the radial equilibrium equation 7.4 in terms of absolute velocity at the nozzle exit:

$$c_2 \cos\alpha_2 \frac{d(c_2 \cos\alpha_2)}{dr} + \frac{c_2 \sin\alpha_2}{r} \frac{d}{dr}(rc_2 \sin\alpha_2) = 0 \tag{7.73}$$

from which, assuming α_2 constant:

$$c_2 \frac{dc_2}{dr} + \frac{c_2^2 \sin^2\alpha_2}{r} = 0 \tag{7.74}$$

Rearranging the above gives:

$$\frac{dc_2}{dr} = -\frac{c_2 \sin^2\alpha_2}{r} \implies \frac{dc_2}{c_2} = -\frac{\sin^2\alpha_2}{r} dr \tag{7.75}$$

Integrate both sides from midspan radius r_m to r:

$$\int_{c_2(r_m)}^{c_2(r)} \frac{1}{c_2} dc_2 = -\sin^2\alpha_2 \int_{r_m}^{r} \frac{1}{r} dr \tag{7.76}$$

i.e.

$$\ln\left[\frac{c_2(r)}{c_2(r_m)}\right] = -\sin^2\alpha_2 \ln\left[\frac{r}{r_m}\right] \implies c_2(r) = c_2(r_m)\left(\frac{r_m}{r}\right)^{\sin^2\alpha_2} \tag{7.77}$$

otherwise, in terms of velocity components:

$$c_{z2}(r) = c_2(r)\cos\alpha_2; \quad c_{z2}(r) = c_{z2}(r_m)\left(\frac{r_m}{r}\right)^{\sin^2\alpha_2} \tag{7.78}$$

$$c_{u2}(r) = c_2(r)\sin\alpha_2; \quad c_{u2}(r) = c_{u2}(r_m)\left(\frac{r_m}{r}\right)^{\sin^2\alpha_2} \tag{7.79}$$

Since $u = \omega r$, at the rotor inlet the flow angle β_2 can be determined:

$$\beta_2(r) = \arctan\left(\frac{c_{u2}(r) - \omega r}{c_{z2}(r)}\right) \tag{7.80}$$

Figure 7.6 illustrates the dependence of the dimensionless axial velocity ratio $\frac{c_{z2}}{c_{z2m}}$ and the circumferential velocity ratio $\frac{c_{u2}}{c_{u2m}}$ on the normalized radial position $\frac{r}{r_m}$ for different inlet flow angles $\alpha_2 = 50°, 60°, 70°$. The corresponding variation of the rotor inlet flow angle β_2 is shown on the right axis.

At the rotor outlet, the solution of the radial equilibrium may be obtained for three cases:

- Free vortex rotor at outlet, i.e. $rc_{u3} = C_3$.
- De-swirled rotor, i.e. $\alpha_3 = 0$.
- Constant rotor work output, i.e. $L(r) = L$ is independent of r.

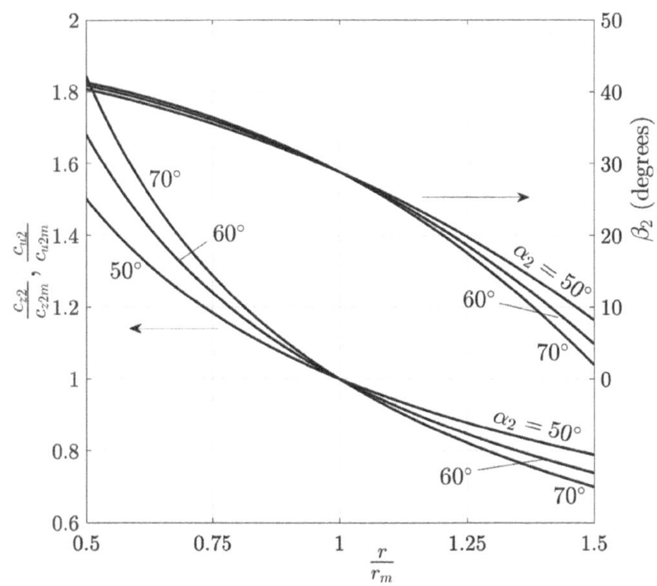

Figure 7.6 Spanwise variation of non-dimensional velocities and flow angle β_2 in a turbine stage with $\alpha_2 = \text{const}$.

Free vortex rotor at outlet. The solution for the flow at the exit of the rotor is given by computing dh_3^0/dr recognizing that $dh_2^0/dr = 0$:

$$\frac{dh_3^0}{dr} - 0 = \frac{dL}{dr} = -\omega\frac{d}{dr}\left[rc_{u2} + rc_{u3}\right] = -\omega c_{u2}(r_m)r_m^{\sin^2\alpha_2}\left(1 - \sin^2\alpha_2\right)r^{-\sin^2\alpha_2} \tag{7.81}$$

from which the radial equilibrium equation applied to station 3 gives ($\frac{d}{dr}(rc_{u3}) = 0$):

$$c_{z3}\frac{dc_{z3}}{dr} = -\omega c_{u2}(r_m)r_m^{\sin^2\alpha_2}\left(1 - \sin^2\alpha_2\right)r^{-\sin^2\alpha_2} \tag{7.82}$$

The final solution for c_{z3} is obtained for $\sin\alpha_2 \neq 1$:

$$c_{z3}(r) = \sqrt{C - 2\omega c_{u2}(r_m)r_m^{\sin^2\alpha_2}r^{1-\sin^2\alpha_2}} \tag{7.83}$$

while:

$$\beta_3 = \arctan\left(\frac{c_{u3}+u}{c_{z3}}\right),\ \alpha_3 = \arctan\left(\frac{c_{u3}}{c_{z3}}\right) \tag{7.84}$$

$$\varepsilon_r(r) = 1 - \frac{rc_{u2} - C_3}{2\omega r^2} \tag{7.85}$$

Figure 7.7 shows the spanwise distribution of quantities obtained in the free vortex rotor at the outlet.

De-swirled rotor. When $c_{u3} = 0$, the expression of $c_{z3}(r)$ is the same as in equation 7.83. Further quantities are calculated by enforcing $c_{u3} = 0$, i.e. $C_3 = 0$:

$$\beta_3 = \arctan\left(\frac{u}{c_{z3}}\right),\ \alpha_3 = 0 \tag{7.86}$$

Constant rotor work output. We need to find c_{u3} such that $L = \omega r(c_{u3} + c_{u2}(r))$ is independent of r, i.e. $dL/dr = 0$:

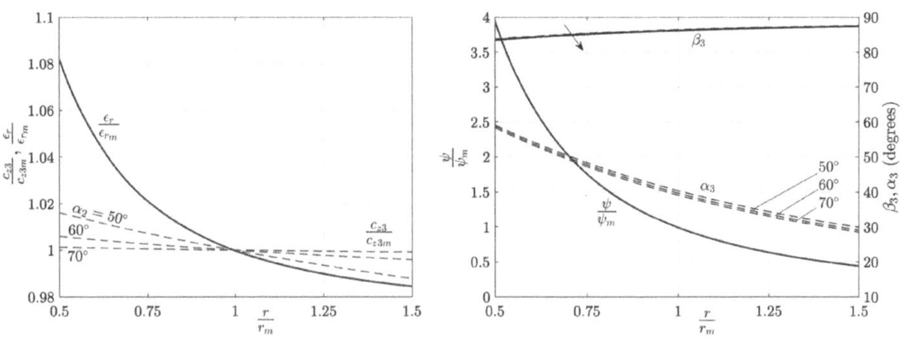

Figure 7.7 Spanwise variation of non-dimensional velocities, non-dimensional degree of reaction, non-dimensional work coefficient, and flow angles in a turbine stage with $\alpha_2 =$ const. and free vortex rotor at outlet.

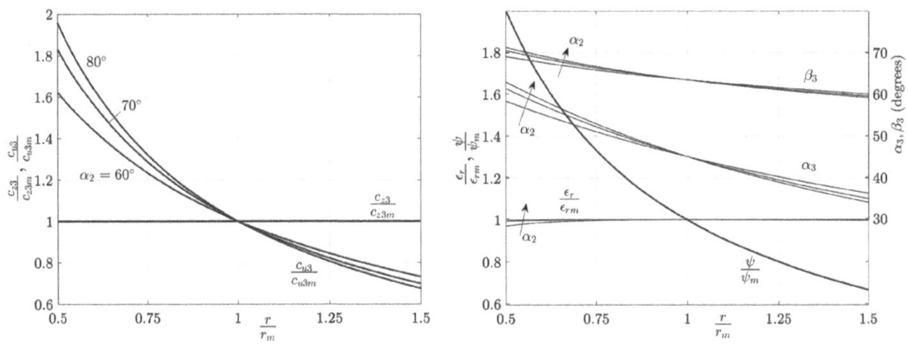

Figure 7.8 Spanwise variation of non-dimensional velocities, non-dimensional degree of reaction, non-dimensional work coefficient, and flow angles in a turbine stage with $\alpha_2 = $ const. and $L = $ const.

$$\frac{d}{dr}(rc_{u3}) = -\frac{d}{dr}(rc_{u2}) \tag{7.87}$$

Since:

$$\frac{d}{dr}(rc_{u2}) = c_{u2}(r_m)\left(\frac{r_m}{r}\right)^{\sin^2\alpha_2}(1+\sin^2\alpha_2) \tag{7.88}$$

from equations 7.87 and 7.88, integrating both sides yields:

$$c_{u3}(r) = -c_{u2}(r_m)\frac{(1+\sin^2\alpha_2)}{1-\sin^2\alpha_2}\left(\frac{r_m}{r}\right)^{\sin^2\alpha_2} \tag{7.89}$$

Next, from:

$$c_{z3}\frac{dc_{z3}}{dr} + \frac{c_{u3}}{r}\frac{d}{dr}(rc_{u3}) = 0 \tag{7.90}$$

after some steps:

$$c_{z3}(r) = \sqrt{2}c_{u2}(r_m)\frac{(1+\sin^2\alpha_2)}{\sqrt{\sin^2\alpha_2(1-\sin^2\alpha_2)}}\left(\frac{r_m}{r}\right)^{\sin^2\alpha_2} \tag{7.91}$$

Figure 7.8 shows the spanwise distribution of quantities obtained in constant rotor work output.

7.3.3 PRACTICAL ISSUES ON VORTEX LAWS IN TURBINES

As for compressors, also in turbines free-vortex design blade turning varies significantly along the span and results in highly twisted blades with large differences between hub and tip incidence angles. In contrast, constant-reaction vortex laws produce nearly symmetric inlet and outlet velocity triangles, which lead to limited blade twist and more uniform turning.

In practical turbine design, a mean-span velocity diagram corresponding to approximately 50% reaction is often selected to maximize aerodynamic efficiency. This results in near-impulse conditions at the hub and highly reaction-loaded flow near the tip. Such vortex distribution laws are designed to minimize secondary flow losses by reducing radial pressure gradients and ensuring optimal loading across the span.

Mechanical and manufacturing constraints can limit the application of ideal vortex profiles. For example, the use of constant-chord blades, favored for their structural simplicity and ease of cooling, can deviate significantly from the ideal flow dictated by radial equilibrium. This mismatch often leads to increased endwall losses and reduced spanwise efficiency due to suboptimal incidence and diffusion characteristics near the hub and casing.

7.4 PROBLEMS

1. An axial compressor rotor is designed under the assumption of free vortex flow with uniform inlet total enthalpy. The design parameters are $\psi_{\text{tip}} = 0.33$, $r_{\text{tip}} = 0.8\,\text{m}$, $r_{\text{hub}} = 0.48\,\text{m}$, $N = 3500\,\text{rpm}$, $\phi = c_m/u_{\text{tip}} = 0.55$, and $c_{u1,\text{tip}}/u_{\text{tip}} = 0.15$. Determine the tangential velocities $c_{u2,\text{tip}}$ and $c_{u2,\text{hub}}$ after the rotor.

   [141 m/s, 235 m/s]

2. An axial compressor rotor is designed under the assumption of free vortex flow with uniform inlet total enthalpy. The tip radius is $r_{\text{tip}} = 1000\,\text{mm}$, the rotational speed is $N = 3000\,\text{rpm}$, and the flow coefficient is $\phi = 0.5$. The specific work coefficient at the tip is $\psi_{\text{tip}} = 0.4$, and the non-dimensional tangential velocity at the inlet tip is $c_{u1,\text{tip}}/u_{\text{tip}} = 0.1$. Determine the hub radius r_{hub} such that the relative flow angle at hub exit $\beta_{2,\text{hub}}$ is zero. Also, calculate the degree of reaction at both the hub and tip, as well as the relative flow angles $\beta_{1,\text{tip}}$, $\beta_{1,\text{hub}}$, $\beta_{2,\text{tip}}$, and $\beta_{2,\text{hub}}$.

   [0.71 m, 0.7, 0.4, 61°, 49°, 45°, 0°]

3. Determine the hub and tip radii of an axial compressor rotor designed using the free vortex criterion. The rotor is positioned downstream of a set of inlet guide vanes (IGVs) that also follow the free vortex criterion and receive calm air from an open chamber. The specific work coefficient at the tip is $\psi_{\text{tip}} = 0.4$. The rotational speed is $N = 3000$ rpm, and the flow coefficient is $\phi = 0.5$. The target conditions are on the volumetric flow rate at rotor inlet $\dot{m}/\rho_1 = 68\,\text{m}^3/\text{s}$ and on the degree of reaction at the mean radius $\varepsilon_r(r_m) = 0.5$.

   [66 mm, 120 mm]

4. An axial flow pump rotor handles water ($\rho = 1000\,\text{kg/m}^3$) entering without preswirl $c_{u1} = 0$. The pump's head is $h = 8\,\text{m}$ and the flow rate is $Q = 1.55\,\text{m}^3/\text{s}$. The rotational speed is $N = 540\,\text{rpm}$, and the expected flow coefficient is $\phi = 0.5$. Determine the blade angles of the rotor at both the hub and tip under the following assumptions: (i) a free vortex distribution for the tangential velocity component c_u at the rotor exit, (ii) a blade tip speed of $u_t = 24\,\text{m/s}$, (iii) a linear variation in cascade solidity with radius given by $\sigma = 1.7 - 0.1r$ r in meters), and (iv)

blade sections following a parabolic-arc camberline with $a/\ell = 0.4$. The expected hydraulic efficiency of the rotor is $\eta_{\text{hyd}} = 0.92$.

[53.5°, 58.6°, 63.8°, 65.9°]

5. The shroud of an existing axial flow pump has an internal diameter of $d_e = 1.2$ m and is set to accommodate a retrofitted rotor. Both the original and retrofitted rotors must ensure a theoretical head of $h_t = 13.54$ m and a flow rate of $Q = 16.05\,\text{m}^3/\text{s}$ at a rotational speed of $N = 450$ rpm. The original rotor features a hub-to-tip diameter ratio of 0.6 and a blade arrangement based on the free vortex design criterion. The new rotor, designed using the forced vortex criterion, may have a different hub diameter, potentially targeting a flow coefficient of $\phi = 0.7$. Determine the relative flow angles upstream and downstream of both rotors. Additionally, compare and comment on the differences in blade geometry between the two designs.

[Original rotor: $\beta_{1,2-\text{hub}} = 37.4°, 22.4°$, $\beta_{1,2-\text{tip}} = 51.9°, 46.75°$; retrofitted rotor: $\beta_{1,2-\text{hub}} = 37°, 35.4°$, $\beta_{1,2-\text{tip}} = 55.6°, 43.7°$]

6. A 0.6 m-diameter axial flow compressor stage handling 32 kg/s of air has a hub-to-tip diameter ratio 0.75 and is to be designed with a constant nominal degree of reaction equal to 50% using a first power vortex law. The aimed polytropic efficiency of the rotor is 0.95. Air at the rotor inlet has $T_1^0 = 300$ K, $p_1^0 = 110$ kPa and $\rho_1 = 1.25$ kg/m^3. The blade velocity is 300 m/s, and the total temperature rise across the rotor is $\Delta T_0 = 16.1$ K. Determine the axial velocity profile before and after the rotor. The air properties are $c_p = 1.005$ kJ/(kg·K) and $\gamma = 1.4$. (Assume $p_2 \approx p_2^0$ and $T_2^0 \approx T_2$).

7. A repeated reaction turbine stage handling combustion gases $\gamma = 1.33$ and $R = 290$ J/(kg·K)) is to be designed for an expansion ratio of 5 $\pi_e = p_1^0/p_3^0$ based on the free vortex criterion. The flow coefficient is 0.54, and the rotor operates at an angular speed of 1400 rad/s. The expected stage polytropic efficiency is 0.88, with a total inlet temperature of $T_1^0 = 650$ K. Determine the minimum hub diameter that satisfies these design constraints while ensuring that the absolute velocity angle at the hub exit of the nozzle vanes is $\alpha_{2,\text{hub}} = 66°$. Additionally, calculate the relative velocity angles at the hub.

[0.47 m, 21.5°, 21.5°]

NOTE

1. We refer to this quantity as *nominal* since the *actual* degree of reaction is a function of both c_{z1} and c_{z2}, which are no longer equal.

8 Three-Dimensional Flow Losses

In the previous chapters, we have developed an understanding of the behavior of flow in turbomachinery under simple conditions. We concentrated on two-dimensional models of cascades since they allowed for the analysis of losses without the additional complication of three-dimensional effects.

The limitations of ideal flow and the purely two-dimensional hypothesis will be removed in this chapter in order to examine the more complex loss mechanisms that result from the actual fluid flows in turbomachines being essentially three-dimensional. In contrast to simpler models, real machines exhibit flow phenomena that are far from two-dimensional, including boundary layer growth, tip leakage, spanwise variations, and secondary flows. A thorough understanding of the non-uniform flow distribution in the radial and circumferential directions, as well as the three-dimensional interactions between the rotor and stator, is necessary to properly characterize the performance of turbomachinery.

In essence, endwall effects and tip leakage flow form three-dimensional losses. Unlike the two-dimensional losses observed in cascades, these three-dimensional losses are closely tied to the actual geometry of the machine and its working conditions. This often results in physical models that lack generality and theoretical rigor.

To offer a complete view of overall losses and an accurate prediction model of machine performance, three-dimensional and two-dimensional losses must be combined.

8.1 GENERAL PICTURE OF 3D FLOW LOSSES

Endwall losses and tip leakage losses are the two main types of three-dimensional flow losses (Figure 8.1). It is crucial to understand that these losses are not always easily separable, as they are frequently connected and impacted by the dynamics of the flow as a whole.

- **Endwall losses**. Occurring in the boundary layer regions near the casing and hub, endwall losses are closely linked to the interaction between the boundary layer over the endwalls and the bladings. The thickening of the boundary layer along the walls and its interaction with the main flow cause flow blockage and additional frictional losses, which are exacerbated by the curvature and pressure gradients near the walls. These losses, which result in passage and horseshoe-shaped vortices, are caused by pressure differentials and flow deflection within the blade passages. The result is a considerable loss of energy and a decline in efficiency, particularly in the vicinity of the hub and shroud areas.

 DOI: 10.1201/9781003619956-8

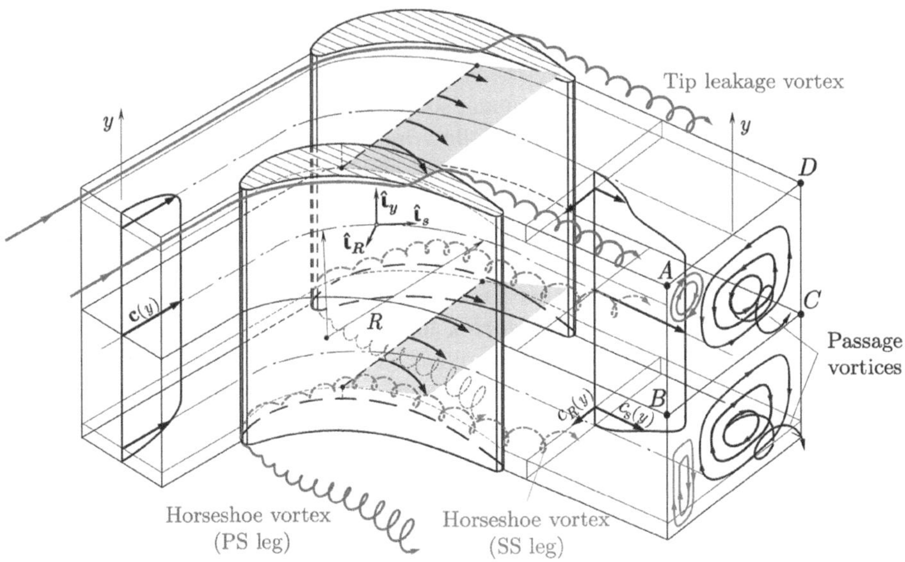

Figure 8.1 General picture of flow losses in a rotor passage.

- **Tip leakage losses.** In rotor blades, a finite pressure difference occurs between the pressure and suction sides, which drives flow across the blade tip, forming a leakage vortex causing substantial energy loss—especially in high-pressure stages. The resulting vortex structures interfere with secondary flows and boundary layers and add to the overall complexity of loss in the flow field.

The nature of these losses is strongly influenced by the compressibility of the flow. As the Mach number relative to the blade approaches unity, compressibility effects become increasingly significant and additional losses emerge as a result of the formation of shock waves. These shocks not only cause a direct loss of energy but also intensify the already complex three-dimensional flow structures, further contributing to dissipation.

An additional complexity arises when the turbomachine operates off-design, which greatly impacts the magnitude of the losses.

8.2 ENDWALL LOSSES

Endwall losses are often referred to as *secondary losses* not because they are of secondary importance, but because they are associated with secondary flows that are distinct from the primary one. Despite their designation, these losses are very significant due to their impact on machine performance.

Secondary flow losses are primarily driven by the interaction between the pressure gradients across the blade passages and the three-dimensional nature of real viscous flows. These losses can be broken down into two main components: *passage vortex losses* and *stagnation point* (or *horseshoe*) *vortex losses*. Other sources of secondary flows may generate additional vorticity, but these are typically of minor significance.

- **Passage vortex losses**. The dominant source of secondary flow losses is the formation of passage vortices. As the flow moves through the blade passage, the pressure difference between the pressure and suction sides induces a swirling motion. This vortex forms near the endwalls and grows in strength as it moves downstream, entraining low-energy fluid and creating regions of high dissipation. Passage vortices are especially detrimental near the hub and casing, where they disrupt the main flow and increase mixing losses.
- **Stagnation point (horseshoe) vortex losses**. Stagnation or horseshoe vortices form around the leading edge of the blade as the flow approaches the stagnation point. The flow splits into two streams, one moving along the pressure side and the other along the suction side of the blade. This creates a vortex pair that wraps around the blade's leading edge and moves downstream. These vortices, while initially small, interact with the passage vortices and contribute to the overall secondary flow losses by increasing flow separation and boundary layer thickness near the endwalls.
- **Viscous losses on hub and casing**. These losses are caused by the development of boundary layers on the hub and casing surfaces, which grow thicker as the flow progresses through the machine. The frictional resistance caused by these boundary layers results in a loss of flow momentum and energy, contributing to a reduction in overall efficiency. On the hub, viscous losses can be exacerbated by the presence of secondary flows and vortex formations that interact with the boundary layer. Similarly, on the casing, the development of a boundary layer and its interaction with the flow can cause additional pressure drops and energy dissipation.

Each of these secondary flow mechanisms is interconnected, and their combined effect significantly alters the flow field within the blade passage, thus leading to increased energy losses.

8.2.1 PASSAGE VORTEX

To understand the development of passage vortex losses, we can draw an analogy to the behavior of an incompressible viscous flow moving through a curved bend, as shown in Figure 8.1. Consider a fluid particle entering a curved channel with a uniform velocity, denoted as c_1. The particle is assumed to follow a curved path with a radius of curvature R, mimicking the streamline within a blade passage in a turbomachine.

For simplicity, we assume that the flow is incompressible and that the meridional velocity component, c_{m1}, remains constant along the entire flow path. This is reasonable outside the boundary layer where the flow remains largely inviscid. Using the equilibrium equation along direction R, we obtain:

$$\frac{1}{\rho}\frac{\partial p}{\partial R} = \frac{|\mathbf{c}|^2}{R} \tag{8.1}$$

where $|\mathbf{c}|$ is the magnitude of the velocity of the fluid particle tangent to its streamline.

This equation implies that the centrifugal effect, $|\mathbf{c}|^2/R$, balances the pressure gradient $\frac{\partial p}{\partial R}$ in the same direction. In this scenario, the flow outside the boundary layer follows this equilibrium, which leads to a pressure distribution across the channel. As the velocity $|\mathbf{c}|$ decreases within the boundary layer, the same pressure gradient $\frac{\partial p}{\partial R}$ must still be maintained (since the pressure remains uniform within the layer). Equation 8.1 allows a fluid particle inside the boundary layer to satisfy this by following a path with a smaller radius of curvature R.

Hence, fluid particles in the upper and lower boundary layers tend to move towards the convex side of the channel, resulting in a spiral motion of the flow at the exit of the bend, with the directions indicated in Figure 8.1. Since these directions are opposite on the lower and upper walls, two secondary vortices with opposite curls will form.

To prove this, let us consider a steady Euler flow along a curved streamline with radius curvature R. Assume a velocity field given by $\mathbf{c} = |\mathbf{c}|\hat{\mathbf{i}}_s = c_s\hat{\mathbf{i}}_s$:

$$(\mathbf{c} \cdot \nabla)\mathbf{c} = -\frac{1}{\rho}\nabla p \tag{8.2}$$

Expanding this into curvilinear coordinates, we have two components:

$$c_s\frac{\partial c_s}{\partial s} = -\frac{1}{\rho}\frac{\partial p}{\partial s}, \quad \text{along streamline } s \tag{8.3}$$

$$\frac{c_s^2}{R} = \frac{1}{\rho}\frac{\partial p}{\partial R}, \quad \text{perpendicular to the streamline along } R \tag{8.4}$$

The full Euler equation becomes:

$$c_s\frac{\partial c_s}{\partial s}\hat{\mathbf{i}}_s + \frac{c_s^2}{R}\hat{\mathbf{i}}_R = -\frac{1}{\rho}\left(\frac{\partial p}{\partial s}\hat{\mathbf{i}}_s + \frac{\partial p}{\partial R}\hat{\mathbf{i}}_R\right) \tag{8.5}$$

In such a case, the vorticity $\nabla \times \mathbf{c}$ has 0 value component along $\hat{\mathbf{i}}_s$:

$$\Omega_s = 0 \tag{8.6}$$

If we now assume the presence of a boundary layer in the y-direction and the effect of centripetal acceleration, the velocity field becomes

$$\mathbf{c} = c_s(y)\hat{\mathbf{i}}_s + c_R(y)\hat{\mathbf{i}}_R \tag{8.7}$$

where both c_s and c_R are functions of y, and ∇p remains the same across the boundary layer. The vorticity vector becomes:

$$\Omega = \left(\frac{\partial c_R}{\partial y}\right)\hat{\mathbf{i}}_s + \left(\frac{c_s}{R}\right)\hat{\mathbf{i}}_y - \left(\frac{\partial c_s}{\partial y}\right)\hat{\mathbf{i}}_R \tag{8.8}$$

as c_R does not depend on s and there is no explicit variation of c_s with R. This concludes the proof and confirms that, along the curved bend, a vorticity of local intensity $\frac{\partial c_R}{\partial y}$ will form which will eventually accumulate and lead to a finite circulation (or vortex strength)[1]:

$$\Gamma = \int_S \left(\frac{\partial c_R}{\partial y}\right) dA \tag{8.9}$$

The vorticity magnitude of the secondary vortices requires detailed knowledge of the velocity distribution in the boundary layer and the flow conditions through the channel, which can be obtained through boundary layer theory, numerical simulations, or experimental data. Therefore, its analytical determination, although obtainable using Stokes' theorem, is impractical.

Since the early stages of compressor and turbine analyses, it has been a common practice to account for the effects of secondary flows by introducing an additional equivalent drag coefficient. This approach allows secondary losses to be modeled and incorporated into overall performance assessments. The following empirical expression:

$$C_{D,s} = 0.018 C_L^2 + 0.020 \frac{s}{H} \tag{8.10}$$

captures the primary dependencies of secondary losses on flow deflection and blade aspect ratio, where C_L is the lift coefficient, s is the blade spacing, and H is the blade height. This formula highlights that secondary losses are more pronounced in blades with high camber, where flow deflection is greater, and in configurations with lower aspect ratios, where $\frac{s}{H}$ is larger. Consequently, secondary flows tend to be more intense for highly loaded blades with low aspect ratios.

8.2.2 HORSESHOE VORTEX LOSSES

The formation of the horseshoe vortex begins as the incoming boundary layer near the endwall, which has less momentum than the free-stream flow, encounters the stagnation region at the leading edge of the blade. The flow around the leading edge is subjected to a strong pressure gradient and intense curvature changes, causing the flow to separate and turn into vortices under a similar physical mechanism that governs the formation of the passage vortex. These vortices move along the blade passage and interact with the main flow, contributing to secondary flow structures such as the passage vortex.

The two vortex legs behave differently as they travel along the blade passage. The leg on the suction side tends to stay close to the blade surface, while the leg on the pressure side, due to the strong pitch-wise pressure gradient, is pushed toward the suction side of the adjacent blade. As the blade's turning and loading increase, this separated vortex leg is redirected toward the suction surface earlier in the flow. Once it reaches the suction surface, the fluid is pushed upward to higher spans and carried along the blade by the mainstream flow. This process continues until it reaches the trailing edge, where the horseshoe vortex creates a region of low total pressure near the junction of the suction surface and the endwall. The interaction between the horseshoe vortex and the main flow induces a complex three-dimensional flow pattern, which leads to additional aerodynamic losses. A simplified model is frequently used to estimate horseshoe vortex losses by analyzing the behavior of the boundary layer as it encounters an obstacle mimicking a blade leading edge. The model simplifies the geometry by considering a flat plate onto which a cylindrical leading edge is mounted. Assuming a fully turbulent boundary layer developing on the flat plate, the boundary layer thickness δ at a streamwise location x is given by Schlichting's empirical formula[2]:

$$\delta(x) = 0.37x\left(\frac{x\,U_\infty}{\nu}\right)^{-1/5} \tag{8.11}$$

where U_∞ is the freestream velocity and ν is the kinematic viscosity of the fluid. By integrating the momentum deficit caused by the separation at the bottom of the cylinder and vortex formation, the model estimates the total pressure loss induced by the horseshoe vortex as:

$$\frac{\Delta p_H^0}{\frac{1}{2}\rho U_\infty^2} \approx 0.054 \tag{8.12}$$

This value implies that the horseshoe vortex contributes a loss equivalent to about 5.4% of the local dynamic pressure; therefore, its direct contribution to total aerodynamic losses is modest in comparison to other sources.

8.2.3 VISCOUS LOSSES ON HUB AND CASING

Increased flow resistance and flow blockage result from the thickening of the boundary layer that forms along the endwalls (Figure 8.2). This causes an increase in frictional resistance and restricts the effective flow area, thereby increasing energy losses.

These losses are further intensified by the curvature of the endwalls and the presence of pressure gradients, particularly near the walls. The combination of adverse pressure gradients and wall curvature promotes the formation of secondary flows, such as vortices, which further disturb the main flow and contribute to energy dissipation. The interaction between these vortices and the boundary layer heightens the overall complexity of the flow in these regions, leading to more significant endwall losses and reduced aerodynamic performance.

8.2.3.1 Endwall Flow Blockage

Flow blockage generally refers to the reduction of the effective flow area caused by the thickening of the boundary layer. This phenomenon is especially significant in compressors, where boundary layer growth is exacerbated by adverse pressure gradients that naturally arise during the compression process. In multistage compressors, the accumulation of boundary layer growth and secondary flows exacerbates flow blockage, restricting the available flow area and degrading performance along the flow path.

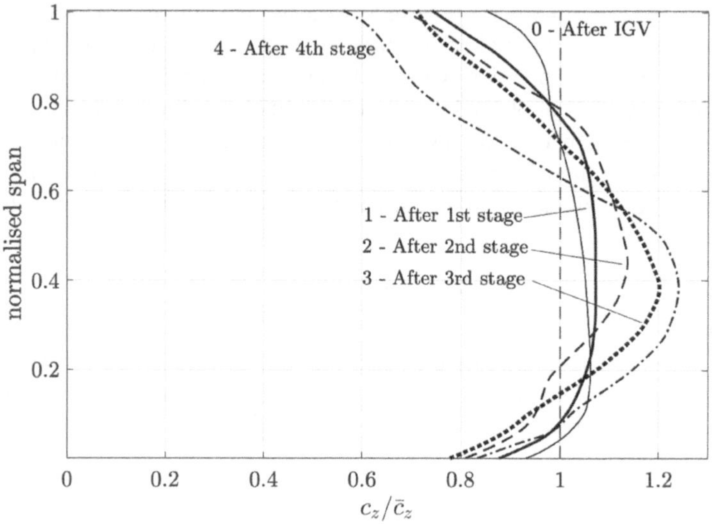

Figure 8.2 Spanwise profiles of normalized axial velocity $c_z/\bar{c}_z$ at various axial locations in a multi-stage axial compressor.

To account for these effects, the concepts of *blockage factor* and *work done factor* are introduced.

Blockage factor. The blockage factor ξ is defined as the ratio of the reduced effective flow area to the nominal, or geometric, area. It can be expressed mathematically as:

$$\xi = \frac{A_{\text{blocked}}}{A_{\text{nominal}}} \tag{8.13}$$

To determine the effective flow area A_{blocked}, we subtract the area occupied by the boundary layers from the nominal flow area:

$$A_{\text{blocked}} = A_{\text{nominal}} - \Delta A_{\text{boundary}} \tag{8.14}$$

where $\Delta A_{\text{boundary}}$ is the reduction in area due to boundary layers, which can be computed using semi-empirical models. The effective flow area A_{blocked} is in fact the portion of the nominal flow area that remains available for mass flow after accounting for the displacement effect of boundary layers. As such, it represents the area that an equivalent inviscid flow would occupy to carry the same mass flow rate as the actual viscous flow.

Work done factor. In addition to blockage, endwall boundary layer growth in compressors causes the axial velocity to be not uniform along the blade height. As a consequence, the axial velocity profile becomes increasingly peaked as the flow moves from the first to subsequent stages. By the fourth stage, the profile generally stabilizes, showing minimal changes beyond this point. The *work done factor* λ_w measures the actual work capacity of a stage compared to its ideal value, as predicted by the Euler equation. In particular, it is defined as the ratio of the actual work done by the stage to the ideal work done that would be achieved if no blockage existed:

$$\lambda_w = \frac{L_{\text{actual}}}{L_{\text{ideal}}} \Longrightarrow L_{\text{actual}} = \lambda_w u \Delta c_u \tag{8.15}$$

The work done factor typically ranges from 0.96 in the first stage to about 0.85 in the fourth and subsequent stages.

8.3 TIP LEAKAGE VORTEX LOSSES

A leakage flow enters the tip gap from the pressure side, separates from the blade tip and forms a jet with a contraction coefficient of around 0.6 (thus forming a *vena contracta*). This phenomenon occurs both in compressors and turbines, though it is far more intense in the latter. The leakage jet mixes with the main flow near the suction surface, creating a vortex that contributes to the losses (Figure 8.1). The total pressure drop is difficult to compute without simplistic assumptions. In the case of incompressible flow, we can estimate the velocity of the flow through the tip clearance v_{tip} based on the pressure difference Δp across the blade and the fluid density ρ:

$$\frac{1}{2}\rho v_{\text{tip}}^2 = \Delta p \tag{8.16}$$

where ρ is the fluid density, v_{tip} the velocity of the flow through the tip clearance and Δp the pressure difference between the pressure side and suction side of the blade. Rearranging this equation to solve for v_{tip} and introducing a discharge coefficient C_d, which accounts for the flow losses (typical values are less than 1, often around 0.8 for compressors and 1 for turbines), we get:

$$v_{\text{tip}} = C_d \sqrt{\frac{2\Delta p}{\rho}} \tag{8.17}$$

The velocity v_{tip} can be used in conjunction with the Borda–Carnot equation to estimate the loss of head due to mixing and expansion in the tip gap. The Borda–Carnot equation for total pressure loss Δp_L^0 with the contraction coefficient $\chi \approx 0.6$ is given by:

$$\Delta p_L^0 = \rho \frac{(1-\chi^2)v_{\text{tip}}^2}{2} = (1-\chi^2)C_d^2 \cdot \Delta p \tag{8.18}$$

Although similar in principle to turbines, leakage flow in unshrouded compressor blades differs due to thinner blades and larger tip gaps. In compressors, the leakage jet often does not reattach to the blade within the tip gap, leading to lower pressure recovery. The mixing of the leakage flow with the mainstream results in significant total pressure losses, particularly in decelerating flow regions near the suction surface.

8.4 SEMI-EMPIRICAL CORRELATIONS FOR 3D FLOW LOSSES

Semi-empirical correlations for three-dimensional (3D) flow losses in compressor and turbine stages are widely used in turbomachinery design and analysis. These correlations attempt to describe the complexities of 3D flows by relying on approximate or schematic representations of real-flow physics. As a result, they often lack theoretical rigor and are not grounded in first-principles analysis. However, despite their simplified nature, loss correlations are extensively employed by manufacturers due to their proven ability to provide reasonable performance predictions for both on- and off-design operating conditions.

This approach remains the only practical method for obtaining rapid assessments of turbomachinery performance, making it invaluable for a variety of applications such as gas path design, preliminary optimization, and performance analysis. Several formulations for loss correlations exist in the open literature, incorporating both algebraic models and look-up tables derived from empirical data. Although the number of models is vast, this discussion will focus on a selected subset, recognizing that none of the available correlations can be universally "best." Each has its own advantages and limitations, and their use must be carefully and critically evaluated based on the specific requirements of the application at hand.

8.4.1 COMPRESSORS

One of the most commonly used sets of formulations is summarized by Tournier and El-Genk (2010),[3] who collected and revisited previous approaches suggested by various experts. All losses are quantified in terms of total pressure loss coefficients between inlet (1) and outlet (2) of a generic blading, i.e.:

$$Y = \frac{p_1^0 - p_2^0}{p_1^0 - p_1}, \text{ for stators} \quad Y = \frac{p_{w1}^0 - p_{w2}^0}{p_{w1}^0 - p_1}, \text{ for rotors} \tag{8.19}$$

8.4.1.1 Endwall Losses: Viscous Losses on Hub and Casing

Endwall losses are often assessed by assuming incompressible flow in high-radius-ratio compressors with repeating stages by analyzing the efficiency loss due to the effects of the boundary layer near the hub and casing. These losses are related to the key properties of the endwall boundary layers, in particular to the displacement thickness, δ^*.

Although representative values of δ^* are deduced from measurements of flow profiles in several low-speed multistage compressors, no universal correlation has proven to be sufficiently reliable for low-to-moderate aspect ratios (down to approximately unity). Some authors suggest the following correlation:

$$Y_{h,c} = 0.0146 \frac{\ell}{H} \left(\frac{\cos\beta_1}{\cos\beta_2} \right)^2 \tag{8.20}$$

where H is the blade height. An analogous expression can be found for stators.

8.4.1.2 Endwall Losses: Passage and Horseshoe Vortex Losses

An overall loss coefficient due to flow vorticity in rotors is given as:

$$Y_s = 0.018\sigma_R \frac{\cos^2\beta_1}{\cos^3\beta_\infty} C_L^2 \tag{8.21}$$

where the ideal blade lift coefficient, C_L, is expressed as:

$$C_L = C_{L,R} = \frac{2\Delta w_u}{w_\infty \sigma_R} \tag{8.22}$$

Analogous expressions can be found for stator blades.

8.4.1.3 Tip Leakage Vortex Losses

Due to the scarcity of data for compressors, a loss factor Y_{TC} can be calculated using the correlations developed for turbines, where two loss components are identified, namely the loss due to tip leakage and the loss within the tip clearance:

$$Y_{TC} = Y_{\text{tip}} + Y_{\text{gap}} \tag{8.23}$$

where:

$$Y_{\text{tip}} = 1.4\, K_E \sigma_R \frac{\varepsilon}{H} \frac{\cos^2\beta_1}{\cos^3\beta_\infty} C_L^{1.5} \tag{8.24}$$

$$Y_{\text{gap}} = 0.0049\, K_G \sigma_R \frac{\ell}{H} \frac{\sqrt{C_L}}{\cos\beta_\infty} \tag{8.25}$$

ε being the tip clearance, and, for mid-loaded compressor blades ($x_A/\ell = 0.5$), $K_E = 0.5$ and $K_G = 1.0$. Once again, analogous expressions can be found for stator blades.

8.4.2 TURBINES

Again, one of the most comprehensive and still widely used sets of correlations is those summarized by Tournier and El-Genk. 3-D effects are incorporated into endwall losses (primarily related to secondary flows outside the tip region) and tip clearance vortex losses, which affect the rotor only.

8.4.2.1 Endwall Losses

Endwall losses are estimated in terms of pertinent total pressure losses to be computed for both nozzles and rotors:

$$Y_{\text{end},R} = F_{AR} \cdot \frac{0.038 + 0.41 \cdot \tanh(1.2\delta^*/H)}{\sqrt{\cos\xi} \cdot (\cos\beta_1/\cos\beta_2) \cdot (\cos\beta_2/\cos\xi)^{0.55}} \tag{8.26}$$

where the aspect ratio factor F_{AR} is a function of the blade aspect ratio as:

$$F_{AR} = \left(\frac{\ell}{H}\right)^{0.55}, \quad \text{when } H/\ell < 2.0 \tag{8.27}$$

$$F_{AR} = 1.36604 \cdot \left(\frac{\ell}{H}\right), \quad \text{when } H/\ell > 2.0 \tag{8.28}$$

In equation 8.26, the boundary layer displacement thickness at the inlet endwall, δ^*, is calculated as:

$$\delta^* = \frac{0.0463\, l_{\text{ref}}}{(\rho_1 w_1 l_{\text{ref}}/\mu_1)^{0.2}} \tag{8.29}$$

where the reference length, l_{ref}, is taken as half the blade axial chord.

8.4.2.2 Tip Clearance Vortex Losses

The loss coefficient due to the formation and dissipation of the tip clearance vortex, Y_{TC}, is calculated as previously shown:

$$Y_{TC} = Y_{\text{tip}} + Y_{\text{gap}}, \tag{8.30}$$

where

$$Y_{\text{tip}} = 1.4 K_E \sigma_R \frac{\varepsilon}{H} \frac{\cos^2\beta_2}{\cos^3\beta_\infty} C_L^{1.5} \tag{8.31}$$

and

$$Y_{\text{gap}} = 0.0049\, K_G \sigma_R \frac{\ell}{H} \frac{\sqrt{C_L}}{\cos\beta_\infty} \tag{8.32}$$

For mid-loaded turbine blades, $K_E = 0.5$ and $K_G = 1.0$, and for front- or aft-loaded blades, $K_E = 0.566$ and $K_G = 0.943$.

8.5 OVERALL STAGE LOSSES

8.5.1 COMPRESSORS

Figure 8.3 presents a performance efficiency plot of a 3D compressor stage, illustrating the impact of different loss mechanisms on total stage isentropic efficiency $\eta_{is,\text{tot}}$ as a function of the normalized flow coefficient ϕ/ϕ^*. Annulus losses in this context are interpreted as viscous losses arising from the formation and evolution of boundary layers on both the hub and shroud surfaces. These losses are primarily influenced by the flow coefficient, which serves as a proxy for the flow rate. As the flow rate increases, viscous losses tend to rise approximately with the square of the velocity, similar to what is observed in pipe flow. This explains why efficiency initially remains high at moderate flow rates but begins to decline at higher flow rates due to the increasing impact of annulus losses.

Secondary or end-wall losses include both tip-clearance losses and losses due to secondary flow structures such as passage and horseshoe vortices. Since these losses are, at a first approximation, dependent on blade loading, they tend to decrease with increasing flow rate. This behavior is attributed to the reduction in pressure ratio at higher flow rates, which in turn weakens the pressure difference driving secondary flows. Consequently, as the flow rate increases, secondary losses diminish, contributing to an initial improvement in efficiency before viscous losses begin to dominate.

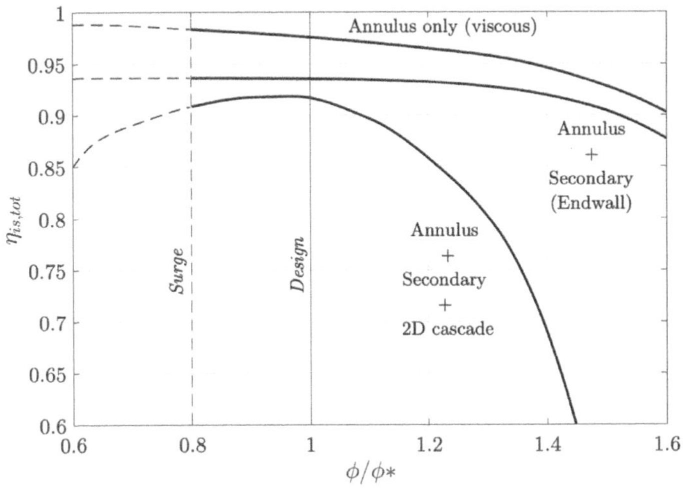

Figure 8.3 Overall compressor stage losses and pertinent breakdown as functions of flow coefficient.

Profile or two-dimensional cascade losses follow the typical parabolic trend observed in compressor cascade analysis, where efficiency peaks at the design condition or slightly below it. This is consistent with classical aerodynamic theory, which suggests that an optimal incidence angle leads to minimal profile losses. The surge condition, in some cases, is close to the design condition in terms of mass flow rate, and according to the figure, it is expected to occur at approximately 20% lower flow than the design condition $\phi/\phi^* \approx 0.8$.

8.5.2 TURBINES

Figure 8.4 presents the total pressure loss coefficient Y_{tot} as a function of incidence angle i in a turbine stage and exemplifies the contributions of the different loss mechanisms. The lowest curve represents the 2D cascade losses, which follow a typical parabolic trend exhibiting a minimum loss near zero incidence and increasing non-symmetrically for positive and negative incidences. These losses are primarily driven by the blade profile and boundary layer development along the airfoil and behave as expected from classical cascade aerodynamics. Since turbine airfoils typically have relatively high camber, losses at highly positive incidence angles are significantly higher than those at corresponding negative incidences. This is due to the stronger adverse pressure gradients that develop in the former case, which leads to earlier flow separation and greater aerodynamic dissipation.

The middle curve, labeled "Endwall (no Annulus)," accounts for secondary flow and endwall-related losses. Endwall losses respond to changes in incidence more gradually than 2D profile losses, which exhibit abrupt shifts. At low angles of incidence, the losses are essentially constant; however, they start to rise more sharply when the incidence rises.

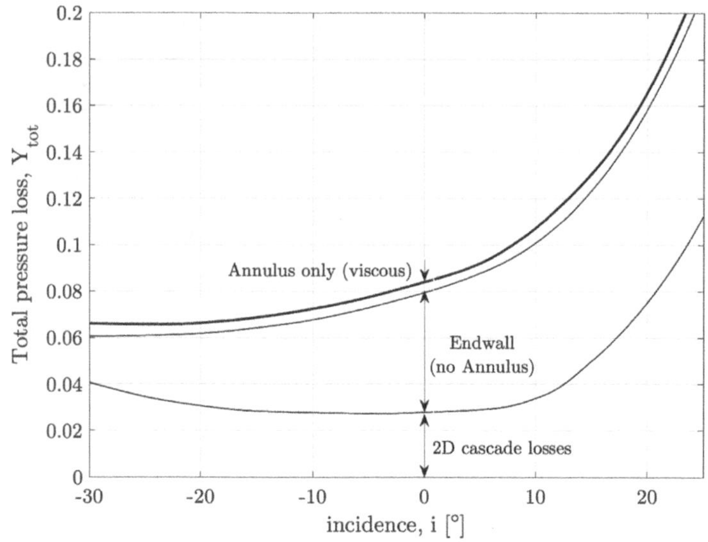

Figure 8.4 Overall turbine stage losses and pertinent breakdown as functions of flow incidence.

The uppermost curve, "Annulus only (viscous)," represents the total loss, incorporating both annulus viscous effects and endwall losses. In this case, annulus losses refer to those that result from boundary layer formation and growth along the hub and shroud surfaces. As expected, as long as the mass flow rate remains fixed, these losses stay relatively steady when the incidence varies. However, their combined impact becomes more noticeable when added to secondary flow and profile losses. This effect is especially evident at high incidence angles, where total aerodynamic losses rise sharply.

8.6 PROBLEMS

1. At its mean radius, a free-vortex air compressor rotor operates with an inlet absolute flow angle of $\alpha_1 = 0°$, an inlet relative flow angle of $\beta_1 = 68°$, and an outlet relative flow angle of $\beta_2 = 50°$. The relative exit velocity is $w_2 = 85$ m/s, the airfoil chord length is $\ell = 0.04$ m, and the blade height is $H = 0.085$ m. The cascade solidity is $\sigma_R = 1.1$. Assuming incompressible air with density $\rho = 1.22$ kg/m^3, the total pressure loss coefficient is given by $\zeta_R = 0.045$. Compute the ratio of the drag coefficient due to secondary flow losses to the 2-D cascade profile drag coefficient, $\frac{C_{D,s}}{C_{D,R}}$.

[2.26]

2. In the z-r plane, the hub and shroud contours of an axial flow compressor are given by r_h [m] = $-0.76z_h^2 + 1.095z_h$ and r_s [m] = $0.8e^{-0.07z_s}$, where z_h and z_s are in meters. The rotor of the first stage has its leading and trailing edges at

$z_1 = 0.3$ m and $z_2 = 0.4$ m, respectively. Assuming a blockage factor $\xi = 0.97$ from the leading to the trailing edge of the rotor, determine the ratio of the average meridional velocities c_{m2}/c_{m1}. The rotor is designed following the free vortex criterion and provides a density ratio of $\rho_2/\rho_1 = 1.1$.

[0.98]

3. An axial compressor operates with four stages under steady conditions. The blade speed is $u = 300$ m/s, with axial velocities of $c_m = 180$ m/s in the first stage and $c_m = 200$ m/s in the fourth stage. The inlet and outlet relative velocity angles for both stages are $\beta_1 = 40°$ and $\beta_2 = 25°$, respectively. The isentropic efficiency of both stages is $\eta_s = 0.88$, and the inlet stagnation temperatures are $T_1^0 = 288$ K for the first stage and $T_1^0 = 335$ K for the fourth stage. The pressure ratios across a single stage are $\pi_s = 1.38$ for the first stage and $\pi_s = 1.32$ for the fourth stage. Determine the work done factor and specific work for both the first and fourth stages. Assume $c_p = 1004.5$ J/(kg·K) and $\gamma = 1.4$.

[0.94, 0.85, 31.7 kJ/kg, 31.6 kJ/kg]

4. Consider a repeated 0.5-reaction compressor stage with an inlet temperature of 350 K. The rotor and stator blade heights are both 30 mm, while the chord lengths are 50 mm. The solidity of the rotor is 1.2, and for the stator, it is 1.3. The inlet and outlet flow angles for the rotor are 56° and 40°, respectively. The tip clearance is 0.5 mm. The lift coefficients for the rotor and stator are 0.9 and 0.85, respectively. The meridional velocity is given as 60 m/s. Evaluate the sum of endwall losses, passage vortex losses, and tip leakage losses for both the rotor and stator in the form of total pressure loss coefficients. Assume $c_p = 1004.5$ J/(kg·K) and $\gamma = 1.4$.

[Rotor: 0.06, Stator: 0.08]

5. A low-speed, repeated 0.5-reaction compressor stage operates with an inlet temperature of 400 K and pressure of 180 kPa. The rotor and stator blade heights are 30 mm, with chord lengths of 50 mm. Rotor and stator solidities are 1.0 and 1.5, respectively. Their cascade loss coefficients are 0.025 and 0.022. The rotor's inlet and outlet flow angles are 60° and 45°, with a tip clearance of 1.8% of blade height. Lift coefficients are 0.9 for the rotor and 0.85 for the stator. The meridional velocity is 60 m/s. Determine the stagnation pressure at the stator outlet for both actual non-isentropic and ideal isentropic compression. Assume $c_p = 1004.5$ J/(kg·K) and $\gamma = 1.4$, with constant inlet density throughout the stage.

[189.7 kPa, 191.3 kPa]

6. A 0-reaction turbine stage handles combustion gases entering at 800 K and 2050 kPa. The nozzle has a blade height of 0.08 m, chord length of 0.055 m, stagger angle of 40 degrees, solidity of 1.7, inlet flow angle $\alpha_1 = 0°$, exit flow angle $\alpha_2 = 75°$, and 2-D cascade total pressure loss coefficient of 0.065. The rotor has a chord length of 0.035 m, stagger angle of 10 degrees, solidity of 1.5, tip clearance of 0.002 m, with inlet flow angle $\beta_2 = 45°$, exit flow angle $\beta_3 = 45°$, blade lift coefficient $C_L = 2.4$, and 2-D cascade total pressure loss coefficient of 0.075. The absolute velocity at the nozzle inlet is c_1 120 m/s, with specific heat capacity of

1120 J/(kg·K) and isentropic exponent of 1.33. Dynamic viscosity is determined using Sutherland's law:

$$\mu(T) = \mu_0 \frac{\left(\frac{T}{T_0}\right)^{3/2} (T_0 + S)}{(T + S)}$$

where $\mu_0 = 1.716 \cdot 10^{-5}$ Pa·s is the reference viscosity at $T_0 = 273.15$ K, and $S = 110.4$ K is Sutherland's constant. Determine overall nozzle and rotor total pressure loss coefficients.

[0.081, 0.169]

NOTES

1. Stokes' theorem is used to relate the circulation around a closed contour to the integral of the vorticity over the enclosed surface. The circulation Γ around a closed contour C (such as ABCD in the passage of Figure 8.1) is given by the line integral of the velocity field $\vec{V}$ along that contour. Stokes' theorem states that this line integral is equal to the surface integral of the vorticity $\vec{\Omega}$ over the area A enclosed by the contour:

$$\Gamma = \oint_C V \cdot ds = \int_A (\nabla \times V) \cdot dA = \int_A \Omega \cdot dA$$

This circulation represents the strength of the secondary vortices.

2. Schlichting, H., and Gersten, K., *Boundary-Layer Theory - With contributions from Egon Krause and Herbert Oertel Jr.*, Ninth Edition, Springer-Verlag Berlin Heidelberg (2017).
3. Tournier, J., and El-Genk, M. S., "Axial Flow, Multi-stage Turbine and Compressor Models," *Energy Conversion and Management*, 51(1): 16–29 (2010).

9 Axial Flow Compressors

9.1 INTRODUCTION

In the previous chapters, we examined the flow characteristics of axial compressor stages, progressing from 1D to 3D analysis.

Now we will build on that knowledge to explore how a system of stages, i.e. a multistage axial compressor, functions. In practice, almost all axial compressors are multistage, as this configuration is essential to achieve the pressure ratios required in real applications (Figure 9.1). This chapter will focus on how multiple stages work together and the factors that affect the performance of the entire compressor system.

The transition from stage-level to multistage analysis introduces several key topics of interest that are essentially dependent on *stage matching*.

Stage matching requires aerodynamic compatibility above all, i.e. the flow leaving one stage should enter the next stage at the appropriate velocity and angle, maintaining optimal velocity triangles at all radii in the presence of all 3D flow effects. It also requires that pressure rise, specific work, and mean flow angles be well distributed throughout the stages to prevent any stage from being overloaded or underloaded in both design and off-design. Furthermore, axial velocity must be consistent as it passes through multiple stages to prevent choking or flow separation. Stage matching also plays a role in maintaining the stability margins of the compressor. A mismatch between stages can make the system more susceptible to local surge or stall events, which can propagate through the entire compressor. This problem is particularly relevant during start-up and when the machine is forced very fast from its nominal working point. Finally, stage matching must account for mechanical considerations,

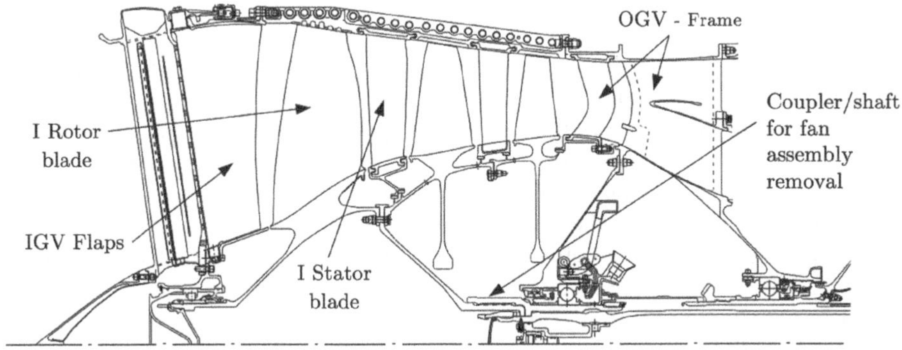

Figure 9.1 The U.S. F110-GE-132 multistage compressor (or LP fan) used in the F-16E/F aircraft by Lockheed Martin. (Wadia, A. R. and James, F. D., "F110-GE-132: Enhanced Power Through Low-Risk Derivative Technology," *ASME Transactions, Journal for Turbomachinery*, 123: 544–551 (2001). Reproduced from ASME with permission.)

DOI: 10.1201/9781003619956-9

such as blade loading, rotor-stator aero-mechanical interactions, and vibrations. Mismatched stages can result in uneven loading on blades or other components, leading to increased mechanical stress and fatigue.

9.2 THE EFFECTS OF THROUGH FLOW BLOCKAGE

Since the early days of compressor development, it has become clear that mid-stage performance differs significantly from the cleaner, more ideal conditions found in the front stages. As air passes through the compressor, the boundary layers along the walls thicken rapidly during the first few stages, leading to flow blockage and a more non-uniform axial velocity profile. In fact, the axial velocity profile becomes increasingly skewed, with a faster flow in the center and a slow flow near the walls.

We have already seen that, to account for these effects, the work done factor λ_w is used:

$$\lambda_w = \frac{h_3^0 - h_1^0}{u\Delta c_u} \tag{9.1}$$

where the numerator accounts for the actual total energy rise in the flow, and the denominator is the work spent in a rotor blading subject to clean, undistorted flow.

In the frontal stages, the work done factor is close to 1, which indicates minimal degradation in performance. By the fourth stage, the work done factor drops to about 0.9 and, after the 10th to 12th stage, the work done factor stabilizes around 0.85, where no further significant boundary layer growth establishes.

The implications involve compensatory blade design interventions: mid- and rear-stages are designed with lower work coefficients to account for the reduced energy transfer, while blade shapes, loading, and spacing are optimized to improve performance despite non-uniform conditions.

9.3 THE COMPRESSOR WORKING LINE

The compressor *working line* (also known as the *operating line*) is a graph on the compressor performance map (Figure 9.2) that shows the actual points at which the compressor operates during various conditions, such as at different speeds or power settings. It is a representation of the relationship between the corrected mass flow rate and the pressure ratio at which the compressor operates in real-world applications. In particular, the working line represents a series of stable operating points for a given set of conditions (rotational speed, inlet temperature, and pressure).

In the compressor map, the surge line indicates the minimum flow rate at which the compressor can operate stably without surging. Speed lines are curves representing the relationship between the corrected mass flow rate and the total pressure ratio at different corrected rotational speeds. The speed lines show increasingly negative slopes as the flow rate increases. The maximum flow rate that the compressor can handle for each speed line before experiencing a choked flow (where the sonic velocity is reached at the narrowest point) is the choking mass flow.

The working line must remain between the surge line and the choking limit. If the operating point approaches the surge line, there is a risk of instability, while

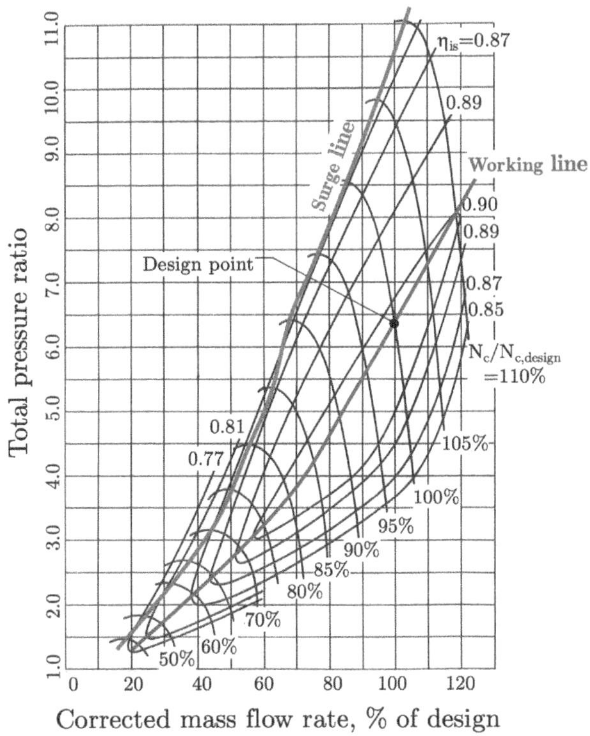

Figure 9.2 A typical compressor map.

approaching choke results in reduced efficiency and flow choking. In addition, the working line will typically traverse multiple speed lines in such a way that optimal performance is obtained.

In general, the working line moves rightward as the compressor moves toward higher mass flow rates and leftward as mass flow decreases. At higher power outputs or during demanding conditions (e.g. takeoff in aircraft engines), the working line shifts closer to choking, whereas at lower loads or idle conditions, it approaches the surge line.

9.4 COMPRESSOR STABILITY: STALL AND SURGE

Stall in a compressor is a critical condition that can manifest itself in a variety of components, including an isolated blade, a cascade or a stage, or even the entire compressor. The concept of stall is borrowed from studies on isolated airfoils and 2D cascades, where it is linked to an unstable behavior of the boundary layer, especially on the suction side, which leads to massive flow separation accompanied by a sudden loss in lift.

Although some clear analogies with airfoils exist, the exact causes behind stall in multistage compressors remain elusive. These systems are far more complex than single-bladed or 2D cascade setups. The airflow does not just move cleanly over one row of blades as it twists, shifts, and interacts with multiple components across stages. Furthermore, the influence of other parameters, such as blade tip clearance, inlet flow distortions, and subtle pressure feedback between stages, makes the picture become even more ambiguous.

The characteristic compressor map indicates either: (i) the minimum corrected mass flow for which steady and stable operation is guaranteed on a certain speed line or, alternatively, (ii) the maximum total pressure that the compressor can deliver for a given corrected flow without entering into stall. These limits are typically defined through experimental tests.

The stall margin is typically expressed as follows:

$$SM_1(\%) = \frac{|\dot{m}_{\text{stall}} - \dot{m}_{\text{operating}}|}{\dot{m}_{\text{operating}}} \cdot 100 \tag{9.2}$$

$$SM_2(\%) = \frac{p^0_{\text{stall}} - p^0_{\text{operating}}}{p^0_{\text{operating}}} \cdot 100 \tag{9.3}$$

The first definition is common in applications where the stability and efficiency of the pressure ratio are crucial, such as in high-performance aerospace engines. In these cases, the minimum required stall margin is from 10% to 15%.[1] The second definition is applied in industrial compressors, where a common value of the minimum stall margin range is from 15% to 20%.

Surge is a more complex phenomenon than stall and usually follows it under particular system operations. Surge is characterized by a complete breakdown of flow within a compressor system. When a compressor reaches its limit, e.g. due to throttling, the delivery pressure collapses as reverse flow rate develops. This is succeeded by the reestablishment of forward flow and the subsequent build-up to peak delivery pressure, which is followed by the collapse back into a cyclical pattern. Mechanical failure can occur rapidly as a result of these violent pressure and temperature fluctuations. Depending on a system's capacities and wave transit times, the frequency of this cycle can vary from less than 1 cycle per second in test plants with long duct systems to nearly 10 cycles per second in aircraft.

As mentioned previously, conventional cascade stall limits are empirically based and two-dimensional, making them somewhat unrelated to the actual stalling process in actual compressor stages. Research on straight cascades indicates that stall behavior is strongly influenced by end effects, with local corner-flow separations caused by the migration of end-wall boundary layer fluid onto the blade's convex surface. It is likely that in actual compressor blade rows, stall also initiates at the blade ends. Stall in compressor stages often develops into rotating stall cells, a phenomenon extensively studied but not yet fully understood.

Various theories predict the tendency for rotating stall cells to form, but they fail to adequately predict their propagation speed, size, and number. These theoretical treatments are typically two-dimensional and based on small perturbations, while in reality, rotating stall cells involve flow separations and reverse flow. For instance, severe forms of rotating stall have been observed, involving the centrifuging of air in stalled rotor blade channels and strong reverse flow at their tips.

The initial onset of stall does not automatically lead to surge. It is well established that stable rotating stall can occur in the early stages of high-pressure ratio compressors when operating at significantly reduced speeds without developing surge.

9.5 THE BASICS OF STAGE MATCHING

9.5.1 NOMINAL STAGE CHARACTERISTICS

Under design conditions, stage matching is primarily achieved by sequentially stacking the nominal properties of each stage, beginning with the first. Each stage's outlet conditions are determined using the relevant equations and the inlet thermo-fluid dynamic properties. The outlet conditions of one stage then serve as the input for the next stage, with this process continuing until the final stage is reached.

Under the assumptions that:

- the compressor geometry is known in its entirety; in particular, the rotor and stator hub and tip radii, which define the meridional flow path, are specified at both the leading and trailing edges of each stage, as well as that all blading details of each stage are given, e.g. assume a circular-arc camber line;
- the axial velocity c_m is constant throughout each stage but can change from one stage to another;
- each rotor's vortex law is such that a constant spanwise specific work is provided;
- the inlet total pressure and temperature, rotational speed, aerodynamic blockages at the rotor inlet and exit are given as boundary conditions;
- the stage characteristics are assumed to be known in terms of $(\psi, \eta_s) = f(\phi)$ (for example as in Figure 9.3).
- the behavior of each stage can be described by analyzing the flow at the mean diameter (mean-line approach) along with proper corrections to account for 3D effects and losses;
- the mean radius $r_m = D_{1,m}$ is constant;
- the stalling point of any stage is defined as the point at which the slope of $\psi = f(\phi)$ characteristic reaches zero;
- choking at any stage occurs when the stage flow coefficient reaches its maximum in the $\psi = f(\phi)$ characteristic;
- the degree of reaction is constant and equal to 0.5 at mean radius;

each i-th stage (see Figure 9.4) can be solved using the following equations. An iterative process allows for the verification that the provided geometry and boundary conditions meet a desired distribution of design flow coefficient ϕ^*. A pseudocode of the matching procedure can be described as follows.

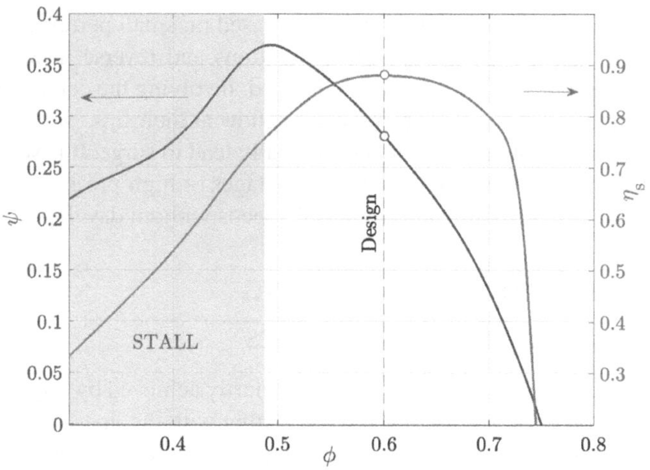

Figure 9.3 Sample compressor stage characteristics.

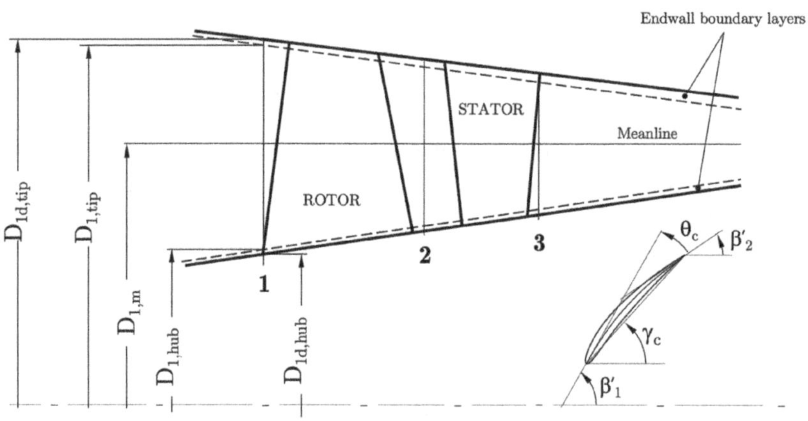

$$\xi_1 = \frac{(D_{1,tip}^2 - D_{1,hub}^2)}{(D_{1d,tip}^2 - D_{1d,hub}^2)}$$

Figure 9.4 Schematic of the meridional view of an axial compressor stage.

1. **Initialization**
   - Define gas properties, e.g.:

$$\gamma = 1.4, \quad R = 287\,\text{J/(kg·K)}, \quad c_p = 1005\,\text{J/(kg·K)}$$

   - Set inlet conditions:

$$T_1^0, \quad p_1^0, \quad \dot{m}, \quad u, \quad \cos\alpha_1$$

- Define profile parameters (stagger angle ξ and camber angle θ_c for all stages.
- Set the target flow coefficient ϕ^*.
- Initialize the initial area distribution $A_1(i)$ for each stage.
- Define the work done factor $\lambda_w(i)$ as a function of stage number.

2. **For each stage $i = 1$ to Z_s,**
   - If $i = 1$, set initial conditions for the first stage:

$$T_1^0 = T_1^0(1), \quad p_1^0 = p_1^0(1)$$

   Otherwise, set inlet conditions from the previous stage's exit values:

$$T_1^0 = T_3^0(i-1), \quad p_1^0 = p_3^0(i-1)$$

   - Set the degree of reaction $\varepsilon_r = 0.5$.
   - Iteratively adjust area $A_1(i)$ for each stage to meet target ϕ^*:
     - Set an initial Mach number Ma_1.
     - Solve the implicit equation for Ma_1 from the continuity equation:

$$\frac{\dot{m}\sqrt{RT_1^0}}{p_1^0 A_1 \sqrt{\gamma}} = \text{Ma}_1 \cos\alpha_1 \left(1 + \frac{\gamma-1}{2}\text{Ma}_1^2\right)^{-\frac{\gamma+1}{2(\gamma-1)}}$$

     This equation is solved numerically using a method like Newton-Raphson or bisection.
     - Calculate stage entry temperature, pressure, and density:

$$T_1 = \frac{T_1^0}{1 + \frac{\gamma-1}{2}\text{Ma}_1^2}$$

$$p_1 = \frac{p_1^0}{\left(1 + \frac{\gamma-1}{2}\text{Ma}_1^2\right)^{\frac{\gamma}{\gamma-1}}}$$

$$\rho_1 = \frac{p_1}{RT_1}$$

     - Calculate axial velocity c_m:

$$c_m = \frac{\dot{m}}{\rho_1 A_1}$$

     - Calculate tangential velocity c_{u1}:

$$c_{u1} = c_m \tan\alpha_1$$

     - Calculate relative velocity at inlet w_{u1} and flow angles:

$$w_{u1} = u - c_{u1}, \quad \beta_1 = \tan^{-1}\left(\frac{w_{u1}}{c_m}\right)$$

- Calculate blade angles using profile parameters θ_c and ξ:

$$\beta_1' = \xi + \frac{\theta_c}{2}, \quad \beta_2' = \xi - \frac{\theta_c}{2}$$

- Calculate deviation δ:

$$\delta = m\frac{\theta_c}{\sqrt{\sigma}},$$

- Calculate exit flow angle β_2:

$$\beta_2 = \beta_2' + \delta$$

- Calculate exit velocities and angles:

$$w_{u2} = c_m \tan\beta_2, \quad c_{u2} = u - w_{u2}, \quad \alpha_2 = \tan^{-1}\left(\frac{c_{u2}}{c_m}\right)$$

- Calculate flow coefficient ϕ and work coefficient ψ:

$$\phi = \frac{c_m}{u}, \quad \psi = \frac{c_{u2} - c_{u1}}{u}$$

- If ϕ is within tolerance of ϕ^*, stop iteration; otherwise, adjust area A_1:

$$A_1 = A_1 \cdot \frac{\phi}{\phi^*}$$

- Calculate the stage total temperature rise and pressure ratio:
  - Use the map for ψ and η_s as a function of ϕ (Figure 9.3), same for all stages
  - Calculate total temperature after the stage:

$$T_3^0 = T_1^0 + \lambda_w(i) \cdot \psi \cdot \frac{u^2}{c_p}$$

  - Calculate the stage pressure ratio π_s:

$$\pi_s = \left(1 + \eta_s \cdot \frac{(\gamma - 1) \cdot \psi \cdot u^2}{\gamma \cdot R \cdot T_1^0}\right)^{\frac{\gamma}{\gamma-1}}$$

  - Update total pressure:

$$p_3^0 = p_1^0 \cdot \pi_s$$

3. **Final output** (see Figure 9.5):
   - Adjusted area distribution A_1 for each stage.
   - Final flow coefficient ϕ per stage.
   - Stage pressure ratio and entry Mach number.

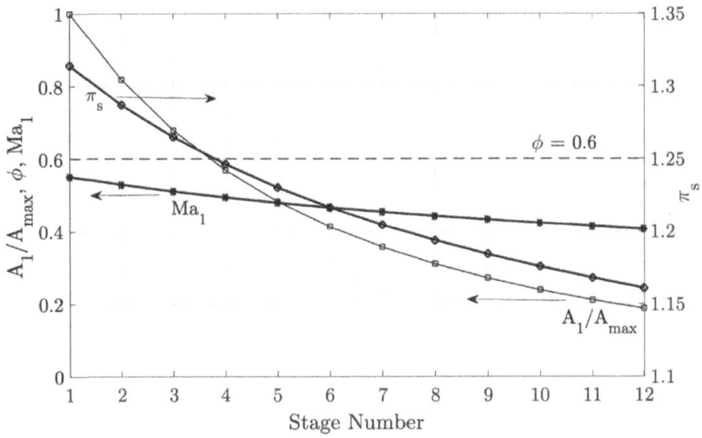

Figure 9.5 Stage quantities in a sample 12-stage compressor.

Once the flow area distribution is determined, the geometric areas A_{d1} can be obtained:

$$\frac{A_1}{A_{d1}} = \xi_1 = \frac{A_{d1} - \int_{hub} \delta^* \, dr - \int_{shroud} \delta^* \, dr}{A_{d1}}$$
$$= 1 - \frac{\iint_{hub} \left(1 - \frac{\rho \, c_{m1}}{(\rho \, c_{m1})_{inviscid}}\right) r dr d\theta}{A_{d1}} \tag{9.4}$$
$$- \frac{\iint_{shroud} \left(1 - \frac{\rho \, c_{m1}}{(\rho \, c_{m1})_{inviscid}}\right) r dr d\theta}{A_{d1}}$$

where the blockage factor ξ_1 can be modeled as a function of the stage index i using an exponential decay law:

$$\xi_1(i) = \xi_{1,0} \cdot e^{-\alpha \, i} \tag{9.5}$$

and $\xi_{1,0}$ is the initial blockage factor at the first stage (typically close to 1) and α is a decay constant that controls the rate of decrease of the blockage factor per stage. This value typically ranges between 0.02 and 0.05, depending on compressor design and operating conditions.

Although the procedure outlined above is a well-established method for predicting the overall compression ratio, it is not suitable for calculating overall efficiency as it does not account for flow losses specifically.

9.5.2 OFF-DESIGN STAGE CHARACTERISTICS

It is mandatory that all stages have comparable stall margins at nominal operation. A non-uniformity in this regard often triggers a stall in the rear stages at reduced rotational speeds.

9.5.2.1 Frontal and Rear Stages

Under off-design conditions, particularly at reduced rotational speeds, the compressor cannot sustain the same mass flow due to working line limitations. In this situation, stage matching remains a challenge because the flow characteristics in each stage deviate from the design point. As speed and mass flow decrease, the total pressure ratio drops, while the relative velocities between stages must adjust to meet local conditions defined by blade geometry. This leads to a lower density ratio across the compressor, which is most noticeable in the later stages. In fact, the density ratio decreases more significantly than the mass flow while the rear stages must still handle the flow passed through the earlier stages. As a consequence, because of stages' limited ability to compress the fluid, the axial velocity tends to increase from the front to the rear stages.

To prove this, let us consider that, for a compressor of given geometry, the pressure ratio π_c and mass flow $\dot{m}$ at different speeds N can be scaled using affinity laws. In particular, π_c scales with the square of speed:

$$\pi_c \propto \left(\frac{N}{N_{\text{design}}}\right)^2 \tag{9.6}$$

while mass flow scales linearly with speed:

$$\dot{m} \propto \frac{N}{N_{\text{design}}} \tag{9.7}$$

The stage density ratio, ρ_r, i.e. the ratio of outlet density ρ_{out} to inlet density ρ_{in}, approximately follows the same tendency as pressure ratio[2], i.e.:

$$\rho_r \propto \left(\frac{N}{N_{\text{design}}}\right)^2 \tag{9.8}$$

thus the density ratio decreases quadratically with speed, while mass flow decreases linearly. As a consequence, at reduced rotational speeds, the rear stages of the compressor experience higher axial velocities compared to the front stages. Thus, rear stages are subjected to lower incidence angles in rotors, and therefore are prone to choke, while front stages, with lower axial velocities and increased incidence angles, tend to approach stall (see Figure 9.6, where a sample result is provided for a ten-stage compressor with nominal $\phi = 0.4$).

At higher rotational speeds, the pressure ratios exceed the design values, causing a reversed mismatch. This increases the likelihood of stall in the rear stages. Such a phenomenon can trigger premature stall in upstream stages so as to initiate surge.

When the rear stages are choked and the frontal stages are stalled (left of Figure 9.6), a severe imbalance in the flow occurs throughout the compressor. The rear stages operate near their maximum capacity with sonic flow at the blade tips or throats, which limits further mass flow through these stages. Since airflow cannot increase beyond this limit, the rear stages are essentially bottlenecked. This creates

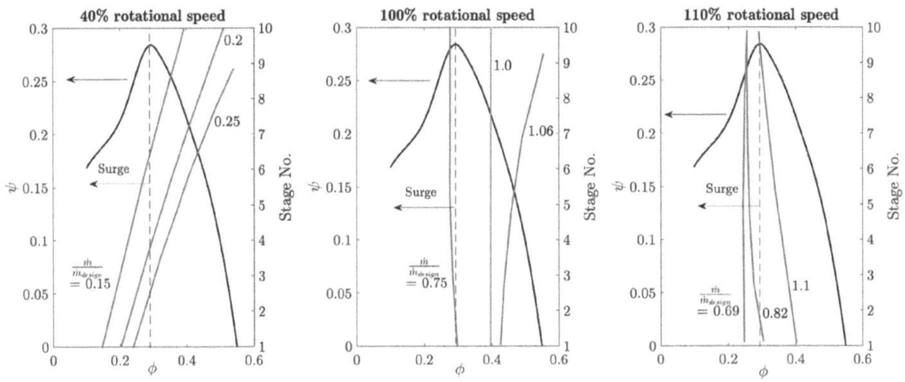

Figure 9.6 $\psi - \phi$ characteristics of a ten-stage compressor at different rotational speeds and mass flows.

a flow mismatch with the front stages. In contrast, frontal stages experience flow separation due to high incidence angles caused by the reduced flow demand from the rear. The airflow is disrupted so as to cause pressure fluctuations and, eventually, stall. This scenario can be particularly dangerous because the flow instability in the central stages can propagate upstream, further aggravating the stall condition and creating a vicious cycle: the choked rear stages prevent enough flow from moving downstream, exacerbating the stalled conditions in the front stages. This can lead to unsteady operation, high vibrations, and severe compressor surge, i.e. a complete breakdown of the flow through the compressor.

When the frontal stages are close to choke (right of Figure 9.6) and the rear stages are stalled, the fluid dynamics is quite different, but still problematic. The front stages operate at their sonic limits due to elevated ingested mass flow at higher-than-nominal rotational speed, allowing no further increase in mass flow through them. This high-pressure, high-velocity flow progresses downstream, creating high-flow demand on the rear stages. As the rear stages receive this high-pressure flow, they are forced to operate under conditions beyond their design capabilities. The incidence angles on the rear-stage blades increase leading to flow separation and stall there. The rear stages cannot handle the airflow, causing pressure drops and localized flow reversal. In this case, the inability of the rear stages to handle the incoming flow creates a back-pressure effect on the front stages. While the front remains choked, the mismatch in downstream pressure could worsen the rear-stage stall and potentially result in compressor surge.

9.6 IMPROVEMENTS IN SURGE CHARACTERISTICS

The key to enhancing the behavior of a compressor towards surge is to conduct a systematic analysis of the stage characteristics. The goal is to ascertain which stage, or stages, have already reached stall or are currently approaching it and to evaluate the compatibility of the remaining stages with requirements on stall margin. Upon identification of stall-causing stages, the surge line can be improved by implementing numerous modifications. Ideally, these modifications will result in the relocation of stall points to higher pressure ratios and lower mass flows. A standard approach is to decrease the stalling mass flow of those stages where instability is predicted, without affecting the pressure ratios at stall or higher flows. This enables other stages to achieve higher pressure ratios and lower flow coefficients.

To this purpose, several strategies and modifications can be employed. The following is a discussion of key factors.

9.6.1 EFFECT ON PERFORMANCE OF VARYING INLET GUIDE-VANE STAGGER AND VARIABLE STATOR VANES

Inlet guide vanes (IGVs) regulate the angle at which airflow enters the first compressor stage; in fact, the performance and surge behavior of the compressor can be significantly influenced by changing the stagger of the IGVs. In particular, by increasing their stagger angle, the flow incidence on the compressor blades is altered in a way that a reduction in mass flow and an improvement in the pressure ratio at lower speeds are achieved. This mitigates the surge in the initial stages.

In addition to using IGVs, variable stator vanes (VSVs) are commonly implemented in high-pressure compressor stages. In this case, all stator vanes in the compression stages adjust their stagger angle as a function of the mass flow rate and rotational speed according to a desired schedule. This is beneficial, in particular for later stages, for maintaining optimal flow characteristics (Figure 9.7) and to increase the stall margin.

As an example, in many aircraft engines, IGVs are placed in front of the first of the low-pressure stages, while VSVs are used in the early stages of the high-pressure stages.

9.6.2 BLEED VALVES

Intermediate bleeding can help prevent choke in the rear stages by redistributing airflow. During startup, the compressor operates at low speeds where the mass flow is still stabilizing. By bleeding off air, the amount of airflow passing through the later stages is reduced, which avoids approaching choke conditions (Figure 9.8). In addition, bleeding off some air between stages makes it possible for the compressor to reduce the overall pressure rise (not shown in Figure 9.8), which decreases the likelihood of reaching the surge line. This ensures smoother flow through the compressor stages and prevents backflow and oscillations.

9.6.3 USE OF MULTI-SPOOL CONFIGURATIONS

Multi-spool designs, where different compressor stage groups (low-pressure spool compressor, or LPC, high-pressure spool compressor, or HPC) are driven by separate movers, provide flexibility in managing surge and choke across various conditions (right in Figure 9.8).

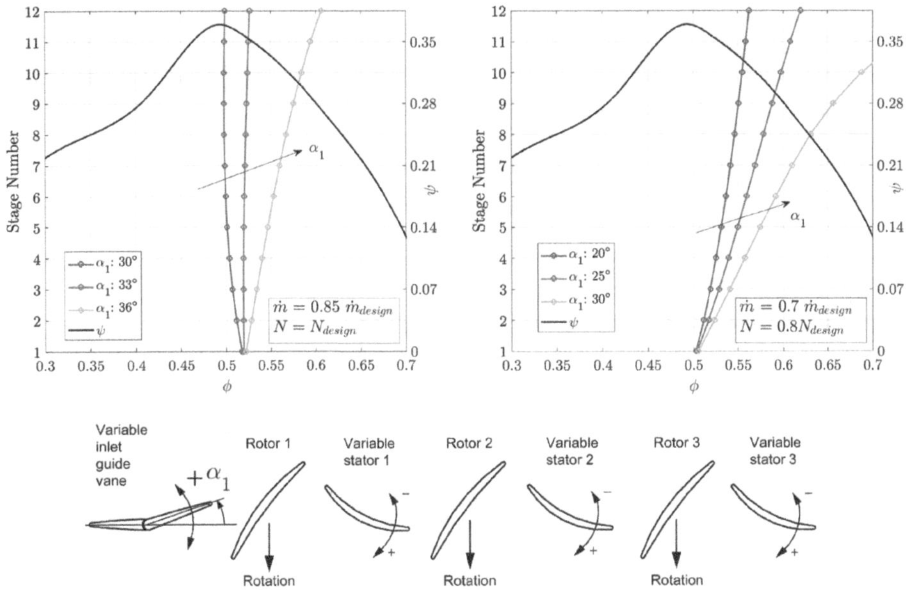

Figure 9.7 Effect of combined IGV and VSV outlet angle variation in a 12-stage compressor at the nominal speed and 85% of nominal flow rate. All vanes are assumed to produce the same flow deviation from 30° to 36°.

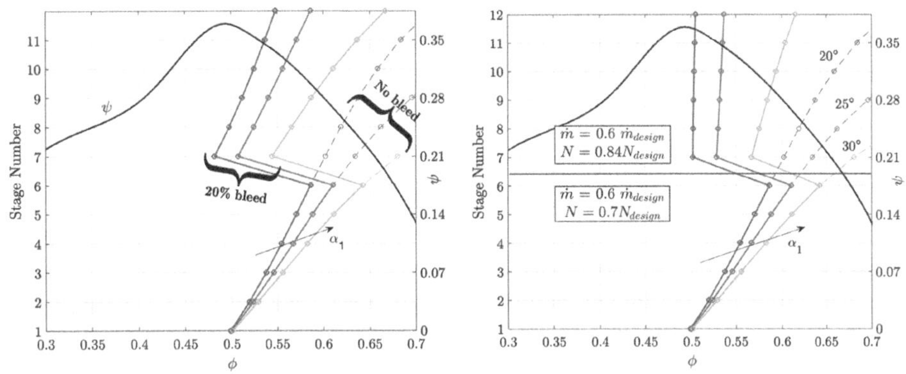

Figure 9.8 Combined effect of α_1 variation, bleed and 2-spool architecture (right) in a 12-stage compressor.

By spinning the larger-diameter, cooler LPC at a lower rpm and the smaller, hotter HPC at a higher rpm, the engine achieves better aerodynamic stage on-off design matching and higher compression efficiency.

In fact, independent spools provide significant benefits in off-design operation and responsiveness. At part-power or other off-design conditions, a multi-spool compressor can adjust each spool speed to match the required airflow while maintaining higher efficiency and smoother stage matching across the compressor. As a matter of fact, a multi-spool architecture mitigates stage mismatch issues that could cause front stages to stall or rear stages to stop absorbing torque due to nearly null values of ψ (windmilling) at low speeds (left of Figure 9.6).

The transient response is also improved. The smaller high-pressure spool has lower rotational inertia, so it can accelerate or decelerate more quickly. This results in a faster throttle response of the engine and better adaptability to rapid changes in power demand.

9.7 PRACTICAL ASPECTS OF LPC AND HPC STAGES

Although both LPCs and HPCs are axial compressors, their aerodynamic designs diverge due to differences in required pressure ratio, size, and operating conditions.

9.7.1 NON-DIMENSIONAL STAGE PARAMETERS

Modern HPCs often employ a transonic first stage rotor to achieve a large pressure rise per stage ($\approx$1.6–1.8), whereas rear HPC and all LPC stages may operate in a high-subsonic regime to ensure efficiency and stability. Supersonic relative flows are usually confined to the front stages of the HPC or the fan, and are carefully managed with blade shaping (such as swept leading edges) to minimize shock intensity and associated losses.

Blade geometry adaptations differ between subsonic and transonic stages (Figure 9.9). Subsonic blades typically have rounded leading edges and moderate camber, focusing on diffusion without flow separation. Transonic rotor blades, by contrast, often feature thin, cambered- or even S-shaped profiles to handle shock-boundary layer interactions. The presence of a terminal normal shock on a transonic rotor limits its achievable pressure rise, so that designers must position the shock near the trailing edge and control the downstream diffusion to avoid separation.

The stage reaction is also tuned differently: transonic stages often favor a slightly higher degree of reaction to reduce the diffusion burden on the stator in the high Mach environment. Despite the additional losses, a well-designed transonic stage can attain stage work coefficients higher than those of a purely subsonic stage, allowing fewer stages for a given pressure ratio. For example, an advanced core compressor can reach an HPC work coefficient of about $\psi_{\text{HPC}} = 0.55 - 0.60$ (for the first stage). In contrast, an LPC stage loading in the same engine may attain $\psi_{\text{LPC}} = 0.38 - 0.40$. In other words, an HPC stage is attributed roughly 45% more specific work than an LPC stage in the same engine.

The flow coefficient trend is roughly the inverse: HPC stages tend to have lower ϕ, meaning that they operate at a smaller meridional velocity relative to the rotor speed. This is partly because the HPC flow path is smaller in diameter and higher in density (due to pre-compression by the LPC/fan), and because variable stators often restrict flow at entry to the HPC at off-design. In contrast, the LPC front stages (especially the fan) process very large airflows at a moderate rotational speed, implying

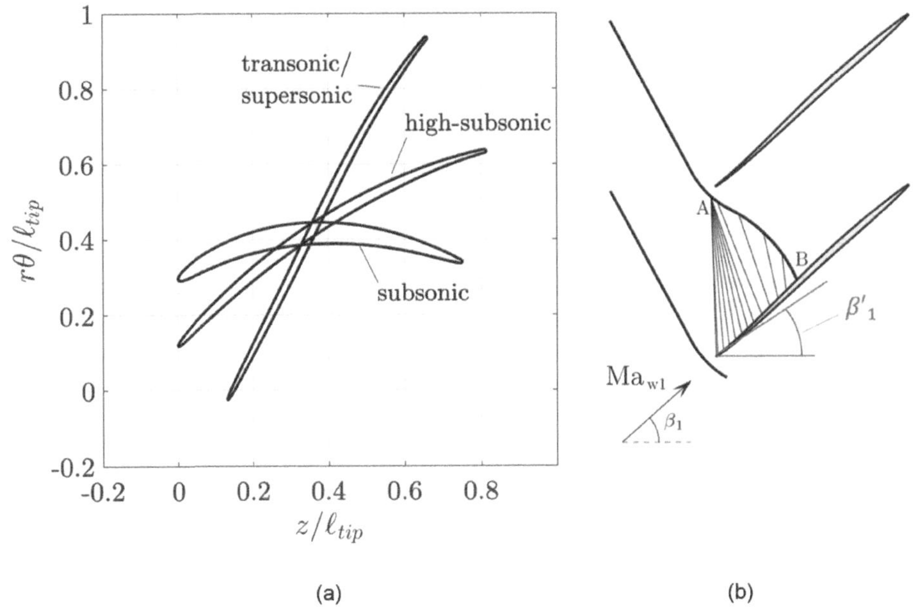

Figure 9.9 Blade shapes and shock interaction in compressible turbomachinery flow. (a) Blade geometries classified by flow regime—subsonic, high-subsonic, and transonic/supersonic—shown in normalized cylindrical coordinates. (b) Schematic of oblique shock and expansion wave interactions in supersonic flow over a blade section.

a relatively higher ϕ. For instance, a large fan might have $\phi \approx 0.6$–0.7 at tip (very high flow rate per tip speed), while an HPC mid-stage might run at $\phi \approx 0.4$–0.5.

The allowable diffusion factor DF for an HPC is typically in the range of 0.5–0.6 for a single rotor blade (a much higher value compared to subsonic stages). LPC stages, on the other hand, can use more conservative DF values (<0.45) to maximize efficiency.

However, it is important to note that while HPC stages have higher loading on average, not all HPC stages are equally loaded - typically, the front stages of the HPC carry the highest load, with a gradual reduction toward the rear stages. For example, a high-pressure core might have a first-stage pressure ratio $\approx$ 1.6–1.8, while later subsonic stages might only contribute $\approx$ 1.15–1.3 each. LPC stages, which are larger in diameter and often subsonic, usually have more uniform moderate loading (e.g. $\approx$ 1.25 per stage) and do not reach the extreme of HPC front stage loading.

9.7.2 STAGE COUNT AND PRESSURE-RATIO SCALING

Because of the blade loading differences described, the allocation of pressure ratio and number of stages between the LPC and HPC is a critical design choice. The overall pressure ratio OPR of a multistage compressor having Z_s stages is the product of the stage pressure ratios:

$$\Pi = \pi_1 \cdot \pi_2 \cdots \pi_{Z_s}$$

where π_i refers to stage i.

As a result of the above mentioned limits on π_i for both LPC and HPC, the pressure ratio split between LPC and HPC is fundamentally governed by the following considerations.

Ideally, each spool operates at its peak efficiency point, which often means dividing the overall pressure ratio such that both LPC and HPC have reasonable stage loadings. An unfair split (e.g. overburdening the HPC with too much pressure rise) could make the HPC stages excessively loaded and prone to stall, whereas underloading the HPC and pushing a higher ratio to the LPC might force the large-diameter LPC stages into an inefficient regime or even transonic inlet conditions at the fan. The optimum split often emerges from cycle optimization: for a given overall pressure ratio, there is a certain intermediate pressure (after the LPC) that maximizes total compressor efficiency.

As mentioned, there is an important trade-off between stage loading and stage count: one could achieve a given overall pressure ratio with fewer, highly loaded (high-ψ) stages or more, lightly loaded stages. Fewer stages mean lighter weight and shorter compressor length, but each stage must run at a higher pressure ratio, which can incur efficiency penalties. More stages allow each to run at a more optimal (lower) pressure ratio for high efficiency, but the additional blades add weight, cost, and complexity. Designers therefore optimize stage count for minimum stage stacking loss and acceptable weight. Conversely, distributing the pressure ratio over more stages can raise efficiency but with diminishing returns – beyond a certain point, adding stages yields only marginal gains in polytropic efficiency while, at the same time, significantly increasing mass and mechanical complexity.

Modern HPCs (Figure 9.10) use transonic front stages to maximize $\Delta h^0_{\text{stage}}$ for the first few stages, and employing many stages to accumulate a high total Π. For example, an advanced core compressor may achieve 22 total pressure ratio with 11 stages, corresponding to an average $\pi \approx 1.32$ per stage, but with the first stage doing about 1.65 and the later stages progressively less.

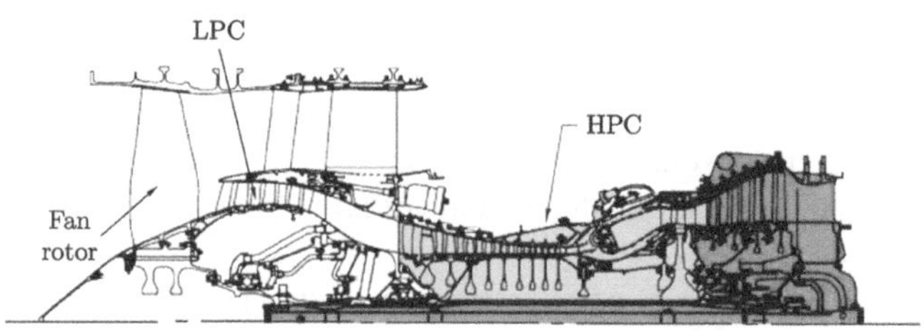

Figure 9.10 The two-spool CFM56 turbofan engine. (Atasoy, V. E., "Improved Modelling of Low-Pressure Rotor Speed in Commercial Turbofan Engines: A Comprehensive Analysis of Machine Learning Approaches," *Energy* 312: 133571 (2024). Reproduced from Elsevier with permission.)

Table 9.1
Compressor Characteristics of Modern Commercial High-Bypass Turbofan Engines

Engine	OPR	HPC PR	Stages (Fan/LPC/HPC)	Cruise rpm (LP/IP/HP)	Diameter (Fan/LPC/HPC) (m)
GE9X (GE)	60:1	27:1	1/3/11	2,355/–/9,561	3.40/~1.70/~0.89
Trent XWB (Rolls-Royce)	50:1	–	1/8/6	2,700/8,200/12,600	3.00/–/–
PW1100G-JM (P&W)	~50:1	–	1/3/8	~13,000/~20,000	2.06/–/–
LEAP-1A (CFM)	40:1	22:1	1/3/10	3,894/–/19,391	1.98/–/–
GEnx-1B (GE)	43.8:1	23:1	1/3/10	2,560/–/11,377	2.82/–/–

Table 9.2
Compressor Characteristics of Modern Military Turbofan Engines

Engine	OPR	HPC PR	Stages (Fan/LPC/HPC)	Cruise rpm (LP/IP/HP)	Diameter (Fan/LPC/HPC) (m)
F135-PW-100 (P&W)	28:1	6:1	3/6	~10,000/~20,000	1.16/–
F414-GE-400 (GE)	30:1	–	3/7	~13,360/~16,810	0.88/–
EJ200 (EuroJet)	26:1	–	3/5	~13,000/~17,000	0.74/–
M88-2 (Safran)	24.5:1	–	3/6	~12,000/~16,000	0.70/–
RD33 (Klimov)	21:1	–	4/9	–/–	1.04/–

In contrast, an LPC booster could aim for a much lower total pressure ratio (e.g. 3–5) over 3–5 stages, averaging ≈1.2–1.3 per stage, which is well within subsonic capability. The fan itself (considered the first stage of an LPC in a large civil turbofan) typically has a pressure ratio of ≈1.5 for high bypass engines, although with a large diameter and one or two splitters or stators to diffuse flow.

In military applications, such as in 5th generation fighter jets, LPCs employ multistage fans featuring three or four stages, along with 5-9 HPC stages (Tables 9.1 and 9.2).

9.7.3 OPERATING RANGE, SURGE MARGIN, AND EFFICIENCY TRENDS

9.7.3.1 HPC

HPCs, with many highly loaded stages, tend to be more prone to flow instability (stall/surge) if not carefully managed. The combination of high stage pressure ratios and many stages in series means that an HPC's stall margin can be relatively narrow. If one stage encounters flow separation, it can rapidly trigger a cascade of stalls through upstream stages (a rotating stall or surge event). HPCs are often the limiting component for low-flow stability. Indeed, without control measures, an HPC designed for maximum pressure ratio at full speed might surge violently at lower speeds or throttle settings. To maintain an adequate surge margin across the flight envelope, HPCs universally employ variable geometry and bleed control.

- Variable Stator Vanes (VSVs) are used on many or all stator rows of the front stages of an HPC. By rotating the stator vanes, the incidence angle on the downstream rotor is controlled, effectively tailoring the compressor's flow capacity as the speed changes. For example, the GE CF6-80E1 HPC has variable stator vanes on every stage of the high-pressure compressor to optimize airflow and prevent stall. At low speeds (low mass flow), VSVs are closed (reduced stagger) to restrict the flow and keep the front stages from stalling; at high speed, they open up to allow more flow. This scheduling maintains the HPC's operating point near its design incidence across speeds.
- Bleed valves are often used in the mid or rear stages of an HPC (and/or at the LPC outlet) during transients and low-speed operation. By bleeding off some flow from middle stages, the effective flow through the front stages is increased, moving away from surge at low speeds. As one source notes, introducing a bleed in the middle/rear of the compressor increases airflow in the front at lower engine speeds. This helps avoid back-pressure buildup that could stall the front stages. For instance, a common design is the 2.5-stage bleed in two-spool engines – a bleed valve at the LPC exit (station 2.5) that opens during acceleration or idle to unload the HPC entrance. Similarly, an HPC might have bleed valves that dump air overboard or recirculate it during starting and low rpms to keep the compression system stable.

Thanks to these measures, an HPC can operate from idle (low ϕ) to max power (high ϕ) without crossing the surge line. However, even with such controls, HPCs typically have a smaller relative surge margin than fans or LPCs. The numerous stages of an HPC create a steep pressure rise curve that drops off sharply near surge. Designers sometimes quantify surge margin as a percentage of pressure or flow, e.g. an HPC might have $\approx$10-20% stability margin in flow at a given speed, whereas a fan might have more. The HPC surge margin is often the critical factor determining the minimum idle speed of the engine (to avoid HPC surge, a certain core speed is required).

9.7.3.2 LPC

The LPC (booster and fan) spool generally benefits from a different stability characteristic. The fan, which has only one (or a few) stage(s), features a fairly broad operating range by itself, but is coupled with the downstream HPC via the flow. The booster typically has no variable geometry (aside from possibly an inlet guide vane). Its surge margin must be sufficient on its own or aided by the same bleed that helps the HPC. In practice, the LPC is usually not the limiting factor for surge because if the HPC is stable, the LPC (pumping into the HPC) will also be in a stable regime.

The LPC's fewer stages and lower pressure ratio per stage give it inherently a wider stall margin. Moreover, being on a separate shaft, the LPC can decouple to some extent – at low thrust, the LPC (fan) speed can be kept higher relative to the HPC, avoiding LPC stall even if core flow is reduced.

9.7.4 EFFICIENCY TRENDS

In terms of efficiency trends, both LPCs and HPCs are designed for high efficiency at their design point, but their efficiency behavior off-design can differ:

- HPC polytropic efficiency at design is typically very high (in the order of 93%–95%). Despite the challenges of shocks and high loading, modern airfoil design and three-dimensional optimization have yielded HPC efficiencies comparable to or better than older low-speed compressors. However, when operating away from design (say at part-speed or part-flow), HPC efficiency can drop off more sharply. Variable stators mitigate this by keeping incidence near optimal, but there are still losses due to blade profiles not being ideally angled at all conditions. Furthermore, running many stages at off-design incidence and Reynolds number can incur additional loss. The rear stages of an HPC at low flow may operate at very low incidence (or even negative incidence if VSVs are not adopted in those stages), which is suboptimal for efficiency.
- LPC (booster) polytropic efficiency is usually quite high at design as well (often in the 90%–92% range). Having fewer stages, its contribution to overall work is smaller, but it is crucial to overall efficiency since it processes a large mass flow. Off-design, an LPC spool's efficiency is maintained relatively well if the fan speed can adjust to demand. The booster stages typically see more uniform incidence changes (since usually no VSVs are there, and the whole booster operates along a single speed-line). The efficiency of the fan, in particular, is highest near its design pressure ratio and drops if the fan is operated at too high (risk of stall) or too low (near choke) incidence. Still, fans are often designed with some margin such that at cruise (which is partial power for turbofans) they operate efficiently at a slightly off-design condition.

A notable efficiency concern for HPCs is the effect of blade aspect ratio and tip clearances in later stages. As the flow is compressed, the blade height decreases toward the rear of the compressor. This results in very short blades in the last stages of an HPC. These blades have a low aspect ratio (short span relative to chord length) and a tip clearance that might be a significant fraction of the span. Tip leakage losses and secondary flows thus consume a larger fraction of the work in those stages, tending to reduce the stage efficiency in the rearmost stages. An LPC, having larger blades, suffers relatively less tip clearance penalty (clearance might be a few percent of span for a fan blade vs. $> 10\%$ for a last-stage HPC blade). HPC designers address this with advanced tip clearance control (tight seals, maybe even active clearance control by cooling the casing) and by sometimes not pushing the last stage to a high pressure rise to avoid high losses. In fact, in some high-pressure compressors, the last stage is deliberately lightly loaded, acting more as a flow straightener than a high-work stage, because an overloaded last stage would have very poor efficiency and could jeopardize stability. The net effect is that the incremental efficiency of each successive HPC stage may diminish; the overall HPC efficiency is a weighted result of all stages. High-loading front stages might have slightly lower efficiency due to shocks, middle stages the highest efficiency, and small rear stages moderate efficiency due to clearance and Reynolds number effects.

Efficiency improvements over time have increased both LPC and HPC performance. For example, the introduction of 3D blading (swept and leaned stators) and optimized airfoil profiles has given stage efficiency gains of 1-2% per stage in some cases. Non-axisymmetric endwall contouring (discussed below) further reduces losses on hub/casing regions, benefiting mostly the HPC where secondary flows are stronger. As a numerical example, some studies report that careful endwall contouring reduced loss in a cascade by 18% and improved peak efficiency by 0.45%. These gains, multiplied over many HPC stages, have allowed modern compressors to achieve polytropic efficiencies above 92%–94% even at very high pressure ratios. LPCs, particularly propulsive fans, have also been able to improve efficiency, thanks to features such as swept blades and composite materials allowing thinner and more aerodynamic shapes. But fans face a trade-off with noise, which sometimes leads to designs that are not purely optimized for efficiency but also account for acoustic considerations.

9.7.5 INTER-SPOOL DYNAMICS

In acceleration, if an HPC slows relative to the LPC, the HPC could surge since the LPC is forcing more flow rate into it. Conversely, if the fan/LPC is choked or held back by an inlet throttle, such as a closed variable inlet or small nozzle, the HPC might overspeed and surge. Engine controllers carefully schedule fuel and variable geometry to keep both spools on a safe operating line. Some advanced engines also incorporate a variable area fan nozzle (VAFN) to modulate the fan/LPC operating line. By enlarging the exhaust nozzle area at low thrust, the engine effectively allows the fan to operate at a lower pressure ratio, increasing the flow margin of the fan. As the engine accelerates, the nozzle closes to increase the fan pressure ratio for best performance.

9.8 EXTERNAL CAUSES OF MISMATCHING

Compressor stage mismatching occurs when the stages of a compressor operate under different conditions, leading to inefficiencies, performance degradation, or even operational instability. These mismatches can be caused by external factors that disrupt the intended design and uniform operation of the compressor.

9.8.1 INLET FLOW DISTORTIONS

Flow distortions are a common external cause of stage mismatching in compressors. The efficiency of the frontal stage is significantly influenced by the characteristics of the air or gas that enters the system. For example, upstream components, including filters, elbows or, more generally, ducts, may induce a non-uniform inlet velocity and pressure profile. A significant metric for assessing these distortions is DC_{60}, an industry-standard parameter that quantifies the extent to which the total pressure varies across different regions of the inlet face.

$$DC_{60} = \frac{p^0_{avg} - p^0_{60}}{q_{avg}} \tag{9.9}$$

where p^0_{60} is the average total pressure across the entire inlet area, p^0_{min} the mean total pressure in the worst 60° sector (out of the full 360° circle), and q_{avg} the dynamic pressure that is averaged over the inlet area. The DC_{60} is calculated by partitioning the inlet face into sectors.

In industry practice, a DC_{60} of about 10% is often treated as a practical upper limit for safe engine operation in civil applications. Distortions below that (i.e. in the single-digit percentage range) are considered manageable with minimal performance loss, whereas exceeding approximately 10%, DC_{60} starts to significantly erode the surge margin. Large civil turbofans expect very low distortion in service as they allow only DC_{60} in the order of 0.03–0.05 (3%–5%) at most, and even that only during extreme crosswind or off-design conditions. In normal cruise, DC60 is basically 0% for a well-designed intake. Thus, any large civil engine is only validated against small distortion levels, and the surge margin deterioration for even 5%–10% distortion is carefully analyzed.

Military engines often have S-duct or chin intakes and, therefore, tolerate higher admissible distortion indices by design. This often corresponds to DC_{60} values in the 0.2+ range (20%+) during worst-case maneuvers.

A temperature distortion index (TDI) parallels the pressure metrics: it quantifies the variation in total temperature at the aerodynamic interface plane AIP, i.e. the station right at the compressor inlet. In mathematical terms, TDI can be expressed as:

$$\text{TDI} = \frac{T^0_{max} - T^0_{min}}{T^0_{avg}} \tag{9.10}$$

or a sector-based difference similar to the DC but for temperature. In practice, temperature distortion often has both circumferential and radial components, so standards sometimes include a combined temperature distortion descriptor. Engines are generally less tolerant to temperature non-uniformity because hot streaks can raise blade metal temperatures and upset compressor stall margin.

In Table 9.3, admissible distortion levels are summarized according to SAE International (formerly the Society of Automotive Engineers), Mil Spec, a specification from the United States Department of Defense, and other company standards.

Table 9.3
Summary of Permissible Inlet Distortion Indices by Standard or Engine Type

Standard - Source	Distortion Metric	Max Permissible Value	Notes
SAE ARP1420 (Guideline)	DC_{60} – 60° pressure distortion	~0.10 (10%) (industry practice)	ARP1420 provides methodology only; ~10% DC60 is treated as a practical upper limit for safe engine operation.
MIL-E-5007D (Mil Spec)	Overall total pressure distortion = $(p^0_{max} - p^0_{min})/p^0_{avg}$	0.30 (30%) at high power	Engine must sustain 30% pressure drop across the inlet face under high power. Lower limits apply at idle. Accounts for several military intakes.
Rolls-Royce Civil (typical)	DC_{60}	~0.03–0.05 (3%–5%) (design target)	Not published publicly; varies by engine type. Very low intake distortion expected. Minimal surge margin allocated for distortion tolerance.
Rolls-Royce Military (typical)	DC_{60}	~0.15–0.25 (15%–25%) (required operability)	Higher distortion levels permitted. Engine and intake are co-designed to ensure margin under worst-case operational conditions.
Temperature Distortion (general)	TDI = $(T^0_{max} - T^0_{min})/T^0_{avg}$ or ΔT pattern	No single value – combined with pressure limits	For example, MIL-E-5007D reduces the allowable pressure distortion from 30% when temperature distortion (TDI) is present. Engines often tested under combined distorted conditions.

9.8.2 DOWNSTREAM OPERATING CONDITIONS AND BACK PRESSURE

Deviations in downstream conditions, especially changes in back pressure, are an additional major external factor contributing to stage mismatch. A compressor can be driven toward surge conditions by an increase in the back pressure, which is frequently the result of increased blockages in downstream components. This has an impact on the later stages, as they are pushed to manage lower mass flow rates, whereas the early stages may still be operating at or near their design points.

Conversely, the compressor might be forced into choke conditions when back pressure decreases, a condition that can be the consequence of unexpected decreased downstream resistance. In this case, the compressor has to handle a greater volume of mass flow than intended, resulting in the early stages being overloaded while the later stages continue to operate efficiently.

9.8.3 LOAD VARIATIONS AND OPERATIONAL TRANSIENTS

Variations in load are common in jet engines and industrial gas turbines as a result of operational transients, including modifications to throttles or changes in the energy demand of power generation systems. These changes may result in temporary mismatches, as the new operating conditions are adjusted by different stages at different time rates.

Furthermore, during idle or low-power operation, the compressor may be operated under partial load conditions, resulting in underloading of certain stages while others remain more heavily loaded. The results of this imbalance in aerodynamic loading are a mismatch in performance across the stages, particularly in the later stages that are more susceptible to mass flow fluctuations.

A FADEC (Full Authority Digital Engine Control) computerized system can help avoid excessively rapid transients, particularly in aircraft gas turbines. FADEC monitors engine parameters in real time and implements closed-loop feedback control to dynamically adjust settings, thus promoting operational stability. The effects of throttle regulation, especially during manual operations of the pilot, are precisely filtered to ensure smooth acceleration and deceleration, while pre-programmed performance maps are employed to optimize fuel flow rate as a function of instantaneous engine parameters. As such, FADEC systems are fundamental in preventing the occurrence of surge and stall.

9.8.4 UPSTREAM AND DOWNSTREAM COMPONENTS

The performance of upstream and downstream components can heavily influence the compressor's stage matching. Poorly designed *upstream ductwork* or diffusers can introduce turbulence, flow separation, or distorted pressure profiles, which affect how each stage processes the incoming flow.

In some cases, mismatching between the compressor and turbine, such as in gas turbines or turbochargers, can lead to imbalances. For example, if the turbine cannot extract enough power to drive the compressor at the required speed, certain compressor stages may operate at off-design conditions, leading to mismatching, particularly at non-optimal load conditions.

9.8.5 SYSTEM INTEGRATION AND MISALIGNMENT

Mismatching of stages can also be caused by mechanical issues associated with system integration and alignment. Vibrations and uneven rotor speeds may be introduced by mechanical misalignment between the compressor and other system components, including couplings and bearings. This results in performance variations between stages due to the uneven aerodynamic loading on the blades.

Furthermore, system resonances, which arise when specific frequencies in upstream or downstream components induce resonances in the compressor, can result in airflow disturbances entering the compressor (see Chapter 13).

9.8.6 FOULING AND INLET CONTAMINATION

Stage performance can be reduced over time as a result of external contamination, such as dust, debris, or moisture entering the compressor. Compressor blades become fouled due to inlet air contamination, which diminishes their aerodynamic efficiency: this frequently leads to mismatches, as the distribution of contaminants is often unpredictable and determines the extent to which certain stages are impacted more than others.

Particles in the few-micron size range tend to adhere to blade surfaces and cause fouling (e.g. from fine dust, sand, and other soil particles), whereas much larger particles (above $\approx$10–20 μm) more often cause erosion rather than fouling. These include salt and corrosive particles, which can cause erosion or material degradation in marine or industrial environments, particularly in the later stages, as a result of the higher pressure and temperature involved there, causing metal components to be more susceptible to chemical reactions like oxidation or corrosion.

Beyond external sources, the engine itself can sometimes be a contributor to fouling. A common example is an internal oil leak at the compressor bearing seals: oil can mist into the compressor airflow and coat the blades, providing a sticky substrate for dirt to adhere to. Over time, this results in heavy, black oily deposits deep in the compressor.

Fouling changes the blade shape and increases surface roughness, reducing the compressor's airflow and pressure-generating capability. The engine is then operating off its design point, that is, the compressor delivers less air to the combustor, and its isentropic efficiency drops. For example, controlled experiments with fouled blades have shown substantial performance declines: a rough fouling layer $\approx$0.025 mm thick has been shown to cause about a 9% drop in compressor pressure ratio and a 6 percentage-point loss in efficiency.

As a consequence, since a fouled compressor is less efficient, the engine must burn more fuel to achieve the commanded power or thrust. Fouling-induced efficiency loss manifests as higher exhaust gas temperature (EGT) and turbine inlet temperature for the same engine output. Operators may notice that over time an engine requires a higher turbine temperature or fuel flow to reach a certain compressor speed or thrust. The increased firing temperature erodes the engine's margin to operating limits and can accelerate hot-section component wear.

Compressor fouling can ultimately lead to compressor stability issues by moving the compressor's operating point closer to the surge line. As fouling deposits restrict flow and alter blade aerodynamics, the compressor's surge margin is reduced. If fouling is severe and uneven (e.g. one quadrant of the compressor is more fouled than another), it can even induce flow distortion that precipitates a rotating stall or surge event.

A combination of good inlet air filtration, regular compressor washing (on-line and off-line), use of protective coatings, and good operating practices ensures that fouling and inlet contamination do not severely compromise engine performance.

9.9 REGULATION

At constant rotational speed, there are two primary methods to regulate the mass flow through a compressor: *inlet valve regulation* and *outlet valve regulation*. While both techniques influence the operating point on the compressor map, they do so in fundamentally different ways.

Recalling the definition of corrected mass flow $\dot{m}_c$:

$$\dot{m}_c = \dot{m}\frac{\sqrt{\frac{T^0_{\text{in}}}{T_{\text{ref}}}}}{\frac{p^0_{\text{in}}}{p_{\text{ref}}}} = \dot{m}\frac{\sqrt{\theta_{\text{corr}}}}{\delta_{\text{corr}}} \tag{9.11}$$

allows us to discuss the cases.

9.9.1 INLET VALVE REGULATION

In the first scenario, mass flow is modulated by partially closing the inlet valve. This introduces a restriction at the compressor inlet, causing the inlet total pressure p^0_{in} to drop. As a result, δ_{corr} decreases. At the same time, the actual mass flow $\dot{m}$ is also reduced due to the throttling effect imposed by the valve.

Substituting into the definition of corrected mass flow, we see that both $\dot{m}$ and δ_{corr} decrease. Because these changes tend to offset one another in a balanced way, the corrected mass flow $\dot{m}_c$ also decreases, but in a relatively smooth and proportional manner. This controlled reduction allows the compressor to shift to lower flow operating points while maintaining stable operation. Consequently, regulation of the inlet valve results in a moderate decrease in the stall margin.

9.9.2 OUTLET VALVE REGULATION

When regulating the compressor using the outlet valve, the inlet conditions (i.e. p^0_{in} and T^0_{in}) remain unchanged, meaning δ_{corr} stays constant. However, the outlet pressure p^0_{out} increases as the valve limits the airflow downstream, leading to an increase in the overall pressure ratio across the compressor. As the outlet valve obstructs the flow, the actual mass flow $\dot{m}$ decreases due to the resistance created downstream of the compressor, but the corrected mass flow $\dot{m}_c$ decreases more sharply. This is because: (i) $\delta_{\text{corr}} = \frac{p^0_{\text{in}}}{p_{\text{ref}}}$ stays constant (since p^0_{in} does not change), and (ii) the actual mass flow $\dot{m}$ decreases due to the increased resistance at the outlet.

As a result, the corrected mass flow $\dot{m}_c$ decreases more significantly compared to inlet regulation because $\dot{m}$ is being reduced while δ_{corr} remains constant. Since the corrected mass flow $\dot{m}_c$ decreases more rapidly in outlet valve regulation, it moves closer to the stall flow rate $\dot{m}_{\text{stall}}$, reducing the stall margin and increasing the risk of compressor stall.

9.9.3 A SPECIAL CASE: REGULATION IN AIRCRAFT ENGINES

Aircraft engines provide a useful illustration of outlet flow control as their exhaust nozzle can serve as a downstream flow restriction. In this situation, the dynamics of the corrected mass flow $\dot{m}_c$ is controlled by the same relationship:

$$\dot{m}_c = \dot{m}\frac{\sqrt{\theta_{\text{corr}}}}{\delta_{\text{corr}}} \tag{9.12}$$

In the case of exhaust throttling, the total inlet pressure p^0_{in} remains virtually constant, which means that δ_{corr} does not change. The back-pressure is raised by a partially closed exhaust nozzle's downstream limitation, which lowers the real mass flow $\dot{m}$ through the engine. Since δ_{corr} remains constant while $\dot{m}$ decreases, the corrected mass flow $\dot{m}_c$ experiences a more significant drop. This is similar to what is observed in compressor outlet valve control: the operating point of the engine moves toward lower corrected mass flow values on the compressor map, closer to the surge line, thus lowering the stall margin.

Under circumstances where the compressor is already running close to its stability limits, this decrease in stall margin increases the likelihood of surge or stall. For this reason, exhaust throttling is not usually preferred as a means for fine control of engine performance. Instabilities in the compressor stages are more likely for the engine given the unequal decline in $\dot{m}_c$ compared to $\dot{m}$.

An overexpanded nozzle raises a related issue, especially if the nozzle throat's sonic conditions are compromised. A rapid rise in backpressure experienced by the compressor can result from this, which could cause negative effects like flow separation and compressor stall. These consequences not only impair performance but also threaten engine operability and safety in a significant way.

9.10 PROBLEMS

1. Given the compressor map shown in Figure 9.2, provide an estimation of the stall margin indicators at 100% and 60% $N_c/N_{c,\text{design}}$.

   [62%, 32%; 23%, 21%]

2. A 4-stage axial flow compressor handles air with a mass flow rate of $\dot{m} = 48.3$ kg/s at an inlet total temperature of $T^0_1 = 300$ K and an inlet total pressure of $p^0_1 = 101.3$ kPa. At the mean diameter, the blade speed is 200 m/s. The working fluid properties are: specific heat ratio $\gamma = 1.4$, gas constant $R = 287$ J/(kg·K). Stage-specific parameters include rotors designed with airfoils having circular-arc camber lines, camber angle $\theta_c = 10.5°$ for all stages, a stagger angle $\xi_R = 36.9°$ for all stages, solidity =1 and a constant inlet flow angle $\alpha_1 = 30°$. The stators are mirrored to the rotors in such a way that a repeated configuration occurs. The initial area distribution is $A_1 = [0.28, 0.24, 0.20, 0.18]$ m^2. The stage work done factors are given as $\lambda_w = [1, 0.95, 0.93, 0.92]$. Each compressor stage has the following efficiency function:

$$\eta_{is_c} = -38.695\phi^3 + 50.299\phi^2 - 18.868\phi + 2.5$$

The objective is to determine the adjusted area distribution required to achieve a target flow coefficient of $\phi^* = 0.52$ for all stages.

$[(0.3473, 0.2927, 0.2492, 0.2143)\,\mathrm{m}^2]$

3. A four-stage axial-flow compressor operates with an inlet total temperature of 300 K and an inlet total pressure of 101.3 kPa. The mass flow rate is 48.3 kg/s, and the blade peripheral speed is 250 m/s. The inlet flow angle is 20°. The flow area distribution across stages is 0.32, 0.27, 0.22, and 0.21 m^2. The work done coefficient values are 1.0, 0.95, 0.93, and 0.92. The rotor profile has circular-arc camber lines, and rotor cascade solidity remains constant at 1.2 for all stages. The initial camber and stagger angles are 14° and 36.9°, respectively, for all stages. The objective is to determine the rotor profile parameters, specifically the camber and stagger angles, such that the work coefficient for each stage remains below 0.4. Additionally, the total pressure ratio across all stages must be computed. (Assume the working fluid has a specific heat ratio of 1.4, a gas constant of 287 J/(kg·K), and a specific heat capacity of 1005 J/(kg·K)).

[Stage 1: $\theta_c = 12.2°, \xi_R = 37.4°$; Stage 2: $\theta_c = 10.9°, \xi_R = 37.8°$; Stage 3: $\theta_c = 14°, \xi_R = 36.9°$; Stage 4: $\theta_c = 5.3°, \xi_R = 40.6°$; $\pi_{\text{overall}} = 2.46$]

4. An axial compressor features eight stages of constant mean diameter, operating with air as the working fluid. Inlet conditions include a total temperature of 300 K, a total pressure of 101.3 kPa, an inlet flow angle $\alpha_1 = 35°$ and a mass flow rate of 75 kg/s. The peripheral speed at the mean radius is 250 m/s and the rotational speed is 3600 rpm. Each stage is repeated and is characterized by rotor airfoils with predefined camber angles increasing linearly from 10.5° (first stage) to 11.2° (last stage) and a constant stagger angle of 36.9°. All airfoils have circular camber lines. The flow area distribution per stage is set at 0.5589, 0.49808, 0.4457, 0.40086, 0.3622, 0.3287, 0.2993, 0.2737 square meters. Determine the inlet and outlet diameters based on the flow area, while ensuring compliance with aerodynamic constraints on the Mach number, work coefficient, and diffusion factor for each stage. (Assume $\gamma = 1.4$, $R = 287$ J/(kg·K) and the following distribution of the work done factor: $\lambda_w = [1\ 0.95\ 0.93\ 0.92\ 0.91\ 0.9\ 0.9\ 0.9]$).

[Entry section: 1.39 m, 1.26 m; Outlet section: 1.36 m, 1.29 m]

5. An axial compressor with 12 stages operates under nominal conditions with a mass flow rate of $\dot{m} = 55$ kg/s. The air properties are defined by a specific heat ratio $\gamma = 1.4$ and a specific gas constant $R = 287$ J/kg·K. The inlet conditions include a total temperature of $T_1^0 = 288$ K and a total pressure of $p_1^0 = 101.3 \cdot 10^3$ Pa. The compressor runs at a rotational speed of 3200 rpm, with a peripheral speed of $u = 280$ m/s at the mean radius. The first rotor is preceded by an IGV ring that sets an inlet flow angle of $\alpha_1 = 30°$. The rotor parameters are indicated below:

Stage	θ_c (deg)	χ_R (deg)	A_1 (m^2)	λ_w
1	10	36.8	0.3114	1.00
2	10	36.8	0.2689	0.95
3	10	36.8	0.2336	0.93
4	10	36.8	0.2045	0.92
5	10	36.8	0.1800	0.90
6	10	36.8	0.1593	0.89
7	10	36.8	0.1417	0.88
8	10	36.8	0.1266	0.87
9	10	36.8	0.1136	0.86
10	10	36.8	0.1023	0.85
11	10	36.8	0.0924	0.85
12	10	36.8	0.0838	0.85

Rotor solidity is 1 for all stages. The stator vanes are mirrored relative to the rotor blades, maintaining repeating velocity triangles across the stages. All airfoils have circular camber lines. Stage stall is predicted when its work coefficient $\psi = \Delta c_u/u$ exceeds 0.43. When the compressor operates along the same speed line at 95% of the nominal flow rate, determine which stage is most likely to stall.

[Stage 3]

6. At takeoff (sea level), a 12-stage axial flow compressor of an aircraft engine handles 103.8 kg/s of air and delivers a pressure ratio of 27 at a rotational regime of 4500 rpm. At the compressor inlet, the stagnation properties are $p_1^0 = 101.02$ kPa and $T_1^0 = 303.15$ K, and the blade speed at midspan is 385 m/s. At takeoff, the compressor features the following geometrical characteristics: IGVs provide an exit flow angle = 15°, while stage data are given in the following Table.

Stage	D_i (m)	D_e (m)	λ_w	ξ_R [°]
1	1.5199	1.7485	1.00	51.77
2	1.5398	1.7286	0.95	50.66
3	1.5564	1.7120	0.93	49.44
4	1.5711	1.6973	0.92	48.05
5	1.5841	1.6844	0.90	46.45
6	1.5960	1.6725	0.89	44.61
7	1.6070	1.6614	0.88	42.49
8	1.6171	1.6513	0.87	40.01
9	1.6265	1.6419	0.86	37.12
10	1.6352	1.6332	0.85	33.73
11	1.6433	1.6251	0.85	29.76
12	1.6510	1.6174	0.85	25.04

The rotors blades have $\theta_c = 12.3°$, $\alpha_1 = \alpha_3 = 15°$ and $\sigma_R = 0.95$ for all stages. The stator vanes are mirrored relative to the rotor blades, maintaining repeating velocity triangles across the stages. All stages' airfoils have circular-arc camber lines. Assume air as an ideal gas ($\gamma = 1.4, R = 287$ J/(kg·K)). Also, assume stage isentropic efficiency to follow $\eta_{is} = f(\phi(i)) = -38.695\phi^3(i) + 50.299\phi^2(i) - 18.868\phi(i) + 2.5$. Verify that the compressor can guarantee the expected pressure ratio at takeoff. At cruise (altitude 10,668 m), the inlet stagnation properties change to $p_1^0 = 38.136$ kPa and $T_1^0 = 261.98$ K, the corrected mass flow rate increases by 3.6%, while the compression ratio and corrected speed drop by 10% and by 16% respectively, all compared to the corresponding values at takeoff. The stages are highly loaded and expected to stall when ψ>0.5. Determine IGV/VSV setting angles in cruise which prevent the compressor from going into stall while assuring the required performance.

$$[\alpha_1 = \{15.00, 15.02, 15.61, 16.47, 17.38, 18.17, 18.78, 19.17, 19.43, 19.66, 20.08, 20.95\} \text{ degrees}]$$

NOTES

1. High-performance applications (e.g. jet engines) tend to have smaller margins, around 10%–12%, to maximize efficiency. However, maintaining a minimum of 10% is vital to ensure safe operation during transient conditions (like takeoff or maneuvering).
2. The stage density ratio can be calculated as follows:

$$\rho_r = \pi_c \cdot \frac{T_{\text{out}}^0}{T_{\text{in}}^0} = \pi_c \cdot \pi_c^{\frac{\gamma-1}{\gamma e_c}} = \pi_c^{1+\frac{\gamma-1}{\gamma e_c}}$$

Using the binomial expansion for small $x = \frac{\gamma-1}{\gamma e_c}$, and expanding π_c^x for small x:

$$\pi_c^x \approx 1 + x \cdot \ln(\pi_c)$$

Thus:

$$\pi_c^{1+\frac{\gamma-1}{\gamma e_c}} \approx \pi_c \cdot \left(1 + \frac{\gamma-1}{\gamma e_c} \cdot \ln(\pi_c)\right) \approx \pi_c$$

when e_c is in the order of 0.9 and π_c values are lower than 2, as commonly found in axial compressor stages.

10 Axial Flow Turbines

10.1 INTRODUCTION

Unlike compressors, where overcoming adverse pressure gradients causes significant aerodynamic issues, turbines operate under favorable pressure gradients, which simplifies certain aspects of turbine aerodynamics. Aerodynamic design in turbines is therefore usually more straightforward, while the main challenges are in other fields—particularly in thermal management and aeromechanics.

High inlet gas temperatures in multistage turbines (see Figure 10.1) cause significant stress on the materials used in the high-pressure stages, therefore raising notable issues regarding structural integrity and lifetime, in addition to efficiency. Enduring such extreme circumstances without compromising performance creates challenging issues regarding thermal stresses, cooling technologies, and material selection. On the other hand, low-pressure stages are more susceptible to aeromechanical issues due to their higher aspect ratio and thinness, which makes bladings more susceptible to vibration-related problems.

Knowing that turbines run with various types of fluids, including steam, in this chapter, we will focus on turbines operating with single-phase flow and configurations involving gases that can be approximated by the ideal gas model.

10.2 GENERAL CHARACTERISTICS

In modern gas turbines, the rotor blades are mounted on a disk that is attached to the main shaft, while the nozzle guide vanes are usually fixed to the casing (Figure 10.1). Turbine blades are attached to their disks using dovetail- or fir-tree-shaped roots fitting into corresponding slotted disk cavities, an arrangement ensuring a secure mechanical interlock and distributing stress evenly while allowing for thermal expansion. A similar design features the nozzle guide vanes, which are often mounted in segments within the turbine casing, once again making it possible for thermal expansion and ease of assembly or replacement.

Interstage labyrinth seals are typically incorporated along the shaft and disk, as well as between consecutive stages, to minimize leakage in the clearance gaps. Shrouds are often used to improve mechanical integrity and reduce clearance gap leakage, especially in situations where maintaining very small clearances is impractical. Some shrouds could also have seal fins to increase leakage control. A comparable concept operates in multistage axial-flow turbines used in aeroengines, but with more focus on efficiency, weight savings, and high performance. Operating at significantly higher rotational speeds, these turbines usually need more sophisticated materials to withstand high temperatures and stresses.

To avoid overheating and guarantee structural integrity during high-temperature operation, complicated cooling systems are also frequently included in the blade

DOI: 10.1201/9781003619956-10

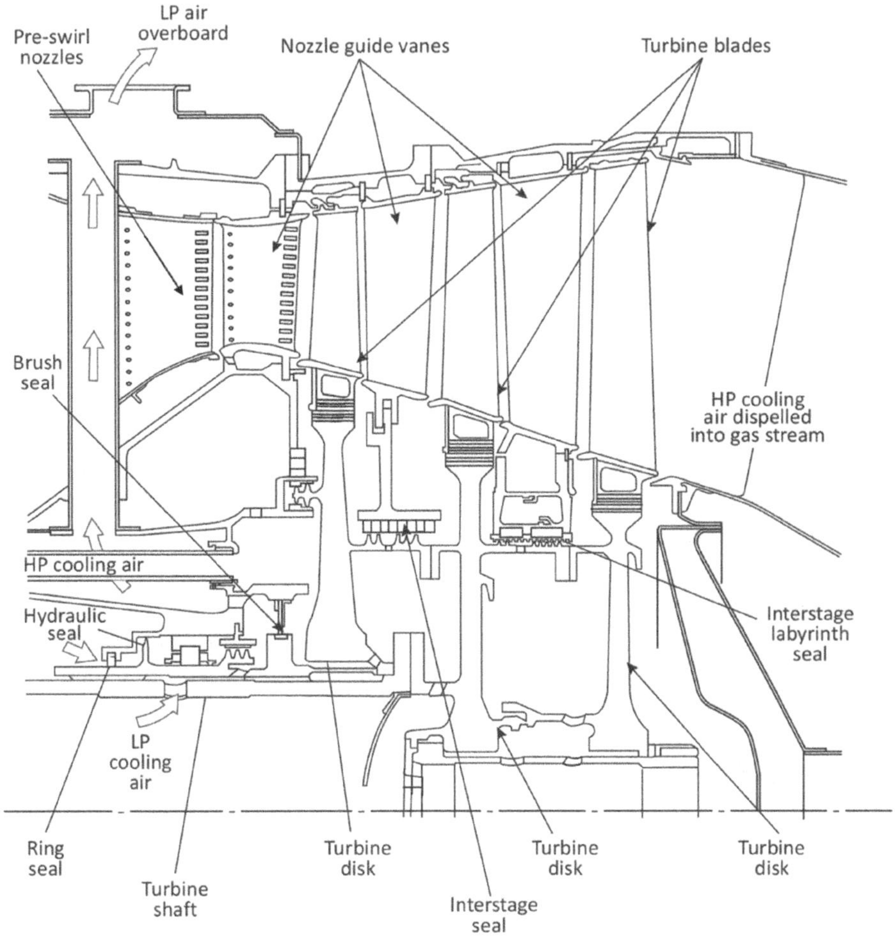

Figure 10.1 Cross-sectional view of an axial gas turbine.

and turbine stages. Extracted from earlier compressor stages, high-pressure air is channeled through internal passages, especially cooling the turbine blades, disks, and other hot components. To lower blade temperature, HP air flows through the turbine blades and finally dissipates into the gas stream. Cooling the turbine shaft and disks is done using low-pressure air entering via dedicated channels, which guarantee operational and structural integrity.

10.3 TURBINE MAP

A performance map is a graphic representation of the operational characteristics of a given turbine across varying conditions of mass flow rate, rotational speed and pres-

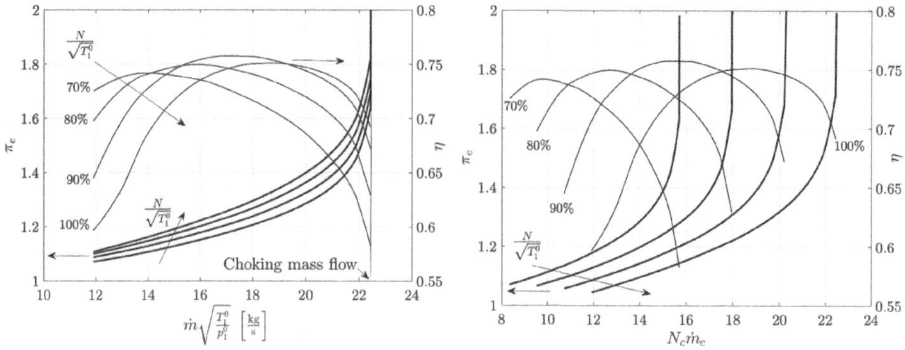

Figure 10.2 Turbine maps.

sure ratio (Figure 10.2). The left-hand-side figure presents the map using pressure ratio π_e, i.e. the expansion ratio, as a function of the corrected mass flow rate for various corrected rotational speeds $\frac{N}{\sqrt{T_1^0}}$. A turbine's adiabatic efficiency η is plotted as contour lines within the map. Each curve corresponds to a specific percentage of the normalized rotational speed, starting from 70% to 100%. The curve shapes show that, as mass flow increases, the pressure ratio rises until a point of choking, where further increases in mass flow cannot be realized independently from the pressure ratio. The choking mass flow is typically achieved when the entry nozzle chokes, which makes the choking mass flow independent of the rotational speed.

In the right-hand side of the figure, the x-axis has been transformed to represent the product of the corrected mass flow rate and the corrected rotational speed, denoted as $N_c\dot{m}_c$. Such representation offers a clearer and more intuitive understanding of the turbine's operational boundaries and speed lines as the spread between each speed line is more apparent and the shape of the efficiency contours becomes clearer. For example, at 90% speed, the curves exhibit a more distinct peak efficiency region that allows better insight into optimal turbine operation.

10.4 FAMILY OF TURBINE TYPES

Turbine stages can be systematically analyzed and categorized using the *Smith's diagram* (Figure 10.3), where the stage work coefficient ψ is plotted against the flow coefficient ϕ for several stage configurations, with parametric contours representing different levels of isentropic efficiency η. The diagram serves both as an informative tool and a guide for the preliminary selection of turbine stage configurations that maximize performance.

As stated in the previous chapters, turbine stages can generally be divided into reaction stages and action stages, each of which occupies distinct regions on Smith's diagram corresponding to different operating conditions and performance goals.

In reaction stages, the enthalpy drop is distributed between the rotor and stator and balanced to obtain the best isentropic efficiencies. For this purpose, it is essen-

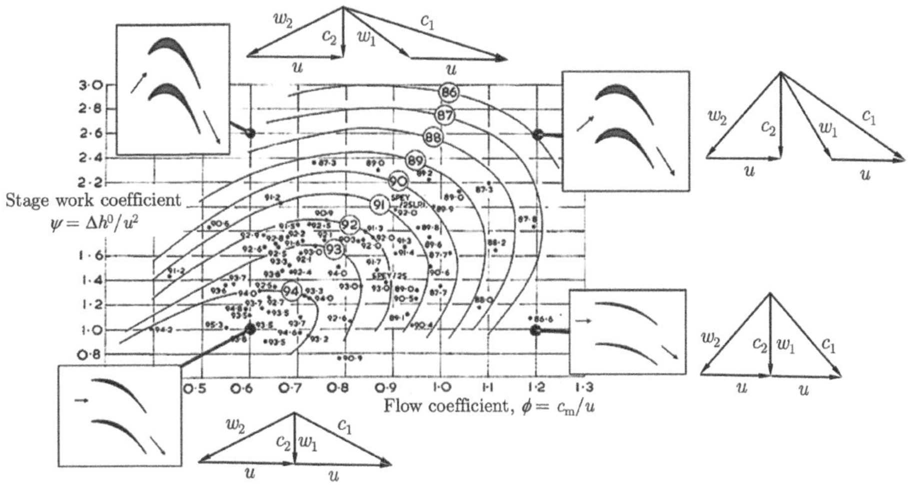

Figure 10.3 Smith's diagram.

tial to keep both the work coefficient and the flow coefficient values within tight, ideal ranges, possibly around $\phi = 0.6$ and $\psi = 1.0$. Increasing ψ while maintaining relatively high values for efficiency requires ϕ to rise almost proportionally with ψ.

In action stages, the rotor blades primarily redirect the flow without causing significant pressure variations, while the nozzle guide vanes account for the majority of the enthalpy drop. As is known, these stages are designed to increase energy transfer per stage, albeit at the expense of reduced efficiency, by allowing for a higher work coefficient ψ at a reduced ϕ.

Action stages are preferred in the initial high-pressure turbine sections because they offer better mechanical strength, lower sensitivity to tip clearance losses, and simpler blade design under extreme conditions. In contrast, reaction stages are more efficient at lower pressures and with larger blade radii, making them ideal for low-pressure turbine stages where the gas has expanded. Thus, turbines typically start with action stages for durability and shift to reaction stages for efficiency as pressure decreases and blade size increases.

10.4.1 GENERIC STAGE ANALYSIS

Assume that the hub and tip diameters remain nearly constant throughout a stage, with the hub-to-tip diameter ratio approaching one. As a result, radial velocity components between the annular walls can be neglected, leading to no tangible variations in flow properties from root to tip. Ultimately, this makes it possible to adopt a mean line approach with adequate confidence for preliminary stage performance assessment. Under these assumptions, a calculation procedure similar to the one described for compressor stages can be adopted. However, in the preliminary analysis of turbine stages, it must be recognized that endwall blockage is generally less pronounced than in compressors, and therefore, the work done factor can be assumed to be unitary.

A pseudocode of the matching procedure can be described as follows.

1. Define gas properties: γ, R, $c_p = \frac{\gamma R}{\gamma - 1}$
2. Set inlet conditions: T_i^0, p_i^0, $\dot{m}$, u, α_1,[1] α_2
3. Define rotor profile parameters: stagger angle ξ_R and camber angle θ_c for all stages. For the sake of simplicity, assume constant values for each stage and a repeated stage layout, i.e. $\alpha_1(i) = \alpha_3(i)$ for stage $i > 1$.
4. Define flow area distribution A for each stage.
5. For each turbine stage:
    - Set a tentative value for the flow coefficient of each stage: ϕ^*.
    - Assume a tentative value of stage efficiency η_{is} based on ϕ from the Smith's diagram. Assume a tentative value of total pressure loss coefficient of nozzle rows π_N[2]:
    - If $i = 1$, set:

$$T_1^0 = T_i^0,\ p_1^0 = p_i^0$$

otherwise, update:

$$T_1^0 = T_3^0(i-1),\ p_1^0 = p_3^0(i-1)$$

    - Solve for Mach number Ma_2 at the nozzle outlet using:

$$f(\text{Ma}_2) = \text{Ma}_2 \cos\alpha_2 \left(1 + \frac{\gamma - 1}{2}\text{Ma}_2^2 \cos^2\alpha_2\right)^{-\frac{\gamma+1}{2(\gamma-1)}} - \frac{\dot{m}\sqrt{RT_1^0}}{p_1^0 A_1 \sqrt{\gamma}} = 0$$

    - Compute temperature, pressure, and density at the nozzle outlet:

$$T_2 = \frac{T_1^0}{1 + \frac{\gamma-1}{2}\text{Ma}_2^2},\ p_2 = \frac{\pi_N p_1^0}{\left(1 + \frac{\gamma-1}{2}\text{Ma}_2^2\right)^{\frac{\gamma}{\gamma-1}}},\ \rho_2 = \frac{p_2}{RT_2}$$

    - Compute axial velocity (constant within each stage):

$$c_m = \frac{\dot{m}}{\rho_2 A}$$

    - Compute velocity components at the rotor inlet:

$$c_{u2} = c_m \tan\alpha_2,\ w_{u2} = c_{u2} - u$$

    - Compute relative flow angle at rotor inlet:

$$\beta_2 = \tan^{-1}\left(\frac{w_{u2}}{c_m}\right)$$

- Compute flow deviation[3]:

$$\delta\,[^\circ] = \frac{(AVDR)^3\left(\frac{1}{\sigma_R}\right)^{1.1}(\beta_2\,[^\circ]+\beta_3'\,[^\circ])^{2.25}}{\xi_R\,[^\circ]^{1.45}\left(\frac{t_m}{\ell}\right)^{0.3}\left(22^\circ+0.22\cdot\beta_2'\,[^\circ]^{1.64}\right)}$$

where, other than known symbols, all angles are in degrees, and t_m/ℓ is the blade maximum thickness-to-chord ratio. The Axial-Velocity-Density-Ratio $AVDR = \frac{\rho_2 c_{m2}}{\rho_3 c_{m3}}$ is usually set to one unless more accurate information is available.
- Compute exit flow angles:

$$\beta_3 = \beta_3' - \delta$$

- Compute velocity components at the rotor exit:

$$w_{u3} = c_m \tan\beta_3,\; c_{u3} = w_{u3} - u$$

- Update flow coefficient, compute work coefficient, and stage reaction:

$$\phi = \frac{c_m}{u},\; \psi = \frac{c_{u2}+c_{u3}}{u},\; \varepsilon_r = 1 - \frac{c_{u2}-c_{u3}}{2u}$$

- If $|\phi - \phi^*| <$ tolerance, exit loop.
- Otherwise, update ϕ: $\phi^* \leftarrow \phi$. Update $\eta_{is} = f(\phi, \psi)$.

6. Compute stage pressure ratio and expansion ratio:

$$\pi_s = \left(1 - \frac{(\gamma-1)\psi u^2}{\eta_{\text{is}}\gamma R T_1^0}\right)^{\frac{\gamma}{\gamma-1}},\; \pi_e = \frac{1}{\pi_s}$$

7. Compute stage exit total temperature and total pressure:

$$T_3^0 = T_1^0 - \frac{\psi u^2}{c_p},\; p_3^0 = p_1^0 \pi_s$$

8. Compute overall expansion ratio:

$$\pi_{e,\text{overall}} = \prod_i \pi_e$$

Figure 10.4 shows an exemplary distribution of quantities along a four-stage turbine of prescribed geometry and design characteristics.

10.5 TURBINE COOLING

The primary objective of turbine cooling is to reduce the surface temperature of critical turbine components without significantly impacting performance. This allows the turbines to operate at higher entry gas temperatures, which directly improves the thermal efficiency of the engine cycles in which turbines work.

There are various methods of cooling turbine blades. The most commonly employed techniques include:

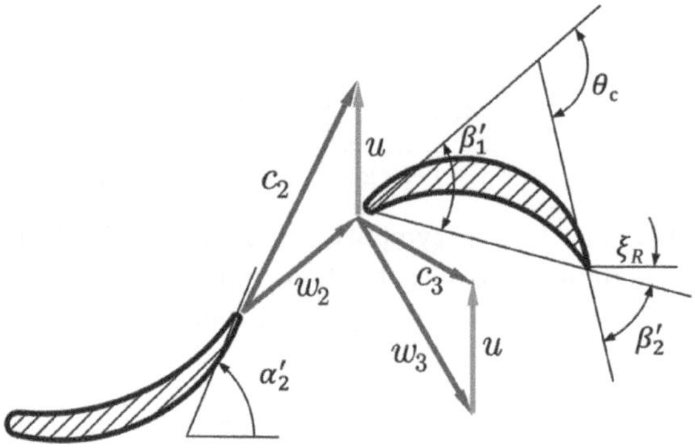

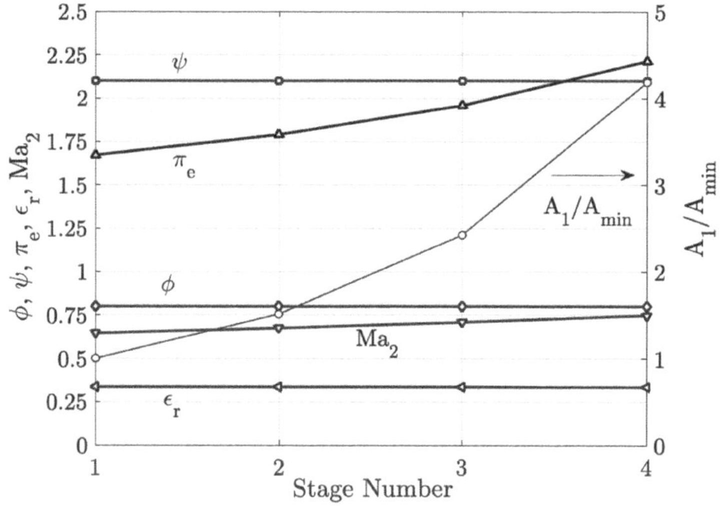

Figure 10.4 Stage characteristics of a sample four-stage turbine. ϕ, ψ, η_{is} selected according to Smith's diagram.

1. Internal cooling (convection + impingement cooling).
2. Film cooling.
3. Transpiration cooling.
4. Thermal Barrier Coatings (TBCs).

10.5.1 INTERNAL COOLING (CONVECTION)

In internal cooling, air is extracted from the compressor section and routed through the internal passages of the turbine blades (Figure 10.5). The cooling air flows through serpentine channels or ribbed passages inside the blade and removes heat by convection. The heat transfer rate in internal cooling is primarily governed by the Nusselt number Nu, which is a function of the Reynolds number Re and the Prandtl number Pr.

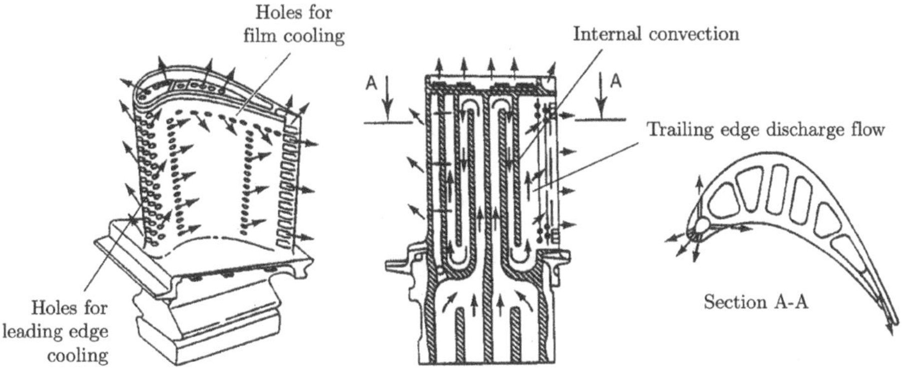

Figure 10.5 Turbine blade cooling.

Newton's law determines the heat transfer rate per unit area, or heat flux, in the absence of substantial compressibility effects (Ma < 0.3):

$$\dot{q} = \alpha(T_g - T_w) \tag{10.1}$$

where α is the convective heat transfer coefficient, T_g the coolant gas temperature and T_w the surface temperature of the blade.

When Ma > 0.3 the impact of high-velocity flows on heat transfer is non-negligible, as high-speed convection involves conversion of mechanical energy into thermal energy, leading to significant temperature fluctuations within the fluid. In these cases, Newton's law is written taking into account the so-called adiabatic wall temperature, T_{aw} of gases as (Figure 10.6):

$$\dot{q} = \alpha(T_{aw} - T_w) \tag{10.2}$$

where T_{aw} is defined as a function of static and total temperatures by using a recovery factor r:

$$T_{aw} = T_g + r\left(T_g^0 - T_g\right) \tag{10.3}$$

r being the recovery factor, i.e. the proportion of free-stream kinetic energy converted into thermal energy as the flow decelerates over an adiabatic surface. In turbulent flows ($0.7 < \text{Pr} < 0.85$), $r \approx \text{Pr}^{1/3}$. Therefore, in high-speed flows, the proper temperature difference driving heat transfer is $(T_{aw} - T_w)$. This implies that the direction of heat transfer—whether heat flows to or from the surface—depends on whether the surface temperature is higher or lower than the adiabatic wall temperature (Figure 10.6).

For temperatures of air in the order of 1000°C, Pr remains largely constant across a wide range of pressures and is around 0.83[4], therefore $\text{Pr}^{1/3} \approx 0.94$. As a consequence, for the sake of simplicity, in the following, we will use the total temperature in lieu of the adiabatic wall temperature:

$$\dot{q} = \alpha(T_g^0 - T_w) \tag{10.4}$$

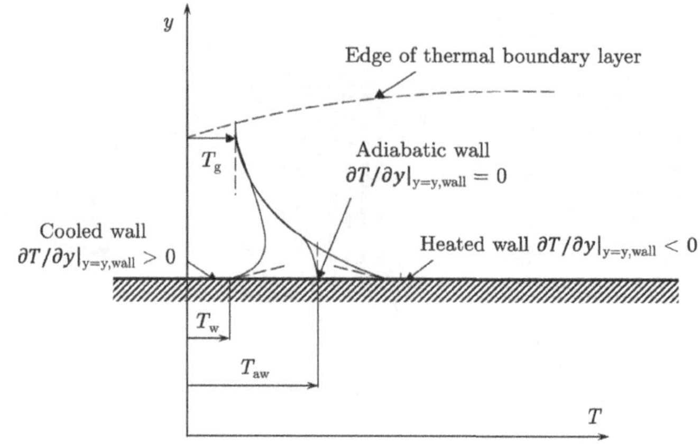

Figure 10.6 Schematic representation of the thermal boundary layer and temperature gradient behavior near a flat wall. The diagram illustrates different wall conditions: cooled wall $\frac{\partial T}{\partial y} > 0$, adiabatic wall $\frac{\partial T}{\partial y} = 0$, and heated wall $\frac{\partial T}{\partial y} < 0$.

The convective heat transfer coefficient, α, is related to the Nusselt number by the equation:

$$\alpha = \frac{\text{Nu} \cdot k}{D} \tag{10.5}$$

where k is the thermal conductivity of the cooling air and D is the hydraulic diameter of the internal cooling channels.

For a turbulent flow inside the cooling channels, the Nusselt number can be calculated using the well-known Dittus-Boelter correlation:

$$\text{Nu}_c = 0.023\,\text{Re}_c^{0.8}\text{Pr}_c^{0.4} \tag{10.6}$$

where $\text{Re}_c = \frac{\rho_c V_c D_c}{\mu_c}$, $\text{Pr} = \frac{c_{pc}\mu_c}{k_c}$ and V_c is the velocity of cooling air having viscosity μ_c. The correlation performs relatively well for internal passages of circular cross sections of diameter D_c. Also, it still provides reasonably accurate predictions for non-circular passages when the hydraulic diameter D_h (in lieu of D_c) of the cross section is used. Rotational effects on cooled rotor blades should also be accounted for using correction factors in the Dittus-Boelter correlation.

In advanced aircraft turbines, because of extremely high entry gas temperatures, simple convective cooling alone has proven inadequate to keep the blade material within acceptable temperature limits. To address this, impingement cooling is commonly applied along with convection to cool both the leading edge and mid-chord sections of the blade, thus enhancing the effectiveness of internal cooling. Specific correlations for the heat transfer coefficient in impingement cooling can be found in the literature.

In order to fully characterize internal convective cooling, we can adopt a very simplified approach based on the model shown in Figure 10.7. Here, the hollow turbine vane is idealized as a double-walled channel, where hot combustion gases

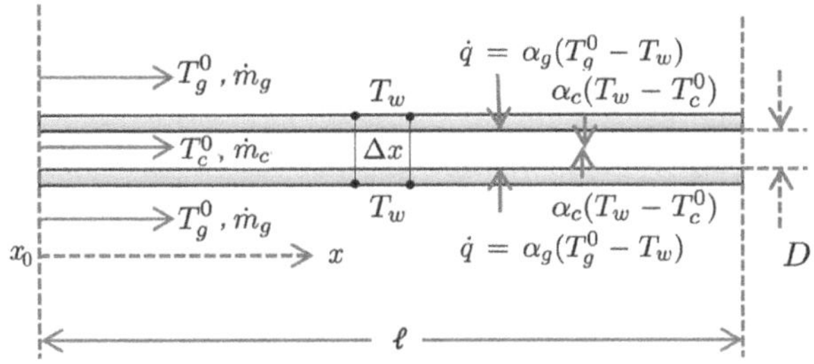

Figure 10.7 Schematic representation of convective cooling in a double-walled channel.

$T_g^0, \dot{m}_g$ flow through the outer sections, while cooling air $T_c^0, \dot{m}_c$ flows through the central passage, where the distance between the walls is $D = D_h$. Convective heat transfer inside the vane is modeled as a pipe flow, while outside the blade, it can be described by correlations of turbulent flow over a flat plate at constant temperature. The local Nusselt number correlation is as follows:

$$\mathrm{Nu}_g = 0.0292\,\mathrm{Re}_{gx}^{0.8}\,\mathrm{Pr}_g^{0.33} \tag{10.7}$$

where x is the local coordinate along the blade chord, $\mathrm{Re}_{gx} = \frac{\rho_g c_g x}{\mu_g}$ is the x-based Reynolds number of the hot flow having c_g velocity.

Neglecting axial heat conduction and assuming zero blade wall thickness to ignore pertinent conduction leads to $\dot{q} = \alpha_g(T_g^0 - T_w) = \alpha_c(T_w - T_c^0)$, which, upon substitution of equations 10.6 and 10.7, leads to the following equation for wall temperature as a function of x:

$$T_w(x) = \frac{0.0292\frac{k_g}{x}\mathrm{Re}_{gx}^{0.8}\mathrm{Pr}_g^{1/3}T_g^0 + 0.023\frac{k_c}{D_h}\mathrm{Re}_c^{0.8}\mathrm{Pr}_c^{0.4}T_c^0(x)}{0.0292\frac{k_g}{x}\mathrm{Re}_{gx}^{0.8}\mathrm{Pr}_g^{1/3} + 0.023\frac{k_c}{D_h}\mathrm{Re}_c^{0.8}\mathrm{Pr}_c^{0.4}} \tag{10.8}$$

The latter equation can be solved as a function of x assuming the worst-case scenario, where T_g^0 does not change with x (i.e. hot gases do not cool during the flow expansion within the blade channel) and gases and coolant flow in a cocurrent way. Summing the heat fluxes from both sides of the blade into the i-th increment of x, i.e. Δx = blade length/number of finite number elements, according to the following equation, gives:

$$2\dot{q}(x) = \frac{\dot{m}_c c_{pc}}{H\Delta x}\left(T_{c,i+1}^0 - T_{c,i}^0\right) \tag{10.9}$$

Substituting $\dot{q}(x)$ and solving for $T_{c,i+1}^0$, the increase in cooling temperature flow is obtained:

$$T_{c,i+1}^0 = T_{c,i}^0 + \frac{2H\Delta x}{\dot{m}_c c_{pc}} \cdot 0.0292\frac{k_g}{x}\mathrm{Re}_{gx}^{0.8}\mathrm{Pr}_g^{1/3}\left[T_g^0 - T_w(x)\right] \tag{10.10}$$

where H is the blade height. The initial boundary condition is given by the known value of T_c^0 at $i = 1$. Using the preceding equation makes it possible to estimate the

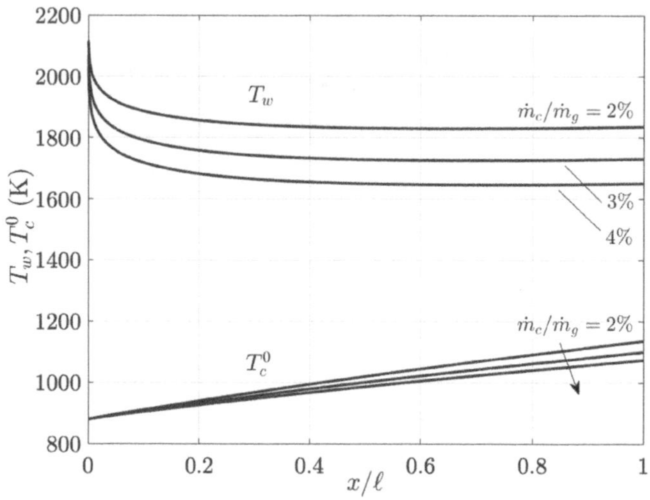

Figure 10.8 Convective cooling: wall temperature T_w and coolant total temperature T_c^0 distribution along the chordwise location of a turbine vane for different coolant-to-hot gas mass flow ratios $\dot{m}_c/\dot{m}_g$.

temperature profile from the internal side as well as the pertinent mass flow rate of the coolant gas needed to maintain the skin of the outer metal at a given temperature.

Figure 10.8 shows the distribution of wall temperature T_w and coolant temperature T_c along the chordwise direction of a turbine vane under varying coolant-to-hot gas mass flow ratios $\dot{m}_c/\dot{m}_g$. As expected, higher coolant flow rates result in lower wall temperatures, thus improving cooling efficiency. It is evident that pipe flow convection cooling alone is insufficient to lower the wall temperature significantly below 1650 K when the combustion temperature is 2280 K, and the coolant-to-hot gas mass flow ratio is below 4%. Incorporating fin-type structures inside the blade can enhance cooling efficiency; however, convection cooling alone is generally inadequate to maintain the blade material at an acceptable temperature.

10.5.2 INTERNAL PLUS FILM COOLING

Film cooling involves the injection of cooler air onto the blade surface to create a protective layer (film) between the hot gases and the blade. Film cooling is used together with internal cooling. Cooling air is ejected through small holes on the surface of the blade, forming a cooling film that reduces the convective heat transfer from the hot gas to the blade surface (Figure 10.5).

To apply conventional slot cooling correlations, individual holes must be treated as an equivalent continuous slot. The total coolant flow rate, $\dot{m}_c$, is split into two parts (Figure 10.9). A fraction ν of the coolant flow is directed internally through the blade for convective cooling. To model the latter, the equations of the preceding section can be applied using the proper adjustments. In particular, the updated Reynolds number

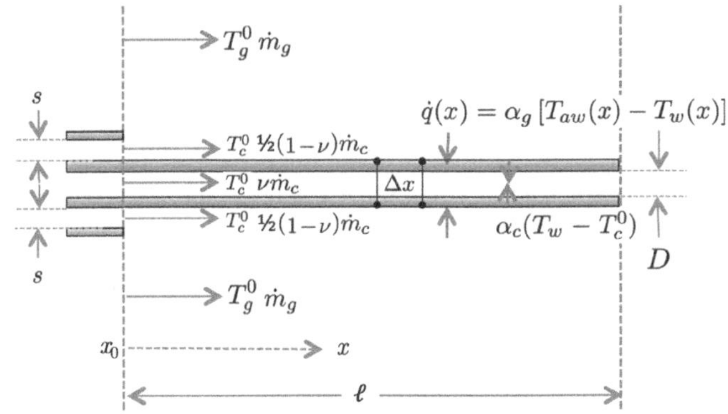

Figure 10.9 Schematic representation of convective cooling with film cooling injection.

for internal pipe flow is computed as follows:

$$\mathrm{Re}_c = \frac{\rho_c V_c' D_c}{\mu_c} = \frac{\nu \dot{m}_c D_c}{A_c \mu_c} \tag{10.11}$$

The remaining fraction of coolant flow, $1 - \nu$, is used for film cooling. By balancing internal convective cooling with external film cooling, the turbine blade can be maintained at a safe operating temperature. Turbulent film cooling correlations adjust for heat transfer changes that occur as a result of variations in both velocity and temperature near the surface. Focusing first on velocity, the momentum ratio between the coolant and the hot gas flow is defined as:

$$\varphi = \frac{\rho_c V_c}{\rho_g c_g} \tag{10.12}$$

10.5.2.1 Film Flow: Boundary Layer Flow Regime

The boundary layer regime will occur for values between $0.5 < \varphi < 1.3$, represented by the following correlation:

$$\mathrm{Nu}_g = 0.069\,\mathrm{Re}_s^{0.7} \tag{10.13}$$

where the Reynolds number is computed based on the slot width s as follows:

$$\mathrm{Re}_s = \frac{\rho_c V_c s}{\mu_c} = \frac{(1-\nu)\dot{m}_c s}{A_s \mu_c} \tag{10.14}$$

where A_s is the area of the slot.

Using the latter correlation, the heat flux outside the vane can be computed as a function of the axial coordinate x by neglecting axial conduction and considering zero wall thickness:

$$\dot{q}(x) = \alpha_g\left[T_{aw}(x) - T_w(x)\right] = 0.069\frac{k_g}{x}\left(\mathrm{Re}_s \cdot \frac{x}{s}\right)^{0.7}\left[T_{aw}(x) - T_w(x)\right] \tag{10.15}$$

where the adiabatic wall temperature T_{aw} is used instead of the total gas temperature T_g^0 as in film cooling the recovery factor can no longer be considered around unity due to the presence of cool flow attached to blade surface, which alters the temperature distribution in the wall's thermal boundary layer.

By neglecting axial conduction and assuming zero wall thickness for simplicity, the wall temperature equation as a function of axial position is obtained by equating the convective and film heat fluxes:

$$T_w(x) = \frac{0.069\frac{k_g}{x}\left(\mathrm{Re}_s\frac{x}{s}\right)^{0.7} T_{aw}(x) + 0.023\frac{k_c}{D_h}\mathrm{Re}_c^{0.8}\mathrm{Pr}_c^{0.4}T_c^0(x)}{0.069\frac{k_g}{x}\left(\mathrm{Re}_s\frac{x}{s}\right)^{0.7} + 0.023\frac{k_c}{D_h}\mathrm{Re}_c^{0.8}\mathrm{Pr}_c^{0.4}} \tag{10.16}$$

To properly estimate the adiabatic wall temperature, a film-cooling effectiveness parameter η_x is defined as follows:

$$\eta_x = \frac{T_g^0 - T_{aw}(x)}{T_g^0 - T_c^0} \tag{10.17}$$

such that:

$$T_{aw}(x) = T_g^0 - \eta_x(T_g^0 - T_c^0) \tag{10.18}$$

When $\eta_x = 1$, the wall temperature is equal to the coolant stagnation temperature $T_{aw} = T_c^0$, indicating *perfect cooling*, where the film cooling fully protects the surface. Opposite to this is the case where $\eta_x = 0$, that is, when the wall temperature equals the stagnation temperature of the hot gas $T_{aw} = T_g^0$, meaning no cooling effect. As a general case $0 < \eta_x < 1$, where the surface experiences *partial cooling*.

Different correlations are available for both the near-slot region and the region farther downstream from the slot. The near-slot region is characterized by $0 < \frac{x}{s} < 150$, and the correlation is given by:

$$\eta_x = 0.6\left(\frac{x}{\varphi \cdot s}\right)^{-0.3}\left(\mathrm{Re}_s\frac{\varphi\mu_c}{\mu_g}\right)^{0.15}, \quad \text{if}\, \eta_x \leq 1\,\text{or} \tag{10.19}$$

$$\eta_x = 1.0, \quad \text{otherwise} \tag{10.20}$$

Data for this correlation are limited to $0.5 < \varphi < 1.3$, $0.8 < \frac{\rho_{sc}}{\rho_{sg}} < 2.5$, and $0.0019 < s < 0.0064\,\mathrm{m}$. The correlation for the region defined by $\frac{x}{s}>150$ is expressed as:

$$\eta_x = 3.68\left(\frac{x}{\varphi \cdot s}\right)^{-0.8}\left(\mathrm{Re}_s\frac{\mu_c}{\mu_g}\right)^{0.2} \tag{10.21}$$

10.5.2.2 Film Flow: Wall Jet Regime

For $\varphi>1.3$, a wall jet regime is estimated to occur where a high-velocity jet of the coolant is established, which is directed parallel to the blade surface. In these conditions, the following correlation applies:

$$\mathrm{Nu}_{gx} = \frac{h_g x}{k_g} = 0.10\mathrm{Re}_s^{0.8}\left(\frac{x}{s}\right)^{0.44} \tag{10.22}$$

The convective heat transfer per unit area at any axial position along the outer surface of the blade is represented as follows for the wall jet regime:

$$\dot{q}(x) = \alpha_g[T_{aw}(x) - T_w(x)] = 0.10\frac{k_g}{x}\mathrm{Re}_s^{0.8}\left(\frac{x}{s}\right)^{0.44}[T_{aw}(x) - T_w(x)] \quad (10.23)$$

Once again, assuming negligible axial conduction and zero wall thickness, the equation for the wall temperature along the axial position becomes:

$$T_w(x) = \frac{0.10\frac{k_g}{x}\mathrm{Re}_s^{0.8}\left(\frac{x}{s}\right)^{0.44}T_{aw}(x) + 0.023\frac{k_c}{D_h}\mathrm{Re}_c^{0.8}\mathrm{Pr}_c^{0.4}T_c^0(x)}{0.10\frac{k_g}{x}\mathrm{Re}_s^{0.8}\left(\frac{x}{s}\right)^{0.44} + 0.023\frac{k_c}{D_h}\mathrm{Re}_c^{0.8}\mathrm{Pr}_c^{0.4}} \quad (10.24)$$

The adiabatic wall temperature T_{aw} is determined using film-cooling effectiveness for the regime where $0 < x/s < 150$, $1.3 < \varphi < 4$, and $x/(\varphi \cdot s) < 8$:

$$\eta_x = 1.0 \quad (10.25)$$

For the regime $0 < x/s < 150$, $1.3 < \varphi < 4$ and $8 < x/(\varphi \cdot s) < 11$, the correlation becomes:

$$\eta_x = \left(0.6 + 0.05\frac{x}{\varphi \cdot s}\right)^{-1} \quad (10.26)$$

For the regime $0 < x/s < 150$, $1.3 < \varphi < 4$, and $x/(\varphi \cdot s) > 11$, the effectiveness is defined as:

$$\eta_x = 0.7\left(\frac{x}{s}\right)^{-0.3}\left(\mathrm{Re}_s\frac{\mu_c}{\mu_g}\right)^{0.15}\varphi^{-0.2} \quad (10.27)$$

For the far downstream region where $x/s > 150$, the correlation provided in equation 10.21 is reused.

The increase in coolant flow temperature $T_{c,i+1}$ between increments Δx for a given axial distance is calculated using the following relationship, incorporating the heat flux $\dot{q}(x)$ for the boundary layer regime when $0.5 < \varphi < 1.3$:

$$T_{c,i+1}^0 = T_{c,i}^0 + \frac{2H\Delta x}{\dot{m}_c c_{pc}} \cdot 0.069\frac{k_c}{x}\left(\mathrm{Re}_s \cdot \frac{x}{s}\right)^{0.7}[T_{aw}(x) - T_w(x)] \quad (10.28)$$

In the wall jet regime where $\varphi > 1.3$, the corresponding temperature increase for the coolant flow is expressed as:

$$T_{c,i+1}^0 = T_{c,i}^0 + \frac{2H\Delta x}{\dot{m}_c c_{pc}} \cdot 0.10\frac{k_c}{x}\mathrm{Re}_s^{0.8}\left(\frac{x}{s}\right)^{0.44}[T_{aw}(x) - T_w(x)] \quad (10.29)$$

These equations are integrated numerically to compute heat flux $\dot{q}(x)$, coolant flow temperature $T_c(x)$, adiabatic wall temperature $T_{aw}(x)$ and wall temperature $T_w(x)$ along the blade surface.

The left-hand side of Figure 10.10 shows the wall temperature distribution along the axial position of a turbine blade for different coolant-to-hot-gas mass flow rate ratios $\dot{m}_c/\dot{m}_g$. The data correspond to a cooling configuration in which 75% of the

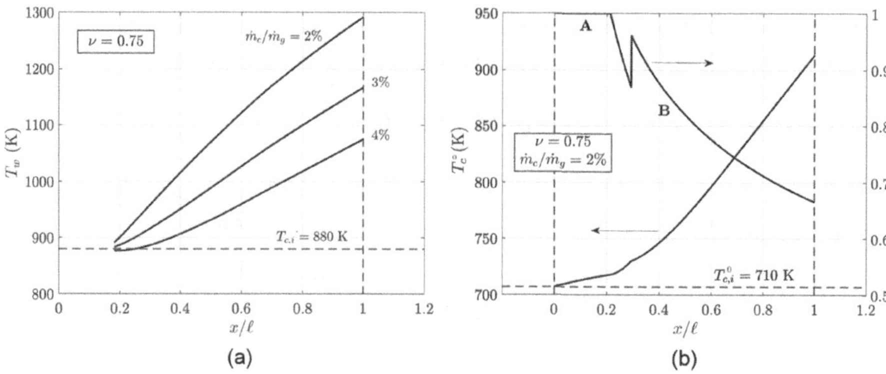

Figure 10.10 Convective plus film cooling for $\nu = 0.75$: (a) sample wall temperature T_w as a function of axial position in a boundary layer flow regime for different coolant-to-hot-gas mass flow ratios $\dot{m}_c/\dot{m}_g$; (b) cooling air total temperature T_c^0 and film-cooling effectiveness η_x in wall jet regime, zone **A**, $x/(\varphi \cdot s) < 8$, near-slot region), zone **B**, $x/(\varphi \cdot s) > 8$, fully developed film regime).

coolant is used for internal cooling and 25% for film cooling, with an inlet gas temperature of $T_g = 2280$ K and an inlet coolant total temperature of $T_{c,i}^0 = 880$ K. An effective reduction in temperature from 2280 K significantly reduces the heat flux and the resulting wall temperature. The right-hand side of Figure 10.10 illustrates the variation in the total temperature of the cooling air T_c^0 and the efficiency of the film cooling η_x along the normalized stream-wise coordinate x/ℓ. In Zone A, where $x/\ell < 0.22$, the cooling flow is in the wall jet regime, which means the coolant stays attached to the wall with minimal mixing with the main hot gas. The cooling effectiveness η_x is around 1, indicating excellent cooling performance. As the flow progresses into Zone B, where $x/\ell > 0.22$, the coolant has mixed more substantially with the surrounding hot gas, thus transitioning into a fully developed film cooling layer. The cooling effectiveness η_x decreases to 0.67 as the cooling effect is reduced due to enhanced mixing with hot gases, while the cooling air temperature T_c^0 shows a gradual increase as the coolant gains heat from the surrounding gas.

The flat plate film and convection cooling correlations presented above effectively capture the cooling trends but may be overly optimistic due to their idealized assumptions. In reality, injecting coolant through discrete holes rather than a continuous slot enhances turbulence and heat transfer, and the latter is further amplified by the positive pressure gradients within turbine blades that thin the boundary layers.

10.5.3 TRANSPIRATION COOLING

Transpiration cooling involves a porous surface through which the cooling air seeps. Unlike conventional film cooling, air flows through the porous material, therefore allowing for a more evenly distributed cooling layer.

Typically, Darcy's law governs the heat transfer process in transpiration cooling:

$$\mathcal{V} = -\frac{K_p}{\mu}\nabla p \tag{10.30}$$

where $\mathcal{V}$ is the seepage velocity, K_p the permeability of the porous material, μ the dynamic viscosity of the cooling air and ∇p the pressure gradient across the porous medium. The characteristics of the porous material and the flow conditions determine the general cooling efficacy.

Today, the leading edge and midsection of turbine vanes and rotor blades are often cooled by transpiration. Usually, metal foams or ceramic matrix composites (CMCs) serve as porous materials. CMCs are preferred for their high-temperature resistance, low density, and capability to withstand thermal stress. Recognizing that including the porous material in the blade is more usual in cutting-edge designs employing advanced manufacturing methods such as 3D printing, both integral porous structures and inserts are employed.

10.5.4 THERMAL BARRIER COATINGS (TBCs)

Thermal Barrier Coatings (TBCs) are used to protect the turbine blade surface by applying a layer of material with low thermal conductivity. In this way, the heat transfer to the blade material is reduced, which allows for higher operating temperatures in the turbine stage.

The heat flux through the TBC layer can be described by Fourier's law of conduction:

$$\dot{q} = -k_{\text{TBC}}\frac{\Delta T}{\Delta x} \tag{10.31}$$

where k_{TBC} is the thermal conductivity of the TBC material, ΔT is the temperature gradient across the TBC and Δx is the thickness of the TBC layer.

Advanced ceramic oxides are the most commonly used solutions for TBCs due to their superior thermal insulation characteristics (thermal conductivity 1.0 to 1.5 W/(m·K)), high melting point (approximately 2,700°C), and phase stability at high temperatures.

Manufacturing techniques such as Electron Beam Physical Vapor Deposition (EB-PVD) and Air Plasma Spraying (APS) allow precise control of coating properties, therefore ensuring good adhesion, low thermal conductivity, and high-temperature resistance.

TBCs are often used in combination with internal and external cooling methods to maximize thermal protection.

10.6 THERMAL EFFICIENCY OF THE COOLED TURBINE

In a cooled turbine, the presence of several gas streams changes the definition of efficiency relative to the adiabatic one. The interaction of multiple streams causes entropy to rise from frictional and mixing losses, while the main gas stream expands through both a nozzle and a rotor. Coolant air streams are also added that mix with the main stream and cause energy transfer and loss creation.

The thermal efficiency is expressed as the ratio of the net useful power output to the isentropic power ideally available, incorporating both primary and cooling flows:

$$\eta_{is} = \frac{\dot{m}_g(\Delta h^0)_g + \sum_j \dot{m}_{c,j}(\Delta h^0)_j - P_c}{\dot{m}_g(\Delta h^0)_{g,s} + \sum_j \dot{m}_{c,j}(\Delta h^0)_{j,s}} \tag{10.32}$$

where $\dot{m}_g$ is the mass flow rate of the primary gas stream and $(\Delta h^0)_p$ the pertinent actual total enthalpy drop, $\dot{m}_j$ the mass flow rate of individual j−th coolant streams undergoing $(\Delta h^0)_j$ enthalpy drop. P_c is the cumulative parasitic power required to pump the coolant flows in the turbine rotors. At the denominator $(\Delta h^0)_{g,s}$ and $(\Delta h^0)_{j,s}$ are the ideal isentropic enthalpy drops for the primary and coolant flows, respectively.

The expansion of both the primary gas flow and all injected coolant air is computed from the turbine inlet total pressure to the exit total pressure assuming that the coolant and primary flows are fully mixed and exit the turbine at the same total temperature. To allow a fair comparison between a cooled and an uncooled turbine, the coolant is assumed to be at the same pressure as the primary stream at the turbine inlet and is considered to expand through the turbine along with the combustion products.

The isentropic enthalpy change for each coolant stream is given by:

$$(\Delta h^0)_{j,s} = c_{pc} T_{1c}^0 \left[1 - \left(\frac{p_3^0}{p_1^0}\right)^{(\gamma-1)/\gamma}\right]_j \tag{10.33}$$

where T_{1c}^0 is the stagnation temperature of the coolant at the turbine inlet, while p_1^0 and p_3^0 are the total pressure at the j−th stage inlet, and exit, respectively.

The actual enthalpy change for a coolant stream j is calculated as:

$$(\Delta h^0)_j = c_{pc}(T_{1c}^0 - T_3^0)_j \tag{10.34}$$

where, due to mixing, T_3^0 is the stagnation temperature at the turbine exit, common to both the coolant and the primary gas.

In practical turbine configurations, the coolant mass flow entering the j-th rotor stage (i.e. at station 2) is typically injected near the hub and initially travels in a predominantly radial direction. As a result, the tangential component of the coolant's absolute velocity at entry is small or approximately equal to the blade speed at the hub:

$$c_{u2} \approx u_{2,\text{hub}} \tag{10.35}$$

As the coolant passes through the rotor passage, it is swept outward and mixed with the main flow, ultimately leaving the rotor at the mean radius (station 3). At this point, the coolant acquires a significant tangential velocity component, denoted by $c_{u3,m}$, and moves at the mean blade speed $u_{3,m}$. The coolant thus gains angular momentum at the expense of turbine work. Consequently, the change in specific total enthalpy associated with the coolant stream is expressed as follows:

$$P_c = \dot{m}_{c,R}\left(u_{3,m}\cdot c_{u3,m} - u_{2,\mathrm{hub}}^2\right) \tag{10.36}$$

which directly quantifies the rotor work required to accelerate the coolant from its initial low-momentum state at the hub to its final energetic state at the rotor exit.

The final expression of the thermal efficiency of the cooled turbine is therefore (assuming just one cooled rotor):

$$\eta_{is} = \frac{\dot{m}_g(\Delta h^0)_g + \sum_j \dot{m}_{c,j}c_{pc}(T_{1c}^0 - T_3^0)_j - \dot{m}_{c,R}\left(u_{3,m}\cdot c_{u3,m} - u_{2,\mathrm{hub}}^2\right)}{\dot{m}_g(\Delta h^0)_{g,s} + \sum_j \dot{m}_{c,j}c_{pc}T_{1c}^0\left[1-\left(\frac{p_3^0}{p_1^0}\right)^{(\gamma-1)/\gamma}\right]_j} \tag{10.37}$$

10.7 PRACTICAL ASPECTS OF HPT AND LPT STAGES IN GAS TURBINES

Multistage axial turbines in gas turbine engines are conventionally divided into high-pressure turbine (HPT) and low-pressure turbine (LPT) sections (Figure 10.11), each customized to different thermodynamic and flow requirements. The HPT follows the combustor and typically has a small number of highly-loaded stages operating in the transonic/supersonic regime. The LPT follows, expanding the already cooled gas over more stages at a lower pressure ratio per stage, usually with fully subsonic flows.

HPT stages often run at the practical tip speed limit and large flow turning. In contrast, LPT stages process a much larger volumetric flow rate at a lower pressure

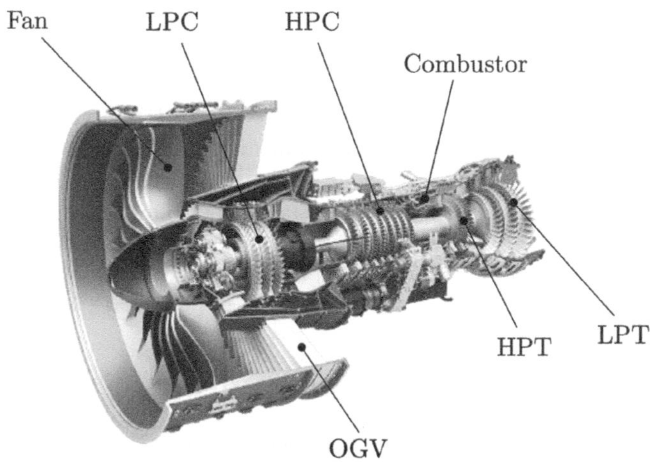

Figure 10.11 Cross-sectional diagram of a high-bypass turbofan engine.

ratio per stage, so they typically use lower loading but a higher flow coefficient ϕ to accommodate the axial flow. Typical HPT designs use $\phi \sim 0.3$–0.6, $\psi \sim 1.5$–2.0, whereas LPT designs often have $\phi \sim 0.7$–1.0 and $\psi \sim 0.5$–1.0. Furthermore, HPT stages, which require internal and film cooling, are often designed with lower ϕ and degree of reaction ε_r to reduce relative velocities and heat transfer rates.[5]

Generally, as the Mach number increases, HPT stage losses increase sharply due to shock formation. Transonic HPT blades produce oblique shocks, often at the trailing edge, which dominate losses in those stages. LPT stages, in contrast, remain subsonic and avoid shock losses, yielding higher static pressure recovery. Thus, a fundamental trade-off is found where HPT pushes the efficiency peak with a high enthalpy drop with some loss of stage efficiency, whereas LPT achieves higher passage efficiency per stage, but requires many stages to achieve the desired total expansion.

10.7.1 STAGE COUNT AND PRESSURE RATIO SCALING

The number of turbine stages and the per-stage pressure ratio differ significantly between high-pressure turbines and low-pressure turbines due to distinct thermodynamic and mechanical constraints.

In commercial aeroengines, the overall expansion ratio across the HPT is typically moderate (e.g. 4–7 in modern turbofans), but this is achieved using very few stages. In practice, single-stage HPTs are feasible for a pressure ratio up to about 5–6. Beyond this, either additional HPT stages or split-spool architecture is required. LPT stages (usually 4–7) are more lightly loaded and require less cooling, but need careful aerodynamic optimization to meet overall performance targets. Published design data and cycle studies indicate typical LPT stage pressure ratios on the order of 1.15–1.40 per stage. For example, the five-stage LPT of the NASA/GE EEE program achieved >91% efficiency at the design point, consistent with per-stage pressure ratios of around ≈1.2. Similarly, the six-stage Trent XWB LPT has an overall expansion that works out to roughly the same range per stage, and the three-stage PW1000G LPT–which is much more loaded–operating closer to the upper end (≈1.4 per stage).

Modern day fighter jets have a single-stage cooled HPT (Figure 10.12) because the latter is one of the most expensive parts of the engine. Single-stage LPTs were the norm until a few years ago, when additional stages were added for improved fuel burn. Today, almost all engines have single-stage LP turbines, except for the General Electric F110, the Pratt & Whitney F135, which have two-stage LP turbines, and the General Electric/Rolls-Royce F136, which features a three-stage LPT, possibly due to a higher fan pressure ratio than the other engines.

10.7.2 OPERATING RANGE AND CONSTRAINTS

HPT and LPT sections have different operating envelopes.

HPT stages are designed for one or two high-load operating points (takeoff/climb) with relatively limited part-speed flexibility. Because an HPT stage often runs near

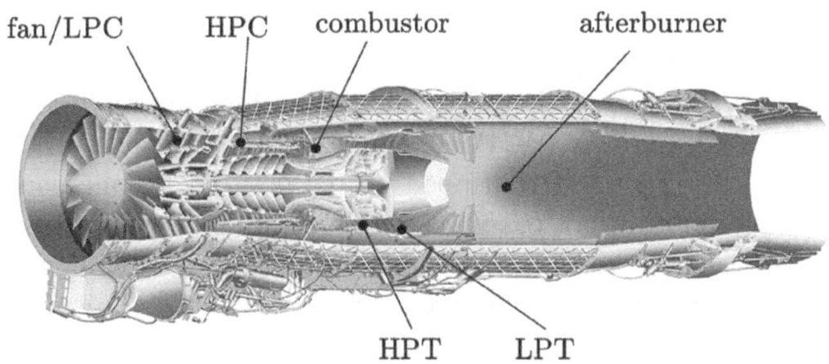

Figure 10.12 The Eurojet EJ200 aircraft engine. (Lane, R. J., and Behenna, J., "EJ200—The Engine for the New European Fighter Aircraft," *Journal of Engineering for Gas Turbines and Power*, 113: 25–32 (1991). Reproduced from ASME with permission.)

choked flow at design, its off-design mass flow and turn can only vary within a narrow band.

LPT stages usually span a wide range of mass flows (idle to takeoff) smoothly. These stages must tolerate large variations in inlet conditions (from high-mass-flow, low-pressure-ratio at takeoff to low-mass-flow, higher-pressure-ratio at cruise) without compromising their flow stability. In an LPT, choking is generally not reached in normal operation, but aerodynamic loading can cause separation near the endwall due to low Reynolds numbers at cruise. LPT blades often feature low-solidity and high-lift profiles, which make them susceptible to separation when off-design incidence changes. Additionally, an LPT turbine must cope with the aerodynamics of a heavily expanding flow: the axial-velocity profile tends to have large gradients (low near hub and tip due to leakage and secondary flows) that can degrade performance. Finally, LPT stages may operate at much lower chord Reynolds at cruise, requiring careful profile design to avoid laminar separation. These differences mean that the map and control strategy of each section must be designed for its own stable region.

10.7.3 TEMPERATURE LIMITS AND COOLING

One of the starkest contrasts between HPT and LPT is temperature and cooling.

In terms of turbine inlet temperature, the highest values are found in military aero-engines. Publicly available data demonstrate that it is consistently between 1450°C and 1580°C in fourth-generation aircraft engines. However, much higher values are observed in more recent engines, with particular emphasis on Pratt & Whitney F135, where it is nearly 2000°C. As a result, HPT blades are made from advanced high-temperature materials (single-crystal or directionally solidified Ni-based superalloys).

As the metal temperature of an HPT blade is usually limited to ≈1100°C–1200°C, nearly 15%–20% of compressed air is bled for cooling from the high-pressure compressor. Typical cooling schemes include serpentine passages with rib turbulators, showerhead or film-hole injection on the pressure and suction sides, trailing-edge slot or pin cooling, as well as tip cooling, be it of the squealer or convection type. As noted above, these features allow high work extraction but reduce stage efficiency. The practical challenges are immense: cooling geometries must be manufacturable (often via casting or additive methods), durable under high thermal gradients, and effective at varying engine conditions. For example, coolant injected through film holes at low blow ratios can be rapidly stripped off, so pressure balance and hole shaping are critical. In addition, hot gas ingestion in rotating cavities (disk cavities, tip gaps) must be prevented. Controlling the thermal stress is a challenge: cooling creates temperature gradients that can fatigue the material, and any mismatch in coolant distribution (e.g. blocked hole) can cause local hot spots.

LPT blades see much lower temperatures (often <800°C–900°C at rotor inlet) and operate at much lower pressures. In many engines, LPT blades (particularly mid-to-late stages) are uncooled or only minimally cooled, e.g. tip or trailing-edge cooling may be used in very high-power turboprops. The materials can therefore be lighter: for example, titanium or titanium-aluminide alloys are common for LPT rotors and stators, taking advantage of lower density and adequate strength at moderate temperatures.[6]

10.7.4 EFFICIENCY TRENDS ACROSS REGIMES

Stage efficiency varies with the flow coefficient, loading, and Mach number. Both HPT and LPT are designed to peak near their design flow conditions, with efficiency dropping off at off-design.

Beyond indications from the Smith diagram, which remains indeed fundamental for preliminary sizing, HPT efficiency is somewhat penalized by cooling and shock losses; typical first-stage efficiencies might be in the upper-80%s to mid-90%s at design. LPT efficiency can reach similar or slightly higher levels, often in the mid 90%s.

As operating conditions vary, efficiency decreases more sharply as losses increase. For example, transonic HPT stages typically see a steep drop in efficiency if exit Mach increases because of enhanced shock/boundary-layer interaction. LPT stages, running at high ϕ, may actually see efficiency degrade if ϕ drops (e.g. at part load) because stall losses increase. In general, both spools have flat bottom efficiency curves: HPT is optimized at high loads, often near 100% speed, while LPT is designed for cruise/runaway conditions with a broad plateau. Unsteady effects and Reynolds number variations also affect performance; e.g. at low rotational speed, LPT transition may delay, and the separation may grow, further reducing efficiency.

10.8 PROBLEMS

1. A 4-stage, low-reaction, repeated-stage axial turbine handles combustion gases with properties: $\gamma = 1.40$, $R = 287$ J/(kg·K) (thus $c_p = \dfrac{\gamma R}{\gamma - 1} \approx 1004$ J/(kg·K)). The inlet total conditions are $T_1^0 = 650$ K and $p_1^0 = 1.5 \times 10^6$ Pa. The turbine operates at a constant peripheral speed (mean radius) of $u = 180$ m/s and at a rotational speed of 4500 rpm. The mass flow rate through the machine is $\dot{m} = 50$ kg/s. The

stator vanes set the absolute exit flow angle to $\alpha_2 = 60°$. Each stage has isentropic efficiency $\eta_{is,c} = 0.925$. The target flow coefficient is $\phi_{\text{target}} = 0.80$ at each stage. The rotor profiles use circular-arc camber lines with camber angle 95° and stagger angle 18°. The blade chord is $\ell = 0.025$ m, the maximum thickness-to-chord ratio is $t_{\max}/\ell = 0.2$, and the chord-normalized maximum camber is $a_r/\ell_r = 0.5$. These quantities are constant across all stages. At midspan, rotor solidity is $\sigma_R = 1.2$, and the area–velocity–density ratio is AVDR = 1.0. The nozzle-outlet (stator exit) flow passage areas for the four stages are:

$$A_1 = \{0.13836,\ 0.24000,\ 1.08000,\ 1.44000\}\ \text{m}^2$$

Determine the characteristics of all five stages, including velocity triangles at each stage, pressure and temperature variations across the stages, work output, and degree of reaction per stage. Finally, determine the overall expansion ratio of the turbine.

[π_e =4.85]

2. A single turbine nozzle vane has constant height $H = 55$ mm and constant chord $\ell = 40$ mm. It handles $\dot{m}_g = 2$ kg/s of hot gases $\gamma_g = 1.33, c_{pg} = 1120$ J/(kg·K), $k_g = 0.164$ W/(m·K) having $T_g = 2200$ K, $p_g = 30$ bar and Mach number $\text{Ma}_g = 0.4$. The cooling air ($\gamma_c = 1.4, c_{pc} = 1004.5$ J/(kg·K), $k_c = 0.053$ W/(m·K)) flows in internal passages of area $A_c = 5.0265\text{e-}05\ m^2$, at $T_c = 880$ K, $p_c = 30$ bar and Mach number $\text{Ma}_c = 0.35$. Determine the minimum wall temperature on the turbine vane as a result of convective cooling. Also, determine the maximum coolant total temperature.

[1706 K, 1022 K]

3. A turbine vane is subjected to high-temperature combustion gases and a cooling flow to regulate its surface temperature. The combustion gases have a total pressure of $p_g^0 = 3.04$ MPa and a total temperature of $T_g^0 = 2280$ K, with a thermal conductivity of $k_g = 0.164$ W/(m·K), a ratio of specific heats $\gamma_g = 1.3$, a Mach number $\text{Ma}_g = 0.6$, and a dynamic viscosity $\mu_g = 7.41 \cdot 10^{-5}$ Pa·s. The cooling flow has the same total pressure of $p_c^0 = 3.04$ MPa and a total temperature of $T_c^0 = 880$ K. Its thermal conductivity is $k_c = 0.0553$ W/(m·K), the ratio of specific heats is $\gamma_c = 1.4$, the Mach number is $\text{Ma}_c = 0.3$ and the dynamic viscosity is $\mu_c = 3.89 \cdot 10^{-5}$ Pa·s. The vane has a height of $H = 0.05$ m, a width of $W = 0.026$ m and a chord length of $\ell = 0.05$ m. The specific gas constants for the combustion gases and the coolant are $R_g = 292$ J/kg·K and $R_c = 287$ J/kg·K, respectively. The coolant mass flow is 2% of the hot gas one, and is injected into the boundary layer through a slot of width $s = 0.001$ m with a fraction of $\nu = 0.75$ of the total cooling flow dedicated to internal convection. Determine the wall temperature distribution along the vane surface considering both convective and film cooling effects and, in particular, compute the maximum wall temperature.

[908 K]

4. A turbine blade operates in a high-temperature environment, requiring effective cooling for thermal protection. The combustion gases have a total pressure of

4 MPa and a total temperature of 2250 K, with a thermal conductivity of 0.164 W/(m·K), a specific heat ratio of 1.3, and a Mach number of 0.3. The cooling flow, injected through discrete film cooling holes, has the same total pressure but a lower total temperature of 850 K, a thermal conductivity of 0.0553 W/(m·K), a specific heat ratio of 1.4, and a Mach number of 0.6. The turbine vane channel measures 0.045 m in length, 0.06 m in width, and 0.05 m in height, with 75% of the coolant used for internal convective cooling and the rest for film cooling through 0.8 mm slots. The gas constants for combustion gases and cooling air are 292 J/(kg·K) and 287 J/(kg·K), respectively. The dynamic viscosity is determined using Sutherland's law, with reference values of $2.25 \cdot 10^{-5}$ Pa·s, 300 K, and 194 K (respectively, the reference viscosity, the reference temperature and the Sutherland constant) for combustion gases, and $1.7894 \cdot 10^{-5}$ Pa·s, 288.15 K, and 110.4 K for cooling air. Assess film cooling efficiency along the vane length and determine the maximum wall temperature as well as the maximum coolant total temperature.

[1044 K, 861 K]

NOTES

1. $\alpha_1 = 0$ for the first stage.
2. According to empirical data, the total pressure loss coefficient for nozzle rows, π_N, increases with the nozzle outlet angle α_2 due to greater flow turning and associated losses. For well-designed, subsonic nozzle cascades, typical values are approximately $\pi_N \approx 0.03$ for $\alpha_2 = 10°$, rising to about 0.035 for $\alpha_2 = 20°$, 0.040 for $\alpha_2 = 30°$, 0.045 for $\alpha_2 = 40°$, 0.050 for $\alpha_2 = 50°$, and reaching 0.055 at $\alpha_2 = 60°$.
3. This correlation, proposed by Islam and Sjolander, gives more accurate results compared to Carter's.
4. Approximate values for air properties at 1000°C:

   - Dynamic viscosity: $\mu = 54.9 \cdot 10^{-6}\,\text{Pa} \cdot \text{s}$
   - Thermal conductivity: $k = 0.076\,\text{W/(m·K)}$
   - Specific heat: $c_p = 1148\,\text{J/(kg·K)}$

   Using these values, we can calculate the Prandtl number:

$$\text{Pr} = \frac{c_p \mu}{k} = \frac{1148\,\text{J/kg·K} \cdot 54.9 \cdot 10^{-6}\,\text{Pa} \cdot \text{s}}{0.076\,\text{W/(m·K)}} = 0.829$$

5. While higher relative velocities increase the convective heat transfer coefficient, they also raise the heat flux, thus increasing the thermal load on the blade. Reducing relative velocities lowers external heat flux, minimizing coolant requirements and improving overall turbine efficiency.
6. Ni-based alloys are still used in the last few LPT stages of very high-power engines where the exhaust temperature is still high). Example: the GE's GEnx turbofan uses gamma-TiAl for its 5th–7th LPT stages.

11 Centrifugal Compressors

11.1 FUNDAMENTAL COMPONENTS

The geometry of impellers of a centrifugal compressor can vary according to the design objectives (Figure 11.1). *Radial blades* ($\beta_2' = 90°$), offer a balance between work transfer and aerodynamic efficiency. *Backward-curved blades* ($\beta_2' < 90°$) are preferred to radial ones when higher efficiency and broader operational range are required. These blades reduce outlet flow velocity and, consequently, facilitate more effective pressure recovery in the diffuser. Ultimately, *forward-curved blades* (β_2'>90°), although capable of higher work transfer, often generate excessive outlet flow velocities, which can lead to inefficiencies in downstream components, particularly in the diffuser.

Diffusers can be either vaneless or vaned. Vaneless diffusers are useful in high-pressure ratio applications, where near-supersonic velocities are easily achieved, as they allow gas to expand freely without being constrained by vanes, thus reducing the risk of flow separation or choking at high Mach numbers. Furthermore, the absence of vanes makes it possible to widen the operating range of the compressor without compromising off-design efficiency. In contrast, vaned diffusers optimize pressure recovery and overall peak efficiency, but at the cost of poorer off-design performance. To achieve the best efficiency, the clearance between the impeller and diffuser vanes must be precisely managed to prevent aerodynamic disturbances and unsteady effects.

The casing, which provides structural integrity and contains internal components of the compressor, is designed according to the specific operational needs of the compressor (Figure 11.2). Horizontally split casings offer easier access to internal components including the impellers, bearings, and seals. This design is particularly useful in large multistage compressors where regular maintenance and inspections are necessary. Often for high-pressure applications, barrel-type casings are used as they can safely regulate extreme pressures. Designed for easy disassembly, these

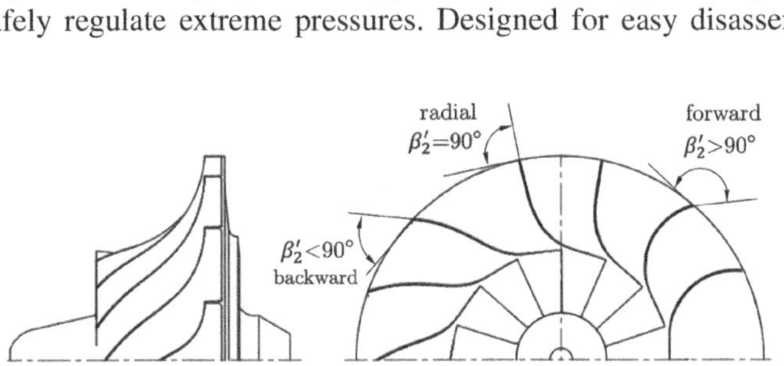

Figure 11.1 Schematic representation of different centrifugal compressor impeller blade configurations based on exit blade angle β_2', illustrating backward-curved, radial, and forward-curved blades.

DOI: 10.1201/9781003619956-11

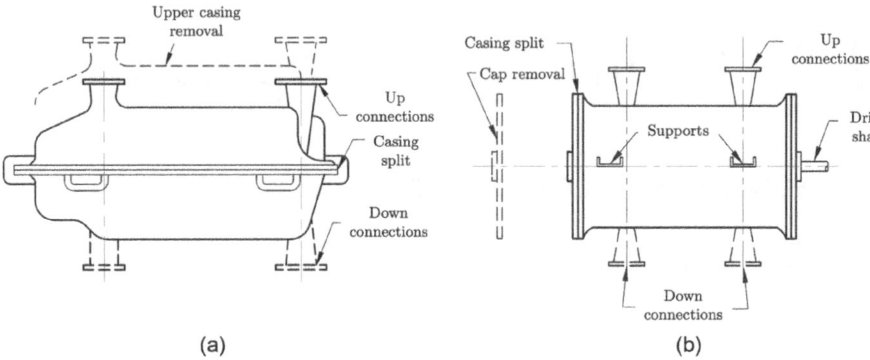

Figure 11.2 Horizontally split casings (a) and barrel-type casings (b) for centrifugal compressors.

casings allow the internal components to be removed for maintenance without requiring complete disassembly of the whole unit.

Using multiple stages, multistage compressors (Figure 11.3) achieve a desired pressure ratio higher than the one achievable with a single stage. Every stage is made up of a return channel, all meant to sequentially compress the gas, a diffuser, either bladed or open, and an impeller. Often, back-to-back impellers (Figure 11.5b) are used to counteract the axial thrust forces generated during compression, thus enhancing the mechanical stability of the compressor. In some applications where a cold sink is usable, intercooling between stages is a common method to decrease gas temperature; this lowers the power required for compression and improves its efficiency.

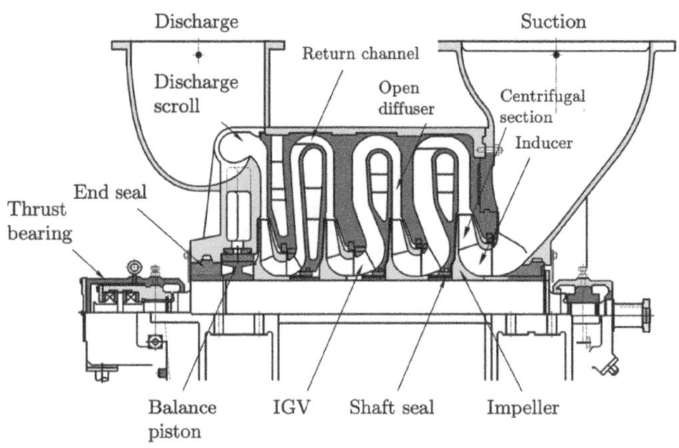

Figure 11.3 Sectional view of a multi-stage centrifugal compressor.

11.2 APPLICATIONS

Centrifugal compressors are employed in many industrial processes, particularly in the petrochemical sector and gas turbine applications, and, more generally, in all cases where stage pressure ratio is relatively high compared to the mass flow rate.

Although they feature inherently higher stall margins compared to axial flow stages, some limitations persist in their stable operating range, which spans 45%–90% of the rated mass flow rate. Operation outside this window can negatively impact efficiency, particularly when running at partial loads. Their high rotational speeds, reaching up to 140,000 rpm for aircraft and spacecraft applications, also present engineering challenges. Commercial units operate below 20,000 rpm, but even at these speeds, issues such as bearing lubrication, vibration, and rotor balancing must be carefully managed.

Within the process gas industry, centrifugal compressors handle a wide variety of gases across numerous applications. These compressors (Figure 11.4) are designed with relatively low pressure ratios per stage, typically in the range of 1.8–3.5. This allows for a much wider operating range between the surge and choke limits.

In the gas turbine sector, centrifugal compressors are widely deployed across a range of applications. Small gas turbines under 5 MW output commonly incorporate single-stage centrifugal compressors capable of producing pressure ratios up to 6 (left of Figure 11.5). As power outputs increase, configurations become more complex, with aeroderivative turbines potentially utilizing multistage arrangements combining axial and centrifugal compressor stages. Even smaller microturbine units below 350 kW frequently employ radial compressors and turbines in a compact package.

11.3 DETAILED DISCUSSION OF COMPONENTS

11.3.1 INLET GUIDE VANES

Inlet guide vanes (IGVs) are positioned at the compressor inlet to adjust airflow before it enters the impeller (Figure 11.4, list item no. 2). IGVs control the flow direction by imparting a tangential velocity, known as *prewhirl*, to the incoming air. Prewhirl can either aid or hinder compressor performance depending on whether the swirl direction is in alignment with the impeller's rotation.

IGVs can be part of either *single-entry* or *double-entry* systems (Figure 11.5). In double-entry systems, the incoming airflow is split into two streams, reducing the inducer's tip Mach number. While effective, double-entry systems are often more complex to integrate.

Prewhirl can be categorized as either *positive* or *negative*. Positive prewhirl, which aligns with the direction of impeller rotation, reduces energy transfer by lowering the Euler head. Negative prewhirl, on the other hand, increases energy transfer by opposing the impeller's rotation. Prewhirl also affects the relative Mach number $\frac{w_1}{a_1}$ at the impeller inlet. In fact, positive prewhirl lowers the relative Mach number, while negative prewhirl increases it.

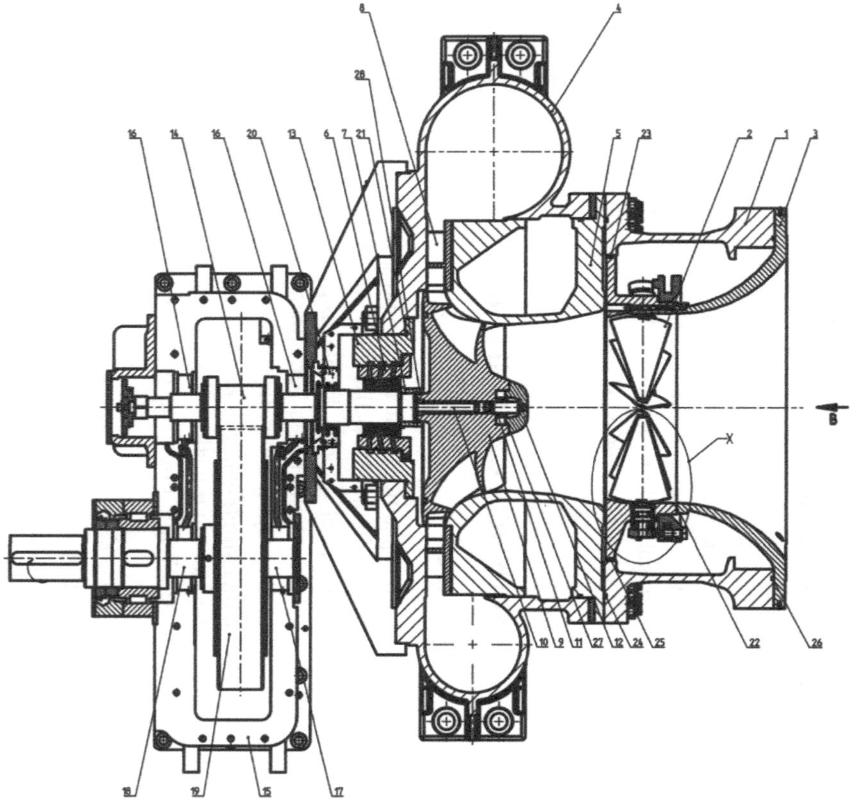

Pos. Item No.	description	QTY	material
1	casing for IGV	1	G X5CRNI13-4 QT2
2	IGV	11	G X5CRNI13-4 QT1
3	ring for IGV	1	G X5CRNI13-4 QT2
4	volute casing	1	G X5CRNI13-4 QT2
5	inlet casing	1	G X5CRNI13-4 QT2
6	dry gas seal	1	Eagle Burgmann
7	locknut	1	X3CrNi13-4
8	diffuser	1	X6 CRNITI 18-10
9	impeller	1	X5CRNICUNB16-4
10	stud	1	X3CRNI13-4
11	locknut	1	X3CRNIMO13-4
12	impeller cap	1	X5CRNICUNB17
13	seal casing	1	G X5CRNI13-4 QT2
14	pinion shaft	1	56 NICRMOV 7
15	gear box	1	Siemens
16	radial tilting pad bearing	2	-
17	journal bearing	1	-
18	bearing	1	-
19	gear shaft	1	56 NICRMOV 7
20	seal pinion shaft	1	-
21	o-ring	1	FPM-Viton
22	o-ring	1	FPM-Viton
23	o-ring	1	FPM-Viton
24	o-ring	1	FPM-Viton
25	o-ring	1	FPM-Viton
26	o-ring	1	FPM-Viton
27	o-ring	1	FPM-Viton
28	o-ring	1	FPM-Viton
29	rotary shaft seal	11	PTFE Compound Nr. 27
30	rotary shaft seal	11	PTFE Compound Nr. 27
31	rolling contact bearing	22	Edelstahl
32	connecting piece	1	-

Figure 11.4 Sectional view of a centrifugal compressor with corresponding parts list.

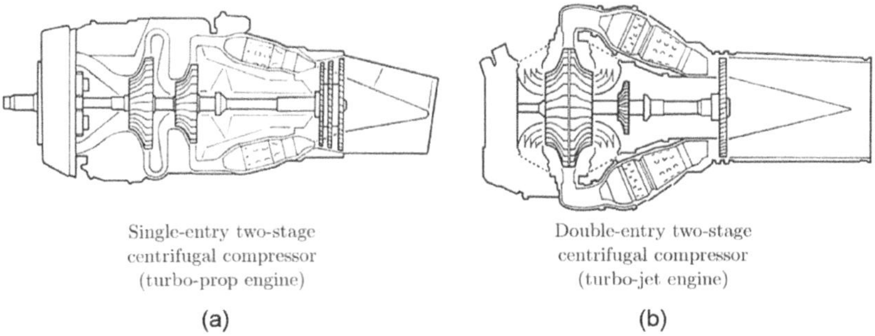

Figure 11.5 Single-entry (a) and double-entry (b) two-stage centrifugal compressors used in turbo-prop and turbo-jet engines. (Courtesy of Rolls-Royce plc.)

IGVs can also throttle the mass flow rate without altering the compressor's rotational speed. This allows compressors to operate more efficiently over a range of flow rates and prevents operational issues like surging or stalling.

11.3.2 IMPELLER ENTRY: THE INDUCER

The inducer increases the angular momentum of the incoming fluid without changing its radius of rotation, guiding the flow from the axial to radial direction. At the inducer tip, controlling the relative Mach number is critical to avoid shock waves and flow separation. When the relative Mach number approaches or exceeds 1, shock losses occur, leading to loss of efficiency and possible entry choke.

To avoid this, splitter blades can be used that are placed in a recessed axial location with respect to the main blades (Figure 11.6). This makes it possible to enlarge the entry flow passage and therefore increase throat dimension while improving flow uniformity in the farther downstream flow.

The stable operating range between stall and choke depends heavily on the inducer characteristics. Key elements in the design process include the hub and shroud radii, as well as the shroud blade angle $\beta'_{1,s}$, which determine the distribution of the relative Mach number in station 1.

The minimum hub radius, $r_{1,h}$, is typically established based on considerations regarding the necessary cross-sectional area of the shaft for torque transmission.

On the other hand, the radius of the inducer shroud, $r_{1,s}$, indicates where the blade velocity, $u_{1,s}$ and the corresponding relative velocity, $w_{1,s}$, reach their maximum. By decreasing this radius, the peripheral speed at the tip of the blade is reduced, leading to a decrease in the relative Mach number $\mathrm{Ma}_{w1,s}$. However, this adjustment also results in a smaller flow area, $\pi(r_{1,s}^2 - r_{1,h}^2)$, and increased meridional velocity, $c_{m1,s}$, highlighting the need for a balanced approach.

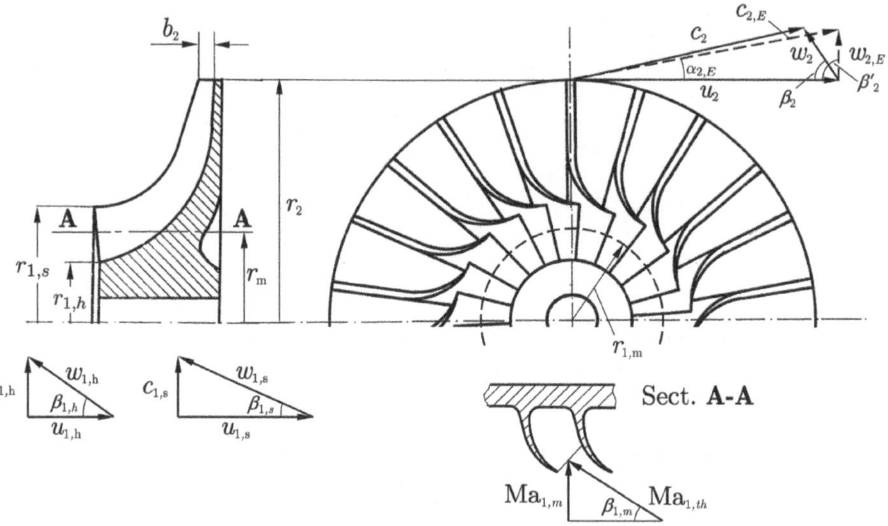

Figure 11.6 Geometric and velocity characteristics of a radial compressor impeller.

11.3.2.1 Minimization of Relative Mach Number at Shroud

Let us analyze the relationships among the various velocities illustrated in the inlet velocity triangle assuming zero prewhirl:

$$w_{1,s}^2 = c_{m1,s}^2 + u_{1,s}^2 = c_1^2 + \omega^2 r_{1,s}^2 \tag{11.1}$$

or, in terms of Mach numbers:

$$\mathrm{Ma}_{w1,s}^2 = \mathrm{Ma}_1^2 + \frac{\omega^2 r_{1,s}^2}{\gamma R T_1} \tag{11.2}$$

Now let us introduce the expression of a nondimensional mass flow rate[1] for the impeller having outer radius r_2 and rotational speed ω:

$$\frac{\dot{m}}{\rho_1^0 \omega r_2^3} \propto \dot{m}_c = \frac{\dot{m}}{\rho_1^0 a_1^0 \pi r_2^2} = \frac{\mathrm{Ma}_1}{\left(1 + \frac{\gamma-1}{2}\mathrm{Ma}_1^2\right)^{\frac{1}{\gamma-1}}} \cdot \left(\frac{a_1^0}{a_1}\right)^{-1} \cdot \frac{r_{1,s}^2}{r_2^2}\left(1-\nu^2\right) \tag{11.3}$$

where $\nu = r_{1,h}/r_{1,s}$, while a_1^0 and a_1 are the stagnation and the static speeds of sound, respectively. Given that:

$$\frac{a_1^0}{a_1} = \left(1 + \frac{\gamma-1}{2}\mathrm{Ma}_1^2\right)^{\frac{1}{2}} \tag{11.4}$$

substituting equation 11.4 into equation 11.3, we get:

$$\dot{m}_c = \mathrm{Ma}_1 \left(1 + \frac{\gamma-1}{2}\mathrm{Ma}_1^2\right)^{-\frac{\gamma+1}{2(\gamma-1)}} \frac{r_{1,s}^2}{r_2^2}\left(1-\nu^2\right) \tag{11.5}$$

from which:

$$\frac{r_1^2}{r_2^2} = \dot{m}_c \cdot (1-\nu^2)^{-1} \cdot \frac{\left(1+\frac{\gamma-1}{2}\mathrm{Ma}_1^2\right)^{\frac{1+\gamma}{2(1-\gamma)}}}{\mathrm{Ma}_1} \tag{11.6}$$

Substituting this into 11.2, after some steps we get:

$$\mathrm{Ma}_{w1,s}^2 = \mathrm{Ma}_1^2 + \frac{\mathrm{Ma}_u^2 \dot{m}_c}{1-\nu^2}\frac{1}{\mathrm{Ma}_1}\left(1+\frac{\gamma-1}{2}\mathrm{Ma}_1^2\right)^{\frac{3\gamma-1}{2(\gamma-1)}} \tag{11.7}$$

where $\mathrm{Ma}_u = \omega r_2/(1+\frac{\gamma-1}{2}\mathrm{Ma}_1^2)^{\frac{1}{2}} = u_2/a_1^0$ is the tangential impeller Mach number. The plot of equation 11.7 is given in Figure 11.7: to enhance the impeller's performance and avoid excessive shock losses at the inducer inlet, one effective strategy is in fact to minimize the relative Mach number $\mathrm{Ma}_{w1,s}$ for a given $\frac{\mathrm{Ma}_u^2 \dot{m}_c}{(1-\nu^2)}$. This provides optimal values of both Ma_1 and inducer inlet angle $\beta_{1,s}$.

The continuity equation ultimately yields the actual value of the inducer eye.

The arguments given above about the choice of Ma_1 must be complemented with the necessity of ensuring that a reasonable margin exists towards choke. Therefore, recognizing that in a choked inducer $\mathrm{Ma_{th}} = \mathrm{Ma}^* = 1$, from continuity – under the assumption of isentropic flow from station 1 to station *th* at the mean diameter $D_m = \sqrt{0.5(D_{1,s}^2 + D_{1,h}^2)}$ –:

$$A_1^* \sin\beta_{1,m} \frac{\mathrm{Ma}_{w1,m}}{\left(1+\frac{\gamma-1}{2}\mathrm{Ma}_{w1,m}^2\right)^{\frac{1+\gamma}{2(\gamma-1)}}} = A_{th}^*\left(\frac{2}{\gamma+1}\right)^{\frac{1+\gamma}{2(\gamma-1)}} \tag{11.8}$$

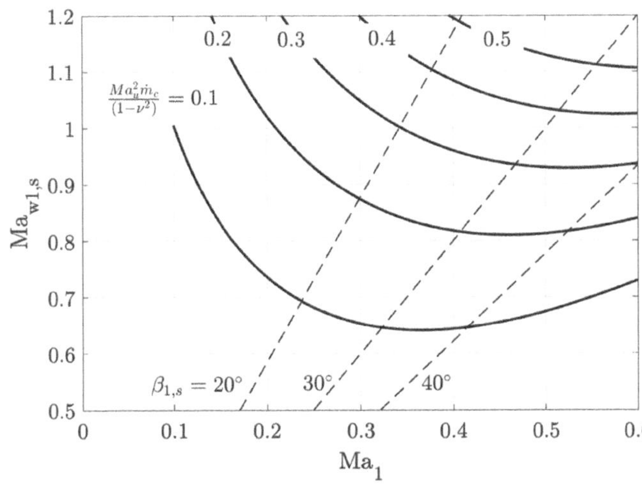

Figure 11.7 Relative Mach number at the inducer shroud as a function of the axial Mach number for different values of the mass flow parameter. Dashed lines indicate constant inducer blade angles at the shroud.

from which the critical area ratio can be estimated:

$$\left(\frac{A_{th}}{A_1}\right)^* = \frac{\xi_1 \sin\beta_{1,m} \cdot \mathrm{Ma}_{w1,m}}{\varkappa\left(\frac{2}{\gamma+1}\right)^{\frac{1+\gamma}{2(\gamma-1)}} \cdot \left(1+\frac{\gamma-1}{2}\mathrm{Ma}_{w1,s}^2\right)^{\frac{1+\gamma}{2(\gamma-1)}}} \tag{11.9}$$

where $\varkappa$ can be set at perhaps 1.2 (due to blade blockage) and ξ_1 is a flow blockage factor which can be given a value of 0.97.

11.3.3 IMPELLER DISCHARGE: THE CENTRIFUGAL SECTION

For a given impeller diameter, the most important parameter in analyzing the discharge section is the blade sweep angle β_2'.

Backswept blades, characterized by $\beta_2' < 90°$, result in a comparatively lower exit flow Mach number Ma_2 than radial blades, which, as noted earlier, leads to milder diffusion in downstream stator components, that is, the volute or diffuser. Additionally, back-swept blades reduce the work input to the flow, which in turn implies lower compression ratios and allows increased stall margins. They also help decrease blade loading due to a smaller blade-to-blade flow deflection $(\beta_2 - \beta_1)$ relative to radial configurations.

However, this design choice comes with trade-offs. To achieve the same pressure ratio as for radial blades, backswept blades require a higher tip speed, which increases centrifugal stress at the blade root.

Another key parameter is the rotational speed ω, especially for a fixed impeller diameter. Higher values of ω require the use of high-speed prime movers or gearboxes for speed multiplication, which add complexity and cost.

If the impeller diameter is a design variable, increasing it generally leads to a higher overall compressor size and greater rotational inertia, factors that are especially important in applications such as turbochargers, where response time is critical.

To analyze the impeller discharge, let us start from the outlet Mach number:

$$\mathrm{Ma}_2 = \frac{c_2}{a_2} = \frac{c_{u2}}{a_1^0 \cos\alpha_2}\frac{a_1^0}{a_2^0}\frac{a_2^0}{a_2} = \frac{c_{u2}}{a_1^0 \cos\alpha_2}\left(\frac{T_1^0 T_2^0}{T_2^0 T_2}\right)^{1/2} \tag{11.10}$$

where

$$\frac{T_2^0}{T_1^0} = 1 + (\gamma-1)\psi\mathrm{Ma}_u^2 \tag{11.11}$$

and, in a zero prewhirl flow:

$$c_{u2} = \psi u_2 \tag{11.12}$$

It follows from 11.10 that:

$$\mathrm{Ma}_2 = \frac{\psi\mathrm{Ma}_u}{\cos\alpha_2}\sqrt{\frac{1+\frac{(\gamma-1)}{2}\mathrm{Ma}_2^2}{1+(\gamma-1)\psi\mathrm{Ma}_u^2}} \tag{11.13}$$

Solving for Ma_2 gives:

$$\mathrm{Ma}_2 = \frac{\psi \mathrm{Ma}_u}{\cos\alpha_2}\left[1+(\gamma-1)\psi \mathrm{Ma}_u^2 - \frac{(\gamma-1)}{2}\left(\frac{\psi \mathrm{Ma}_u}{\cos\alpha_2}\right)^2\right]^{-\frac{1}{2}} \tag{11.14}$$

The work coefficient ψ can be written as[2]

$$\psi = \frac{\sigma_{slip}}{1+\left(\frac{\tan\alpha_{2,E}}{\tan\beta_2'}\right)} \tag{11.15}$$

where the slip factor can be estimated using the Wiesner formula:

$$\sigma_{slip} = 1 - \frac{\sqrt{\sin\beta_2'}}{Z^{0.70}(1-\phi\cot\beta_2')} = 1 - \frac{\sqrt{\sin\beta_2'}\cdot(\tan\alpha_{2,E}+\tan\beta_2')}{Z^{0.70}\cdot\tan\beta_2'} \tag{11.16}$$

since:

$$\phi = \frac{\tan\alpha_{2,E}\cdot\tan\beta_2'}{\tan\alpha_{2,E}+\tan\beta_2'} \tag{11.17}$$

Given that:

$$\sigma_{slip} = \frac{\tan\alpha_{2E}}{\tan\alpha_2} \tag{11.18}$$

it follows that, from equation 11.14, $\mathrm{Ma}_2 = f(\mathrm{Ma}_u, \alpha_{2,E}, \beta_2', Z)$. The plot of this function is provided in Figure 11.8 for sample values of $\alpha_{2,E} = 25°$ and $Z = 30$. Similar plots can be drawn by varying the values of angles and the number of blades. To this extent, however, the dependence on $\alpha_{2,E}$ is less apparent with respect to that on β_2', therefore only parametric values of β_2' are displayed in the figure.

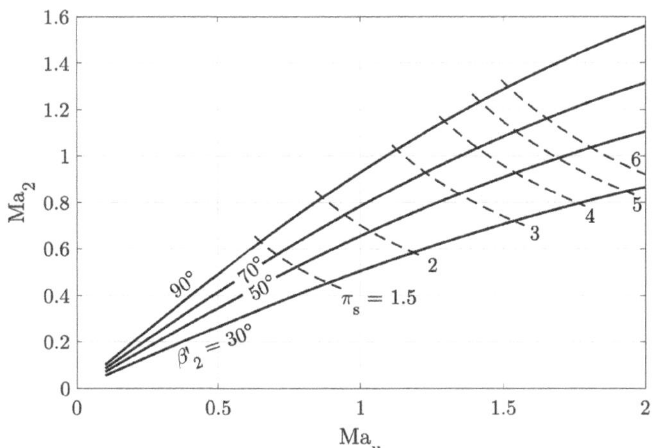

Figure 11.8 Variation of Ma_{a2} as a function of Ma_u for $\alpha_{2,E} = 25°$ and $Z = 30$, with parametric curves for different values of β_2'. Dashed lines represent contours of constant stage pressure ratio π_s.

In Figure 11.8, parametric values of the compression ratio:

$$\pi_s = \left[1+\eta_s(\gamma-1)\psi \mathrm{Ma}_u^2\right]^{\frac{\gamma}{\gamma-1}} \tag{11.19}$$

are given assuming $\eta_s = 0.8$.

From the figure, it becomes apparent that increasing the backward sweep of impeller blades reduces the discharge Mach number for a given impeller speed. To maintain the desired pressure ratio despite the reduced Mach number, the impeller speed Ma_u must be increased, which in turn raises the discharge Mach number. For instance, at a pressure ratio of 3, decreasing the backward sweep angle from 90 to 50° lowers the discharge Mach number from about 1.03 to ≈ 0.7. To compensate, the impeller speed needs to increase.

Although a reduction in the discharge Mach number can be beneficial in lowering flow losses and improving efficiency, boosting the impeller speed also increases the inducer blade speed. This can lead to mechanical stress and potential flow separation in the inducer region. The inlet relative Mach number is influenced by factors such as the inlet flow angle β_{1s} and the radius ratio r_{1s}/r_2, both of which need careful consideration to ensure optimal inducer performance. Thus, while backward sweep can improve performance, it requires balancing the effects of increased impeller speed to avoid negative impacts on overall design.

The blade exit height b_2 and the blade number Z must be determined in a way that two conditions are fulfilled: (i) the continuity equation holds and (ii) a condition is enforced such that the minimum relative velocity always remains positive within the blade-to-blade channel.

From continuity:

$$\dot{m} = \rho_1 \pi r_{1s}^2\left(1-\frac{r_{1h}^2}{r_{1s}^2}\right) c_{m1} = \rho_2 2\pi r_2 b_2 c_{m2}\xi_2 \tag{11.20}$$

the following relationship is obtained:

$$\frac{b_2}{r_2} = \frac{1}{2\xi_2}\frac{\rho_1}{\rho_2}\frac{r_{1s}^2}{r_2^2}(1-\nu^2)\frac{c_{m1}a_1^0}{a_1^0 c_{m2}} \tag{11.21}$$

where ξ_2 is the flow blockage at the impeller exit station caused by developed boundary layer on impeller internal surfaces. Next, recognizing that $\frac{\rho_1}{\rho_2} = \frac{\rho_1}{\rho_1^0}\frac{\rho_1^0}{\rho_2}$, where:

$$\frac{\rho_1^0}{\rho_1} = \left(1+\frac{\gamma-1}{2}\mathrm{Ma}_1^2\right)^{\frac{1}{\gamma-1}} = \left(1+\frac{\gamma-1}{2}\mathrm{Ma}_{w1}^2\sin^2\beta_1\right)^{\frac{1}{\gamma-1}} \tag{11.22}$$

and:

$$\frac{\rho_1^0}{\rho_2} = \frac{p_1^0}{p_2^0}\cdot\frac{p_2^0}{p_2}\cdot\frac{T_2^0}{T_1^0}\cdot\frac{T_2}{T_2^0} = \left(1+\frac{\gamma-1}{2}\mathrm{Ma}_2^2\right)^{\frac{1}{\gamma-1}}\pi_s^{\frac{\gamma-1}{\gamma\eta_p}-1} \tag{11.23}$$

it is finally found that:

$$\frac{b_2}{r_2} = f\left(\xi_2, \sigma_{slip}, \eta_p, \nu, \beta_{1s}, \beta_2', \alpha_{2E}, \mathrm{Ma}_u, \frac{r_{1s}}{r_2}\right) \tag{11.24}$$

where

$$\frac{c_{m1}}{a_1^0} = \frac{c_1}{a_1^0} = \frac{u_{1s}}{a_1^0}\tan\beta_{1s}, \quad \frac{c_{m2}}{a_1^0} = \frac{c_2}{a_1^0} = \frac{u_2}{a_1^0}\tan\alpha_2 \tag{11.25}$$

On the other hand, the number of blades should be selected to ensure that the relative velocity does not reach zero in the $r-\theta$ plane within the blade-to-blade surface. A zero relative velocity would indicate potential flow separation on the pressure side of the blade. A practical approach to enforcing this constraint is to rely on the simplified model already used in centrifugal pumps, such that:

$$Z \geq \frac{\pi}{\phi\left(\frac{1}{\tan^2\beta_2'} + 1\right)} \tag{11.26}$$

11.3.4 VANED DIFFUSER

Designing and analyzing vaned diffusers for centrifugal compressors presents significant challenges due to the complex fluid dynamics involved. By focusing on subsonic configurations, we make a first-order approximation that treats the flow as one-dimensional and steady. In addition, the fluid is assumed to be inviscid, and the flow is adiabatic.

The diffuser profiles are typically defined using a circular-arc camber line with a large radius of curvature. In fact, as the radius approaches infinity, this configuration simplifies to a straight-channel diffuser. An arbitrary thickness distribution along the camber line can be specified.

The geometry of the diffuser can be described using parameters such as radius ratios $\frac{r_3}{r_2}$ and $\frac{r_4}{r_2}$, curvature radius ratio $\frac{R_c}{r_2}$, vane angle α_3', thickness t and number of vanes Z', all of which allow for precise control over the diffuser shape, channel width and curvature. The boundary conditions for the analysis include the stagnation temperature T_2^0 and pressure p_2^0, the mass flow rate $\dot{m}$, the inlet flow angle α_2, and the impeller Mach number Ma_u.

Diffuser channel flow geometry is defined from impeller exit (subscript 2), vaned diffuser leading edge (subscript 3), diffuser throat (subscript th), and diffuser exit (subscript 4). The values of relevant section areas from station 2 to station 3 are determined as follows:

$$A_2 = 2\pi r_2 b_2 \sin\alpha_2 \frac{1}{Z'} \tag{11.27}$$

$$A_3 = b_3 h_3 = \begin{cases} b_3\left[2\pi r_3/Z'\sin(\alpha_3') + \frac{2\pi r_2 b_2 \sin(\alpha_3')/Z' - t}{\cos(\alpha_3 - \alpha_3')}\right] & \text{if } \alpha_3 \geq \alpha_3', \\ b_3\left[\begin{array}{l} 2\pi r_3/Z'\sin(\alpha_3) + 2\pi r_3/Z'\cos(\alpha_3)\tan(\alpha_3' - \alpha_3) \\ -\frac{t}{2\cos(\alpha_3' - \alpha_3)} \end{array}\right] & \text{if } \alpha_3 < \alpha_3' \end{cases} \tag{11.28}$$

where

$$\tan\alpha_3 = \tan\alpha_2 \frac{\rho_2}{\rho_3}\frac{b_2}{b_3} \tag{11.29}$$

It is worth noting that from equation 11.28, α_3 can be calculated only when the thermodynamic status of the flow at station 3 has been defined. This, in turn, depends on the knowledge of A_3 and therefore again of α_3. The problem can be solved by iterative calculations.

The diffuser throat area is measured perpendicularly to the local tangent of the profile camber:

$$A_{th} = b_3\left(\frac{2\pi r_{th}}{Z'}\sin\alpha_{th} - t_{th}\right) \tag{11.30}$$

Downstream of the diffuser throat, the flow is confined within the blade channel. From the throat to the diffuser exit, the passage areas are measured perpendicularly to the channel mean line according to the following relation:

$$A(r) = \frac{2\pi r}{Z'}\sin\alpha - t \tag{11.31}$$

where

$$\alpha = \arccos\left(\frac{r^2 + R_c^2 - i^2}{2rR_c}\right) \tag{11.32}$$

Once the channel geometry has been defined, the isentropic flow within the diffuser is computed using one-dimensional gas dynamics for adiabatic flow, while the Mach number distribution is calculated using the following nonlinear equation:

$$\dot{m} = A(r)\sqrt{\frac{\gamma}{R}}\frac{p_2^0}{\sqrt{T_2^0}}\mathrm{Ma}(r)\left(1+\frac{\gamma-1}{2}\mathrm{Ma}(r)^2\right)^{-\frac{\gamma+1}{2(\gamma-1)}} \tag{11.33}$$

while the relationships among fluid quantities and section areas can be obtained using the Hugoniot equations (see Appendix A.4):

$$\frac{dA}{A} = \frac{dc}{c}(\mathrm{Ma}^2-1);\ \frac{dp}{p} = -\gamma\left(1-\frac{1}{\mathrm{Ma}^2}\right)\frac{dA}{A};\ \frac{d\rho}{\rho} = -\left(1-\frac{1}{\mathrm{Ma}^2}\right)\frac{dA}{A} \tag{11.34}$$

Diffuser performance can be evaluated using a total pressure loss coefficient:

$$\Omega = \frac{p_3^0 - p_4^0}{p_3^0 - p_3} \tag{11.35}$$

and a pressure recovery coefficient:

$$C_{p,r} = \frac{p_4 - p_3}{p_3^0 - p_3} \tag{11.36}$$

The actual performance of a diffuser depends strongly on the entropy generation within the diffuser channel. These can be calculated considering the following components of losses:

- incidence losses;
- blade surface profile losses;
- blade wake losses.

Incidence losses can be evaluated using the following:

$$\Delta p^0_{\text{inc}} = f_{\text{inc}} \rho_3 \frac{\Delta c_u^2}{2} \tag{11.37}$$

where $\frac{\Delta c_u^2}{2}$ is the loss of the tangential specific kinetic energy of the flow:

$$\frac{\Delta c_u^2}{2} = \frac{1}{2} c_{m3}^2 \left(\frac{1}{\tan \alpha_3} - \frac{1}{\tan \alpha_3'} \right)^2 \tag{11.38}$$

and f_{inc} is a parameter depending on the actual working condition of the diffuser: when $\alpha_3 = \alpha_3'$ the value of 0 can be assumed (i.e. no loss), when $\alpha_3 < \alpha_3'$ the value of 0.9 is suggested, whereas for $\alpha_3 > \alpha_3'$ some value between 0.5 and 0.6 is recommended.

The blade surface profile and wake losses can be evaluated by computing the entropy generation rate per unit area within the wall and the wake boundary layers. This calculation is quite complex and goes beyond the scope of this book. An approximate evaluation for relatively thin airfoils can be done assuming that the total pressure loss is proportional to $\text{Re}_s^{-1/6}$, where s is the curvilinear coordinate of the meanline from station 2 to station 4 in Figure 11.9.

Once the losses have been calculated, the following loss coefficients can be determined:

$$\zeta_{\text{inc}} = \frac{\Delta p^0_{\text{inc}}}{p_3^0 - p_3}, \quad \zeta_{BL} = \frac{\Delta p^0_{BL}}{p_3^0 - p_3}, \quad \zeta_{\text{wake}} = \frac{\Delta p^0_{\text{wake}}}{p_3^0 - p_3} \tag{11.39}$$

such that:

$$\Omega = \zeta_{\text{inc}} + \zeta_{BL} + \zeta_{\text{wake}} \tag{11.40}$$

The method described provides good accuracy when the flow is high-subsonic (up to $\text{Ma}_u = 0.8$) and close to the design point (Figure 11.10). It maintains acceptable precision for $\pm 5°$ in the incidence angle variation. Beyond this limit, compressibility effects start to be significant, and performance prediction is less accurate.

11.3.5 VOLUTE AND DISCHARGE SYSTEM

The volute is designed to guide high-speed gas from the impeller and direct it into the discharge system while converting its kinetic energy into pressure.

A single-volute design offers simplicity and is generally effective for compressors operating within a narrow range of flow rates. However, it can lead to unbalanced radial forces, which may cause vibration and stress on bearings at off-design conditions. To address this, double-volute designs use two parallel volute chambers, balancing radial forces and providing more stability across a range of operating conditions, although they are more complex to manufacture and occupy more space.

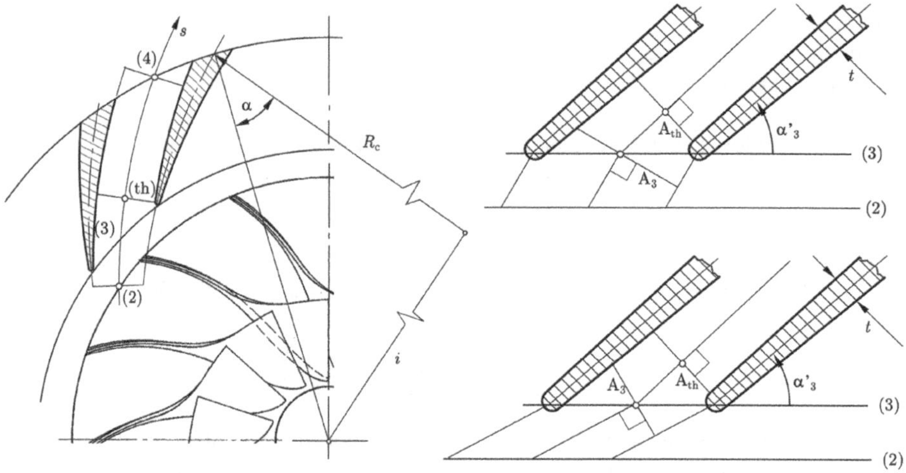

Figure 11.9 A vaned diffuser for a centrifugal compressor.

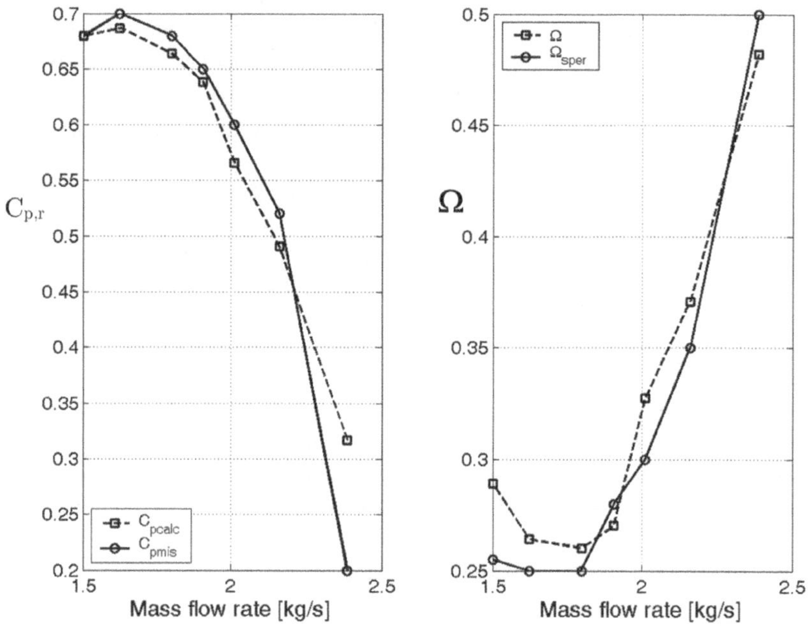

Figure 11.10 Pressure recovery coefficient and total pressure loss coefficient as a function of the mass flow rate in a vaned diffuser: comparison between experimental and calculated data.

In high-performance applications, vaned volutes are sometimes used, incorporating guide vanes at the volute inlet to improve alignment between the gas flow and the volute walls. This configuration enhances efficiency, but can be sensitive to flow variations, leading to possible efficiency losses under design conditions.

The volute's cross-sectional area is a fundamental design aspect, expanding gradually to reduce gas velocity and increase pressure as it moves towards the discharge. This increase in area aligns with the principles of angular momentum conservation, minimizing fluctuations in velocity that could otherwise create pressure drops. In addition, the tongue or cutwater, a feature where the flow exits the volute into the discharge system, is meticulously positioned to avoid abrupt flow direction changes that would introduce turbulence, noise, and pressure losses.

The discharge system, which channels compressed gas from the volute outlet to downstream equipment, is integral to maintaining the desired pressure and flow. Key components in the discharge system include the discharge nozzle, check and isolation valves, and connecting piping. The shape and position of the discharge nozzle at the volute outlet guide the gas flow out of the compressor and into the connected piping system, where it is often directed through diffusers to further reduce velocity and increase pressure. Check and isolation valves help prevent backflow during shutdown and facilitate maintenance, while acoustic treatments in the piping reduce noise and vibration, ensuring stable gas flow into downstream processes.

Volute flow is characterized by the inlet conditions from the impeller exit, which include radial velocity c_{m2}, tangential velocity c_{u2}, and total pressure p_2^0. These quantities may have circumferential non-uniformities caused by the impeller-volute interaction.

The first loss component is the meridional velocity dump loss Δp_m^0, which accounts for the complete loss of kinetic energy associated with the radial velocity component:

$$\Delta p_m^0 = \frac{1}{2}\rho c_{m2}^2 \tag{11.41}$$

The friction losses, Δp_f^0, are calculated based on the throughflow velocity $c_u{}^3$ in the volute passage, using a friction factor f dependent on the Reynolds number and surface roughness:

$$\Delta p_f^0 = f\frac{l_p}{D_h}\frac{1}{2}\rho c_u^2 \tag{11.42}$$

where l_p represents the path length or flow distance that the fluid travels within the volute passage and D_h is the hydraulic diameter of the volute flow passage cross-section.

As the tangential velocity decreases from the inlet to the outlet ($c_{u2}>c_{u3}$), the swirling flow decelerates, resulting in tangential velocity dump losses Δp_D^0:

$$\Delta p_D^0 = \frac{1}{2}\rho(c_{u2} - c_{u3})^2 \tag{11.43}$$

Finally, exit cone losses Δp_{EC}^0 are modeled as gradual expansion losses due to area increase:

$$\Delta p_{EC}^0 = f_{EC}\frac{1}{2}\rho(c_{u4} - c_{u3})^2 \tag{11.44}$$

where f_{EC} is typically around 0.15 for well-designed volute exit cones.

The total pressure loss coefficient ω_T and the static pressure recovery coefficient C_p' in the volute can then be calculated from the inlet and exit conditions using the total pressure loss $\Delta p_{\text{total}}^0 = \Delta p_m^0 + \Delta p_f^0 + \Delta p_D^0 + \Delta p_{EC}^0$:

$$\Omega_v = \frac{p_2^0 - p_4^0}{p_2^0 - p_2} \tag{11.45}$$

$$C_{p,r}' = \frac{p_4 - p_2}{p_2^0 - p_2} \tag{11.46}$$

Although a one-dimensional volute performance prediction model provides a useful first estimate, it has several inherent limitations due to the assumptions made. More advanced models are needed to accurately capture the complex three-dimensional swirling flow within the volute passage.

11.4 COMPRESSOR MAP

While the basic principles are similar for both centrifugal and axial compressors, there are some notable differences in their compressor maps due to the fundamental differences in their flow physics and geometry (Figure 11.11).

Compared to axial compressors, centrifugal compressors usually have a broader operating range in terms of mass flow rates because the centrifugal force field allows them to run at lower flow coefficients without stalling. In addition, the surge line is usually more vertical, thus leading to a fairly constant surge mass flow rate across a wide range of pressure ratios. Furthermore, in contrast to the sudden choke limit seen

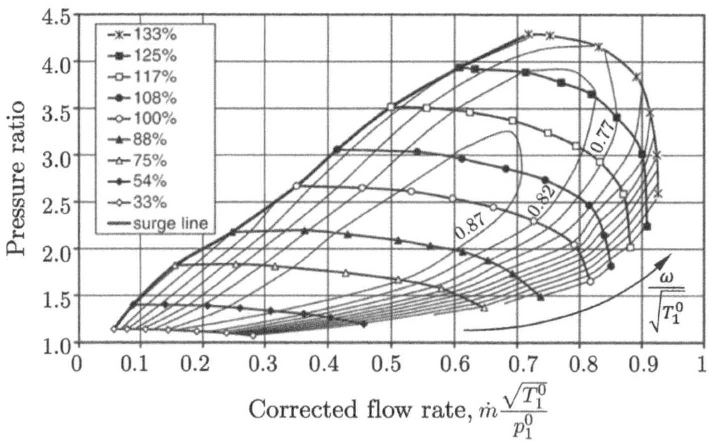

Figure 11.11 Sample map of a centrifugal compressor.

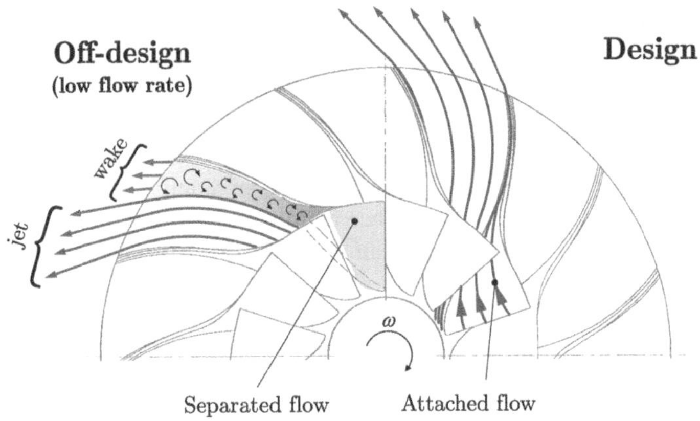

Figure 11.12 Comparison of flow behavior in a centrifugal compressor rotor passage under off-design (low flow rate) and design conditions.

in the axial stages, centrifugal compressors experience a more gradual trend towards choke.

11.4.1 OFF-DESIGN FLOW DYNAMICS

One important aerodynamic feature is the *jet-wake* flow pattern that emerges from the impeller at off-design conditions (Figure 11.12). This pattern is most pronounced at lower flow rates when the inducer operates at high incidence: part of the flow may stall on the suction side of the blades, creating a thick wake, while the rest of the passage still accelerates the flow, creating a jet. In a vaneless diffuser, the jet-wake will gradually mix out circumferentially, but if a vaned diffuser is present, the impeller wakes and jets impinge on the diffuser vanes periodically, causing localized high incidence and low momentum unsteady flows, respectively. This unsteady interaction can lead to periodic pressure fluctuations and can promote diffuser flow separation under some conditions. At off-design conditions, especially near surge, these impeller–diffuser interactions become stronger, contributing to overall losses, and can even trigger operational instability such as a rotating stall.

Centrifugal impellers also experience *secondary flows* due to centrifugal and Coriolis forces acting on the boundary layer. As fluid moves through the impeller passage, low-momentum fluid from the blade surfaces (particularly the shroud side in an unshrouded impeller or the hub/shroud boundary layers in a shrouded impeller) tends to migrate under pressure gradients. This can lead to phenomena such as a build-up of low-momentum fluid on one side of the passage (e.g. toward the pressure side or toward the shroud) forming secondary flow vortices. For example, in a backswept impeller, the combination of Coriolis force and adverse pressure gradient can drive a secondary flow that accumulates near the blade suction side at the shroud, contributing to the wake. These secondary flows reduce the effective flow area and can cause early flow separation if the passage loading is too high. At low flow off-design, secondary flows are more pronounced because the flow is more three-dimensional and separations can occur at the endwalls. On the other hand, at high flow near choke, secondary flows may be somewhat less troublesome than in axial machines because the through-flow is very high, and this can reenergize boundary layers.

For a *vaned diffuser*, off-design incidence can cause the diffuser blades to stall. At low flow rates, the flow entering the diffuser has a large angle relative to the diffuser vane angle, leading to a positive incidence and likely flow separation on the diffuser vane suction side (Figure 11.9). This can result in diffuser stall before the impeller itself stalls. In fact, in many cases, the surge of a centrifugal stage with vaned diffuser is precipitated by the diffuser stalling first. The diffuser stall might initially be localized in one or a few passages, but if it progresses, it actually affects all the flow and causes the entire stage to surge. In contrast, a vaneless diffuser cannot stall in the same discrete way, but it can still experience flow reversal or rotating flow instabilities under heavily throttled conditions. At high flow, a vaned diffuser can choke at its throat if the Mach number is too high.

In single-stage compressors, the *volute* can also influence off-design performance. At very low flows, the volute may not be able to efficiently redirect the flow towards the exit, and the flow might recirculate in the volute. In extreme cases, flow might even reenter the diffuser from the volute during surge cycles. Typically, volutes are designed to be large enough not to choke or cause separation within the normal operating range, but poorly designed volutes can reduce the stable range.

11.5 REGULATION TECHNIQUES

Effective regulation is necessary to maintain optimal efficiency and pressure ratio while avoiding surge. The most widely used techniques include throttling, variable inlet guide vanes (IGVs), and variable speed drives (VSDs).

11.5.1 THROTTLING

Throttling is one of the simplest regulation techniques, achieved by adjusting a throttle or control valve located on the discharge or suction side of the compressor. By partially closing a valve, the system inlet or back pressure is altered, thus changing the compressor's operating point.

Throttling at the inlet is often preferred over discharge throttling because of its advantages in maintaining efficiency and reducing required throttling intensity. Suction throttling minimizes the loss in efficiency compared to discharge throttling, as it directly reduces the inlet pressure p_1^0, thus naturally decreasing the mass flow rate. Throttling on the discharge side increases back pressure without affecting p_1^0 and T_1^0, which remain at the design inlet conditions. Consequently, discharge throttling requires more intense throttling action, which increases resistance, raises energy consumption, and leads to efficiency losses.

In fact, for a given corrected speed, $N_c = \frac{N}{\sqrt{T_1^0}}$ and for almost constant inlet Mach number, $\dot{m}$ is a function of the inlet conditions T_1^0 and p_1^0 11.33:

$$\dot{m} = A\sqrt{\frac{\gamma}{R}}\frac{p_1^0}{\sqrt{T_1^0}}\mathrm{Ma}\left(1+\frac{\gamma-1}{2}\mathrm{Ma}^2\right)^{-\frac{\gamma+1}{2(\gamma-1)}} \tag{11.47}$$

When the inlet pressure p_1^0 is reduced via suction throttling, the mass flow rate $\dot{m}$ decreases proportionally. However, the corrected mass flow rate:

$$\dot{m}_c = \dot{m}\frac{\sqrt{T_1^0}}{p_1^0} \tag{11.48}$$

remains nearly constant, which means that the operating point on the compressor map is not significantly altered. When discharge throttling is applied, the corrected mass flow is lower compared to suction throttling for the same actual mass flow, which negatively affects compressor efficiency and decreases stall margin.

11.5.2 VARIABLE INLET GUIDE VANES (IGVS)

When IGVs adjust the airflow angle, the velocity triangle at the impeller inlet changes, altering not only the velocity direction but also, in general, the mass flow rate and compression ratio. In addition, a non-negligible effect is registered on both total and static pressure (Figure 11.13). The relationship between pressure ratio and IGV angle can be expressed as:

$$\frac{p_2^0}{p_1^0} = f(\alpha, \dot{m})$$

where f represents a function of IGV angle α and mass flow rate $\dot{m}$. The relationship between the IGV angle and the inlet air velocity c is given by:

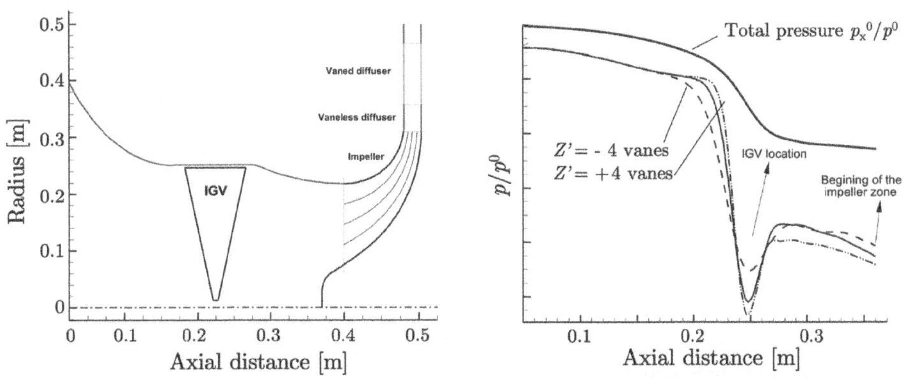

Figure 11.13 Effect of IGV on pressure upstream of the impeller in a centrifugal compressor.

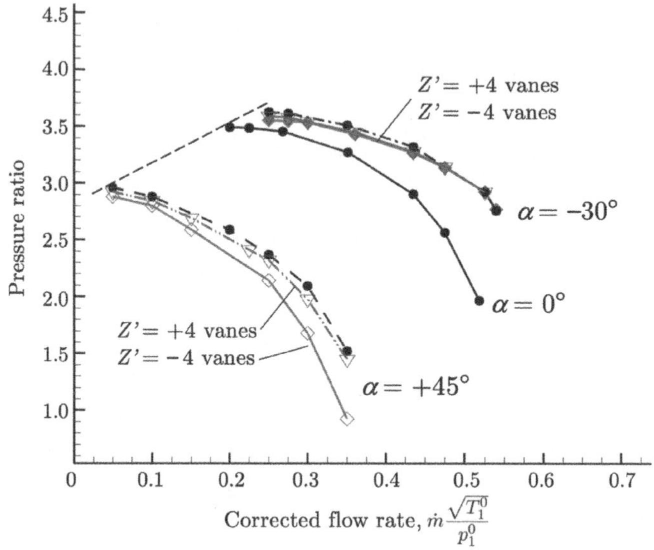

Figure 11.14 Effect of IGV setting on compressor speed line.

$$c = c_{\text{design}} \cos \alpha$$

where c_{design} is the inlet velocity at the design angle of the IGV (typically 0°). An exemplary effect of changing the angle α on a compressor speed line is shown in Figure 11.14. When the IGVs have a positive angle, they introduce prewhirl in the rotor's direction, reducing the work done by the impeller on the fluid and consequently lowering the pressure ratio. This in turn requires a higher meridional flow velocity to maintain the mass flow rate, thereby reducing the choke limit. Conversely, a negative IGV angle introduces prewhirl in the opposite direction, increasing the compressor head and enhancing the choke mass flow rate. A reduced pressure rise associated with a positive IGV angle also allows safer compressor operation at lower flow rates by decreasing the likelihood of stall due to a reduced positive pressure gradient.

In addition to the effects described above, a positive prewhirl decreases the incidence angle at the impeller leading edge, especially at low flow rates, which helps delay separation near surge conditions, thus enhancing stability. Furthermore, the impact of the parameter Z' on the increase in compressor pressure becomes significant with increasing IGV angle α, particularly at $\alpha = +45°$. An increase in Z' creates a higher inlet pressure drop and aligns the impeller flow angle better with the IGV angle, although it also introduces a higher blockage ratio at lower spans. This blockage redirects flow to higher spans, where blockage effects are less pronounced, but large IGV angles can still lead to inlet distortions with impacts on impeller outlet pressure and velocity distribution.

11.5.3 VARIABLE SPEED DRIVE (VSD)

By adjusting rotational speed N using VSD, the compressor performance curve can be shifted to achieve the desired pressure and flow combinations.

The flow rate $\dot{m}$ and the pressure ratio p_2^0/p_1^0 vary proportionally with changes in the compressor speed. In fact, given a speed N, the performance of the compressor can be described by the following affinity laws:

$$\dot{m} \propto N; \frac{p_2^0}{p_1^0} \propto N^2; P \propto N^3$$

11.6 PRACTICAL ASPECTS OF CENTRIFUGAL COMPRESSORS

11.6.1 AEROENGINES

In aerospace gas turbines (e.g. turboprop, turboshaft, and small turbojet/turbofan engines), weight and size are essential, and very high overall pressure ratios are needed for maximizing cycle thermal efficiency. Historically, centrifugal stages were used in early jet engines and continue to be used in *auxiliary power units* (APUs) and smaller engines because a single stage can provide a comparatively large pressure ratio, thus reducing the number of stages needed. For example, many helicopter turboshaft engines (such as the GE T700 shown at the top of Figure 11.15, or the Pratt & Whitney PT6, a variant of which is depicted in the bottom part of the same figure) use a centrifugal impeller as the final stage of compression after some axial stages, contributing a major fraction of the overall pressure ratio. In APUs and small jet engines, often a single centrifugal stage is used alone, as in the Garrett AiResearch GTCP series of APUs.

The design of aerospace centrifugal compressors pushes for high pressure ratio per stage (from 4 to 6) to minimize stage count, and uses high tip speeds enabled by advanced materials such as titanium alloys and manufacturing, which often requires 5-axis milled blading with precise contours. The blades are typically backswept to improve efficiency and surge margin. The efficiencies achieved in aerospace centrifugal compressors are quite high (isentropic efficiency in the high-80% range) since they often operate at high Reynolds numbers and are finely optimized; however, when pushed to very high loading (e.g. in transonic conditions), the efficiency can drop into the 70%s. Aerospace compressors typically operate at a fixed design speed, such that the operating range required is somewhat limited, so designers can optimize the map for a narrower usage band. Surge margin is maintained by careful matching of the centrifugal stage with any axial stages and by using bleed valves at low speeds if necessary.

Finally, because of high rotational speeds, mechanical constraints are significant: impellers must be designed for large stresses and thermal gradients while accounting for pertinent deformations at operating speeds, so that clearances remain acceptable.

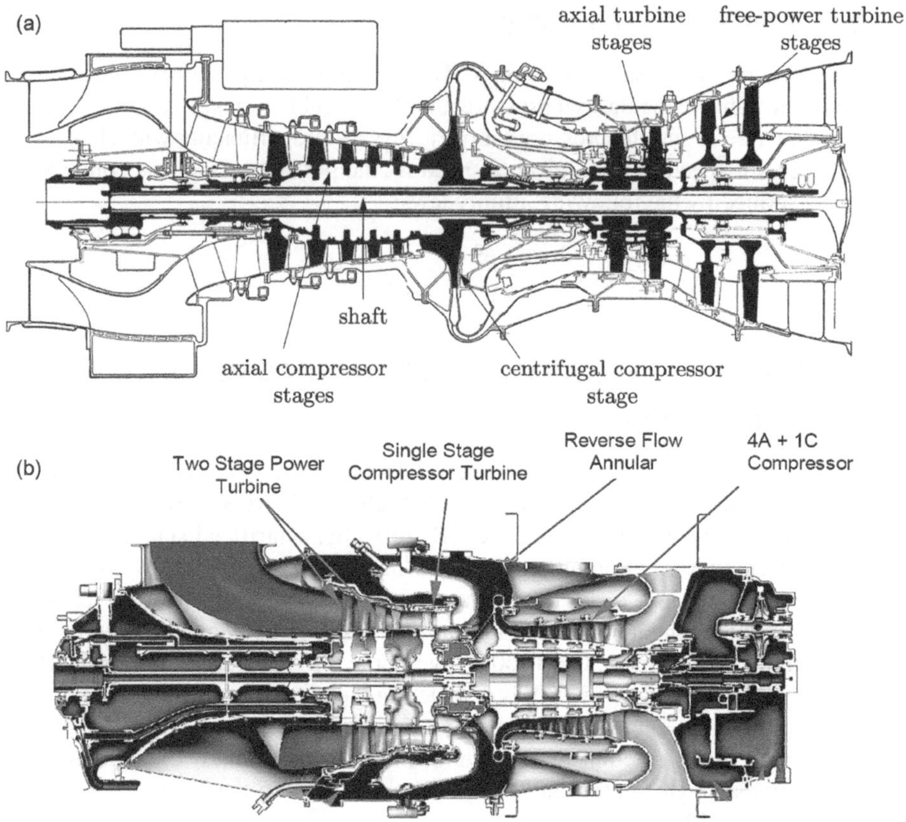

Figure 11.15 (a) Cutaway view of the GE T700 engine. (b) Cutaway view of the Pratt & Whitney PTC-67A. (Smailys, H. *et al.*, "Advanced Technologies Applied to PT6C-67A Tiltrotor Engine Design," *Proceedings of the 57th AHS International Annual Forum*, pp. 1647–1658 (2001). Reproduced from The Vertical Flight Society with permission.)

11.6.2 INDUSTRIAL PROCESS COMPRESSORS

Industrial centrifugal compressors (such as those in refineries, chemical plants, or pipeline booster stations) are usually multistage machines designed for continuous operation at or near their design point. For example, in a pipeline station, the compressor might always run at 100% speed, and the operating point moves along that speed line as the throughput demand changes. If a wider flow range is needed, sometimes a throttle or adjustable inlet guide vane is installed to shift the curve.

Key concerns for these machines are efficiency and reliability. These compressors often handle large flow rates at moderate to high pressure ratios (for example, compressing natural gas from 1 bar to 20 bar might use four or five stages). Efficiency is paramount because these machines consume significant power (tens or hundreds of MW), so even a 1% efficiency gain means big energy savings. As such, industrial compressors use multiple stages with lower per-stage pressure ratios (≈1.5–3.0) to keep each stage in an efficient regime. They frequently incorporate intercooling between stage groups to reduce the work input and manage temperatures (for instance, compressing air in three stages with intercoolers to reach 8 bar, as is common in large air separation plants).

One notable design approach is the integrally geared compressor, where each stage, or pair of stages, runs on its own high-speed pinion gear (Figure 11.16); this allows each impeller to operate at its optimal speed, since in a single-shaft machine

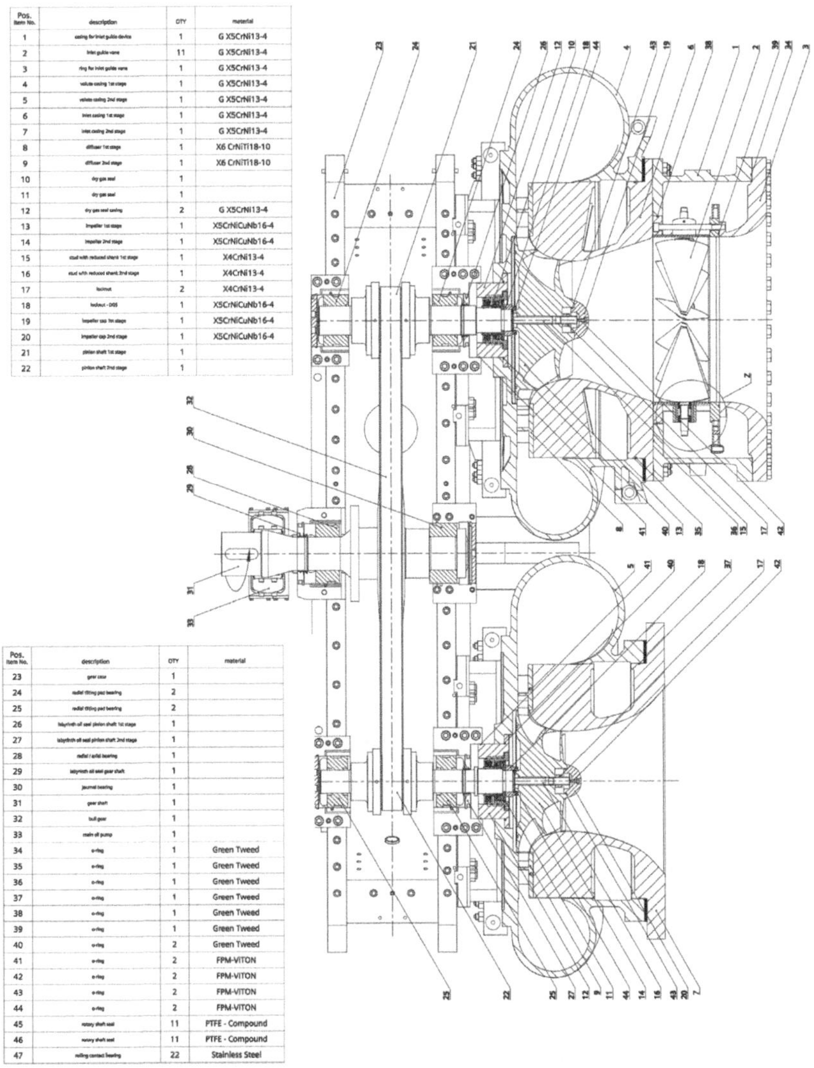

Figure 11.16 A geared centrifugal compressor composed of two stages.

later stages would be oversized or spinning sub-optimally if all on one shaft. By optimizing the speed of each stage, the overall efficiency is improved.

Industrial compressors also often employ open impeller designs with tight clearances or abradable linings to maintain efficiency over time, and coatings to reduce fouling if the process gas is dirty. The vaneless diffuser or a low-solidity vaned diffuser is frequently employed to get a broader stable range, even if it means a slightly lower peak efficiency.

Despite industrial compressors having to often handle a range of conditions, they are not typically required to drastically change flow in nearly real-time, therefore the management of surge margin is relatively straightforward. As a result, many industrial units maintain at least 10%–15% surge margin at normal operation and include an anti-surge control system that continuously monitors conditions and opens a blow-off valve if approaching surge. Sometimes, two recycle loops equipped with two blow-off systems are employed (one for large surge events, one for control during smaller fluctuations), while typically modern compressor installations incorporate surge anticipation by detecting lead indicators like slight pressure fluctuations or changes in shaft vibrations that precede surge.

11.6.3 MICROTURBINES AND TURBOCHARGERS

At the lower size limit, centrifugal compressors in microturbines (30–200 kW range) or automotive turbochargers face a different set of challenges.

These compressors have impeller diameters of a few centimeters and operate at extremely high speeds (100,000–200,000 rpm is common for turbochargers). The small size leads to low Reynolds numbers, which increase friction and heat transfer losses. As a result, the maximum efficiencies of these compressors are comparatively low - for example, a well-designed automotive turbocharger compressor could have a peak efficiency of around 75%–78% and a microturbine compressor around 70%–80%.

They must also operate over a wide range of flows. For a turbocharger, engine operating conditions can vary from near idle (low-flow, low-pressure ratio) to full-throttle (high-flow, high-pressure ratio). Consequently, turbochargers are designed with wide maps, a feature that can be achieved using both blow-off valves and circumferential bleed slots at the inducer to extend low-flow stability. In addition to these precautions, impellers with very low rotational inertia are common, as sustained pumping of a large-inertia rotor against a closed throttle might cause surge.

The pressure ratios for automotive turbos are typically up to 3.0–4.0 for a single stage. Some high-performance turbo systems use two stages sequentially to reach higher boosts. The impellers are usually made of aluminum for low cost and, due to operating temperatures being moderate on the compressor side, are sometimes forged/machined, or even cast in large-volume production.

Microturbine compressors for small power generation use a single centrifugal stage similarly for simplicity and cost, and achieve overall pressure ratios of 3–5. Because these feed a combustor, the pressure ratio directly impacts thermal efficiency, so there is an incentive to maximize it. However, material restrictions, often requiring steel or titanium impellers, and cost mean that they typically stay in the four range at efficiency ≈70%–75%. Some microturbines incorporate a two-stage centrifugal compressor with an intercooler to reach higher overall pressure ratios (≈6–8) if higher thermal efficiency is needed, but this increases cost and complexity. Surge control valves are put in place to avoid flameout.

11.7 PROBLEMS

1. The inducer of a centrifugal compressor impeller ingests an air mass flow rate of $\dot{m}_c = 0.7453$ kg/s at inlet stagnation conditions $p_1^0 = 101.3$ kPa and $T_1^0 = 300$ K. The compressor is designed to minimize the relative inlet Mach number to reduce inducer shock losses. The impeller has an external diameter of $D_2 = 88$ mm and operates at a rotational speed of N = 60,000 rpm. The inducer is constrained by an external diameter of $D_{1,s} = 65$ mm and an internal diameter of $D_{1,h} = 32$ mm. Determine the inlet Mach number Ma_1 and calculate the corresponding inducer blade angle at the shroud β_{1s}, which minimize the relative inlet Mach number.
[0.497, 32.1°]

2. The inducer of a centrifugal compressor impeller receives an air mass flow rate of $\dot{m}_c = 1.4776$ kg/s at inlet stagnation conditions $p_1^0 = 101.3$ kPa and $T_1^0 = 288$ K. The compressor operates at a rotational speed of $N = 50,000$ rpm, with an impeller outer diameter of $D_2 = 95$ mm. The internal inducer diameter is constrained to $D_{1,h} = 40$ mm. Determine the maximum possible inducer outer diameter, $D_{1,s}$, such that the relative Mach number at the inducer, $\mathrm{Ma}_{w1,s}$, does not exceed 1.
[75 mm]

3. An impeller of a centrifugal compressor with radial blades $\beta_2' = 90°$ handles air. The impeller has 16 blades, an isentropic efficiency of $\eta_s = 0.85$ and a flow coefficient of $\phi = 0.466$. Determine the outlet Mach number M_2 and the pressure ratio across the impeller $\frac{p_2^0}{p_1^0}$, under the condition that the tangential Mach number at the impeller tip, Ma_u, does not exceed 1.
[0.97, 2.62]

4. Design a centrifugal compressor impeller for the following operating conditions: the mass flow rate is $\dot{m} = 0.62$ kg/s, the rotational speed is $N = 70{,}000$ rpm, and the maximum blade tip speed is $u_{2,\max} = 500$ m/s. The compressor is required to achieve a total-to-total pressure ratio of $\pi_s = 3.1$, with an inlet stagnation pressure of $p_1^0 = 101.3$ kPa and an inlet stagnation temperature of $T_1^0 = 300$ K. The expected isentropic efficiency of the impeller is $\eta_s = 0.75$.

NOTES

1. The non-dimensional mass flow rate is in fact a corrected mass flow rate.
2. For a zero prewhirl impeller, $\psi = \frac{c_{u2}}{u_2} = \frac{\sigma_{slip} c_{u2,E}}{u_2} = \frac{\sigma_{slip} u_2}{1+\frac{\tan\alpha_{2E}}{\tan\beta_2'}}$, since $c_{u2,E} = u_2 - \frac{c_{m2}}{\tan\beta_2'}$ and $c_{u2,E} = \frac{c_{m2}}{\tan\alpha_{2E}}$, which implies $c_{u2,E} = u_2\left(1 + \frac{\tan\alpha_{2E}}{\tan\beta_2'}\right)$.
3. Since c_m tends to 0 as r increases, we identify the tangential component $c_u(r)$ as the velocity perpendicular to the local cross section of the volute.

12 Centripetal Turbines

In radial inflow turbines, commonly known as centripetal turbines, the working fluid flows radially inwards from the periphery to the center of the rotor. Despite their considerable applications in hydraulic systems, this chapter will concentrate solely on the function of these machines as gas expanders.

Radial inflow gas turbines are extensively utilized in various industries owing to their compact architecture, high efficiency, and capacity to manage significant pressure differentials per stage. Their durable construction enables them to perform effectively in demanding applications, making them ideal for situations where operational reliability and performance are essential. Radial inflow turbines are preferred in situations where energy efficiency and size limitations are critical, including small-scale power generation, industrial process control, and turbocharging internal combustion engines.

12.1 GENERAL CHARACTERISTICS

The primary components along the mean flow path of a radial inflow turbine include (Figure 12.1):

- **Inlet volute**. This is a scroll-shaped casing that evenly directs the incoming high-pressure gas around the circumference of the turbine, guiding it toward a radial

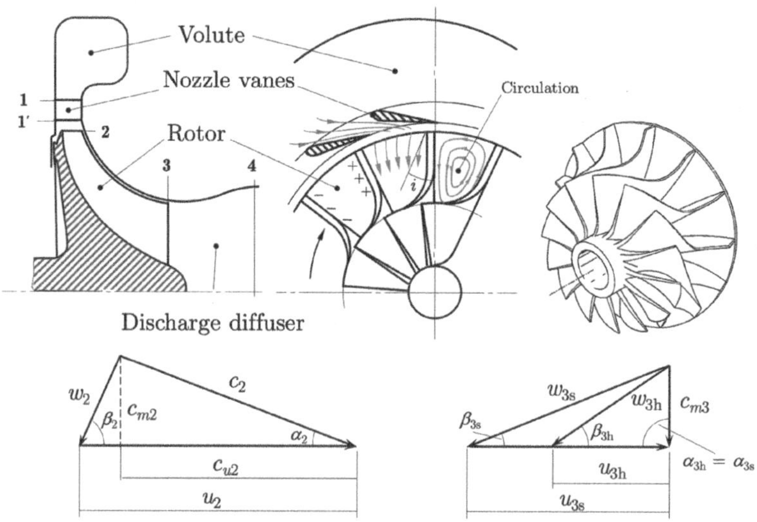

Figure 12.1 Schematic representation of a centripetal turbine geometry and pertinent velocity triangles.

DOI: 10.1201/9781003619956-12

ring of nozzle vanes or, if such a nozzle is not present, directly into the rotor. By distributing the gas flow, the volute reduces pressure and ensures a steady inlet condition as well as mild acceleration. Due to its shape, it also provides an inlet tangential velocity component to the flow.

- **Radial nozzle vanes** (when present). In some configurations, particularly those requiring additional control over velocity and pressure drop, a radial vaned nozzle is positioned immediately downstream of the volute. The nozzle converts the enthalpy energy of the gas into kinetic energy by accelerating the flow before it reaches the rotor. The nozzle blades direct the gas to the rotor blades at an optimal angle to maximize energy transfer and efficiency.
- **Centripetal rotor**. Gas enters the rotor radially, moving inward toward the center of the turbine in a centripetal flow pattern. The rotor blades are typically backward-curved or radially curved.
- **Exhaust diffuser**. The low-pressure expanded gas exits the turbine rotor through the exhaust diffuser, which decelerates the gas flow converting the residual kinetic energy into pressure energy for the purpose of aiding in exhaust system efficiency.

12.2 THE ROTOR

In general terms, the work exchange in the rotor (from station 2 to 3) can be expressed as follows:

$$L = \frac{c_2^2 - c_3^2}{2} + \frac{w_3^2 - w_2^2}{2} + \frac{u_2^2 - u_3^2}{2} \tag{12.1}$$

To maximize L, the following conditions must be satisfied. Firstly, u_2 must be larger than u_3, a condition ensured by the rotor configuration. Furthermore, c_2 must exceed c_3, which implies that the nozzle vanes should provide maximum flow acceleration while minimizing the rotor outlet velocity. Finally, w_3 must exceed w_2, which means that the rotor should accelerate the relative flow. These requirements lead to the velocity diagrams shown in Figure 12.1.

The blade angle β_2' is typically set to 90°, creating radial inflow relative velocity and enabling a blade geometry made of radial fibers. This configuration ensures that the blades are radial in any section taken perpendicular to the rotor axis. Such radial blades achieve a stress condition dominated by tensile loads, allowing for very high peripheral speeds (up to 500 m/s). The absolute outlet velocity angle α_3 is set to 90° to minimize the flow velocity at the blade outlet and reduce kinetic energy losses. A negative exit swirl can sometimes be used to increase specific work output, although this approach necessarily reduces efficiency.

Assuming a constant value for the outlet absolute velocity along the radius, the condition $c_{u3} = 0$ implies that flow turning is maximum at the blade tip. In fact, from the exit triangle shown in Figure 12.1:

$$c_{m3} = \text{const.} \rightarrow r_3 \tan\beta_3 = \text{const.} \tag{12.2}$$

Consequently, the relative velocity is also maximized at the tip radius, potentially causing choking in the exit shroud region. The turbine rotor, in fact, behaves like a convergent nozzle for the relative flow, and the relative Mach number can approach unity at the exit.

A common criterion for selecting rotor dimensions is to use a non-dimensional design approach, where all dimensions are normalized with respect to the rotor radius, while flow quantities are normalized using the peripheral velocity or, in cases emphasizing compressibility, the stagnation speed of sound at station 2.

In single-stage turbine rotors, the total-to-static isentropic efficiency is often used to measure the irreversibility in energy exchange:

$$\eta_{ts} = \frac{h_1^0 - h_3^0}{h_1^0 - h_{3s}} \tag{12.3}$$

or, for an ideal gas:

$$\eta_{ts} = \frac{T_1^0 - T_3^0}{T_1^0 - T_{3s}} \tag{12.4}$$

This metric is usually preferred to total-to-total efficiency because the kinetic energy of the gases at the rotor outlet is considered to be a loss. The total-to-static isentropic efficiency closely correlates with the total-to-static expansion ratio. In fact:

$$\eta_{ts} = \frac{1 - \frac{T_3^0}{T_1^0}}{1 - \frac{T_{3s}}{T_1^0}} = \frac{P_f}{1 - \left(\frac{p_1^0}{p_3}\right)^{\frac{1-\gamma}{\gamma}}} \tag{12.5}$$

where P_f is called *power factor*. The previous equation can be used to determine P_f for a desired $\frac{p_1^0}{p_3}$, provided that η_{ts} is postulated or known.

12.2.1 THE INLET SECTION

The analysis of the rotor begins with determining the optimal flow conditions at the inlet. This involves minimizing the inlet Mach numbers, Ma_2 and Ma_{w2}, for a given power factor P_f and inlet flow angle α_2. In fact, minimizing Ma_2 reduces losses generated within the guide vanes (when present), while minimizing Ma_{w2} mitigates losses within the rotor's inlet section. Achieving these objectives requires a thorough understanding of the role of flow slip at the rotor inlet.

The rotational nature of the relative flow inside the rotor channels generates vorticity, similar to that described for centrifugal pumps or compressor impellers, which alters the flow patterns compared to the ideal Eulerian case. As a result, the relative velocity angle that optimizes aerodynamic efficiency of the inlet rotor section is not 90° but closer to 65°, as confirmed by experimental studies (Figure 12.1), which leads to optimal incidence angles of approximately 25°. In the absence of swirled flow at the rotor outlet, the extracted specific work is:

$$L = \frac{\gamma R}{\gamma - 1}\left(T_1^0 - T_3^0\right) = u_2 c_2 \cos\alpha_2 \tag{12.6}$$

from which:

$$P_f = (\gamma - 1)\cos\alpha_2 \mathrm{Ma}_{2,0}\mathrm{Ma}_u \tag{12.7}$$

where $\mathrm{Ma}_{2,0} = c_2/\sqrt{\gamma R T_2^0}$ and $\mathrm{Ma}_u = u_2/\sqrt{\gamma R T_2^0}$.

On the other hand, from trigonometric considerations:

$$\phi^{-1} = \frac{u_2}{c_{m2}} = \frac{c_{u2} + w_{u2}}{c_{m2}} = \frac{1}{\tan\alpha_2} + \frac{1}{\tan\beta_2} \tag{12.8}$$

therefore:

$$c_2 = \frac{u_2}{\sin\alpha_2}\left(\frac{\tan\alpha_2 \tan\beta_2}{\tan\alpha_2 + \tan\beta_2}\right) \tag{12.9}$$

from which:

$$\mathrm{Ma}_2^2 = \frac{2\,\mathrm{Ma}_u^2 \tan^2\alpha_2 \tan^2\beta_2}{2\sin^2\alpha_2 (\tan\alpha_2 + \tan\beta_2)^2 - (\gamma - 1)\,\mathrm{Ma}_u^2 \tan^2\alpha_2 \tan^2\beta_2} \tag{12.10}$$

To find a simplified implicit equation $f(\mathrm{Ma}_2, \alpha_2, \beta_2, P_f) = 0$, we will systematically combine the given equations and eliminate intermediate variables:

$$\frac{P_f^2 (1 + \frac{\gamma-1}{2}\mathrm{Ma}_2^2)^2}{(\gamma - 1)^2 \cos^2\alpha_2 \,\mathrm{Ma}_2^2} - \mathrm{Ma}_2^2 \sin^2\alpha_2 \left(\frac{\tan\alpha_2 + \tan\beta_2}{\tan\alpha_2 \tan\beta_2}\right)^2 = 0$$

which, along with:

$$\mathrm{Ma}_{w2} = \frac{\mathrm{Ma}_2 \sin\alpha_2}{\sin\beta_2}$$

make it possible to study the influence of α_2, β_2, P_f on Ma_2, Ma_{w2} and Ma_u, as shown in Figure 12.2. For a given power factor P_f, and assuming $\beta_2 = 65°$, the optimal choice for flow Mach numbers Ma_2 and Ma_{w2} occurs at the lowest values of the angle α_2. For instance, when the power factor is 0.3, a subsonic flow is achieved at the vane exit only if the flow angle α_2 is less than 30°. Lower power factors facilitate the development of fully subsonic conditions for both the absolute and relative Mach numbers at the rotor inlet. However, it is important to note that smaller α_2 values restrict the allowed mass flow to the rotor. Consequently, a trade-off is necessary to balance the desired flow rate with the required power extraction.

Once α_2 is identified and β_2 is fixed at 65°, the inlet velocity triangle is fully determined.

The minimum number of rotor blades Z is determined to prevent the relative velocity on the pressure side from becoming zero at the inlet (station 2). The equations used here are simpler than those for pumps and compressors because the relative flow is accelerated within the rotor. This acceleration reduces the impact of the number of blades on performance, even in the case of mild flow separations, which tend to reattach naturally in turbines. Relative flow can be visualized as a superposition of a constant throughflow velocity (independent of θ) and a rotational velocity (Figure 12.3). By imposing $w_{\min} = \bar{w}_r - \omega r \frac{\Delta\theta}{2} = 0$, it follows that:

$$Z = \frac{\pi}{\phi} = \frac{\pi(\tan\beta_2 + \tan\alpha_2)}{\tan\beta_2 \cdot \tan\alpha_2} \tag{12.11}$$

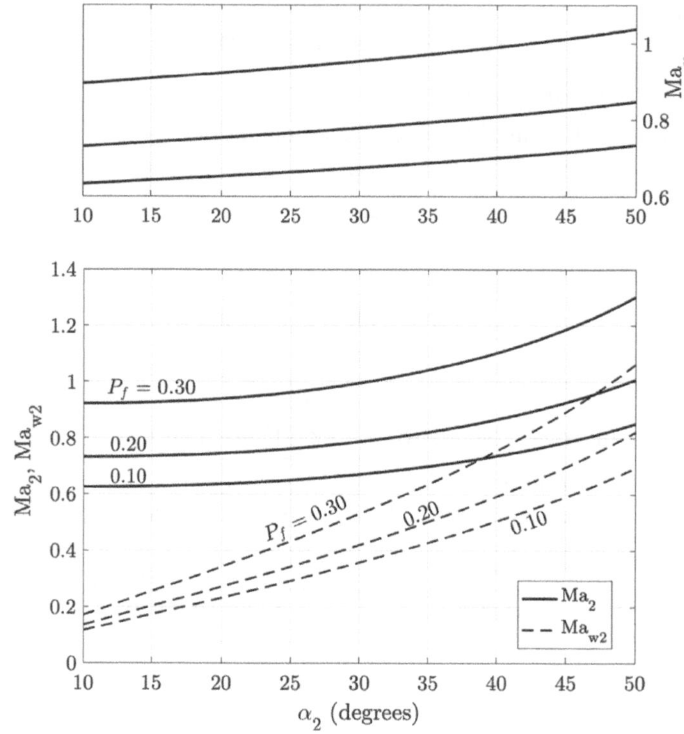

Figure 12.2 Variation in Mach numbers at the rotor inlet as functions of absolute flow angle α_2 for different values of P_f.

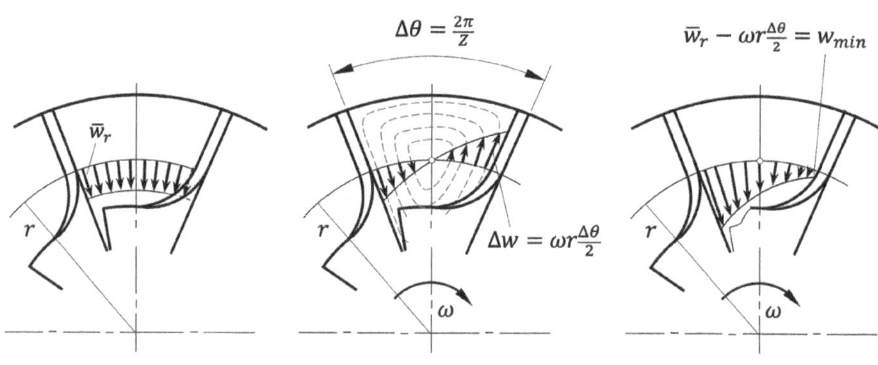

Figure 12.3 Illustration of minimum relative velocity formation in a centripetal turbine blade passage.

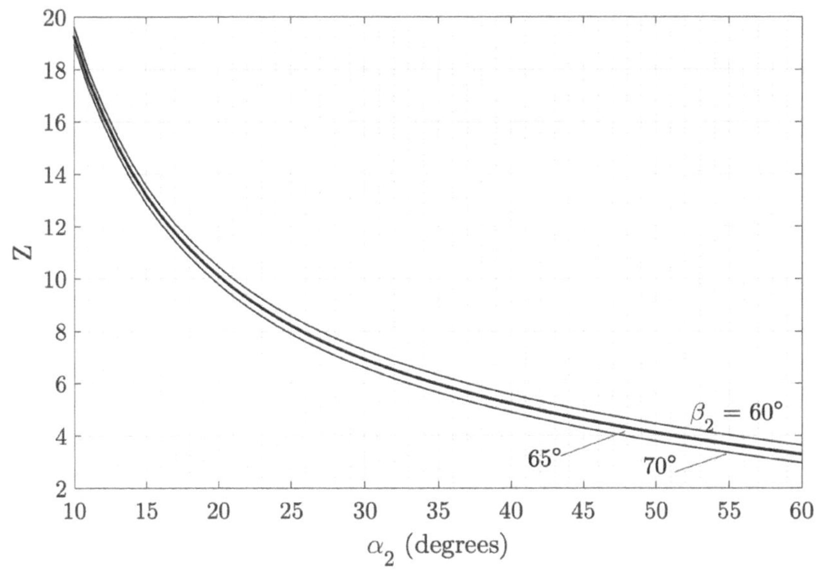

Figure 12.4 Minimum required number of rotor blades Z in a centripetal turbine as a function of absolute flow angle at rotor inlet α_2, for different values of inlet relative flow angle β_2.

A plot of equation 12.11 is shown in Figure 12.4. As can be observed, the influence of β_2 is marginal when β_2 lies within the range from 60 to 70°. In contrast, α_2 has a significant impact on the number of blades, which increases superlinearly as α_2 decreases below 30°, resulting in elevated friction losses. However, it is important to note that such low α_2 values are necessary to minimize the Mach numbers at the rotor inlet (see Figure 12.2). To achieve an optimal solution where overall losses are minimized, a careful balance must be struck regarding α_2.

The flow density in 2 can be computed from the equation of state:

$$\rho_2 = \frac{\pi_N p_1^0}{RT_1^0}\left(1+\frac{\gamma-1}{2}M_2^2\right)^{-\frac{1}{\gamma-1}} \tag{12.12}$$

where $\pi_N = p_2^0/p_1^0$ quantifies the total pressure loss in the nozzle vanes.

The blade height b_2 can finally be determined as:

$$b_2 = \frac{\dot{m}}{\rho_2}\frac{1}{\xi_2 2\pi r_2 \phi \mathrm{Ma}_u \sqrt{\gamma R T_2^0}} \tag{12.13}$$

where, for a radial inflow rotor featuring Z blades with inlet thickness t_2, ξ_2 can be determined as:

$$\xi_2 = 1 - \frac{t_2 Z}{2\pi r_2} \tag{12.14}$$

12.2.2 THE OUTLET SECTION

The rotor outlet (or exducer) can be analyzed in a manner similar to that done for a compressor inducer to minimize the relative Mach number at the shroud exit. In particular, if we define a non-dimensional mass flow rate as:

$$\dot{m}_c = \frac{\dot{m}}{\rho_1^0 a_1^0 \pi r_2^2} = \frac{\rho_3 c_3 r_{3,s}^2 (1-\nu^2)}{\rho_1^0 a_1^0 r_2^2} \tag{12.15}$$

where

$$\nu = \frac{r_{3,h}}{r_{3,s}} \tag{12.16}$$

the relative Mach number $\mathrm{Ma}_{w3,s}$ is computed in a way similar to what was done in Section 11.3.2 (equations 11.3–11.7):

$$\mathrm{Ma}_{w3,s}^2 = \mathrm{Ma}_3^2 + \frac{\dot{m}_c \mathrm{Ma}_u^2}{1-\nu^2} \frac{1}{\mathrm{Ma}_3} \left(1 + \frac{\gamma - 1}{2} \mathrm{Ma}_3^2\right)^{1/2} \left(\frac{T_3^0}{T_1^0}\right)^{-1/2} \frac{p_1^0}{p_3} \tag{12.17}$$

where $\mathrm{Ma}_{w3,s} = w_{w3,s}/\sqrt{\gamma R T_3}$, and recalling equation 12.5:

$$\frac{p_1^0}{p_3} = \left(1 - \frac{P_f}{\eta_{ts}}\right)^{-\gamma/(\gamma-1)}, \quad \frac{T_3^0}{T_1^0} = 1 - P_f \tag{12.18}$$

The plot of equation 12.17 is shown in Figure 12.5. As is evident, for a given power factor P_f and a corrected mass flow $\dot{m}_c$, the hub-to-tip ratio ν can be selected

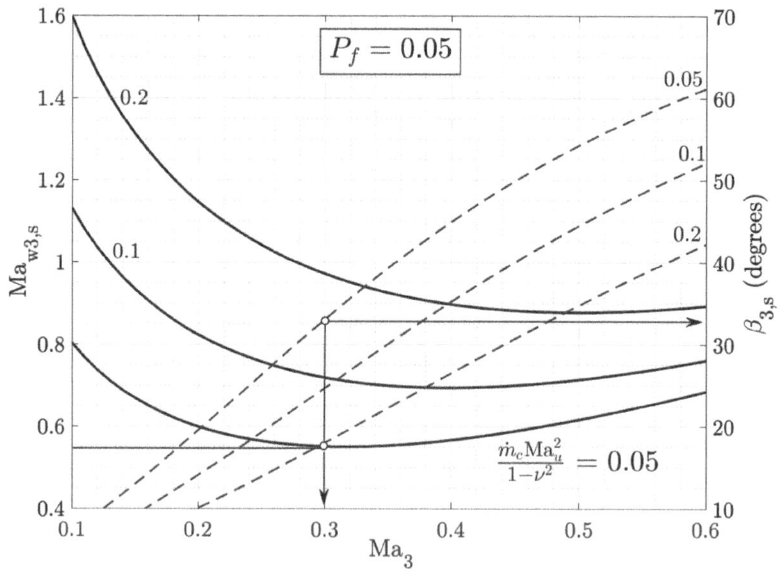

Figure 12.5 Non-dimensional performance chart for the outlet of a centripetal turbine.

to match a specified $\frac{\dot{m}_c \mathrm{Ma}_u^2}{1-v^2}$ (as Ma_u was already determined during the design of the rotor inlet). The resulting $\mathrm{Ma}_{w3,s}$ function (depicted as solid lines in Figure 12.5) exhibits a minimum at a specific value of the outlet Mach number Ma_3. At this minimum, there is a unique value of the rotor blade angle $\beta_{3,s}$, given by:

$$\beta_{3,s} = \arcsin\left(\frac{\mathrm{Ma}_3}{\mathrm{Ma}_{w3,s}}\right) \tag{12.19}$$

This relationship facilitates the optimal design of the rotor outlet section starting from the shroud. For instance, when $P_f = 0.05$ and $\eta_{ts} = 0.85$, the optimal design corresponds to $\mathrm{Ma}_3 = 0.3$, $\mathrm{Ma}_{w3,s} = 0.55$ and $\beta_{3,s} = 33^\circ$.

It is important to note that while a flow slip does occur at the blade outlet, it is almost negligible.

The preliminary design of the rotor can be developed using the equations provided above. In particular, apart from the velocity triangles, the primary dimensions are determined by equation 12.15.

To determine the rotor diameter—an essential parameter from which other dimensions can be proportionally scaled—a useful dimensionless parameter is the specific diameter:

$$D_s = \frac{D_2 \left(\Delta h_s^0\right)^{1/4}}{\sqrt{Q_3}} \tag{12.20}$$

Here, Δh_s^0 denotes the isentropic work output of the rotor, and Q_3 is the volumetric flow rate at the rotor exit, as commonly encountered in turbines. The specific diameter shows a strong correlation with the specific speed, defined as:

$$\omega_s = \frac{\omega \sqrt{Q_3}}{\left(\Delta h_s^0\right)^{3/4}} \tag{12.21}$$

This relationship is supported by the statistical analyses conducted by Balje,[1] as illustrated in Figure 12.6.

12.3 VOLUTE AND INLET SYSTEM

The flow enters the volute generating radial c_m and tangential c_u velocity components. The radial component directs the flow inward, whereas the tangential component transports angular momentum along the periphery of the impeller or the inlet guide vanes, if present.

The volute should be designed to ensure uniform flow at the inlet of the downstream component. However, this ideal must often be compromised due to constraints on size and available volume. As a result, non-uniform flows to the rotor frequently occur, propagating through the rotor and potentially reducing work transfer efficiency.

The tongue extension significantly affects flow uniformity and losses, with azimuth angles of 85°–120° offering a good balance. Shorter tongues are common but

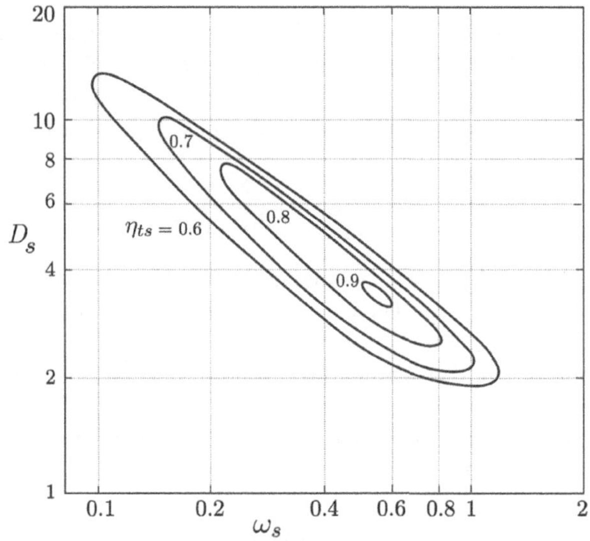

Figure 12.6 The Balje chart for centripetal turbines.

not always optimal, as a complex vortex motion begins within the first 20° of flow turning leading to total pressure losses.

In a vaneless volute, the angular momentum is conserved such that the flow follows a free vortex pattern:

$$rc_u = \text{const.} = K \tag{12.22}$$

The mass flow rate through a cross-section of the volute at azimuthal angle θ is expressed as:

$$\dot{m}_\theta = \rho_\theta A_\theta c_u \tag{12.23}$$

which, for a uniform mass flow distribution, is written as:

$$\dot{m}_\theta = \dot{m}\left(1 - \frac{\theta}{360^\circ}\right) \tag{12.24}$$

Equating the previous two equations:

$$A_\theta = \frac{\dot{m}r}{\rho_\theta K}\left(1 - \frac{\theta}{360^\circ}\right) \tag{12.25}$$

where, assuming a non-isentropic flow within the volute:

$$\rho_\theta = \frac{\pi_v(\theta)p_1^0}{RT_1^0}\left(1 + \frac{\gamma - 1}{2}\text{Ma}_\theta^2\right)^{-\frac{1}{\gamma-1}} \tag{12.26}$$

$\pi_v(\theta)$ being a total pressure loss coefficient, which can be postulated as linearly dependent on θ, such that:

$$A_\theta = \frac{\dot{m}RT_1^0}{\pi_v p_1^0 K}\left(1+\frac{\gamma-1}{2}\mathrm{Ma}_\theta^2\right)^{\frac{1}{\gamma-1}}\left(1-\frac{\theta}{360^\circ}\right) \tag{12.27}$$

The relationship among inlet, exit Mach numbers, and radius ratios is derived from the free vortex equation:

$$c_1 r_1 \sin\alpha_1 = c_2 r_2 \sin\alpha_2 \tag{12.28}$$

expressing the latter in terms of Mach numbers:

$$\frac{r_1}{r_2} = \frac{M_2 \sin\alpha_2}{M_1 \sin\alpha_1}\sqrt{\frac{1+\frac{\gamma-1}{2}M_1^2}{1+\frac{\gamma-1}{2}M_2^2}} \tag{12.29}$$

The volute area ratio can be derived from continuity:

$$\frac{A_1}{A_2} = \frac{\theta_2}{\theta_1}\frac{p_2^0}{p_1^0} \tag{12.30}$$

where θ_1 and θ_2 are the inlet and exit azimuthal angles of the volute, respectively.

The previous equations can be used to establish the shape of the volute for given inlet stagnation gas properties and the desired $\mathrm{Ma}_\theta = \mathrm{Ma}_1 =$ const. For compact designs, a minimum Ma_1 of 0.3 is typical. In gas turbines, low Mach numbers (below 0.15) exiting the combustor can require volute areas that are too large. To address this, nozzle vanes are used to turn and accelerate the gas before the gas enters the rotor.

The angle of flow at the volute exit is determined by the ratio of tangential to meridional velocity:

$$\tan\alpha_2 = \frac{c_{u2}}{c_{m2}} \tag{12.31}$$

and it should be compatible with the required angle for best nozzle vane or rotor operation (see 12.2.1).

12.3.1 DISCHARGE DIFFUSER

The diffuser downstream of a turbine rotor recovers part of the exhausted kinetic energy, thus increasing the static expansion ratio and total-to-static efficiency. This recovery is essential when the rotor exit dimensions are aerodynamically constrained. Diffusers typically feature conical or annular geometries, with performance influenced by geometric parameters (length, area, cone angle, and wall contour) and fluid properties (Mach and Reynolds numbers, turbulence, and swirl).

The design is complex due to the dynamic interactions within the diffuser, including boundary layer growth, turbulence, and low-momentum flow zones. Despite advances, understanding remains limited, relying on empirical data and case studies. The literature highlights progress, but diffuser design remains a trial-and-error process often supported by CFD.

12.4 TURBINE MAP

The following equation for the expansion ratio is derived for a generic turbine stage:

$$\pi_e = \frac{p_1^0}{p_3^0} = \left[1-(\gamma-1)\frac{\mathrm{Ma}_u}{\tan\alpha_2}\left(\frac{\dot{m}}{\rho_2 A_2 a_2^0}\right)\right]^{\frac{1-\gamma}{\gamma\eta_{\mathrm{pol,T}}}} \tag{12.32}$$

where:

$$\frac{\dot{m}}{\rho_2 A_2 a_2^0} = k\cdot\dot{m}_c\left(1+\frac{\gamma+1}{2}\mathrm{Ma}_2^2\right)^{\frac{1}{\gamma-1}} \tag{12.33}$$

k being a constant for a turbine of a given geometry and given working fluid. In this section, $\dot{m}_c$ is defined as the proper corrected mass flow rate, that is:

$$\dot{m}_c = \dot{m}\frac{\sqrt{T_1^0}}{p_1^0} \tag{12.34}$$

In addition:

$$\mathrm{Ma}_2 = \frac{\mathrm{Ma}_u}{\sin\alpha_2}\left(\frac{\tan\beta_2\cdot\tan\alpha_2}{\tan\beta_2+\tan\alpha_2}\right) \tag{12.35}$$

$$\tan\beta_2 = \frac{\mathrm{Ma}_{2,0}\sin\alpha_2}{\mathrm{Ma}_u-\mathrm{Ma}_{2,0}\cot\alpha_2} \tag{12.36}$$

Upon manipulation of previous equations (see Appendix A.5), the following expressions can be derived:

$$\pi_e = \left[1-(\gamma-1)\frac{\mathrm{Ma}_u}{\tan\alpha_2}\cdot k\cdot\dot{m}_c\right]^{\frac{1-\gamma}{\gamma\eta_{\mathrm{pol,T}}}} \tag{12.37}$$

$$\tan\beta_2(\sin\alpha_2+\cot\alpha_2\tan\beta_2) =$$

$$= k\cdot\dot{m}_c\sin\alpha_2\cdot\left[1+\frac{\gamma+1}{2}\cdot\frac{\mathrm{Ma}_u^2}{\sin^2\alpha_2}\cdot\left(\frac{\tan\beta_2\tan\alpha_2}{\tan\beta_2+\tan\alpha_2}\right)^2\right]^{\frac{1}{\gamma-1}} \tag{12.38}$$

These relations can be used to construct the characteristic map of a turbine, expressed as $\pi_e = f(k\cdot\dot{m}_c, \mathrm{Ma}_u, \alpha_2, \gamma, \eta_{\mathrm{pol,T}})$ (Figure 12.7). Specifically, equation 12.37 is illustrated for both cases: when the nozzle vane ring is unchoked (left) and choked (right) at the highest attainable corrected mass flow for a given speed line. In the same figure, the values of total-to-total isentropic efficiency, η_{TT}, are displayed where η_{TT} is modeled as a simplistic decreasing parabolic function of β_2, as this angle serves as a direct indicator of inlet incidence losses occurring at off-design operation.

Choking typically occurs when the stator nozzle throat or the rotor inlet reaches sonic velocity, the mass flow cannot increase further despite the lower backpressure. Operators must ensure that the turbine is sized such that under maximum expected flow, the stage can pass the required flow without destructive overpressure, although choking itself is not detrimental.

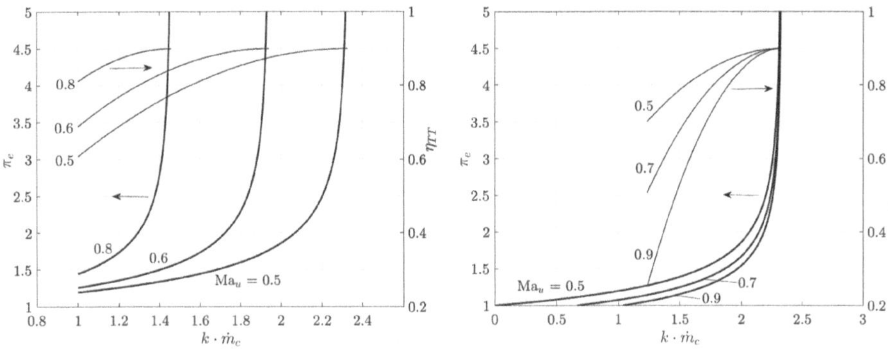

Figure 12.7 Turbine maps. $\alpha_2 = 25°$; nominal $\eta_{\text{pol,T}} = 0.9$.

At the low-flow end, a centripetal turbine does not have a true surge phenomenon in the same severe sense as a compressor. However, there is a practical limit to how far one can reduce mass flow for a given turbine geometry and speed. As the mass flow decreases, the turbine must increase the expansion ratio to produce the required work for whatever load is on the shaft. This leads to ever-increasing incidence angles on the rotor and possibly flow separation. If the flow is reduced too much, the turbine will still pass the imposed flow, but the efficiency drops sharply due to a severe incidence mismatch. In a turbocharger context, for example, the surge line on the turbine map is often defined not by the turbine itself surging, but by the compressor surge: at extremely low engine exhaust flows, the compressor driven by the turbine cannot sustain its flow and surges, which indirectly imposes a limit on how low the turbine flow can go in normal operation. Thus, the left boundary of a turbine's operating map is typically determined by the engine or system rather than by the inherent flow instability in the turbine.

An efficiency contour plot in ψ–ϕ coordinates shows a gently sloping plateau around the optimal $\psi \approx 1$, $\phi \approx 0.25$ and then steep drop-offs beyond $\psi \approx 0.7$ or $\psi \approx 1.2$, and for ϕ below approximately 0.15 or above 0.35.

12.5 REGULATION TECHNIQUES

Aside from conventional throttling using backpressure valves, adjusting the stagger angle of nozzle vanes, i.e. in variable nozzle guide vanes (VNGVs), serves as an effective regulation mechanism. This adjustment alters the flow angle α_2 and the throat area, thus controlling the mass flow rate and energy transfer in the rotor.

In Figure 12.8, the effect of increasing α_2 on turbine performance is shown. By adjusting the vane angle, the incidence angle of the fluid on the rotor blades can be controlled, ensuring optimal aerodynamic efficiency under varying load conditions. At high expansion ratios, the working fluid experiences significant pressure and velocity changes. If α_2 is set to low values under these conditions, the throat area can become excessively restricted, causing the nozzle to approach or reach choking.

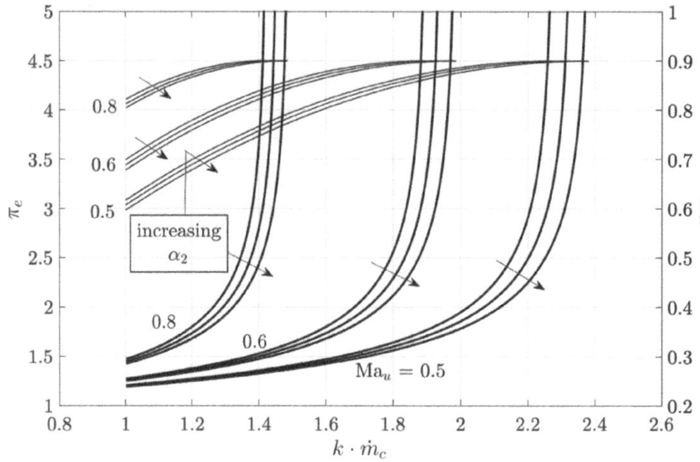

Figure 12.8 Effect of increasing stator exit flow angle α_2 on speed-lines in a turbine.

Operating with low outlet pressure in choked conditions usually induces large total pressure losses in the diffuser downstream of the rotor caused by underexpansion. Choking not only limits the turbine throughput but can also lead to upstream pressure build-up: Localized regions of high Mach numbers within the nozzle can induce shock waves, exacerbating losses, and decreasing overall efficiency. VNGVs are integral to variable geometry turbochargers in internal combustion engines. VNGVs are also a popular choice in radial turbines for power generation.

Another technique consists in bypass regulation by using *wastegate valves*, which involves diverting a portion of the airflow around the turbine. This technique helps to control engine temperatures, manage emissions, and optimize performance during varying operational demands.

12.6 PRACTICAL ASPECTS OF CENTRIPETAL TURBINES

12.6.1 SINGLE- VS. MULTI-STAGE DESIGNS

Most radial turbines are single-stage as this keeps the machine compact and avoids the aerodynamic complexity of turning the flow between stages. Multistage radial inflow turbines have been studied and used in niche cases, for example, in some compressed air energy storage (CAES) expanders and rocket engine turbopumps, but they are comparatively rare. When multiple stages are used, each stage can be smaller and operate at more moderate loading, but the flow must be redirected between stages from axial out of the first rotor and then turn radially into the next, which adds loss and mechanical complexity. Designers generally prefer to extract as much work as possible in one radial stage up to the mechanical limits.

A single-stage radial turbine can typically attain three to six expansion ratios with $\approx$80%–85% efficiency, whereas a two-stage radial turbine could extend beyond 10:1 total expansion if needed, at the cost of greater complexity.

12.6.2 APPLICATIONS

Radial inflow turbines appear in a variety of power and propulsion systems. We will now survey a few key applications, i.e. turbochargers, auxiliary power units, and microturbines, highlighting how performance constraints influence design choices in each.

12.6.2.1 Automotive Turbochargers

Perhaps the most widespread use of radial turbines is in turbochargers for internal combustion engines. In this configuration, a single-stage radial turbine is coupled to a single-stage centrifugal compressor on a common shaft. The turbine expands engine exhaust gases (air mixed with combustion products) to drive the compressor, which in turn boosts the intake air pressure.

Turbocharger turbines operate under challenging conditions: highly pulsating flow (due to the engine exhaust stroke cycle), a very wide range of operating points (from idle to full throttle, low rpm to high rpm), and strict size and cost constraints.

A common approach is to size the turbine for the mid-to-high load point of the engine (e.g. at peak torque rpm) to achieve peak efficiency at that condition, and to accept some efficiency loss at the extremes. Indeed, automotive radial turbines are often tuned for a relatively broad efficiency plateau rather than a sharp peak, since engines frequently operate across a wide range of speeds and loads.

This design philosophy often results in the chosen values of the flow coefficient ϕ and the loading coefficient ψ deviating slightly from their absolute optima (i.e. $\phi \approx 0.25$ and $\psi \approx 1.0$) to widen the operational map. For example, a slightly higher flow coefficient $\phi \approx 0.3$ provides a wider range of permitted mass flow at the cost of a minor reduction in peak efficiency.

Historically, gasoline engine turbochargers employed wastegates for flow control to avoid knocking and prevent turbine overspeed at high loads. Modern diesel engines often utilize variable turbine geometry (VGT or VNT) to improve low-end torque and reduce emissions through better exhaust gas recirculation (EGR) control.

In VGT designs, the stator vanes close at low rpm to accelerate the turbine, and consequently the compressor, more quickly, and open at high rpm to reduce back-pressure. This capability to dynamically adjust the turbine geometry effectively mitigates the trade-off between low-speed boost and high-speed flow capacity.

The limits of the temperature of the material are a critical consideration. Turbocharger turbines experience exhaust gas temperatures ranging from approximately 800°C (in Diesel engines) to over 1000°C (in gasoline engines). Radial turbines operating in this environment are typically manufactured from nickel-based superalloys and produced through investment casting.

Another important constraint is rotor inertia. Smaller in diameter, and therefore lighter, turbines accelerate faster, which reduces turbo lag and improves throttle response. However, reducing the turbine size too far results in excessively high shaft speeds for a given mass flow rate, introducing mechanical limitations such as increased bearing friction and the risk of burst containment failure.

Most automotive radial turbines operate at shaft speeds in the range of 100,000–200,000 rpm.

The pulsating nature of the exhaust flow means that the turbine rarely operates in steady-state conditions. However, average behavior is typically captured using steady-flow performance maps. Some efficiency loss occurs due to cycle averaging, since the turbine cannot continuously adapt to swallow each exhaust pulse at optimal incidence throughout the cycle.

Advanced turbocharger designs consider twin-scroll volutes and pulse-tuned exhaust manifolds to better harness the energy from exhaust pulses and maximize turbine efficiency under transient conditions.

12.6.2.2 Auxiliary Power Units, APUs

Auxiliary power units are small gas turbines on aircraft that provide electrical power and bleed air (for starting engines and air conditioning) when the main engines are off or at idle. Many APUs use a radial compressor and a radial inflow turbine due to their compact size. For example, a common APU might use a centrifugal compressor with a pressure ratio around 4-6 and a single radial turbine expanding the combustor exhaust down to ambient pressure.

An APU usually operates at a constant design speed, optimized for one or a few operating points. It may idle at a lower speed, but for delivering power or bleed air, it runs at or near 100% rated speed. This allows the radial turbine in an APU to be optimized for a specific condition, yielding high efficiency at that point.

If more power is needed than one stage can provide (as in larger APUs), a free power turbine (usually axial) may be added, but the gas generator turbine remains radial.

Variable geometry within the turbine itself is uncommon in APUs because the added complexity and weight are difficult to justify for what is essentially a self-contained unit meant to operate at steady conditions.

The material and stress constraints are similar to those of other turbines. APUs use high-grade alloys and sometimes incorporate blade cooling or thermal barrier coatings if temperatures are high. However, many APUs have turbine inlet temperatures in the range of 600–800°C, which is lower than that of main engines and allows for longer life without the need for sophisticated cooling systems.

The performance of the radial turbine directly affects the APU fuel consumption, since it dictates how much power per unit mass flow can be extracted. A well-designed radial turbine allows the APU to achieve overall thermal efficiencies on the order of 10–15%, which, although seemingly low, is typical for small simple-cycle turbines without heat recovery.

Some modern APUs employ recuperators to improve efficiency. Although this does not fundamentally change the turbine design, it means that the turbine experiences a lower inlet temperature for the same power output (since exhaust heat is reused), which can ease blade stress and extend component life.

12.7 PROBLEMS

1. A radial inflow turbine rotor with $\beta_2' = 90°$ must be designed for a total-to-static pressure ratio of $\frac{p_2^0}{p_3} = 3.5$ and an expected total-to-static efficiency of $\eta_{ts} = 0.78$. The rotor handles 0.8 kg/s of air under inlet conditions $p_2^0 = 3.5$ bar and $T_2^0 = 650°C$. The rotational speed of the rotor is $N = 85{,}000$ rpm and the blade rotor speed can be set at 350 m/s. Conduct the preliminary design of the turbine rotor. Air can be treated as an ideal gas with constant properties.
2. Conduct a preliminary sizing of a radial inflow turbine rotor with $\beta_2' = 90°$, which expands combustion gases ($\gamma = 1.3$, $R = 292$ J/(kg·K)) from a total inlet temperature of $T_2^0 = 400$ K and a total inlet pressure of $p_2^0 = 3.8$ bar to ambient pressure. The goal of the design is to obtain the maximum value of the total-to-static rotor efficiency while accounting for constraints imposed by retrofitting within an existing turbine casing. These constraints include a maximum allowable rotor exit area of $A_3 = 3500$ mm^2, a given rotor diameter of $D_2 = 110$ mm and a prescribed rotor rotational speed of $N = 70{,}000$ rpm.
3. The volute of a centripetal turbine rotor must be designed for the following conditions: the volute inlet has a Mach number $\text{Ma}_1 = 0.36$, total pressure $p_1^0 = 3.5$ bar, total temperature $T_1^0 = 430$ K, and mass flow rate $\dot{m} = 0.9$ kg/s. The radius of the rotor is $r_2 = 45$ mm, and the required inclination of the absolute flow velocity at the rotor inlet is $\alpha_2 = 25°$. Assume that air is the working fluid. The total pressure loss coefficient in the volute is assumed to vary with the azimuthal angle θ as $\pi_v(\theta) = 1 - 0.42 \cdot 10^{-3} \cdot \theta$ (θ in degrees). The Mach number is also assumed to vary linearly from Ma_1 at $\theta = 0°$ to $\text{Ma}_2 = 0.95$ at $\theta = 360°$. Compute the distribution of the area with θ and, in particular, determine the values of the area for $\theta = 0°, 90°, 180°, 270°$ and $360°$.

   $[(22.9; 18.9; 14.2; 8.2; 0) \cdot 10^{-3}\ \text{m}^2]$

NOTE

1. Balje, O. E., *Turbomachines: A Guide to Design, Selection, and Theory*, Wiley, New York (1981).

13 General Principles of Rotor Dynamics in Turbomachinery

13.1 TYPES AND GENERAL CHARACTERISTICS OF ROTORS

Rotor dynamics focuses on the dynamic behavior of rotating shafts and assemblies (collectively called a rotor system or, more simply, a rotor). From our perspective, it is first essential to begin with a distinction between rigid and flexible rotors.

A *rigid rotor* is defined as a rotor that does not experience significant elastic deformation (bending) during operation. This implies that the shaft remains nearly straight under centrifugal and dynamic loads, and its dynamic behavior can be approximated using rigid-body assumptions.

On the other hand, a *flexible rotor* features a shaft undergoing significant elastic deformation (typically bending) within its normal operating speed range.

The distinctive features of the rotor types imply significant differences regarding the dynamic behavior of the rotating shaft with rotational speeds. In particular, we recognize that:

- in rigid rotors, the critical rotational speed, i.e. the speed at which rigid-body modes (typically translational) are excited by imbalance forces, is associated with shaft support flexibility, not shaft bending. This means that, in a single-degree-of-freedom (SDOF), lumped-mass rigid rotor the critical speed is simply:

$$\Omega_c = \sqrt{\frac{k}{m}} \tag{13.1}$$

  where k is the stiffness of the shaft/support system and m is the mass of the rotor. In turbomachinery, usually Ω_c is well above the operational speed range, and resonance is not encountered when rotors are appropriately balanced.[1]
- in flexible rotors, critical speeds are associated with natural flexural shaft frequencies. This means that multiple critical speeds may exist: 1st, 2nd, 3rd..., each corresponding to a distinct bending mode. Mathematically, this implies:

$$\omega \geq \omega_1 \tag{13.2}$$

  where ω is the operational spin speed and ω_1 is the first bending natural frequency (first critical speed). Operation near or above the first critical speed occurs regularly in many turbomachinery applications and causes several effects, such as the

DOI: 10.1201/9781003619956-13

Table 13.1
Comparison of Rigid and Flexible Rotors

Feature	Rigid Rotor	Flexible Rotor
Deflection	Negligible	Significant bending
Critical speeds	Not exceeded	Often exceeded
Analysis method	SDOF, lumped mass	MDOF, modal/FEM
Balancing	1–2 planes, low speed	Multi-plane, high speed
Whirl modes	Not split	FW/BW split (gyros)
Applications	Low-speed machinery	High-speed turbines, compressors

birth of gyroscopic and resonance phenomena. As a consequence, mitigating dynamic phenomena in flexible rotors is much more complex than in rigid rotors and requires multiple strategies.

Table 13.1 summarizes the main characteristics of both rigid and flexible rotors.

Flexible rotors often have high length-to-diameter ratios or high rotational speeds causing slender shaft deflection. In such rotors, imbalance forces can excite bending modes, and simple two-plane balancing at low speed may not be sufficient to control high-speed vibration. Most large turbomachinery rotors (e.g. gas or steam turbine generators, jet engine rotors) are flexible because they spin above their first (and sometimes second) critical speed. These machines are designed to accelerate through critical speed regions quickly and operate in the supercritical regime with acceptable vibration levels.

Excitations of flexible rotors in turbomachinery may originate from a variety of sources, both mechanical and aerodynamic, each capable of inducing dynamic responses depending on the frequency content and interaction with the rotor's natural modes. These excitations can be broadly classified into *mechanical*, *aerodynamic*, *fluid-induced* and *aeroelastic* categories and often act in concert to influence the rotor's behavior.

From a *mechanical* standpoint, the most fundamental source is mass imbalance, where the mass distribution of the rotor is not perfectly symmetric around its axis of rotation. This produces a centrifugal force that acts at the rotor spin frequency, commonly referred to as a 1× excitation. When this excitation aligns with a natural bending frequency of the rotor, it can result in resonance, especially if the rotor operates near its critical speeds. Additionally, misalignment of couplings or shaft segments, shaft bow, and mechanical looseness can introduce periodic forces that contribute to vibratory excitation, often at multiples of the rotational speed (harmonics).

Beyond purely mechanical causes, *aerodynamic* excitations play a significant role in turbomachinery. As the rotor blades interact with stationary components such as vanes or casing features, they experience periodic aerodynamic forces due to pressure fluctuations and wake interactions. These are known as blade passing excitations, and their frequency is determined by the number of vanes and the rotor speed (e.g. a stator with 20 vanes induces a 20· excitation). If the frequency of these aerodynamic forces coincides with a rotor mode, especially a higher-order bending mode, a resonant response can occur.

Another significant category involves *fluid-induced forces*, which emerge from the interaction of the rotor with lubricating films in journal bearings, seals, and dampers. These include effects such as cross-coupled stiffness in seals and fluid-induced instability mechanisms such as oil whirl and oil whip. These phenomena are particularly important in high-speed flexible rotors supported on fluid film bearings, where small lateral displacements of the rotor can induce pressure asymmetries in the fluid film, generating destabilizing forces that act orthogonally to the motion and can lead to self-excited vibrations at subsynchronous frequencies.

Finally, *aeroelastic* forces arising from the coupling between blade vibration and airflow can excite the rotor system indirectly, especially if the blade vibrations transmit forces through the disk into the shaft. In some cases, flutter or other blade instabilities can impose fluctuating torques or lateral loads on the rotor shaft.

13.1.1 GOVERNING EQUATIONS OF MOTION

The vibration of a rotor can be described by differential equations of motion derived from Newton's second law or Lagrange's equations. For a rotor with n degrees of freedom (DOF), the linearized equations of motion about the rotor's equilibrium (steady rotation) can be written in matrix form as[2]:

$$[\mathbf{M}]\,\ddot{\mathbf{q}}(t) + [\mathbf{C}]\,\dot{\mathbf{q}}(t) + [\mathbf{K}]\,\mathbf{q}(t) = \mathbf{Q}(t) \tag{13.3}$$

where $[\mathbf{M}]$ is the mass (inertia) matrix, $[\mathbf{K}]$ is the stiffness matrix (including shaft bending stiffness and any support stiffness) and $[\mathbf{C}]$ is the damping matrix. $\mathbf{q}(t)$ is the vector of generalized coordinates describing the small displacements of the rotor center (e.g. lateral deflections in x and y directions, angular rotations θ, see Figure 13.2) and $\mathbf{Q}(t)$ is the generalized force vector (external excitations). For rotating systems, $[\mathbf{C}]$ generally includes both physical damping and rotational gyroscopic terms.

It is often convenient to separate the gyroscopic effects explicitly. The gyroscopic forces act like velocity-proportional terms that couple orthogonal motions, leading to a gyroscopic matrix $[\mathbf{G}]$ such that the effective damping matrix is $[\mathbf{C}] + \omega[\mathbf{G}]$ (where ω is the constant spin speed). Equation 13.3 can then be expanded to:

$$[\mathbf{M}]\,\ddot{\mathbf{q}} + ([\mathbf{C}] + \omega[\mathbf{G}])\,\dot{\mathbf{q}} + [\mathbf{K}]\,\mathbf{q} = \mathbf{Q}(t) \tag{13.4}$$

For a simple isotropic shaft and disk, $[\mathbf{K}]$ and $[\mathbf{M}]$ are symmetric and $[\mathbf{G}]$ causes coupling between x and y lateral motions. If the rotor spins about a principal axis of inertia, $[\mathbf{G}]$ is proportional to the difference between the polar and transverse moments of inertia of the rotor. As ω increases, the eigenvalues of the homogeneous part of equation 13.3 ($\mathbf{Q} = 0$) become functions of ω, yielding the speed-dependent natural frequencies $\omega_n = f(\omega)$.

13.1.1.1 1- and 2-DOF Jeffcott Rotor

13.1.1.1.1 Conventional Rotor Assembly

As an example case, consider a Jeffcott rotor (top of Figure 13.2)[3] (mass m, shaft stiffness k) about its equilibrium yields (in one lateral direction):

$$m\ddot{y}(t) + c\dot{y}(t) + ky(t) = F(t) \tag{13.5}$$

where $y(t)$ is the lateral (vertical, in this case) displacement of the disk and $F(t)$ is an external force (such as an imbalance force). In the absence of external forces, the free vibration solution is $y(t) = Y\sin(\omega_n t + \phi)$, with $\omega_n = \sqrt{k/m}$ the natural frequency (at $\omega = 0$) (Figure 13.1).

When the disk is not centered, the dynamics are slightly altered due to asymmetry

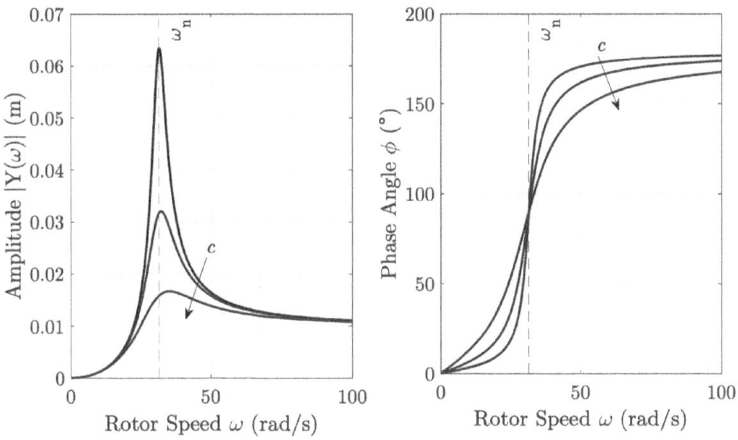

Figure 13.1 Typical solution for the dynamics of 1 DOF Jeffcott rotor with different damping ratios. Left: vibration amplitude vs. rotor speed. Right: phase angle vs. rotor speed.

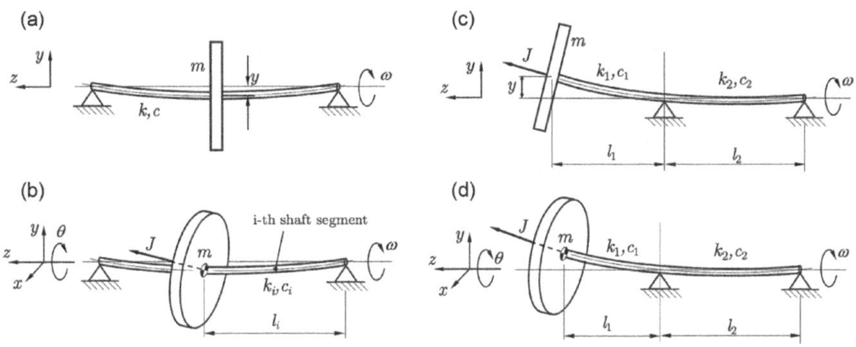

Figure 13.2 (a) A 1-DOF Jeffcott rotor. (b) A 2-DOF Jeffcott rotor with gyroscopic effects. (c) A 1-DOF overhung Jeffcott rotor. (d) A 2-DOF overhung Jeffcott rotor with gyroscopic effects.

in the shaft's flexibility and support conditions. In this case, the equation of motion is again 13.5, where the equivalent transverse stiffness at the disk location is:

$$k = \left(\frac{1}{k_1} + \frac{1}{k_2}\right)^{-1} \tag{13.6}$$

since the shaft behaves like two springs in series with stiffness k_1 and k_2, respectively, and the disk is located at distance l_1 from one end and l_2 from the other. If each span behaves like a simply supported beam:

$$k_i = \frac{48EI}{l_i^3}, \quad i = 1,2 \tag{13.7}$$

where E is the Young modulus and I the second moment of the area of the shaft. For a shaft of circular cross section with diameter d, the second moment of area is:

$$I = \frac{\pi d^4}{64} \tag{13.8}$$

If wobbling of the disk rotor is taken into account, the angular momentum and the angular velocity vectors are no longer parallel and thus induce gyroscopic moments. This often happens when the rotor spins in two perpendicular directions x, y (Figure 13.2b and d), such that equations 13.5 are coupled by gyroscopic terms, producing equations of the form of equation 13.4. Solving the eigenvalue problem at a given rotor spin speed ω, of the form:

$$\det\left([\mathbf{K}] + \lambda^2[\mathbf{M}] + \lambda\omega[\mathbf{G}]\right) = 0 \tag{13.9}$$

yields complex conjugate pairs of eigenvalues λ. These give rise to two families of whirl frequencies: *Forward Whirl (FW)* and *Backward Whirl (BW)*, depending on the direction of precession relative to the shaft rotation. These results reduce to the classical eigenfrequencies when $\omega = 0$.[4]

The general solution of equation 13.3 for forced vibration can be obtained via modal analysis or direct integration. Typically, solutions of the form $\mathbf{q}(t) = \Re\{\mathbf{X}e^{j\omega t}\}$ are assumed for harmonic forcing (with $\Re$ denoting the real part), which leads to solving for complex amplitudes $\mathbf{X}$. The form of the equations and the elements in matrices $[\mathbf{M}], [\mathbf{C}], [\mathbf{K}], [\mathbf{G}]$ can be derived from energy methods and finite element modeling of the rotor-bearing system. For instance, in finite element models (FEM) of rotors, $[\mathbf{M}]$ includes both translational and rotational inertia of shaft and disks, $[\mathbf{K}]$ comes from shaft bending stiffness (and possibly bearing stiffness) and $[\mathbf{G}]$ arises from disk angular momentum.

13.1.1.1.2 Overhung Rotor Assembly

In an overhung Jeffcott rotor as shown in Figure 13.2c, the disk is mounted beyond the two supports. Given that the allowed displacements are in the $x-y$ plane, the system has 2 DOFs. Assume that only the overhung segment contributes to restoring and damping forces at the disk location, while the central shaft section (between the two supports) is dynamically passive in the lumped-parameter model. In this case, the transverse stiffness of the overhung portion of the shaft is:

$$k_1 = \frac{3EI}{l_1^3} = \frac{3\pi E d^4}{64 l_1^3} \tag{13.10}$$

The gyroscopic moments arise due to precession of the spinning disk:

$$\begin{aligned} M_x &= -J\omega\dot{y} \\ M_y &= J\omega\dot{x} \end{aligned} \tag{13.11}$$

The dynamic equations for the disk in the x and y directions, including gyroscopic effects, stiffness, damping, and external forces are:

$$\begin{aligned} m\ddot{x} + c_1\dot{x} + k_1 x - J\omega\dot{y} &= F_x(t) \\ m\ddot{y} + c_1\dot{y} + k_1 y + J\omega\dot{x} &= F_y(t) \end{aligned} \tag{13.12}$$

These are coupled second-order differential equations including gyroscopic terms $\pm J\omega\dot{y}$ and $\pm J\omega\dot{x}$. A sample solution of these equations is given in Figure 13.3. In this lumped-parameter model, the shaft segment between the bearings (with stiffness k_2 and damping c_2 does not appear in the equations of motion, because the disk is not mounted on this segment and does not provide restoring or damping forces at the disk location. Instead, it would contribute to the overall shaft dynamics if a full finite element or distributed-parameter model was adopted.

13.1.1.2 6-DOF Double Jeffcott Rotor

13.1.1.2.1 Equations of Motion

An example of a turbomachinery model using a double Jeffcott rotor is provided in Figure 13.4. The rotors are mounted on a massless, flexible shaft. Each rotor can move laterally in the x and y directions and rotate torsionally (in θ) about the shaft

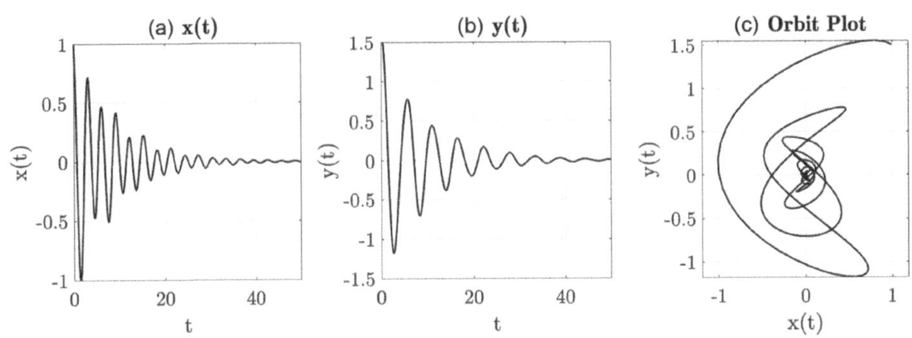

Figure 13.3 2-DOF Jeffcott rotor: (a) and (b), time-domain lateral displacements and (c) orbit of the disk center.

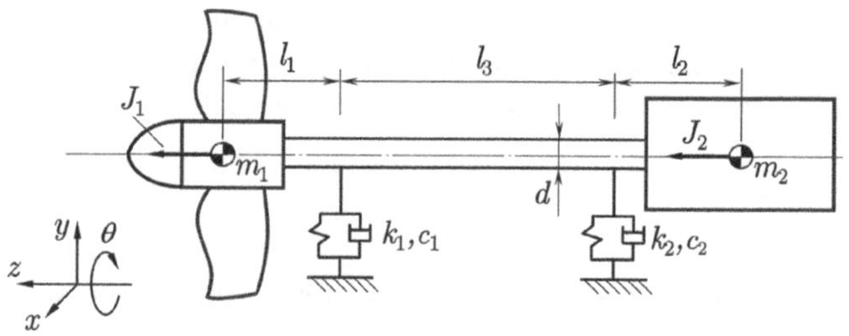

Figure 13.4 Two-disk Jeffcott rotor system mounted on a flexible shaft.

axis. The rotors are connected via the shaft, which provides torsional and lateral stiffness. Each rotor is supported by bearings with lateral stiffness and damping.

The system has six degrees of freedom, defined by the generalized coordinate vector:

$$\mathbf{q}(t) = \begin{bmatrix} x_1(t) \\ y_1(t) \\ \theta_1(t) \\ x_2(t) \\ y_2(t) \\ \theta_2(t) \end{bmatrix} \in \mathbb{R}^6 \tag{13.13}$$

where $x_i(t), y_i(t)$ are the lateral displacements of the rotor i, $\theta_i(t)$ is the torsional rotation of rotor i, $i = 1$ for the left rotor and $i = 2$ for the right rotor.

The mass matrix is diagonal and includes translational and rotational inertias:

$$[\mathbf{M}] = \mathrm{diag}(m_1, m_1, J_1, m_2, m_2, J_2) \tag{13.14}$$

where m_i and J_i are the mass and polar moment of inertia of rotor i.

We assume isotropic (diagonal) viscous damping in the lateral and torsional directions[5]:

$$[\mathbf{C}] = \mathrm{diag}(c_x, c_y, c_{\theta 1}, c_x, c_y, c_{\theta 2}) \tag{13.15}$$

The gyroscopic matrix is skew-symmetric and couples the lateral x and y motions:

$$[\mathbf{G}] = \begin{bmatrix} 0 & G_1 & 0 & 0 & 0 & 0 \\ -G_1 & 0 & 0 & 0 & 0 & 0 \\ 0 & 0 & 0 & 0 & 0 & 0 \\ 0 & 0 & 0 & 0 & G_2 & 0 \\ 0 & 0 & 0 & -G_2 & 0 & 0 \\ 0 & 0 & 0 & 0 & 0 & 0 \end{bmatrix} \tag{13.16}$$

where G_1 and G_2 are the gyroscopic coefficients of the two rotors, $G_1 = J_1$, $G_2 = J_2$.

The stiffness matrix includes the bearing lateral stiffness: $k_{x1}, k_{y1}, k_{x2}, k_{y2}$, a torsional stiffness from the rotor bases $k_{\theta 1}, k_{\theta 2}$, a shaft torsional coupling k_t and a lateral shaft coupling: k_s (same for x and y)[6]:

$$[\mathbf{K}] = \begin{bmatrix} k_{x1}+k_s & 0 & 0 & -k_s & 0 & 0 \\ 0 & k_{y1}+k_s & 0 & 0 & -k_s & 0 \\ 0 & 0 & k_{\theta 1}+k_t & 0 & 0 & -k_t \\ -k_s & 0 & 0 & k_{x2}+k_s & 0 & 0 \\ 0 & -k_s & 0 & 0 & k_{y2}+k_s & 0 \\ 0 & 0 & -k_t & 0 & 0 & k_{\theta 2}+k_t \end{bmatrix} \tag{13.17}$$

Ultimately, the excitation vector $\mathbf{Q}(t) \in \mathbb{R}^6$ may include external forces or torques. For instance, if a sinusoidal force acts on rotor 1 in the x-direction with angular speed Ω:

$$\mathbf{Q}(t) = \begin{bmatrix} F_0 \cos(\Omega t) \\ 0 \\ 0 \\ 0 \\ 0 \\ 0 \end{bmatrix} \tag{13.18}$$

As an example, the equation for the $x_1(t)$ coordinate becomes:

$$m_1 \ddot{x}_1 + c_x \dot{x}_1 + \omega G_1 \dot{y}_1 + (k_{x1}+k_s)x_1 - k_s x_2 = F_0 \cos(\Omega t) \tag{13.19}$$

which shows the coupling between x_1 and x_2 via shaft stiffness k_s and between x_1 and y_1 via the gyroscopic term.

To understand the dynamic behavior of the system, a modal analysis is performed by solving the undamped, unforced homogeneous equation:

$$[\mathbf{M}]\ddot{\mathbf{q}} + \omega[\mathbf{G}]\dot{\mathbf{q}} + [\mathbf{K}]\mathbf{q} = \mathbf{0} \tag{13.20}$$

Assuming a harmonic solution of the form:

$$\mathbf{q}(t) = \mathbf{\Phi} e^{\lambda t} \tag{13.21}$$

and substituting into equation 13.20 yields the quadratic eigenvalue equation:

$$\left(\lambda^2[\mathbf{M}] + \lambda\omega[\mathbf{G}] + [\mathbf{K}]\right)\mathbf{\Phi} = \mathbf{0} \tag{13.22}$$

which provides the complex eigenvalues λ that represent the system's damped natural frequencies and the corresponding eigenvectors $\mathbf{\Phi}$, which represent the mode shapes. In the example of a 6-DOF Jeffcott rotor, the problem manifests 2×6 = 12 values for λ.

In the presence of gyroscopic effects, the eigenvalues appear in complex-conjugate pairs and may split into distinct forward whirl and backward whirl frequencies. The real part indicates damping or growth, and the imaginary part gives the modal frequency.

13.1.1.2.2 Critical Speeds

Operating speeds where excitation frequency coincides with a natural frequency give critical speeds. These are identified from the imaginary parts of λ, converted to Hz:

$$f_n = \frac{|\mathrm{Im}(\lambda)|}{2\pi} \tag{13.23}$$

Critical speeds can be inferred from Figure 13.5 for the case in Figure 13.4. In particular, the signal amplitude is given for each of the six degrees of freedom. A harmonic excitation vector:

$$\mathbf{Q}_0 = \begin{bmatrix} 1 & 1 & 1 & 0 & 0 & 0 \end{bmatrix}^\top \tag{13.24}$$

is applied, meaning the x and y lateral displacements and the torsional rotation of the first rotor are simultaneously excited. The plots compare the amplitude response under increasing damping levels. The plots of $|x_1|$ and $|y_1|$ both exhibit a sharp resonance near 600 rpm referring to the left rotor. As expected, the peak amplitude at resonance decreases with increasing damping, and the resonance becomes wider. The response of θ_1, shown in the top right plot, differs in nature. Unlike lateral displacements, the torsional amplitude does not exhibit a pronounced resonant peak, reflecting the increasing inertial resistance to torsional acceleration as speed increases.

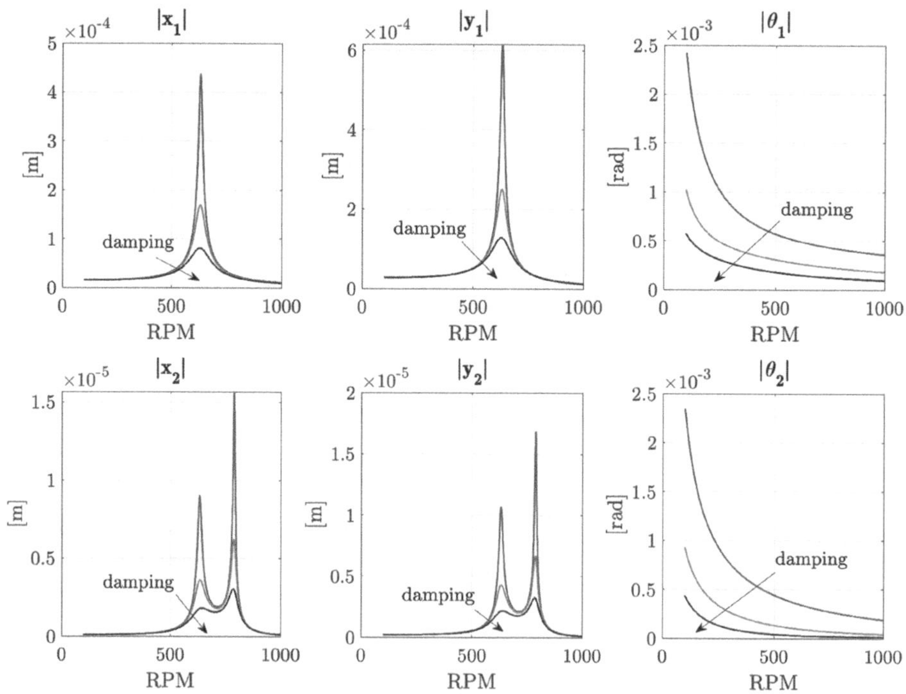

Figure 13.5 Frequency response of a two-disk Jeffcott rotor system.

The inclusion of damping further reduces the amplitude of the torsional vibration, particularly at lower speeds.

The bottom row shows the response of the second rotor, which is not directly excited but is dynamically coupled to the first rotor via the flexible shaft. The responses $|x_2|$ and $|y_2|$ exhibit two distinct resonant peaks. The first peak, located near 500 rpm, is associated with the primary flexural mode shared with the first rotor. The second peak, occurring at a higher rpm ($\approx$ 800 rpm), arises from a torsional-lateral coupled mode, which is enabled by the shaft flexibility and rotor-rotor interaction. The presence of these two peaks is indicative of mode splitting in a coupled multirotor system. The magnitude of the response in x_2 and y_2 is significantly smaller than in x_1 and y_1 due to the indirect nature of the excitation. The response in θ_2 shows a behavior similar to θ_1: a smooth and gradual decay in amplitude with increasing speed, further diminished by damping. Despite the lack of direct excitation, the torsional motion is transmitted from Rotor 1 to Rotor 2 through the shaft.

The phase behavior can be seen in Figure 13.6. The plots of x_1 and y_1 (top left and top center) show the classical behavior of a single degree-of-freedom oscillator: a rapid phase drop of approximately 180° across the resonance region, centered around 600 rpm. This confirms the location of the first critical speed seen in the amplitude plots. The effect of damping is evident in the smoothness and breadth of the transition. For higher damping, the phase transition is more gradual, spreading

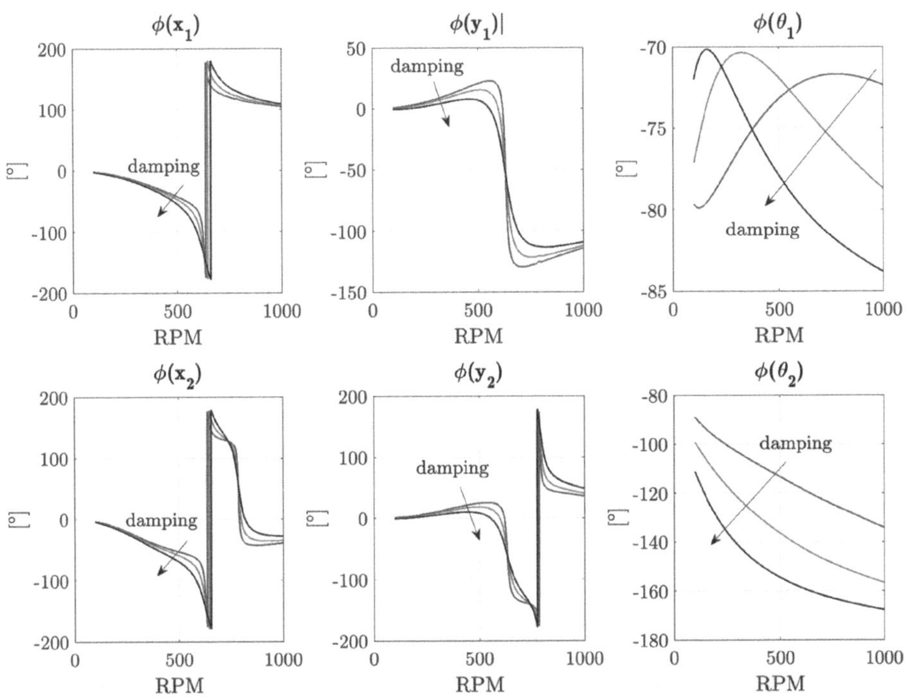

Figure 13.6 Phase angle response of a two-disk Jeffcott rotor system.

over a wider range of speeds. At low speeds, all curves start near 0°, consistent with in-phase response to excitation. As the speed increases beyond resonance, the phase asymptotically approaches −180°, indicating an out-of-phase condition. The phase of θ_1 (top right) exhibits a slow and nonlinear decrease in phase with increasing rpm, which reflects the lack of strong resonance in the torsional DOF and the increasing phase lag due to inertial resistance and damping. The x_2 and y_2 phase plots reveal a more complex behavior, including two distinct phase transitions, one near 600 rpm and the second close to 800 rpm. These correspond to the dual resonance peaks observed in the amplitude plots and reflect the presence of mode splitting due to rotor-to-rotor coupling. Finally, the phase of θ_2 (bottom right) decreases smoothly with rpm, similar to θ_1. Since there is no direct excitation and only torsional coupling via the shaft, the response remains subdued. The influence of damping is again evident in the steepness of the phase lag: higher damping causes the phase to drop more rapidly than in the low damping case.

13.1.1.2.3 Dynamic Response

To simulate the time response, we rewrite the second-order equation into a first-order system:

$$\frac{d}{dt}\begin{bmatrix}\mathbf{q}(t)\\ \dot{\mathbf{q}}(t)\end{bmatrix} = \underbrace{\begin{bmatrix}[\mathbf{0}] & [\mathbf{I}]\\ -[\mathbf{M}]^{-1}[\mathbf{K}] & -[\mathbf{M}]^{-1}([\mathbf{C}]+\omega[\mathbf{G}])\end{bmatrix}}_{[\mathbf{A}]}\begin{bmatrix}\mathbf{q}(t)\\ \dot{\mathbf{q}}(t)\end{bmatrix} + \underbrace{\begin{bmatrix}\mathbf{0}\\ [\mathbf{M}]^{-1}\mathbf{Q}(t)\end{bmatrix}}_{\mathbf{B}(t)} \tag{13.25}$$

This system can be numerically integrated using time-stepping algorithms (e.g. Runge-Kutta methods such as `ode45` in MATLAB). To explore the free response, initial displacement and/or velocity can be specified:

$$\mathbf{q}(0) = \mathbf{q}_0, \quad \dot{\mathbf{q}}(0) = \dot{\mathbf{q}}_0 \tag{13.26}$$

Post-processing of time response includes plotting the time histories $x_i(t), y_i(t)$, $\theta_i(t)$ (Figure 13.7) and orbits (Figure 13.8) in the lateral plane:

$$\text{Orbit: } (x_i(t), y_i(t)) \tag{13.27}$$

The time-domain simulation considers the system in free response (no external forcing) but with initial displacement and velocity applied to Rotor 1. Specifically, the initial condition is given by:

$$\begin{aligned}\mathbf{q}(0) &= \begin{bmatrix}q_1(0) & q_2(0) & q_3(0) & 0 & 0 & 0\end{bmatrix}^\top\\ \dot{\mathbf{q}}(0) &= \begin{bmatrix}\dot{q}_1(0) & \dot{q}_2(0) & \dot{q}_3(0) & 0 & 0 & 0\end{bmatrix}^\top\end{aligned} \tag{13.28}$$

This simulates an initial disturbance in the x, y, and θ of Rotor 1. Each subplot in Figure 13.7 shows the time history of each of the six generalized coordinates. The top row displays the response of Rotor 1. As expected, lateral motions $x_1(t)$ and $y_1(t)$

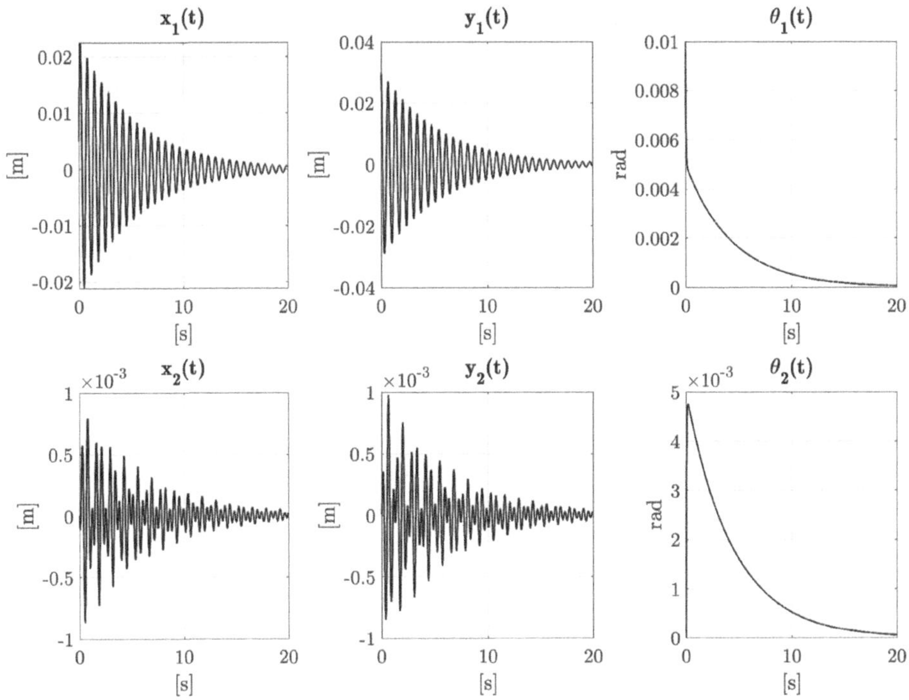

Figure 13.7 Time-domain response of a two-disk Jeffcott rotor system subjected to initial excitation.

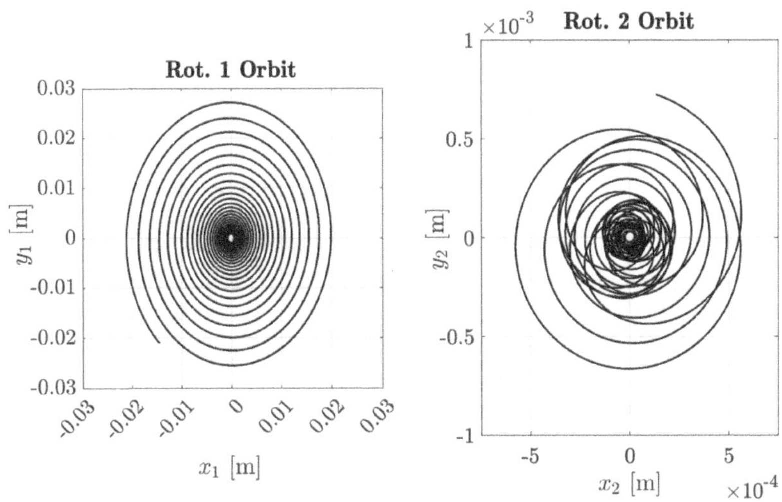

Figure 13.8 Orbit plots of two rotors in a 2-DOF Jeffcott rotor system.

exhibit decaying oscillations from bearing damping and structural damping. The torsional response $\theta_1(t)$ decays smoothly and monotonically due to the dominance of a low-frequency torsional mode and the absence of a strong torsional resonance near the spin speed of the system. The bottom row shows the response of Rotor 2, which starts at rest but becomes excited due to dynamic coupling with Rotor 1 through the flexible shaft. Both $x_2(t)$ and $y_2(t)$ exhibit oscillatory behavior with smaller amplitudes compared to Rotor 1, as excitation is indirect. These responses show mode mixing and interference, evident from the modulated amplitude (beating pattern) as a result of the interaction of multiple closely spaced natural frequencies. The torsional response $\theta_2(t)$ is also induced via shaft coupling and displays an exponential decay similar to $\theta_1(t)$.

The orbit of rotor 1 displays an elliptic spiral that gradually decays toward the origin. This rotor is directly excited in both lateral directions, and the resulting trajectory reveals a dominant forward-whirl pattern. The nearly circular loops suggest balanced mass and stiffness distributions in the lateral directions, while the tight, concentric decay highlights the effectiveness of the system damping in dissipating vibratory energy. The orbit of Rotor 2, shown in the right panel, presents a more intricate picture. Although not directly excited, Rotor 2 responds via dynamic coupling with Rotor 1 through the flexible shaft. Its orbit is characterized by a smaller amplitude and a less uniform spiral, with a noticeably modulated trajectory. This complexity suggests the presence of multiple closely spaced frequency components leading to visible beating patterns.

13.1.1.2.4 Forced Response

The forced response over a range of excitation frequencies can be calculated from:

$$\mathbf{q}(\Omega) = \left(-\Omega^2[\mathbf{M}] + j\Omega([\mathbf{C}] + \omega[\mathbf{G}]) + [\mathbf{K}]\right)^{-1} \mathbf{Q}_0 \tag{13.29}$$

As an example, in Figure 13.9 the forced response of the system in Figure 13.4 is shown. Here we assume the system to be subjected to a harmonic torsional excitation applied at the first rotor. In the upper row of the figure, two distinct resonant peaks are observed in each plot, indicating the presence of two dominant torsional modes due to coupling between the rotors via the shaft. The vertical dashed lines mark the frequency of excitation, which produces the torsional response shown in the lower plots. The time histories reveal the typical signature of resonance, i.e. a steadily increasing amplitude due to the assumed near-zero damping in the system. This unbounded growth is a theoretical artifact of the minimal damping used, intended to emphasize the resonance effect. The waveforms exhibit nearly sinusoidal shapes with slowly increasing envelopes, reflecting the continuous energy transfer from the harmonic torque into the system's torsional mode.

The theoretical framework provided in these sections makes it possible to build highly detailed models and predict the rotor dynamics of very complex systems (Figure 13.10) including multiple Jeffcott rotors and an arbitrarily large number of supports. To address such complexity, the classical double Jeffcott rotor model can

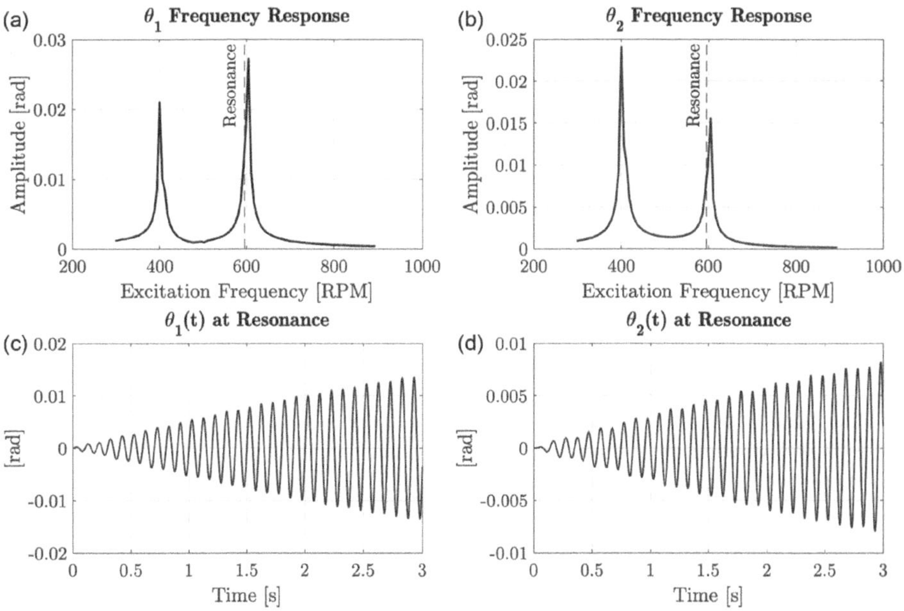

Figure 13.9 Frequency response and time-domain behavior of the angular displacements $\theta_1(t)$ and $\theta_2(t)$ of a two-disk Jeffcott rotor system near resonance. (a) θ_1 Frequency response, (b) θ_2 frequency response, (c) $\theta_1(t)$ at resonance, (d) $\theta_2(t)$ at resonance.

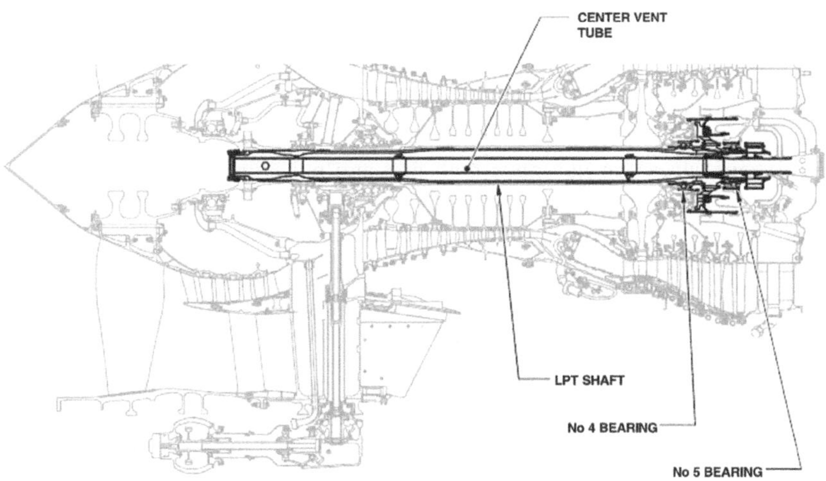

Figure 13.10 Rotor configuration of a modern aircraft engine.

be extended into a multi-degree-of-freedom (MDOF) system by incorporating additional disks and supports. Each disk (such as those representing the LPT and HPT stages) can be modeled as a concentrated mass with its own mass and polar/diametral moments of inertia. The shaft segments between the disks and bearings can be represented as beams.

Support structures, including bearings and seals, can be represented using nonlinear stiffness and damping elements, which may depend on rotor speed, temperature, or misalignment. For example, bearings No. 4 and No. 5 in the figure can be modeled using equivalent stiffness and damping matrices or more advanced hydrodynamic bearing models if needed.

13.2 EXCITATIONS FROM BLADES AND DISK

13.2.1 AERODYNAMIC AND MECHANICAL EXCITATIONS

Blades and disks are subjected to periodic excitations from sources such as passing fluid wakes or inherent manufacturing deviations.

Aerodynamic excitations typically involve integer-order sources, i.e. those sources corresponding to regularly spaced stationary streamlined components immersed in the flow path (e.g. nozzles, struts, stators, IGVs, etc.), which generate a Wake Passing Frequency (WPF). In fact, the periodic interaction of rotating blades with upstream or downstream components generates fluctuating aerodynamic forces that are a function of the discrete number of parts of the component. The frequency of excitation, f_{WPF}, is given by:

$$f_{\mathrm{WPF}} = Z \cdot f_r \tag{13.30}$$

where Z is the number of rotor blades and f_r is the rotor rotational frequency (in Hz, $f_r = \omega/(2\pi)$).

Another typical source of aerodynamic excitation is due to Rotor-Stator Interaction (RSI), whereby the relative motion between consecutive sets of rotating blades and stationary vanes can introduce both pressure pulsations (potential interaction) and wake disturbances (wake interaction). Pressure pulsations arise from the periodic aerodynamic interaction between the rotating and stationary components, particularly from the potential field generated by the blades or vanes. The frequency of these pulsations can be expressed as:

$$f_{\mathrm{potential}} = Z' \cdot f_r \tag{13.31}$$

where Z' is the number of stator vanes. The amplitude of the pressure pulsations is influenced by the blade passing frequency (BPF) and the aerodynamic loading. Assuming a sinusoidal pressure field variation, the pressure fluctuation $p(t)$ at a point can be represented as:

$$p(t) = p_0 + A_0 \cos(2\pi f_{\mathrm{potential}} t + \phi) \tag{13.32}$$

where p_0 is the time-averaged, or mean, pressure, A_0 the amplitude of the pressure fluctuation and ϕ the phase angle. Wake interaction occurs due to the non-uniform flow field behind the rotor blades interacting with downstream stator vanes. The excitation frequency for wake disturbances is determined using approaches based on WPF, while the spatial variation of the wake velocity deficit $\Delta u(x,t)$ can be modeled using Fourier series:

$$\Delta u(x,t) = \sum_{n=1}^{\infty} a_n \cos\left(2\pi n f_{\text{wake}} t - kx + \phi_n\right) \tag{13.33}$$

a_n being the amplitude of the n-th harmonic of the wake velocity, k the wave number related to the spatial periodicity of the wake $k = 2\pi/\lambda$, where λ is the wavelength of the wake and ϕ_n is the phase angle of the n-th harmonic. The interaction of these wakes with the downstream components leads to unsteady aerodynamic forces, which can be quantified using:

$$F_{\text{wake}} = \rho \cdot A \cdot \int_0^{\lambda} \Delta u(x,t) \cdot w(x,t)\, dx \tag{13.34}$$

where A is the blade surface area. Factor $w(x,t)$ is a local aerodynamic weighting function of the downstream component that converts upstream velocity disturbance $\Delta u(x,t)$ into a force contribution at each point x. The combined excitation frequency due to both wake and potential interactions in RSI is often determined by the lowest common multiple (LCM) of the rotor and stator passing frequencies:

$$f_{\text{RSI}} = \text{LCM}(Z \cdot f_r, Z' \cdot f_r) \tag{13.35}$$

The pressure and wake disturbance amplitudes depend heavily on specific geometries and flow conditions, so a direct computation of these amplitudes often requires numerical or experimental analysis.

When aerodynamic excitations are coupled with the structural dynamics of a blade or a set of blades, these can be driven into oscillation at specific frequencies owing to a forced response or to a more complex, self-excited vibration pattern called flutter. Flutter arises when aerodynamic work on the blade reinforces its motion, which can lead to divergence in vibration amplitude and ultimately to rotor failure.

Mechanical excitations can originate from manufacturing tolerances, which can cause slight deviations in blade mass, thus leading to centrifugal forces during rotation. Further causes of mechanical excitations include inertial forces manifesting during start-up, shutdown, and other transient operating conditions.

13.2.2 SYSTEM INSTABILITY: THE CASE OF AERODYNAMIC FLUTTER

Unlike resonance, which requires an external periodic force, flutter is a self-sustaining vibration phenomenon and can occur with little or no direct external time-dependent aerodynamic excitation. In flutter, as the blades vibrate, aerodynamic forces can either dampen or amplify its motion, depending on the phase relationship between the structural deformation of the dynamic system and the aerodynamic loading. If aerodynamic forces feed energy into the system, the vibration amplitude grows uncontrollably. Therefore, to assess the onset of flutter, the work per vibratory cycle done by unsteady aerodynamic forces on the blade+disk assembly can be used. Mathematically, the *work per cycle* is defined as the integral of the instantaneous power, i.e. the product of the instantaneous force $F(t)$ and velocity $\dot{x}(t)$ over one complete vibratory cycle of period $T = \frac{2\pi}{\varpi}$:

$$W_{\text{cycle}} = \int_{t_0}^{t_0+T} F(t) \cdot \dot{x}(t)\, dt \tag{13.36}$$

Assuming a harmonic motion, the displacement and force can be expressed using complex exponentials as:

$$x(t) = \Re\{x_0 e^{i\varpi t}\}, \quad F(t) = \Re\{F_0 e^{i(\varpi t+\phi)}\} \tag{13.37}$$

where x_0 and F_0 are complex amplitudes, while ϕ is the phase difference between force and displacement.

The velocity is given by:

$$\dot{x}(t) = \Re\left\{i\varpi x_0 e^{i\varpi t}\right\} \tag{13.38}$$

The average power transferred over one cycle is:

$$\langle P \rangle = \frac{1}{T}\int_0^T F(t) \cdot \dot{x}(t)\, dt = \frac{1}{2}\varpi\, |F_0|\, |x_0|\, \sin\phi \tag{13.39}$$

Thus, the total work done per cycle becomes:

$$W_{\text{cycle}} = T \cdot \langle P \rangle = \frac{2\pi}{\varpi} \cdot \frac{1}{2}\varpi\, |F_0|\, |x_0|\, \sin\phi = \pi\, |F_0|\, |x_0|\, \sin\phi \tag{13.40}$$

If the force and displacement fields vary spatially over a structure (e.g. along a blade), spatial integration must be carried out. Analytical solutions may be feasible for simple geometries and loading profiles; otherwise, numerical integration methods are typically employed.

The basic criterion for flutter stability is based on the balance between the energy input from the airflow (negative aerodynamic damping) and the energy dissipated by the structural system (structural damping). Since structural damping is often negligible in high-speed turbomachinery applications, we will focus on the aerodynamic damping coefficient δ_{AERO}, defined as:

$$\delta_{\text{AERO}} = \frac{\text{Unsteady aerodynamic work per cycle}}{\text{Average vibratory kinetic energy of the system}} \tag{13.41}$$

The average vibratory kinetic energy of a single degree-of-freedom system is given by:

$$E_{\text{kinetic}} = \frac{1}{T}\int_0^T \frac{1}{2} m\, |\dot{x}(t)|^2\, dt = \frac{1}{4} m\varpi^2 x_0^2 \tag{13.42}$$

Therefore, the aerodynamic damping coefficient becomes the following:

$$\delta_{\text{AERO}} = \frac{W_{\text{cycle}}}{E_{\text{kinetic}}} = \frac{\pi|F_0||x_0|\sin\phi}{\frac{1}{4}m\varpi^2|x_0|^2} = \frac{4\pi|F_0|\sin\phi}{m\varpi^2|x_0|} \tag{13.43}$$

This dimensionless damping parameter characterizes the aerodynamic contribution to the stability of the vibratory system. A negative value of δ_{AERO} indicates that $\cos\phi < 0$, i.e. that the flow is extracting energy from the structure, potentially leading to flutter.

The occurrence of $\delta_{\text{AERO}}>0$ indicates stabilizing aerodynamic damping, while $\delta_{\text{AERO}} < 0$ indicates destabilizing aerodynamic damping, potentially leading to flutter. In this case, the aerodynamic force lags the displacement by more than 90°.

Flutter can occur across a wide range of operating speeds, not just at specific resonant conditions. However, it usually manifests near or beyond a critical speed. It is particularly dangerous because damping mechanisms are often less effective against flutter-induced vibrations compared to resonance.

13.2.2.1 Single Blade Flutter

To study flutter in one turbomachinery blade, we adopt a simplified two-degree-of-freedom model that accounts for structural bending and torsional deformation of the blade. This formulation focuses on self-excited vibrations arising from unsteady aerodynamic loads rather than rotor dynamics or gyroscopic effects.

Assume that the blade is modeled with the following generalized coordinates: $q(t)$, the transverse bending amplitude, representing the first bending mode of the blade, and $\varsigma(t)$ representing the torsional deformation around the elastic axis of the blade. Also, assume that the elastic axis passes through the center of mass to avoid bending-torsion coupling.

The governing equations of motion are:

$$m_{eq}\ddot{q}(t) + k_q q(t) = F(t) \tag{13.44}$$

$$I_\varsigma \ddot{\varsigma}(t) + k_\varsigma \varsigma(t) = M(t) \tag{13.45}$$

where m_{eq} is the equivalent modal mass associated with the first bending mode,[7] I_ς is the torsional moment of inertia, k_q and k_ς are the bending and torsional stiffness, respectively, while $F(t)$ is the aerodynamic force and $M(t)$ is the pitching moment, respectively.

For small oscillations in incompressible flow, we may use Theodorsen's theory to model unsteady aerodynamic loads. CFD can be used as an alternative or complementary tool for more general findings.

To assess flutter stability, we compute the net aerodynamic work done over one vibration cycle using both degrees of freedom. Assuming a harmonic motion:

$$q(t) = q_0 e^{i\varpi t}, \quad \varsigma(t) = \varsigma_0 e^{i\varpi t} \tag{13.46}$$

$$F(t) = F_0 e^{i(\varpi t + \phi_F)}, \quad M(t) = M_0 e^{i(\varpi t + \phi_M)} \tag{13.47}$$

the aerodynamic work per cycle is:

$$W_{\text{cycle}} = \int_0^{2\pi/\varpi} [F(t)\dot{q}(t) + M(t)\dot{\varsigma}(t)]\,dt = \pi\,[F_0 q_0 \sin(\phi_F) + M_0 \varsigma_0 \sin(\phi_M)] \tag{13.48}$$

The average vibratory kinetic energy becomes:

$$E_{\text{kinetic}} = \frac{1}{2} m_{eq} \varpi^2 q_0^2 + \frac{1}{2} I_\varsigma \varpi^2 \varsigma_0^2 \tag{13.49}$$

Therefore, the aerodynamic damping coefficient is:

$$\delta_{\text{AERO}} = \frac{2\pi\,[F_0 q_0 \sin(\phi_F) + M_0 \varsigma_0 \sin(\phi_M)]}{m_{eq}\varpi^2 q_0^2 + I_\varsigma \varpi^2 \varsigma_0^2} \tag{13.50}$$

As already stated, the sign of δ_{AERO} indicates aeroelastic stability: if $\delta_{\text{AERO}}>0$, the unsteady aerodynamic forces dissipate energy and the motion is stable; if $\delta_{\text{AERO}} < 0$, energy is added to the system and the motion may diverge, indicating flutter.

13.2.2.2 Multi-Blade Flutter

In practical turbomachinery, flutter typically involves the interaction of multiple blades. To model this, we extend the 2-DOF single-blade formulation to a cyclic system of N identical blades arranged uniformly around a rotating hub. Each blade is modeled using two degrees of freedom: $q_j(t)$, the transverse bending amplitude of the j-th blade, and $\varsigma_j(t)$, the pitch angle of the j-th blade.

The structural equations of motion for each blade are:

$$m_{eq}\ddot{q}_j(t) + k_q q_j(t) = F_j(t) \tag{13.51}$$

$$I_\varsigma \ddot{\varsigma}_j(t) + k_\varsigma \varsigma_j(t) = M_j(t) \tag{13.52}$$

where $F_j(t)$ and $M_j(t)$ denote the unsteady aerodynamic force and moment acting on blade j. These aerodynamic loads depend not only on the local motion of blade j but also on the motion of adjacent blades due to potential flow and wake effects:

$$F_j(t) = f_F\left(q_j, \dot{q}_j, \varsigma_j, \dot{\varsigma}_j, q_{j\pm1}, \varsigma_{j\pm1}, \ldots\right) \tag{13.53}$$

$$M_j(t) = f_M\left(q_j, \dot{q}_j, \varsigma_j, \dot{\varsigma}_j, q_{j\pm1}, \varsigma_{j\pm1}, \ldots\right) \tag{13.54}$$

Such inter-blade aerodynamic coupling may be derived from empirical data or high-fidelity unsteady CFD simulations.

To reduce the complexity of the full $2N$-DOF system, periodic boundary conditions are imposed (i.e. blade $N+1 \equiv$ blade 1, and the solution is expressed as a traveling wave with inter-blade phase angle (IBPA) Ξ:

$$q_j(t) = \hat{q}(t)e^{i\Xi j}, \quad \varsigma_j(t) = \hat{\varsigma}(t)e^{i\Xi j} \tag{13.55}$$

where $\hat{q}(t)$ and $\hat{\varsigma}(t)$ are the complex time-dependent modal amplitudes common to the whole group of blades:

$$\hat{q}(t) = q_0 e^{i\varpi t}, \quad \hat{\varsigma}(t) = \varsigma_0 e^{i\varpi t} \tag{13.56}$$

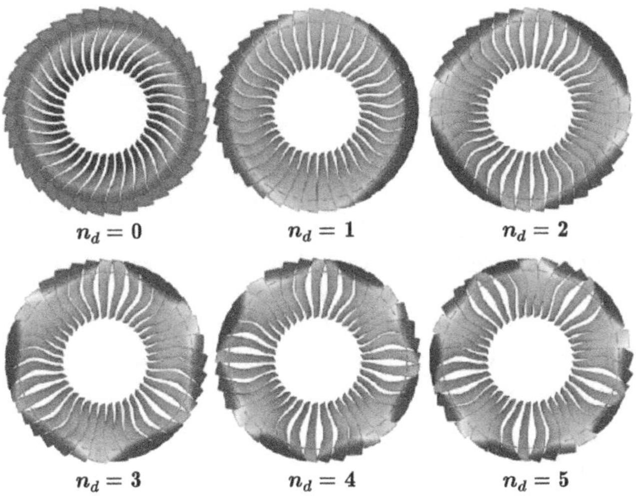

Figure 13.11 Nodal diameters in a bladed disk

IBPA Ξ is related to the nodal diameter n_d (see Figure 13.11) as:

$$\Xi = \frac{2\pi n_d}{N} \tag{13.57}$$

Typical IBPA values describe:

- $\Xi = 0^\circ$: all blades move in phase (synchronous mode),
- $\Xi = 180^\circ$: adjacent blades move in opposite phase (diametrical mode),
- $0^\circ < \Xi < 180^\circ$: traveling wave modes (asynchronous flutter).

This transformation converts the 2N-DOF system into a single 2-DOF system parameterized by Ξ, which greatly simplifies the stability analysis.

Using the same harmonic assumption for forces as in the single-blade case:

$$F_j(t) = F_0 e^{i(\varpi t+\phi_F)} e^{i\Xi j}, \quad M_j(t) = M_0 e^{i(\varpi t+\phi_M)} e^{i\Xi j} \tag{13.58}$$

we compute the total aerodynamic work done over one cycle by summing the contributions of all blades:

$$\begin{aligned} W_{\text{cycle}}^{\text{total}} &= \sum_{j=1}^{N} \int_0^T \left[F_j(t)\dot{q}_j(t) + M_j(t)\dot{\varsigma}_j(t)\right] dt \\ &= \pi N \left[F_0 q_0 \sin(\phi_F) + M_0 \varsigma_0 \sin(\phi_M)\right] \end{aligned} \tag{13.59}$$

and the total kinetic energy of the system is:

$$E_{\text{kinetic}}^{\text{total}} = \frac{N}{2}\left(m_{eq}\varpi^2 q_0^2 + I_\varsigma \varpi^2 \varsigma_0^2\right) \tag{13.60}$$

Hence, the aerodynamic damping coefficient for the blade row is:

$$\delta_{\text{AERO}} = \frac{W_{\text{cycle}}^{\text{total}}}{E_{\text{kinetic}}^{\text{total}}} = \frac{2\pi\left[F_0 q_0 \sin(\phi_F) + M_0 \varsigma_0 \sin(\phi_M)\right]}{m_{eq}\varpi^2 q_0^2 + I_\varsigma \varpi^2 \varsigma_0^2} \tag{13.61}$$

This expression is independent of the number of blades N, as both the total work and the total energy scale linearly with N. Once again, the sign of δ_{AERO} serves as a flutter stability criterion. By evaluating δ_{AERO} across a range of IBPAs Ξ and flow speeds U (or equivalently, for several flow and angular speed regimes), one can construct a flutter stability map and identify the most unstable mode shapes (Figure 13.12). The pseudocode in Algorithm 13.1 describes the procedure to be followed for multiblade flutter analysis in a blade row based on the energy-based methodology described in the present section.

13.2.2.3 Example

To illustrate the modal amplitude q_0, modal function $\hat{q}(t)$, and spatial displacement $q_j(t)$, consider the following configuration:

- Number of blades: $N = 6$
- Nodal diameter: $n_d = 2$

Algorithm 13.1: Multi-Blade Flutter Assessment via Aerodynamic Work and Damping

1: Define number of blades N, define structural parameters m_{eq}, I_ς and ϖ
2: Set vibratory amplitudes q_0, ς_0
3: Set flow velocity or mass flow range: $U = \{U_1, U_2, \ldots, U_k\}$ or $\dot{m} = \{\dot{m}_1, \dot{m}_2, \ldots, \dot{m}_k\}$
4: Set nodal diameters: $n_d = \{0, 1, \ldots, \lfloor N/2 \rfloor\}$
5: **for** each n_d in nodal diameters **do**
6:     Compute IBPA: $\Xi = \dfrac{2\pi n_d}{N}$
7:     **for** each U inflow velocity range **do**
8:         Compute $F_0(\Xi, U)$ and $\phi_F(\Xi, U)$
9:         Compute $M_0(\Xi, U)$ and $\phi_M(\Xi, U)$
10:         Compute aerodynamic work per cycle $W_{\text{cycle}}^{\text{total}}$
11:         Compute total kinetic energy: $E_{\text{kinetic}}^{\text{total}}$
12:         Compute aerodynamic damping coefficient: δ_{AERO}
13:         Store δ_{AERO} for this (Ξ, U)
14:     **end for**
15: **end for**
16: Identify flutter boundary where $\delta_{\text{AERO}} = 0$
17: Assess δ_{AERO} vs. IBPA for each U_k or $\dot{m}_k$
18: Repeat from Step 1 with updated vibration frequency ϖ, if required

---

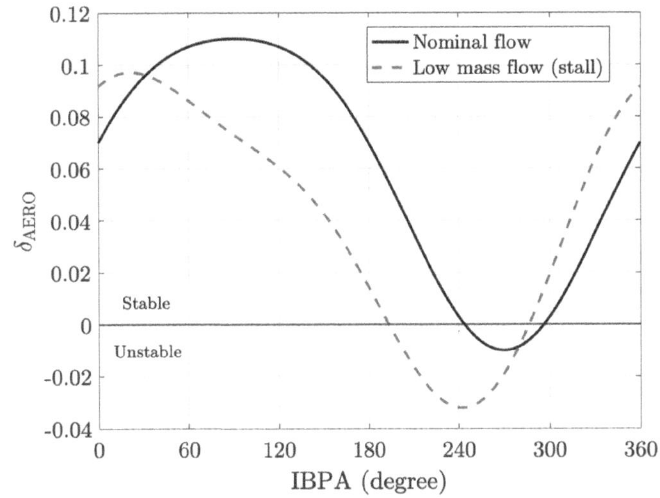

Figure 13.12 Aerodynamic damping vs. IBPA at two operating conditions.

- Vibration frequency: $\varpi = 1000\,\text{rad/s}$
- Time: $t = 0.001\,\text{s}$
- Amplitude: $|q_0| = 10^{-5}\,\text{m}$
- Phase offset: $\theta = \frac{\pi}{6}$

The inter-blade phase angle is:

$$\Xi = \frac{2\pi n_d}{N} = \frac{4\pi}{6} = \frac{2\pi}{3}$$

The shared modal function is:

$$\hat{q}(t) = q_0 e^{i\varpi t} = 10^{-5} e^{i(\varpi t+\theta)} = 10^{-5} e^{i(1.523)}$$

The motion of each blade includes a spatial phase shift:

$$q_j(t) = \hat{q}(t) e^{i\Xi j} = 10^{-5} e^{i(\varpi t+\theta+\Xi j)}$$

The physical displacement is the real part:

$$\Re\{q_j(t)\} = 10^{-5} \cos(\varpi t + \theta + \Xi j)$$

At $t = 0.001\,\text{s}$, $\varpi t + \theta = 1.523\,\text{rad}$. Table 13.2 shows the real part of the displacement for each blade.

13.2.3 PRACTICAL PREDICTION OF BLADE/DISK VIBRATIONS

To predict and mitigate blade/disk vibrations, the following analyses must be performed: (i) modal analysis of rotor-disk assembly using Finite Elements Methods

Table 13.2
Real Displacement of each Blade at $t = 0.001\,s$ with $|q_0| = 10^{-5}\,m$

Blade j	Phase Angle $\phi_j = 1.523 + \frac{2\pi}{3} j$ (rad)	$\Re\{q_{j(t)}\}$ (m)
0	1.523	$4.9 \cdot 10^{-7}$
1	3.617	$-8.89 \cdot 10^{-6}$
2	5.711	$8.37 \cdot 10^{-6}$
3	7.805	$-6.18 \cdot 10^{-6}$
4	9.899	$2.79 \cdot 10^{-6}$
5	11.993	$3.82 \cdot 10^{-6}$

(FEM) to determine the natural frequencies and mode shapes; (ii) forced vibration analysis, again using FEM, to evaluate the system's response to external periodic excitation forces, and (iii) Computational Fluid Dynamics (CFD) and Fluid-Structure Interaction (FSI) to simulate the aerodynamic forces and their coupling with the structural dynamics of the blades. The latter step usually requires significant computational resources. It is fundamental that modal analysis is carried out on a detailed virtual model of the assembly including centrifugal stiffening and elastic properties of blade and disk and, when applicable, assigning the most realistic temperature distribution.

A useful tool to guide the analyses and identify potential resonance conditions is the Campbell, or interference, diagram. It represents the natural frequencies of the blade/disk modes as functions of the rotational speed of the rotor (Figure 13.13). Modes are categorized into bending (e.g. 1B, 2B, and 3B for first, second, and third bending modes) and torsion ones (e.g. 1T, 2T). A series of straight lines, originating from the origin, is also reported, which indicates excitation frequencies proportional to the rotor speed. Each of these lines corresponds to an Engine Order (EO), i.e. to the number of excitations per revolution. For example, 4 EOs correspond to airflow excitations on the rotor assembly caused by 4 upstream struts. Resonance occurs when a natural frequency intersects with an EO line, indicating that the rotor may experience significant vibratory stresses at that specific rotor speed. Interpreting the diagram involves knowledge of geometrical details of the rotor as the number of intersections between EOs and natural frequencies is undoubtedly large. 1 EO, i.e. one excitation per revolution, can be originated by a shaft imbalance misalignment and should not be a problem when the rotor is correctly balanced before full operation. 4 EOs, as in the example discussed, should be carefully looked at alongside its harmonic multiples, i.e. 8 and 12 EOs, to avoid higher-order aerodynamic effects. Furthermore, low EOs (e.g. usually up to 3) are generated by inlet flow distortion at the rotor inlet, e.g. when the machine intake is misaligned with respect to the rotor axis and could be highly dangerous for low frequency modes (points A, B, and C in Figure 13.13). As already mentioned, mid-range EOs (e.g. 4 EOs, 6 EOs) are typically associated with upstream flow disturbances, such as struts, guide vanes, or non-uniform inlet conditions. They need to be carefully inspected as they often interact with the first and second bending modes, which are the most flexible and prone to excitation (points F,

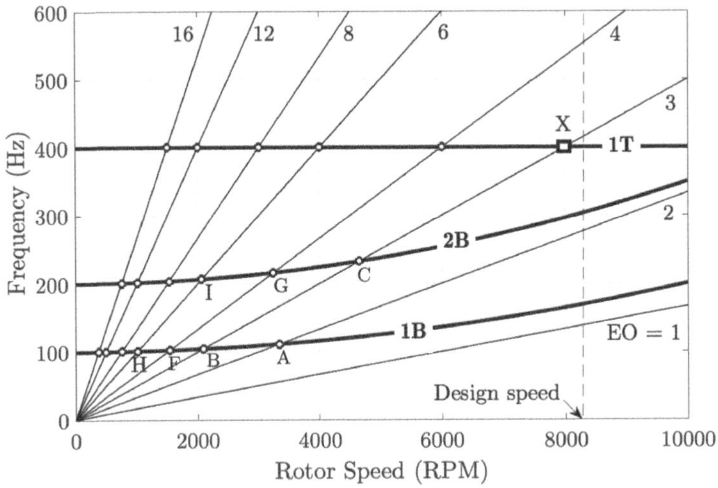

Figure 13.13 The Campbell diagram.

G, H, and I in Figure 13.13); however, for these EOs, a rapid acceleration or deceleration during intersectional rpm provides safety margins against resonance. High EOs (e.g. 12 EOs and above) represent harmonic excitations or higher-order effects. These are less critical than low- or mid-range ones, unless they intersect torsion or higher bending modes at or close to design speeds since in these circumstances a quick acceleration/deceleration might be difficult to achieve due to insufficient torque. In contrast, resonance occurring at low rpms would lie within a transient speed range, decreasing the chances of sustaining a high enough response to cause failure or accumulating enough cycles for failure. Special attention must be given to intersections close to design speeds (point X in Figure 13.13) as they might interfere with stable operation in the neighborhood of the design rpm. Finally, it must be underlined that higher modes (3B, 2T, etc.) are less flexible and harder to excite and are therefore examined after 1B, 2B, and 1T modes to ensure safety in the higher speed range.

In practical bladed systems, the study of forced response and flutter is particularly complex due to the fact that external excitation sources may exhibit arbitrary waveforms and act non-axisymmetrically with respect to the rotor axis, introducing spatially and temporally varying forces that are challenging to characterize and model. This complexity increases in the context of multiblade systems, where the interaction of each blade with the surrounding flow and its neighbors must be considered. Excitations can vary not only in time but also across the circumferential direction, leading to multiharmonic and spatially nonuniform forcing.

Nevertheless, under certain assumptions, such excitations—both aerodynamic and mechanical—can still be represented, at least approximately, by a spatial-temporal Fourier series involving harmonics in time and IBPAs. This representation enables the decomposition of the force acting on each blade into a sum of traveling wave components characterized by distinct nodal diameters and frequencies.

If the unsteady aerodynamic force acting on a given blade is expressed in the form of a Fourier series:

$$F_j(t) = \sum_{n=1}^{\infty} F_n e^{i(\varpi_n t + j\Xi_n + \phi_n)} \tag{13.62}$$

where j is the blade index, ϖ_n the forcing frequency of the n-th harmonic, Ξ_n the inter-blade phase angle (related to the nodal diameter by $\Xi_n = 2\pi n_d/N$), and ϕ_n a phase offset, then the displacement response of each blade can similarly be expressed as a superposition of modal contributions:

$$x_j(t) = \sum_{n=1}^{\infty} x_n e^{i(\varpi_n t + j\Xi_n)} \tag{13.63}$$

where x_n denotes the amplitude of the displacement associated with the n-th harmonic.

When multiple forcing frequencies are present, the assessment of flutter stability in bladed systems becomes considerably more challenging. This complexity is further amplified in multiblade assemblies, where both temporal and spatial interactions must be considered simultaneously. In such scenarios, several key aeroelastic phenomena may arise:

(i) **Modal interactions and coupling**. Different structural modes—especially those with close natural frequencies—can interact through the fluid field, resulting in aeroelastic coupling. This phenomenon can occur across multiple IBPAs. The coupling may either stabilize the system (via aerodynamic energy dissipation) or destabilize it (via energy addition), depending on the relative phasing and distribution of modal energy.

(ii) **Multi-frequency forcing and aerodynamic damping contributions.** Each forcing frequency component introduces an independent unsteady aerodynamic response and contributes to the net aerodynamic damping. For multiblade systems subjected to general, non-axisymmetric forcing, this leads to a composite response formed by a sum of traveling wave components. The total aerodynamic damping can be expressed as:

$$\delta_{\text{AERO}} = \sum_{n=1}^{\infty} \frac{2\pi F_n \sin(\phi_n)}{m\varpi_n^2 x_n} \tag{13.64}$$

where F_n and ϕ_n are the amplitude and phase lag of the unsteady aerodynamic force associated with the n-th harmonic, x_n is the corresponding modal displacement amplitude, and m is the effective modal mass. In multiblade systems, each mode is additionally associated with a spatial pattern (IBPA), and contributions from all nodal diameters must be considered. If the sum of these contributions becomes negative, the system enters an unstable regime, and flutter may develop.

(iii) **Nonlinear effects.** In real turbomachinery applications, nonlinearities can emerge due to large structural displacements, material behavior, or flow phenomena such as shock-induced separation or stall. These nonlinear effects can lead to amplitude-dependent damping, limit-cycle oscillations, or even mode switching, making linear flutter predictions insufficient in many practical scenarios.

13.2.4 PRACTICAL FLUTTER AND VIBRATION PROBLEMS IN LP STAGES

Low pressure (LP) stages, both in compressor and turbine sections, are significantly affected by vibration-related phenomena.

LP compressor stages, particularly fans, are often most vulnerable: inlet distortions (gusts, bleed flows, foreign-object damage) and modern lightweight blade designs make LPC stages highly flutter-sensitive. LP turbine stages can also flutter, particularly in highly-loaded or low-damping designs under off-design, but less often than LPC.

13.2.4.1 Aeroelastic Behavior of LPC

Fan/LPC stages face a high-subsonic to transonic flow regime at blade tip sections, often with significant compressibility and potential shock formation. The fan blades are large in diameter, with a wide chord and a high aspect ratio. These blades encounter strong aeroelastic couplings, especially near stall operating conditions where flow separation can occur. The aerodynamic excitation sources in fans/LPC include inlet distortions (non-uniform inflow), upstream gusts or turbulence, and rotor-stator interactions. Fans in particular often have few blades (e.g. 16–20), so each blade's aerodynamic influence is significant.

Fans/LPCs operate over a wide flow range and are prone to classical and stall flutter. Classical flutter involves coupling of the blade's torsional and flexural modes under steady flow (Figure 13.13), and occurs when aerodynamic excitation at some speed sustains an oscillation (often single-mode). In stall flutter, flow separation due to LPC operation close to surge line causes periodic vortex shedding and, if the shedding frequency approaches blade natural frequency, flutter occurs. In fact, stall flutter is one of the most common failure causes in compressors.

In many flutter cases, the dominant unstable mode is a pure torsion or coupled torsion-bending mode.

For subsonic/transonic flow, classical torsion-bending flutter can appear if blade loading is high. At high Mach numbers (fan tip speeds approaching sonic), shock-related phenomena and choke flutter may also arise.

Differences in blade stiffness, mass, or aerodynamic properties between blades lead to *mistuning*. Structurally, mistuning breaks the cyclic symmetry and causes the traveling wave modes to split and localize, causing vibration to concentrate in a subset of blades. For flutter, a lightly mistuned assembly tends to be more stable than the tuned one in many cases. Physically, mistuning blurs the phase coherence of blade motion that the aerodynamic forces rely on to feed energy into the system. Research has shown that mistuning generally improves the flutter margin by redistributing energy to more stable modes. In a tuned fan, all blade motion is perfectly in phase (for a given nodal diameter pattern), which can maximize the aerodynamic work if that interblade phase is unfavorable; with mistuning, some blades may lag or lead, reducing the net positive work. However, mistuning can also introduce aerodynamic asymmetry, meaning that each blade might experience slightly different flow conditions, potentially creating localized flutter on a single blade if one blade is much weaker or stronger. Intentional mistuning has even been proposed to increase aeroelastic damping.

13.2.4.2 Aeroelastic Behavior of LPT

Low-pressure turbine stages present a different set of conditions. LPT blades can be slender and have a long span, and they may include tip shrouds or interlocking segments that influence mode frequencies and shapes. LPT blades often vibrate in primarily torsional modes due to their low natural bending frequencies (as centrifugal pre-twist tends to stiffen bending). The torsion axis location strongly influences flutter and stability.

Furthermore, many LPT rotors include mechanical coupling (e.g. tip shrouds or fir-tree roots) giving cyclic symmetry (Figure 10.11). A modal analysis of such a design yields complex mode shapes.

LPT blades often see transonic flow near midspan and shocks near certain speeds. Linearized Euler solvers or compressible cascade theory are used to compute the unsteady pressures. Stall-type flutter is less common in expansion flows, but separation-induced forcing (e.g. on the suction side at off-design incidence) can still occur. Also, key findings in the literature note that unsteady wake interactions and mistuning can precipitate LPT flutter off-design.

As in compressors, torsion-flexure coupling and vortex shedding are possible. However, because turbine blades are thinner, less stiff, and operate in lower-density flow, the aerodynamic damping is often even smaller (structural damping dominates). Flutter therefore tends to occur at somewhat different operating points, often associated with large incidence or after-deflection in part-speed conditions. Modern large-diameter LPT blades also have low natural frequencies, increasing flutter susceptibility.

LPT designers often find it sufficient to look at high-frequency torsional modes, whereas LPC analysis may need both torsion and bending. Another difference is that inlet distortions matter less for LPT, though turbine inlet flows from the combustor can be unsteady, but outflow conditions (e.g. entrainment effects) can play a role.

13.3 SHAFT VIBRATIONS DUE TO DYNAMIC LOADS

Shaft vibrations in turbomachines arise from dynamic loads caused by rotor imbalance, geometric asymmetry, or component misalignments. Common causes of imbalance include structural and aerodynamic mistuning, wear and tear, and nonuniform thermal expansion or cooling.

These loads induce unsteady forces and moments that propagate through the shaft system, leading to complex vibrational behavior. For example, rotor imbalance generates centrifugal forces proportional to the imbalanced mass m and eccentricity e:

$$F_{\text{imbalance}} = me\omega^2 \tag{13.65}$$

while asymmetry and misalignment exacerbate these forces, often coupling with shaft dynamics to produce non-synchronous vibrations.

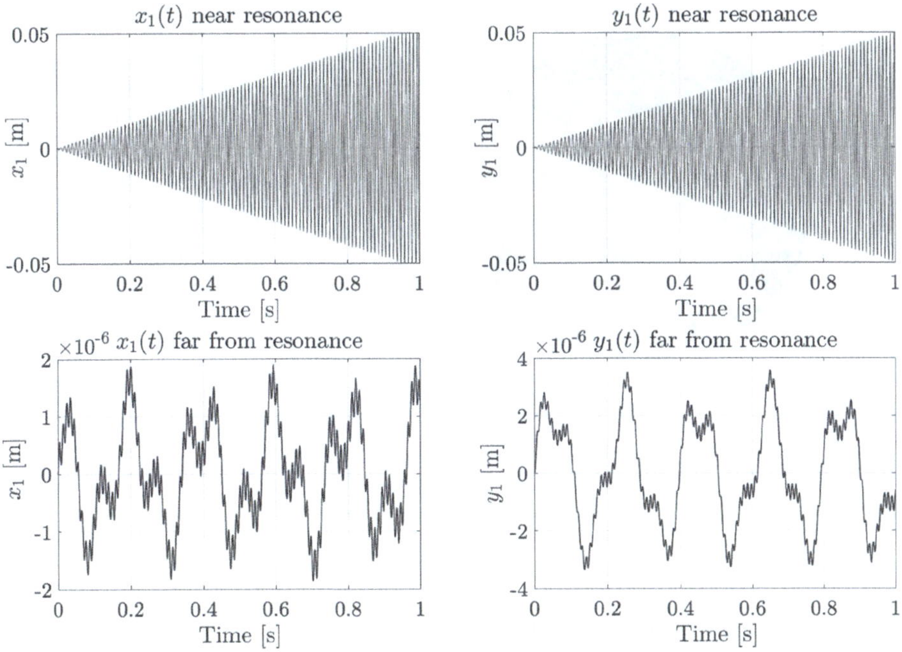

Figure 13.14 Time-domain response of a Jeffcott rotor experiencing force imbalance due to mass eccentricity.

As an example, a force imbalance of the following type:

$$\mathbf{F}_{\text{imb}}(t) = me\omega^2 \begin{bmatrix} \cos(\omega t) \\ \sin(\omega t) \end{bmatrix} \tag{13.66}$$

is applied to the left rotor of the system in Figure 13.4. This force rotates in a circular fashion at the same speed as the rotor, thus causing lateral vibrations typically observed as whirl orbits in the rotor system. To accurately characterize the dynamic behavior of the system, we can identify lateral resonance behavior first using a frequency sweep, where imbalance force is applied at various spin speeds, and the peak steady-state amplitude of the lateral displacement $x_1(t)$ is measured.

The time domain simulations can be conducted later under two conditions (Figures 13.14 and 13.15):

- **Near resonance.** The system response at the identified resonant frequency shows a significant amplitude growth in both $x_1(t)$ and $y_1(t)$. The symmetry in both directions suggests a dominant whirling mode. The orbit of the rotor becomes larger and more circular.
- **Far from resonance.** At much lower spin speed, the response amplitudes remain low (in the order of micro-meters, $\sim 10^{-6}$ m) and the orbit is indeed smaller and more elliptical, thus indicating weak energy transfer to the structure. The

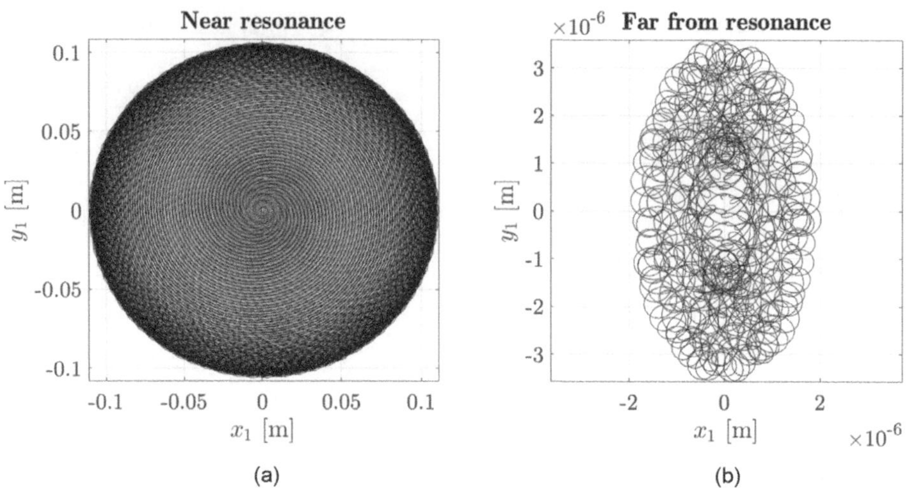

Figure 13.15 Rotor whirl orbits under imbalance: circular orbit at resonance (left) and elliptical orbit far from resonance (right).

amplitude signals appear irregular because of the weak interaction between the excitation frequency and the natural modes of the system.

This comparison highlights the dramatic difference in the behavior of the rotor near and far from resonance. Near resonance, even small imbalance forces can lead to severe vibration due to resonance amplification. Conversely, far from resonance, the same imbalance generates only small, bounded orbits.

13.4 PROBLEMS

1. Consider a rigid rotor of mass m, with transverse moment of inertia J_x and polar moment of inertia J_z, mounted on isotropic elastic supports. The rotor spins at a constant angular speed ω about its geometric z-axis. The supports are located at distances a and b from the rotor's center of mass C, and are modeled as linear springs with stiffness k acting in both the x- and y-directions. Let:

   - $x(t), y(t)$ the transverse displacements of the rotor center in the x- and y-directions;
   - $\theta(t)$ and $\psi(t)$ the angular displacement of the rotor about the x-axis and the y-axis, respectively.

   Derive equations of motion. Prove that in the general case where $a \neq b$, the equations of motion of the system become coupled.

2. A rigid disk is mounted at the center of a simply supported shaft. The shaft has a length $L = 0.6\,\text{m}$ and diameter $d = 0.02\,\text{m}$. The disk has a radius $R = 0.045\,\text{m}$ and mass $m = 9.88\,\text{kg}$. The disk rotates at a constant angular velocity $\omega = 2000\,\text{rad/s}$, inducing gyroscopic effects. The shaft is made of steel, which has a Young modulus $E = 2.1 \cdot 10^{11}\,\text{Pa}$. The shaft stiffness is isotropic and calculated using beam theory:
$$k_x = k_y = \frac{48EI}{L^3}, \quad \text{with } I = \frac{\pi d^4}{64}$$
Viscous damping is included via journal bearings:
$$c_x = c_y = 10\,\text{N s/m}$$
The gyroscopic moment is given by:
$$G = \frac{1}{2} m R^2 \cdot \omega$$
The rotor is initially displaced with:
$$x(0) = 0.001\,\text{m}, \quad y(0) = 0.005\,\text{m}, \quad \dot{x}(0) = \dot{y}(0) = 0$$
Simulate the system dynamics implementing a dedicated code which makes use of a library for the solution of differential equations (e.g. the `ode45` function in MATLAB) and describe the evolution of $x(t)$ and $y(t)$. Identify the forward and backward whirl modes.
3. A rigid disk is mounted at the end of a cantilevered steel shaft in an overhung configuration (see Figure 13.2d). The shaft has a diameter of 0.018 m and an overhung length of 120 mm. The rigid disk has a mass of 0.5 kg, a diameter of 85 mm, and a thickness of 20 mm. The shaft is made of steel with a Young modulus of $2.1 \cdot 10^{11}$ Pa. The disk has an eccentricity of 1 mm. A damping ratio of 0.04 is assumed to model viscous damping in both directions. Compute the first critical speed of the system, simulate the time-domain lateral response at the prescribed rotational speed, and analyze the resulting whirl orbit.
4. A high-speed turbomachinery system features two rigid disks, which are mounted at the ends of a massless, flexible steel shaft supported by journal bearings. The first rotor has a mass of $m_1 = 0.8$ kg, a diameter of $D_1 = 120$ mm, and a thickness of $w_1 = 33$ mm. The second rotor has a mass of $m_2 = 4.0$ kg, a diameter of $D_2 = 75$ mm, and a thickness of $w_2 = 60$ mm. The rotors are connected via a steel shaft with a diameter $d_s = 20$ mm and total length $L = 310$ mm ($L = l_1 + l_2 + l_3$ in Figure 13.4). The Young modulus and shear modulus of the shaft are $E = 210 \cdot 10^9$ Pa and $G = 80 \cdot 10^9$ Pa, respectively. The left bearing has lateral stiffness values $k_{x_1} = k_{y_1} = 1 \cdot 10^5$ N/m, while the right bearing has $k_{x_2} = k_{y_2} = 2 \cdot 10^5$ N/m. The torsional base stiffnesses of the two rotor locations are $k_{\theta 1} = 10$ N·m/rad and $k_{\theta 2} = 20$ N·m/rad. Damping is modeled as viscous and isotropic, with lateral damping coefficients $c_x = c_y = 5$ N·s/m, and torsional damping $c_{\theta_1} = 0.01$ N·m·s/rad and $c_{\theta_2} = 0.02$ N·m·s/rad. The rotor system spins at a constant angular speed of $\omega = 1000 \cdot 2\pi$ rad/s. Compute the undamped natural frequencies of the system.

[(83.96; 244.78; 244.78; 603.46; 603.46; 2064.12) rad/s]

NOTES

1. To balance a rigid rotor dynamically, the rotor is first mounted onto a balancing machine, which may use either soft or hard bearings depending on the sensitivity and calibration method. Once securely mounted, the rotor is spun at low speed, safely below its first critical speed, to avoid resonance effects. Vibration measurements are then taken at two axial planes, typically denoted as plane A and plane B, using sensors placed near the bearings or proximity probes. The appropriate correction weights are placed next in the angular positions needed in each plane to counteract the imbalance.
2. In the present chapter, matrices will be denoted using square brackets, e.g. $[\mathbf{X}]$, to facilitate reading and understanding of equations.
3. A Jeffcott rotor—also known as a de Laval rotor—is a classical theoretical model used to represent the fundamental dynamic behavior of rotating systems. It consists of a massless elastic shaft with a single rigid disk mounted at the midpoint, supported by bearings at both ends. The disk is typically assumed to have mass, but no stiffness and its center of mass may be offset (i.e. imbalanced) from the shaft axis. In the mathematical description of the Jeffcott rotor, it does not matter whether the lateral stiffness comes from the bending of the shaft, from the supports, or both.
4. To avoid confusion, we distinguish between two types of natural frequencies in rotor dynamics:

- ω_0: The natural frequency of the system at *zero spin speed* (i.e. a stationary rotor). It is obtained by solving the standard eigenvalue problem without gyroscopic effects:

$$\det\left([\mathbf{K}] - \omega_0^2[\mathbf{M}]\right) = 0$$

  This frequency serves as a baseline or reference point.
- $\omega_n = f(\omega)$: The *spin-speed-dependent* natural frequency when the rotor is spinning at angular speed ω. Due to gyroscopic effects, each mode generally splits into two distinct frequencies:

$$\omega_n^{\mathrm{FW}} > \omega_0 \quad \text{(Forward Whirl)}, \qquad \omega_n^{\mathrm{BW}} < \omega_0 \quad \text{(Backward Whirl)}$$

5. Damping through the shaft is neglected.
6. Using a single value k_s to represent the lateral stiffness of the shaft is a common and practical simplification in rotor dynamic modeling. This lumped approach is generally appropriate when the shaft is relatively short and stiff, the rotors are significantly heavier than the shaft (allowing the shaft mass to be neglected), and the analysis focuses on the system's low-order dynamics—such as the first bending mode. It is particularly well-suited for preliminary design, modal estimation, and frequency-domain studies around the operating speed range.

However, this simplification becomes less accurate when the shaft is long, slender, or contributes non-negligibly to the system's inertia. In such cases, higher-order bending modes and shaft whirling phenomena become relevant, and a single stiffness term cannot capture the distributed nature of shaft deformation. The approximation also falls short when intermediate supports are present or when there is significant dynamic coupling between torsional and bending motions. For high-fidelity modeling under these conditions, a more detailed representation of the shaft—such as a distributed parameter or finite element model—is recommended.

7. For a blade modeled as a uniform cantilever beam in first bending mode, about 23.6% of the total mass participates in the motion, i.e. $m_{\text{eq}} \approx 0.236 \cdot m_{\text{blade}}$.

14 Materials for Turbomachines

14.1 INTRODUCTION

The selection of materials directly impacts performance, reliability, and operational lifespan of turbomachinery components. As such, material choices must be aligned with the specific functional and operational demands of each application.

This chapter examines the materials commonly used in turbomachinery manufacturing, categorized by application. The approach follows the categories outlined in the second chapter, focusing on material selection as a function of performance requirements and environmental conditions of each of the turbomachines dealt with in this book. Key material groups include metals, ceramics, composites, and polymers, each offering distinct advantages depending on the operating environment.

Additionally, this chapter discusses the primary criteria for material selection, including mechanical properties, thermal performance, resistance to wear and corrosion, manufacturability, and cost considerations. Advances in material technology are explored in the context of modern turbomachinery applications.

14.2 MATERIALS FOR PUMPS

The materials used in pump components are selected based on the fluid being handled, operating conditions, required durability, and cost. Common materials include metals such as cast iron and stainless steel for strength and corrosion resistance, polymers for chemical compatibility, and nickel or titanium for lightweight and specialized applications. Table 14.1 offers a general view of typical materials used for a pump's casing, impeller, and shaft, with indications on comparative price.

14.2.1 IRON-BASED MATERIALS

Iron-based materials are the backbone of pump construction, offering a range of mechanical and corrosion-resistant properties suitable for various applications.

Grey cast iron, characterized by a carbon content $\geq 3\%$, offers excellent machinability, low cost, and sufficient corrosion resistance for neutral to mildly acidic environments. It performs well with fluids featuring a pH range of 6 to 10, making it ideal for handling water and non-aggressive fluids. A notable feature of gray cast iron is its ability to form a protective graphite layer over time during operation, which shields the underlying metal from further corrosion. However, its brittle nature limits its use in high-impact or shock-prone environments, and it is generally avoided in high-temperature or high-pressure applications.

DOI: 10.1201/9781003619956-14

Table 14.1
Pump Materials and Relative Prices

Pump Casing	Pump Impeller and Small Parts	Shaft	Relative Price
Grey cast iron	Grey cast iron	Stainless steel	0.97
Grey cast iron	Red brass	Stainless steel	1.00
Grey cast iron	Acid-resistant steel	Acid-resistant steel	1.08
Spheroidal or nodular cast iron	Acid-resistant steel	Acid-resistant steel	1.15–1.25
Steel casting	Acid-resistant steel	Acid-resistant steel	1.40–1.50
Red brass	Red brass	Stainless or acid-resistant steel	1.65
Acid-resistant steel SIS 2324	Acid-resistant steel SIS 2324	Acid-resistant steel SIS 2324	1.85
High alloy acid-resistant steel	High alloy acid-resistant steel	High alloy acid-resistant steel	2.2
Hastelloy C	Hastelloy C	Hastelloy C	3
Titanium	Titanium	Titanium	10

Spheroidal cast iron increases its ductility compared to gray cast iron, impact resistance, and tensile strength, making it suitable for demanding applications such as high-pressure systems or pump components exposed to mechanical stress. It operates effectively at temperatures up to 300°C, broadening its applicability to hot fluids. Despite these improvements, spheroidal cast iron exhibits reduced corrosion resistance at high flow velocities, where erosion and cavitation may accelerate material degradation. Its ability to handle internal pressures and its compatibility with various protective coatings make it a reliable choice for industrial pumps.

Cast iron pumps are typically heavy, have a rough, grainy surface texture, and usually appear painted or coated to prevent rust. Edges and features might feel coarser or angular as a result of the casting process. Gray cast iron (GG-18 to GG-50) features lamellar graphite, making it brittle with negligible elongation and unsuitable for safety-critical applications like pump casings due to the risk of sudden failure. In contrast, nodular cast iron (GGG-40 to GGG-50) forms spherical graphite grains, offering mechanical properties akin to steel. Although both types show similar corrosion behavior to carbon steels, nodular cast iron (GGG-40) has a lower resistance to erosion corrosion than gray cast iron (GG-25).

14.2.2 STEEL-BASED MATERIALS

Steel-based materials are highly versatile in pump construction, offering a broad range of mechanical properties and corrosion resistance suitable for demanding applications.

Cast carbon steel (e.g. ASTM A216 Grade WCB) is commonly specified for pump casings in higher pressure applications or where impact toughness is needed. WCB cast steel typically has an as-cast tensile strength on the order of 485–655MPa and yield strength around 250 MPa. Compared to gray iron, cast steel provides much greater fracture toughness and can tolerate moderate impact or vibration without cracking. For pump shafts, bolts, and other highly stressed parts, *wrought alloy steels* are typically used. AISI-4140, a Cr-Mo low alloy steel, is a representative shaft material due to its balance of strength and toughness. In normalized or tempered conditions, 4140 steel has a tensile strength of around 650–700 MPa and a yield strength of 415 MPa with 25% elongation. These alloy steels exhibit good high-cycle fatigue resistance; under reversed bending, an endurance limit on the order of 300 MPa can be expected for a 650 MPa UTS steel, which is advantageous for long-life rotating shafts. At elevated temperatures, carbon and low alloy steels gradually lose strength and can undergo creep. Plain carbon steels are generally limited to 400°C service before time-dependent deformation becomes significant. For higher temperature pumps (e.g. in power plant service), low-alloy Cr-Mo steels are used due to their enhanced creep resistance. Steels alloyed with molybdenum (e.g. 0.5%–1% Mo) can maintain usable strength up to 500°C–530°C.

Stainless steel stands out for its ability to resist corrosion, a property derived from its minimum 12% chromium content. Chromium forms a thin, passive oxide layer on the surface, protecting steel from corrosive environments. Stainless steel is particularly effective in chemically aggressive and oxidizing environments, such as those involving strong acids, chlorides or elevated temperatures. Variants with added molybdenum, known as acid-resistant grades, further enhance corrosion resistance against pitting and crevice corrosion, especially in chloride-rich solutions or marine settings. Its toughness and durability also make it a preferred material for pump components subjected to both mechanical stress and chemical exposure. Stainless steels, which include martensitic, austenitic, and duplex types, vary in strength, ductility, and corrosion resistance based on their composition and carbon content. Lower carbon content (<0.07% or <0.03% for high corrosion resistance) improves passivation and reduces the susceptibility to corrosion.

Martensitic steels, typified by AISI 410 (EN 1.4006) or cast equivalents such as CA-15 and CA6NM, have high tensile strength (700–1200 MPa), are suitable for fresh water pump components but are prone to local corrosion, cracking of stress corrosion, and hydrogen embrittlement, limiting their use in chloride-rich environments. Austenitic steels, that is, grades such as AISI 304L and 316L, containing at least 8% nickel, have lower strength (400–650 MPa) but good resistance to uniform corrosion, although they remain vulnerable to local and intergranular corrosion. Duplex steels (e.g. ASTM A890 Grade 3A/EN 1.4462, commonly known as 2205 Duplex), with an austenitic-ferritic microstructure and higher strength (600–800 MPa), offer improved resistance to local corrosion, stress corrosion cracking, and abrasion, making them ideal for harsher environments, particularly in seawater, brine, or other aggressive services where 316 stainless steel may lack sufficient strength or resistance to stress corrosion. Proper alloy composition and heat treatment are critical for optimizing their mechanical and corrosion properties.

Non-alloyed steel, while lacking the corrosion resistance of stainless steel, compensates with superior mechanical strength, making it a reliable choice for pumps designed for high-pressure and hot-liquid applications. It is often used in structural and load-bearing components of pumps where strength and rigidity are critical. However, due to its vulnerability to corrosion, non-alloyed steel is typically employed in conjunction with protective coatings or in applications where the pumped fluid is non-corrosive. The cost effectiveness and machinability of this material make it an attractive option for general-purpose pumps, although it requires careful consideration of environmental conditions to avoid premature degradation.

Steel pumps are typically lighter in comparison to cast iron for the same dimensions and have a smooth, polished, or brushed finish. Mostly non-painted, silvery-toned, they resist rusting under normal conditions. In addition, they feature smoother surfaces, as the material is often machined or polished.

14.2.3 NICKEL-BASED ALLOYS

Nickel-based alloys are premium materials in pump construction, known for their exceptional corrosion resistance and mechanical properties, especially in challenging environments. They find application in pumps handling hot, highly corrosive fluids where even stainless steels would rapidly corrode or lack sufficient strength. These alloys can offer tensile strengths on the order of 700–1000 MPa at room temperature and retain a large fraction of that strength at elevated temperatures (600°C–800°C), along with superior resistance to acid attack and oxidation.

Hastelloy is a family of alloys that is highly resistant to both oxidizing and reducing chemicals, making it a preferred choice for handling strong acids such as hydrochloric and sulfuric acid, as well as oxidizing agents in industrial processes. Hastelloy exhibits remarkable durability under harsh conditions, including elevated temperatures and high-pressure systems. Its ability to resist localized corrosion, such as pitting, crevice corrosion, and stress corrosion cracking, ensures reliability in aggressive chemical environments, particularly those involving chlorides and complex multi-component solutions.

Monel is a nickel-copper alloy, highly effective in marine and saltwater environments, maintaining its structural integrity and corrosion resistance even at high flow velocities. Its resistance to chloride-induced stress corrosion and erosion makes it ideal for pumps exposed to seawater or brine. In addition, Monel has excellent mechanical properties and can withstand thermal and mechanical stresses. It is commonly used for components such as impellers and shafts where strength and corrosion resistance are critical.

Nickel-chromium-iron alloys without molybdenum, such as Inconel 600 (NiCr15) and Inconel 800 (X10NiCrAlTi32-21) are prone to pitting and unsuitable for seawater use.

All nickel-based alloys are often chosen for their ability to withstand environments where traditional materials such as stainless steel or cast iron would degrade rapidly, ensuring long service life and minimal maintenance in most demanding applications. Although costlier than other materials, their superior performance justifies their use in critical systems, such as in the defense, aerospace, chemical, energy, and automotive industries. Blue Origin's BE-4 rocket engine incorporates an Ox Boost Pump (OBP) with key components manufactured using additive manufacturing techniques from Monel. Interstellar Technologies' ZERO rocket engine utilizes a turbopump constructed from nickel alloy, with certain parts made from titanium and aluminum alloys. This combination leverages the heat resistance of nickel alloys and the lightweight properties of titanium to achieve a compact and efficient design capable of operating at rotational speeds up to 40,000 rpm.

Nickel-based alloys are dense materials with a hard surface, resistant to scratches and mechanical wear, and are heavier compared to even some stainless steel pumps of the same size. They often have a dull, silvery-gray color with a slightly yellowish or golden hue in certain light conditions, especially Monel. Components like impellers may have a smooth finish and precise geometry, thanks to finely machined parts.

14.2.4 COPPER-BASED ALLOYS

Copper-based alloys, particularly bronze and brass, are widely used in pump construction due to their excellent corrosion resistance and mechanical properties in specific environments.

Bronze and **brass** are especially effective in seawater and chloride-rich environments, where their natural ability to form protective oxide films provides significant resistance to corrosion. Brass, composed of copper and zinc, is often used in less aggressive conditions and offers good machinability and moderate corrosion resistance. Bronze, typically a blend of copper and tin with additional elements, is more robust and resistant to stress corrosion cracking, making it a preferred choice in marine applications. Both materials are well-suited for pump components like casings, impellers and bearings, where moderate strength and excellent corrosion resistance are required.

Aluminum-bronze alloys contain aluminum as a primary alloying element, significantly enhancing strength, durability, and resistance to cavitation and erosion. They perform well in harsh environments, such as seawater or fluids containing abrasive particles, and are often used in high-performance pump applications. Aluminum-bronze, such as G-CuAl-10Ni, is particularly valued for its ability to resist the combined effects of corrosion, erosion, and wear, making it ideal for impellers and other critical components in demanding operations (tensile strengths up to 600 MPa). Tin bronzes like G-CuSn10 are stable in seawater at low flow rates but have limited ultimate tensile strength (in the order of 200–270 MPa), and yield strength of only 100 MPa, making them unsuitable for high-head applications. Another type is nickel-aluminum bronze (NAB, for example, ASTM B148 C95800), which combines strength with corrosion resistance: typical tensile strength is in the range of 640–700 MPa, with a 0.2% yield strength around 250–300 MPa and an elongation of about 15%.

Together, these alloys provide a versatile and reliable option for pumps operating in environments where other materials might fail due to corrosion or wear, balancing cost, durability, and performance. Brass has a yellowish-golden hue as a result of its zinc content. The surface may tarnish to a dull bronze-like color over time if exposed to air. Bronze is reddish-brown or coppery in shade and darkens to a greenish or brown patina if exposed to environmental conditions like seawater. Aluminum-bronze appears more golden or light bronze in color with a smooth, often polished finish. It may have a slightly more metallic sheen compared to standard bronze.

14.2.5 SPECIALIZED METALS AND NON-METALLIC MATERIALS

Specialized metals and non-metallic materials are used for niche applications where standard materials fail to meet the required performance criteria.

Silicon iron is specifically engineered to withstand highly corrosive environments, such as those involving concentrated sulfuric acid. Its high silicon content provides exceptional chemical resistance, but it comes with limitations, including brittleness and poor performance under high-pressure conditions. As a result, silicon iron is often used in low-pressure systems or as a lining within steel casings to enhance durability.

Titanium alloys (like Ti-6Al-4V) have high specific strength and outstanding corrosion resistance in seawater and chlorine-bearing environments. Ti-alloys used for pump impellers or casings typically exhibit UTS in the range of 800–900 MPa (with 0.2% yield around 800 MPa for Ti-6Al-4V), while commercially pure titanium grades have lower strength (e.g. Grade 3 titanium has UTS $\sim$450 MPa, yield $\sim$380 MPa) but exceptional immunity to many chemicals. The main drawbacks of nickel- and titanium-based alloys are their high cost and, in some cases, difficult manufacturing; therefore, they are reserved for services where more common materials cannot meet the requirements (e.g. pumps for very aggressive acids, high-temperature organic heat transfer fluids, or aerospace turbopumps). A notable example is the Space Shuttle Main Engine (SSME), which employed a Ti-5Al-2.5Sn titanium alloy for several critical components. This alloy was chosen for its optimum performance in liquid hydrogen environments, enabling higher rotational tip speeds and contributing to weight reduction and increased efficiency.

Plastics are increasingly being utilized for their chemical resistance and versatility. Hard plastics like epoxy are favored in static applications where strength and durability are paramount, while thermoplastics such as PTFE (Teflon) excel in dynamic systems requiring flexibility, low friction, and resistance to high-purity chemicals. These materials are lightweight, cost-effective, and resistant to a wide range of corrosive fluids, though they are limited by lower temperature and pressure tolerances compared to metals. For example, PVDF (polyvinylidene fluoride, a fluoroplastic) has a yield strength of about 55 MPa at 23°C, which drops to around 10 MPa by 150 °C. Plastics also tend to creep under sustained loads, which means a pump casing made of plastic can deform over time if operated near its stress limit. As a result, plastic pump components are generally limited to low-pressure and moderate-temperature applications. They often require thicker sections to compensate for lower strength, and fiber reinforcement (e.g. fiberglass) may be used to improve stiffness and creep resistance.

Table 14.2
Compact Overview of Pump Materials

Material Type	Key Features	Applications/Limitations	Tensile Strength (MPa)
Iron-Based	Grey Cast Iron: Low cost, protective graphite layer. Spheroidal Iron: Ductile, handles 300°C.	Grey: Neutral fluids (pH 6–10), brittle under impact. Spheroidal: High-pressure, low corrosion resistance at high flow velocities.	Grey: 150–250 Spheroidal: 400–700
Steel-Based	Stainless Steel: Corrosion-resistant, durable. Non-Alloyed: High strength, prone to corrosion.	Stainless: Marine/chemical environments. Non-Alloyed: Structural parts, non-corrosive fluids.	Stainless: 480–620 Non-Alloyed: 400–550
Nickel-Based	Hastelloy: Resistant to acids, oxidizing agents. Monel: Marine, high-flow saltwater.	Hastelloy: High-temp/pressure systems. Monel: Aerospace pumps, marine use.	Hastelloy: 690–790 Monel: 550–620
Copper-Based	Bronze: Durable, marine corrosion-resistant. Aluminum-Bronze: Cavitation and erosion resistant.	Bronze: Seawater pumps. Al-Bronze: Abrasive, high-performance pumps.	Bronze: 200–300 Al-Bronze: 500–600
Specialized	Silicon Iron: Acid resistant, brittle. Titanium: Lightweight, corrosion-resistant.	Silicon: Low-pressure systems. Titanium: Aerospace, chemical processing.	Silicon Iron: 200–250 Titanium: 900–1200
Non-Metallic	Plastics: Chemically resistant. Rubber/Ceramics: Abrasion and wear resistant.	Plastics: Low temp/pressure. Rubber: Abrasive fluids. Ceramics: Precision parts.	Plastics: 30–50 Ceramics: 200–400

Rubber is a highly adaptable material used in pump components due to its exceptional resistance to abrasion and chemical attack. It is often applied in slurry pumps or as seals and gaskets in corrosive environments.

Ceramics, on the other hand, offer excellent chemical stability and hardness, making them ideal for applications that require wear resistance and precision, such as impellers and liners in highly abrasive or high-temperature systems. However, ceramics are brittle and require careful handling and design considerations to prevent cracking under mechanical stress.

Table 14.2 highlights the materials, their properties and suitable applications/limitations, providing a concise reference for pump material selection.

14.2.6 CAVITATION-RESISTANT MATERIALS

Cavitation leads to a significant reduction in head and overall efficiency, causing the pump to underperform relative to its designed operational curves. This means that while these materials can delay or mitigate the onset of physical damage, they cannot completely eliminate the effects of cavitation.

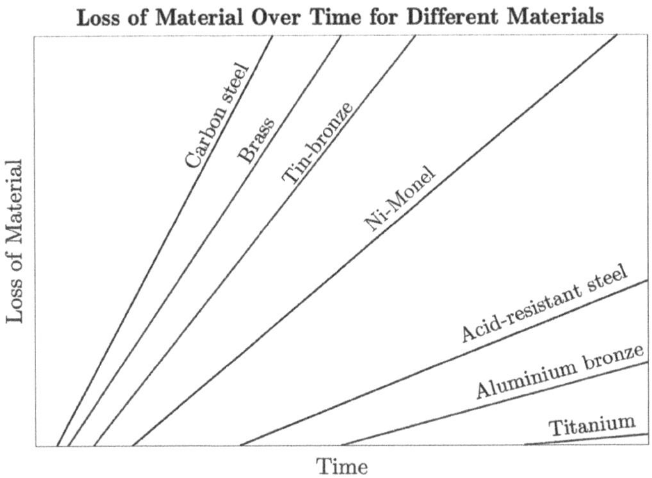

Figure 14.1 Loss of material over time for different materials.

The selection of appropriate materials must be guided by rigorous testing and experimentation to determine their resistance under specific operating conditions. As highlighted in Figure 14.1, the damage process often has a "lag phase", during which visible deterioration is not yet apparent. Understanding this phase through experiments and long-term observation allows engineers to make informed decisions about material suitability and optimize system design to completely reduce cavitation risks. Ultimately, while materials resistant to cavitation enhance durability, addressing the root causes of cavitation, such as optimizing flow design and improving system parameters like NPSH, remains essential to maintain both performance and longevity. In particular, Figure 14.1 illustrates the loss of material for various materials over time under cavitation conditions. The slope of each line reflects the material's cavitation resistance, with lower slopes indicating better performance. Titanium and aluminum bronze are clearly the most resistant to cavitation, while carbon steel and brass are the least resistant.

14.2.7 ADVANCED MATERIALS FOR PUMPS

Traditional materials often fail under high temperatures, pressures, or corrosive environments, making advanced materials essential. Lightweight *composites and polymers* are particularly valuable in sectors like aerospace, offering reduced weight without compromising strength. Advanced materials also provide superior chemical compatibility and wear resistance, essential for handling diverse and abrasive fluids.

High-performance polymers like Polyether Ether Ketone (PEEK), Polyphenylene Sulfide (PPS), and Polytetrafluoroethylene (PTFE) offer exceptional chemical resistance, thermal stability, and low friction, making them ideal for impellers, seals, and wear rings. Advanced ceramics such as silicon carbide, alumina, and zirconia deliver unmatched hardness and wear resistance, excelling in mechanical seals and shafts. Composite materials, including Carbon Fiber Reinforced Polymers (CFRP) and Glass Fiber Reinforced Polymers (GFRP), combine strength and lightweight properties, while specialty metals like duplex stainless steels, nickel-based superalloys, and titanium alloys offer enhanced corrosion resistance and stability under extreme conditions.

Advanced materials often have higher initial costs and require complex manufacturing processes. Ensuring compatibility between different materials and meeting regulatory standards in industries like food and pharmaceuticals can be time-consuming. Despite these hurdles, long-term benefits, including longer lifespans and reduced maintenance, often outweigh upfront costs.

14.3 MATERIALS FOR COMPRESSORS AND HIGH-SPEED FANS

14.3.1 ALUMINUM ALLOYS

Aluminum and its alloys are commonly used in centrifugal compressors for their strength-to-weight ratio, good thermal conductivity, and cost-effectiveness. Less often they are adopted in axial stages due to their relatively poor mechanical strength (Table 14.3). These materials are particularly suited for low- to medium-pressure applications, where lightweight construction is critical to minimize inertia and optimize rotational speeds. Alloys such as 6061-T6, 7075-T6, and 2024-T3 are often selected for their machinability, moderate corrosion resistance (enhanced by coatings), and adequate fatigue strength. However, their usage is generally limited to applications operating below 150°C due to a drop in mechanical properties at elevated temperatures.

14.3.2 TITANIUM ALLOYS

Titanium alloys, such as Ti-6Al-4V, are extensively used in low-pressure compressor blades and fan casings due to their exceptional strength-to-weight ratio (density of 4.43 g/cm^3, ultimate tensile strength of up to 1100 MPa and a yield strength around 930 MPa) and corrosion resistance. High-temperature-resistant alloys, such as Ti-6246 and Ti-834, are used in high-pressure compressor stages and can operate at temperatures up to 500°C. These alloys are essential for reducing weight and improving fuel efficiency. Advances in titanium aluminides (e.g. γ -TiAl) address applications requiring strength at elevated temperatures of up to 750°C. With a low coefficient of thermal expansion ($8.6 \cdot 10^{-6}$/°C) and thermal conductivity of 6.7 W/(m·K), these materials are thermally stable, minimizing the risk of thermal distortion and ensuring reliable operation over a wide temperature range. Their resistance to fatigue and creep under cyclic stresses improves their durability, which is crucial for applications requiring continuous operation. Moreover, the natural passivating oxide layer (TiO_2) that forms on the surface provides excellent resistance to corrosion, protecting the blades from environmental factors like humidity and contaminants present in jet fuel

Table 14.3
Properties of Aluminum Alloys for Centrifugal Compressors

Alloy	Yield Strength (MPa)	Ultimate Tensile Strength (MPa)	Density (g/cm^3)	Key Applications
6061-T6	276	310	2.70	Low-pressure compressors, lightweight systems
7075-T6	500	570	2.81	High-performance compressors, aerospace
2024-T3	345	470	2.78	Aerospace compressors, moderate fatigue resistance
5052-H32	193	228	2.68	Lightweight, corrosion-resistant components
3003-H14	150	180	2.73	Low-strength applications, corrosion resistance

or industrial air streams. Table 14.4 reports a synthesis of the most relevant properties of the Ti-6Al-4V alloy.

While Ti-6Al-4V remains a prevalent choice for compressor blades, several advanced titanium alloys offer enhanced performance for such applications.

Table 14.4
Typical Properties of Ti-6Al-4V Alloy for Compressor Blades

Property	Value
Material Composition	Ti (88%–90%), Al (5.5%–6.75%), V (3.5%–4.5%), Fe (<0.3%), O (<0.2%)
Density	4.43 g/cm^3
Ultimate Tensile Strength (UTS)	950–1100 MPa
Yield Strength (0.2% offset)	880–930 MPa
Elongation at Break	14%–15%
Fatigue Strength (10^7 cycles)	510 MPa
Hardness (Vickers)	349 HV
Melting Point	1660°C
Thermal Conductivity	6.7 W/(m·K)
Specific Heat Capacity	0.56 J/(g·K)
Coefficient of Thermal Expansion	$8.6 \cdot 10^{-6}$/°C
Operating Temperature Range	Up to 400°C
Creep Rate	10^{-7} %/hour at 300°C, 150 MPa
Corrosion Resistance	Excellent due to passivating oxide layer TiO_2

Ti-6Al-2Sn-4Zr-2Mo (Ti-6-2-4-2) exhibits superior high-temperature strength and creep resistance compared to Ti-6Al-4V, making it suitable for components operating at elevated temperatures. Its typical composition includes 6% aluminum, 2% tin, 4% zirconium, and 2% molybdenum.

Ti-8Al-1Mo-1V has been developed for high-pressure compressor blades and offers a favorable balance of strength and fracture resistance. Its composition comprises 8% aluminum, 1% molybdenum, and 1% vanadium.

Titanium Aluminides (TiAl) are intermetallic compounds like gamma titanium aluminide (γ–TiAl), valued for their low density and high-temperature oxidation resistance. They are increasingly used in low-pressure turbine blades and have potential applications in compressor components.

Titanium Matrix Composites (TMCs) incorporate ceramic fibers into a titanium matrix, enhancing stiffness and strength while reducing weight. TMCs are being explored for advanced compressor blades to improve performance.

14.3.3 NICKEL ALLOYS

Nickel-based alloys are particularly suited for the hot sections of compressor blades due to their exceptional ability to withstand extreme conditions of temperature, stress, and corrosive environments. Alloys like Inconel 718 provide excellent thermal stability, high-temperature strength, and resistance to fatigue, creep, and oxidation. Their operating temperature range extends up to 750°C.

The mechanical properties of these alloys further enhance their suitability for hot compressor blades (Table 14.5). With ultimate tensile strength exceeding 1200 MPa

Table 14.5
Typical Properties of Inconel 718 Alloy for Hot Compressor Blades

Property	Value
Material Composition	Ni (50%–55%), Cr (17%–21%), Fe (17%), Nb (4.75%–5.5%), Mo (2.8%–3.3%), Ti (0.65%–1.15%), Al (0.2%–0.8%), C (< 0.08%), Co (< 1%) and trace elements
Density	8.19 g/cm^3
Ultimate Tensile Strength (UTS)	1240 MPa
Yield Strength (0.2% offset)	1035 MPa
Elongation at Break	12%–19%
Fatigue Strength (10^7 cycles)	500 MPa
Hardness (Vickers)	460 HV
Melting Point	1350°C
Thermal Conductivity	11.4 W/(m·K)
Specific Heat Capacity	0.435 J/(g·K)
Coefficient of Thermal Expansion	$13 \cdot 10^{-6}$/°C
Operating Temperature Range	Up to 700°C–750°C
Creep Rate	10^{-7} %/hour at 650°C, 150 MPa
Corrosion Resistance	Excellent resistance to oxidation, pitting and crevice corrosion

and high creep resistance, they maintain their structural integrity under prolonged thermal and mechanical loads. Additionally, their moderate thermal conductivity helps manage heat dissipation effectively, reducing thermal stresses during operation. The coefficient of thermal expansion is carefully balanced to minimize deformation and thermal fatigue in fluctuating thermal environments. The alloy's inherent corrosion resistance protects against aggressive agents such as moisture, salts, and combustion byproducts that are common in engine environments.

14.3.4 CARBON-FIBER-REINFORCED POLYMERS

Carbon-Fiber-Reinforced Polymers (CFRPs) are composite materials made of carbon fibers embedded in a polymer matrix, typically epoxy resin. They are adopted for the cold sections of compressor blades, particularly for large ultra-high-bypass-ratio fans, due to their favorable strength-to-weight ratio (Table 14.6), making them significantly lighter than metallic materials such as titanium. CFRPs exhibit outstanding mechanical properties, including high tensile strength and stiffness, particularly when loads are applied along the direction of the fibers.

In addition to mechanical benefits, CFRPs offer low thermal expansion along the fiber direction, minimizing deformation under temperature variations, which is crucial for the dimensional stability of components. The polymer matrix in CFRPs provides excellent resistance to environmental factors such as moisture, aviation fuels, and corrosive agents, ensuring durability in operational environments. Furthermore, CFRPs are less prone to fatigue compared to metals, making them well-suited for cyclic loading conditions typical of compressor blades. Incorporating a hard mate-

Table 14.6
Typical Properties of CFRP for Cold Compressor Blades (Example Ranges)

Property	Value
Material Composition	Carbon Fibers (50%–70% by volume), Epoxy Resin
Carbon Fiber Tensile Strength	3000–6000 MPa
Carbon Fiber Young's Modulus	200–600 GPa
Density (Composite)	1.5–1.6 g/cm^3
Tensile Strength (CFRP)	600–1500 MPa
Young's Modulus (CFRP)	60–200 GPa
Flexural Strength	700–1500 MPa
Interlaminar Shear Strength	50–80 MPa
Thermal Conductivity	Longitudinal: 20–50 W/(m·K), Transverse: 0.5–1.0 W/(m·K)
Coefficient of Thermal Expansion (CTE)	Longitudinal: -0.1 to 0.5×10^{-6}/°C, Transverse: 30–60×10^{-6}/°C
Operating Temperature Range	−50°C to +120°C
Heat Deflection Temperature	80°C–150°C
Corrosion Resistance	Excellent; inert to chemicals, aviation fuel and salt spray
Fatigue Resistance	Outstanding under high-cycle loading
Impact Strength	Moderate (orientation-dependent)

rial (e.g. titanium or hard metallic coatings such as tungsten carbide or chromium alloys) at the leading edge of a blade is often used as a strategy to improve impact strength.

CFRPs also exhibit good thermal stability for the operating temperatures of cold compressor sections, maintaining their mechanical properties without degradation. Their inherent resistance to corrosion and lack of susceptibility to chemical attacks from aviation environments further enhance their reliability and service life.

14.3.5 STEELS

Steel continues to be used in compressors, particularly in industrial and lower-performance applications, where cost-effectiveness and durability are more critical than weight reduction or extreme thermal performance. Martensitic and stainless steels are commonly employed in cold or low-pressure compressor sections, where operating temperatures and stresses are relatively moderate. In industrial compressors for power generation and processing industries, steel's strength and corrosion resistance make it a practical choice. Additionally, in legacy systems and non-critical components, steel remains valuable due to its availability, affordability, and ease of manufacturing.

14.4 MATERIALS FOR GAS TURBINES

14.4.1 NICKEL SUPERALLOYS

Nickel-based superalloys dominate the turbine blade and vane technology due to their creep resistance, oxidation stability, and high-temperature strength. Innovations in directionally solidified (DS) and single-crystal (SX) superalloys have extended operating limits. For example, CMSX-6 (SX) (Table 14.7) now provides improved creep and thermal fatigue properties, while third-generation SX alloys (e.g. Rene N6) incorporate rhenium to improve high-temperature performance. The absence of grain boundaries in SX alloys and the alignment of grains in DS alloys minimize the structural weaknesses typical of conventional polycrystalline materials, such as sliding or cracking of the grain boundary under stress. Their compositions are carefully tailored with elements such as aluminum and titanium, which form precipitates γ' in the microstructure. This phase γ', comprising up to 70% of the material in some SX alloys, provides resistance to deformation by impeding dislocation motion. Additives like rhenium further enhance creep resistance while chromium contributes to oxidation and corrosion protection.

Table 14.8 provides a comprehensive list including Creep Strength for some of the most commonly used DS and SX superalloys.

14.4.1.1 Refractory High-Entropy Alloys

Refractory High-Entropy Alloys (RHEAs) have emerged as promising materials in turbine blades due to their mechanical and thermostructural properties. These alloys are composed of multiple principal elements, typically including refractory metals

Table 14.7
Properties of CMSX-6 Single-Crystal Nickel Superalloy

Property	Value/Description
Composition	Ni-based alloy with Cr (10%–20%), Al (5%–8%), Ti (2%–5%), Co (10%–15%), Ta (1%–3%), W (1%–4%), Re (0%–3%)
Density	8.7 g/cm^3
Melting Point Range	1230°C–1370°C
Ultimate Tensile Strength (UTS)	1000 MPa at 900°C
Creep Strength	Sustains stress of 250 MPa at 1000°C for over 1000 hours
Fatigue Resistance	High resistance to fatigue cracking; stress cycles exceeding 10^7 at moderate loads
Thermal Conductivity	11 W/(m·K)
Thermal Expansion Coefficient	$14.2 \cdot 10^{-6}$ K^{-1} (20°C–1000°C)
γ' Volume Fraction	70%
Oxidation Resistance	Excellent up to 1050°C with coatings
Microstructure	Single-crystal structure with γ matrix and γ' precipitates

Table 14.8
Creep Strength of Common DS and SX Nickel-Based Superalloys

Alloy Type	Alloy Name	Creep Strength	Test Conditions (°C)
DS	DS GTD-111	230 MPa for 1000 hours	980
DS	René 80	200 MPa for 1000 hours	950
SX	CMSX-4	248 MPa for 1000 hours	1050
SX	CMSX-6	250 MPa for 1000 hours	1000
SX	René N5	260 MPa for 1000 hours	1050
SX	René N6	248 MPa for 1000 hours	1050
SX	PWA 1484	260 MPa for 1000 hours	1093

such as tungsten (W), tantalum (Ta), molybdenum (Mo), niobium (Nb), hafnium (Hf), zirconium (Zr) and vanadium (V), each with an atomic percentage of 5–35 (Table 14.9). This complex composition leads to the formation of simple solid-solution phases, primarily body-centered cubic (BCC) structures, which contribute to their high-temperature performance. In fact, RHEAs maintain superior strength at elevated temperatures. For instance, the VNbMoTaW alloy exhibits a yield strength exceeding 600 MPa at 1400°C, significantly outperforming conventional superalloys like Inconel 718. Table 14.10 includes data on creep strength for some of the most widely used Refractory High-Entropy Alloys (RHEAs).

Table 14.9
Properties of $AlMo_{0.5}NbTa_{0.5}TiZr$ Refractory High-Entropy Alloy

Property	Value/Description
Composition	Al (10%), Mo (12.5%), Nb (25%), Ta (12.5%), Ti (20%), Zr (20%)
Density	7.46 g/cm^3
Melting Point	Estimated >2000°C
Vickers Hardness	678 HV
Compressive Yield Strength (RT)	2466 MPa at room temperature
Compressive Yield Strength (1000°C)	964 MPa
Creep Resistance	Maintains strength with minimal deformation under stress at 1000°C
Thermal Stability	Excellent microstructural stability at high temperatures
Oxidation Resistance	Good due to aluminum content
Microstructure	Body-centered cubic (BCC) structure

Table 14.10
Creep Strength of Common Refractory High-Entropy Alloys (RHEAs)

Alloy Name	Creep Strength	Test Conditions (°C)	Composition (Atomic %)
$AlMo_{0.5}NbTa_{0.5}TiZr$	964 MPa for 1000 hours	1000	Al (10), Mo (12.5), Nb (25), Ta (12.5), Ti (20), Zr (20)
HfNbTaTiZr	840 MPa for 1000 hours	1200	Hf (20), Nb (20), Ta (20), Ti (20), Zr (20)
VNbMoTaW	600 MPa for 1000 hours	1400	V (20), Nb (20), Mo (20), Ta (20), W (20)
NbMoTaW	550 MPa for 1000 hours	1200	Nb (25), Mo (25), Ta (25), W (25)
AlNbTaTiV	900 MPa for 1000 hours	1000	Al (10), Nb (25), Ta (20), Ti (25), V (20)

14.4.2 ADVANCED THERMAL BARRIER COATINGS

Advanced Thermal Barrier Coatings (TBCs, Table 14.11) are used to protect turbine blades from extreme thermal environments. They enhance durability by providing thermal insulation and resistance to oxidation and corrosion. TBCs are primarily applied to high-temperature components made from nickel-based superalloys by means of Air Plasma Spraying (APS) or Electron Beam Physical Vapor Deposition (EB-PVD).

Table 14.11
Types, Composition and Thermal Conductivity of Common TBCs

Type of TBC	Composition	Thermal Conductivity W/(m·K)
YSZ (Yttria-Stabilized Zirconia)	ZrO_2 + 7–8 wt% Y_2O_3	2.6–3.0
Gadolinium Zirconate (GZ)	$Gd_2Zr_2O_7$	1.2–1.6
Yttria-Alumina-Garnet (YAG)	$Y_3Al_5O_{12}$	2.2–2.4
Lanthanum Zirconate (LZ)	$La_2Zr_2O_7$	1.4–1.8
Yttria-Doped Hafnia (YDH)	HfO_2 + Y_2O_3	1.5–2.0
Alumina (Alumina-Based TBC)	Al_2O_3	30–35 (used as bond coat)

Table 14.12
Properties of Common Ceramic Matrix Composites (CMCs)

Type of CMC	Composition	Thermal Conductivity W/(m·K)	Creep Strength	Applications
SiC/SiC	SiC fibers in SiC matrix	10–20	140 MPa at 1200°C	Turbine shrouds, combustor liners
C/SiC	Carbon fibers in SiC matrix	10–12	50 MPa at 1200°C	Brake discs, aerospace thermal shields
Oxide/Oxide	Al_2O_3 fibers in mullite or aluminosilicate matrix	3–5	40 MPa at 1000°C	Turbine exhausts, thermal protection
Si_3N_4/Si_3N_4	Silicon nitride fibers in silicon nitride matrix	15–20	120 MPa at 1200°C	Engine components, aerospace
C/C	Carbon fibers in carbon matrix	20–40	Moderate (use-case dependent)	Rocket nozzles, heat shields

14.4.2.1 Ceramics

Ceramic Matrix Composites (CMCs) are not typically used to construct integral turbine blades due to their inherent brittleness and the extreme mechanical stresses experienced by turbine blades. Instead, CMCs are used more frequently as inserts, shrouds, or liners in specific parts of turbines where high-temperature resistance is critical, but mechanical stresses are relatively lower compared to the blades themselves. In 2021, NASA's Glenn Research Center developed silicon carbide/silicon carbide (SiC/SiC) CMCs capable of withstanding temperatures up to 1482°C. These advancements are part of NASA's Aeronautics research, aiming to enhance turbine engine applications for next-generation ultra-efficient aircraft. Table 14.12 includes types, composition, thermal conductivity, and creep strength of commonly used CMCs.

14.4.3 COMPATIBILITY WITH COOLING TECHNIQUES

The compatibility of turbine materials with cooling techniques is mainly determined by their thermal properties, mechanical strength, and manufacturability (Table 14.13). Nickel-based superalloys are the most compatible with advanced cooling techniques due to their high thermal conductivity, strength at elevated temperatures, and adaptability to manufacturing processes like directional solidification and additive manufacturing. These methods allow for the precise creation of intricate internal convective cooling channels and external film cooling holes.

CMCs have lower thermal conductivity but high-temperature capability. However, their brittleness and anisotropy make them less compatible with internal cooling channel manufacturing. Coated materials like superalloys with TBCs allow the combination of film cooling with advanced manufacturing methods such as laser drilling and additive techniques to create efficient cooling pathways.

RHEAs show promise for operating at temperatures around 1400°C, but their compatibility with cooling techniques is still experimental. Additive manufacturing offers potential for integrating cooling channels, though their low thermal conductivity (10–12 W/(m·K)) and mechanical limitations at ambient temperatures require further research.

CMCs with very low thermal conductivity are not typically used in actively cooled components. They rely on complex machining and surface coatings to enable external cooling in specific applications.

Table 14.13
Compatibility of Turbine Materials with Cooling Techniques

Material Type	Temperature Limit with Cooling	Best Manufacturing Technique for Cooling Channels	Cooling Synergy
Nickel-Based Superalloys	1150°C–1200°C	Directional Solidification (DS), Additive Manufacturing	High
CMC (SiC/SiC)	1482°C	No cooling channels required	Moderate
TBC-Coated Superalloys	1200°C (substrate)	Additive Manufacturing, Laser Drilling	High
RHEAs	1400°C (potential)	Additive Manufacturing (under research)	Experimental
Oxide/Oxide Composites	1200°C (surface)	Not applicable (used in indirect heat exposure areas)	Low
Carbon-Carbon (C/C)	2000°C (inert environments)	Complex Machining, Surface Coating Required	Low

Table 14.14
Materials for HP and LP Stages of Steam Turbines

Stage	Material Type	Composition	Strength (MPa)	Max Temperature (°C)
HP	Inconel 740H	Ni-25Cr-20Co-Ti-Al	800 (at 750°C)	750
HP	P91 Steel	Fe-9Cr-1Mo-V-Nb	600 (at 620°C)	620
HP	Haynes 282	Ni-20Cr-10Co-Mo-Ti	800 (at 750°C)	750
LP	316H Stainless Steel	Fe-16Cr-10Ni-2Mo	250 (at 500°C)	500
LP	17-4 PH Stainless Steel	Fe-15Cr-4Ni-Cu	850 (at 20°C)	480
LP	Martensitic Steel (410SS)	Fe-12Cr-C	450 (at 20°C)	425

14.4.4 MATERIALS FOR STEAM TURBINES

In the high-pressure (HP) stages of steam turbines, materials must withstand relatively high temperatures and very high pressures while maintaining excellent creep strength, corrosion resistance, and thermal stability (in Advanced Ultra-Supercritical, A-USC, Plants, pressure can go up to 350 bar). Table 14.14 provides a list of the most commonly used materials in steam turbines. Nickel-based superalloys, such as Inconel 740H and Haynes 282, are widely used for HP stages, offering operating temperatures up to 750°C due to their resistance to steam oxidation. Advanced ferritic steels such as P91 and P92 are also employed in some HP components, with operating limits around 620°C, providing a balance between cost-effectiveness and performance. In intermediate-pressure (IP) and low-pressure (LP) stages, where temperatures and pressures are lower, materials with moderate strength and good corrosion resistance are preferred. Austenitic stainless steels such as 316H are common in LP sections, particularly in wet steam conditions, with a temperature limit of approximately 500°C. For rotors and blades in LP stages, 17-4 PH stainless steel and martensitic steels such as 410SS are used due to their toughness, erosion resistance, and fatigue properties under cyclic loads. These material choices are critical to achieve efficiency and reliability in steam turbines operating under varying thermal and mechanical demands.

Appendix

A.1 DERIVATION OF THE FIRST LAW OF THERMODYNAMICS IN STAGNATION FORM FOR IDEAL GASES

To prove the validity of $T^0 ds = dh^0 - \frac{dp^0}{\rho^0}$, we will start to elaborate on the known $Tds = dh - \frac{dp}{\rho}$. The static quantities T, p, and ρ can be easily expressed in terms of the corresponding total quantities:

$$T = T^0\left(1+\frac{\gamma-1}{2}\mathrm{Ma}^2\right)^{-1};\ p = p^0\left(1+\frac{\gamma-1}{2}\mathrm{Ma}^2\right)^{\frac{-\gamma}{\gamma-1}};\ \rho = \rho^0\left(1+\frac{\gamma-1}{2}\mathrm{Ma}^2\right)^{\frac{-1}{\gamma-1}}$$

Using the above, the following expression of $Tds = dh - \frac{dp}{\rho}$ can be derived for ideal gases:

$$T^0\left(1+\frac{\gamma-1}{2}\mathrm{Ma}^2\right)^{-1} ds = \frac{\gamma R}{\gamma-1}dT - \left(\frac{dp}{\rho^0}\right)\left(1+\frac{\gamma-1}{2}\mathrm{Ma}^2\right)^{\frac{1}{\gamma-1}} \tag{A.1}$$

Let us start with differentiation on T and p to rearrange the RHS of equation A.1:

$$dT = \left(1+\frac{\gamma-1}{2}\mathrm{Ma}^2\right)^{-1}\left[dT^0 + (\gamma-1)\mathrm{Ma}^2\right] \tag{A.2}$$

$$dT = \left(1+\frac{\gamma-1}{2}\mathrm{Ma}^2\right)^{-1} dT^0 - T^0\left(1+\frac{\gamma-1}{2}\mathrm{Ma}^2\right)^{-2}\left(\frac{\gamma-1}{2}\right)d\left(\mathrm{Ma}^2\right) \tag{A.3}$$

$$dp = \left(1+\frac{\gamma-1}{2}\mathrm{Ma}^2\right)^{-\frac{\gamma}{\gamma-1}} dp^0 - p^0\frac{\gamma}{\gamma-1}\left(1+\frac{\gamma-1}{2}\mathrm{Ma}^2\right)^{-\frac{\gamma}{\gamma-1}-1}\left(\frac{\gamma-1}{2}\right)d\left(\mathrm{Ma}^2\right)$$

Let us substitute equations A.2 and A.3 in equation A.1 and simplify:

$$T^0 ds = \frac{\gamma R}{\gamma-1}dT^0 + \gamma R\mathrm{Ma}^2 - \left(\frac{dp^0}{\rho^0}\right) - \gamma R\mathrm{Ma}^2 = dh^0 - \frac{dp^0}{\rho^0} \tag{A.4}$$

which concludes the proof.

A.2 LIEBLEIN'S METHODS FOR TOTAL PRESSURE LOSS ESTIMATION IN COMPRESSOR CASCADES' WAKES

Assume that at an arbitrary z coordinate, at or downstream of the trailing edge of airfoils, outside the wakes, a constant velocity $c_z = \text{const.}$ is maintained, and the static pressure p remains uniform across the channel spacing s. It is also assumed that the flow angles α of the velocities both in the boundary layer and outside of it are constant. Due to periodicity in the flow in the pitch-wise direction, identical wakes form behind each blade, and the flow is considered incompressible.

Let us introduce the following thicknesses for a two-dimensional boundary layer as displayed in Figure 6.26, where $y = \xi s$:

- the boundary layer thickness δ is the distance from the solid boundary to the point in the flow where the velocity v reaches 99% of the free-stream velocity c. It is mathematically defined as:

$$\frac{\delta}{s} = \xi_{99} \qquad \text{where} \qquad \left.\frac{v}{c}\right|_{\xi=\xi_{99}} = 0.99 \tag{A.5}$$

- the displacement thickness δ^* quantifies the reduction in the mass flow rate due to the presence of the boundary layer, compared to an ideal inviscid flow. It is defined as:

$$\frac{\delta^*}{s\cos\alpha} = \int_0^1 \left(1 - \frac{v(\xi)}{c}\right) d\xi \tag{A.6}$$

- the momentum thickness θ represents the loss of momentum due to the boundary layer, which would have been carried by the ideal inviscid flow:

$$\frac{\theta}{s\cos\alpha} = \int_0^1 \frac{v(\xi)}{c}\left(1 - \frac{v(\xi)}{c}\right) d\xi \tag{A.7}$$

- the energy thickness k measures the reduction in kinetic energy due to the boundary layer, compared to the ideal inviscid flow:

$$\frac{k}{s\cos\alpha} = \int_0^1 \left(\frac{v(\xi)}{c}\right)^2 \left(1 - \frac{v(\xi)}{c}\right) d\xi \tag{A.8}$$

The integrals appearing in these equations are in fact similar to the standard boundary layer thickness parameters, except that the integration path is along y, i.e. normal to the axial direction z instead of being normal to the flow direction, as in conventional 2-D boundary layer theory.

A coefficient for the quantification of total loss between station 1 at the leading edge and a hypothetical far-downstream station 3, where the flow mixing has been completed, can be defined as a function of the velocity in the boundary layer v_3 in station 3 in mixed state:

$$\zeta_c = \frac{p_1^0 - p_3^0}{\frac{1}{2}\rho v_3^2} \tag{A.9}$$

This can be rewritten as:

$$\zeta_c = \frac{p_1^0 - \bar{p}_2^0}{\frac{1}{2}\rho v_3^2} + \frac{\bar{p}_2^0 - \bar{p}_3^0}{\frac{1}{2}\rho v_3^2} \tag{A.10}$$

Recognizing that, since p_2 and α_2 are independent of y:

$$\bar{p}_2^0 = \frac{\int_0^s (p_2(y) + \frac{\rho}{2} v_2^2(y))\, d\dot{m}}{\int_0^s d\dot{m}} = p_2 + \frac{\frac{1}{2}\int_0^s \rho v_2^2(y)\, d\dot{m}}{\int_0^s d\dot{m}} = p_2 + \frac{1}{2}\rho \bar{v}_2^2 \tag{A.11}$$

where $d\dot{m} = \rho v_z dy = \rho v \cos\alpha_2 dy$, equation A.11 becomes:

$$\bar{p}_2^0 = p_2 + \frac{\frac{\rho}{2}\int_0^1 v_2^3(\xi)\, d\xi}{\int_0^1 v_2(\xi)\, d\xi} \tag{A.12}$$

Since the flow outside the wakes is inviscid, the total pressure is conserved from 1 to 2:

$$p_1^0 = p_2 + \rho \frac{c_2^2}{2} \tag{A.13}$$

Equation A.9 can be rewritten as:

$$\zeta_c = \frac{p_1^0 - \bar{p}_2^0}{\frac{1}{2}\rho v_3^2} + \frac{\bar{p}_2^0 - \bar{p}_3^0}{\frac{1}{2}\rho v_3^2} = \frac{1}{v_3^2}\left[c_2^2 - \frac{\int_0^1 v_2^3(\xi)\, d\xi}{\int_0^1 v_2(\xi)\, d\xi}\right] + \frac{\lambda}{v_3^2}\left[\int_0^1 v_2(\xi)\, d\xi\right]^2 \tag{A.14}$$

where:

$$\lambda = \frac{\bar{p}_2^0 - \bar{p}_3^0}{\frac{\rho}{2}\left[\int_0^1 v_2(\xi)\, d\xi\right]^2} \tag{A.15}$$

is termed the mixing loss coefficient.

The mixed-out state 3 is defined by the following conditions:

$$\begin{cases} \dot{m}_2 = \dot{m}_3 \implies \int_0^s \rho v_{z2}(y) dy = \cos\alpha_2 \int_0^s \rho v_2(y) dy = \rho s v_{z3} \quad \text{(continuity)} \\ \int_0^s \rho v_{z2}(y) v_{u2}(y) dy = \dot{m}_3 v_{u3} \quad \text{(y-momentum)} \\ \int_0^s \rho v_{z2}^2(y) dy + p_2 s = \dot{m}_3 v_{z3} + p_3 s \quad \text{(z-momentum)} \end{cases} \tag{A.16}$$

i.e.:

$$\begin{cases} \cos\alpha_2 \int_0^1 v_2(\xi) d\xi = v_{z3} \\ \sin\alpha_2 \frac{\int_0^1 v_2^2(\xi) d\xi}{\int_0^1 v_2(\xi) d\xi} = v_{u3} \\ \int_0^1 \rho v_{z2}^2(\xi) d\xi + p_2 - \dot{m}_3 v_{z3} = p_3 \end{cases} \tag{A.17}$$

from which the total pressure $\bar{p}_3^0$ can be written as:

$$\bar{p}_3^0 = p_3 + \frac{\rho}{2} v_3^3 = p_2 + \rho\cos^2\alpha_2 \left\{\int_0^1 v_2^2(\xi) d\xi - \frac{1}{2}\left[\int_0^1 v_2(\xi) d\xi\right]^2\right\}$$
$$+ \frac{\rho}{2}\sin^2\alpha_2 \left\{\frac{\int_0^1 v_2^2(\xi) d\xi}{\int_0^1 v_2(\xi) d\xi}\right\}^2 \tag{A.18}$$

Ultimately, recognizing that p_2 is independent of ξ, the wake loss coefficient λ can be expressed as:

$$\lambda = I_I - I_{II}^2 + \cos^2\alpha_2 \left[I_{II} - 1\right]^2 \tag{A.19}$$

where:

$$I_I = \frac{\int_0^1 v_2(\xi)^3 d\xi}{\left[\int_0^1 v_2(\xi) d\xi\right]^3} \tag{A.20}$$

$$I_{II} = \frac{\int_0^1 v_2(\xi)^2 d\xi}{\left[\int_0^1 v_2(\xi) d\xi\right]^2} \tag{A.21}$$

All velocities in the above equations can be expressed as functions of the characteristic boundary layer thicknesses using equations A.6 - A.8. In fact, as $v_2(\xi) = c_2 - (c_2 - v_2(\xi))$:

$$\int_0^1 v_2(\xi) d\xi = c_2 - \int_0^1 \left[c_2 - v_2(\xi)\right], d\xi \tag{A.22}$$

and, introducing the wake shape factor $H = \frac{\delta^*}{\theta}$ and the energy factor $K = \frac{k}{\theta}$, it follows that:

$$\int_0^1 v_2(\xi) d\xi = c_2 \left[1 - \frac{\theta}{s\cos\alpha_2} H\right] = c_2 \left[1 - \left(\frac{\theta}{\ell}\right) \frac{\sigma_S}{\cos\alpha_2} H\right] \tag{A.23}$$

$$\int_0^1 v_2^2(\xi) d\xi = c_2^2 \left[1 - \left(\frac{\theta}{\ell}\right) \frac{\sigma_S}{\cos\alpha_2}(1+H)\right] \tag{A.24}$$

$$\int_0^1 v_2^3(\xi) d\xi = c_2^3 \left[1 - \left(\frac{\theta}{\ell}\right) \frac{\sigma_S}{\cos\alpha_2} \frac{3H(H-1)}{3H-1}\right] \tag{A.25}$$

Researchers have demonstrated that, when the wake at station 2 exhibits a power velocity profile similar to turbulent boundary layers, the parameter H remains relatively stable, varying between 1.4 and 2. Using this range, one can plot equation A.14 for different values of α_2, with the independent variable defined as:

$$x = \frac{\theta}{\ell} \frac{\sigma_S}{\cos\alpha_2}.$$

A.3 CALCULATION OF THE NUMBER OF IMPELLER BLADES BASED ON MINIMUM RELATIVE VELOCITY

Starting from the original Euler equation for steady radial flow in the $r - \theta$ plane:

$$w_r \frac{\partial w_\theta}{\partial r} + \frac{w_\theta}{r} \frac{\partial w_\theta}{\partial \theta} + \frac{w_r w_\theta}{r} = -\frac{1}{\rho r} \frac{\partial p}{\partial \theta} - 2\omega w_r \tag{A.26}$$

We calculate $w_\theta = \frac{w_r}{\tan\beta}$ and $\frac{\partial w_\theta}{\partial r}$:

$$w_\theta = \frac{w_r}{\tan\beta} \implies \frac{\partial w_\theta}{\partial r} = \frac{1}{\tan\beta}\frac{\partial w_r}{\partial r} \tag{A.27}$$

as well as $\frac{\partial w_\theta}{\partial \theta}$:

$$w_\theta = \frac{w_r}{\tan\beta} \implies \frac{\partial w_\theta}{\partial \theta} = \frac{1}{\tan\beta}\frac{\partial w_r}{\partial \theta} \tag{A.28}$$

Substituting these into the Euler equation gives:

$$w_r\left(\frac{1}{\tan\beta}\frac{\partial w_r}{\partial r}\right) + \frac{1}{r}\left(\frac{w_r}{\tan\beta}\right)\left(\frac{1}{\tan\beta}\frac{\partial w_r}{\partial \theta}\right) + \frac{w_r\left(\frac{w_r}{\tan\beta}\right)}{r} = -\frac{1}{\rho r}\frac{\partial p}{\partial \theta} - 2\omega w_r \tag{A.29}$$

This simplifies to:

$$\frac{w_r}{\tan\beta}\frac{\partial w_r}{\partial r} + \frac{1}{r\tan^2\beta}w_r\frac{\partial w_r}{\partial \theta} + \frac{w_r^2}{r\tan\beta} = -\frac{1}{\rho r}\frac{\partial p}{\partial \theta} - 2\omega w_r \tag{A.30}$$

Given:

$$p + \frac{1}{2}\rho\left(\left(\frac{w_r}{\tan\beta}\right)^2 + w_r^2 - (\omega r)^2\right) = \text{const.} \tag{A.31}$$

this implies:

$$p = \text{const.} - \frac{1}{2}\left(\left(\frac{w_r}{\tan\beta}\right)^2 + w_r^2 - (\omega r)^2\right) \tag{A.32}$$

Taking the derivative with respect to θ:

$$\frac{\partial p}{\partial \theta} = -\frac{1}{2}\rho\left(\frac{\partial}{\partial \theta}\left(\frac{w_r^2}{\tan^2\beta} + w_r^2\right)\right) \tag{A.33}$$

Since ω is constant with respect to θ and r, the derivative of $-(\omega r)^2$ with respect to θ is zero. Thus, we have:

$$\frac{\partial p}{\partial \theta} = -\rho\, w_r\left(\frac{1}{\tan^2\beta} + 1\right)\frac{\partial w_r}{\partial \theta} \tag{A.34}$$

Substituting $\frac{\partial p}{\partial \theta}$ back into the Euler equation gives:

$$\frac{w_r}{\tan\beta}\frac{\partial w_r}{\partial r} + \frac{1}{r\tan^2\beta}w_r\frac{\partial w_r}{\partial \theta} + \frac{w_r^2}{r\tan\beta} = \frac{w_r}{r\tan^2\beta}\frac{\partial w_r}{\partial \theta} + \frac{w_r}{r}\frac{\partial w_r}{\partial \theta} - 2\omega w_r \tag{A.35}$$

The continuity equation for incompressible flow in cylindrical coordinates is given by:

$$\frac{1}{r}\frac{\partial}{\partial r}(rw_r)+\frac{1}{r}\frac{\partial w_\theta}{\partial \theta}=0 \tag{A.36}$$

Given that $w_\theta = \frac{w_r}{\tan\beta}$, we can substitute this into the continuity equation:

$$\frac{1}{r}\frac{\partial}{\partial r}(rw_r)+\frac{1}{r}\frac{\partial}{\partial \theta}\left(\frac{w_r}{\tan\beta}\right)=0 \tag{A.37}$$

1. The first term remains as is:

$$\frac{1}{r}\frac{\partial}{\partial r}(rw_r) \tag{A.38}$$

2. The second term becomes:

$$\frac{1}{r}\frac{\partial}{\partial \theta}\left(\frac{w_r}{\tan\beta}\right)=\frac{1}{r\tan\beta}\frac{\partial w_r}{\partial \theta} \tag{A.39}$$

Putting these together, we have:

$$\frac{1}{r}\frac{\partial}{\partial r}(rw_r)+\frac{1}{r\tan\beta}\frac{\partial w_r}{\partial \theta}=0 \rightarrow \frac{\partial}{\partial r}(rw_r)+\frac{1}{\tan\beta}\frac{\partial w_r}{\partial \theta}=0 \tag{A.40}$$

which, upon derivation, gives:

$$\frac{\partial w_r}{\partial r}=-\frac{w_r}{r}-\frac{1}{r\tan\beta}\frac{\partial w_r}{\partial \theta} \tag{A.41}$$

Substituting this into the Euler equation, we get:

$$2\omega w_r=\frac{w_r}{r}\left(\frac{1}{\tan^2\beta}+1\right)\frac{\partial w_r}{\partial \theta} \tag{A.42}$$

Substituting $w_r(\theta)=w_{r,\min}+\frac{Z_p}{\pi}(\bar{w}_r-w_{r,\min})\theta$:

$$2\omega\left[w_{r,\min}+\frac{Z_p}{\pi}(\bar{w}_r-w_{r,\min})\theta\right] \tag{A.43}$$

$$=\frac{1}{r}\left[w_{r,\min}+\frac{Z_p}{\pi}(\bar{w}_r-w_{r,\min})\theta\right]\left(\frac{1}{\tan^2\beta}+1\right)\frac{Z_p}{\pi}(\bar{w}_r-w_{r,\min}) \tag{A.44}$$

Substituting $w_{r,\min}=0$:

$$2\omega\frac{Z_p}{\pi}\bar{w}_r\theta=\frac{Z_p^2\bar{w}_r{}^2\theta}{\pi^2 r}\left(\frac{1}{\tan^2\beta}+1\right) \tag{A.45}$$

which, solving for Z_p gives:

$$Z_p=\frac{\omega\pi r}{\bar{w}_r\left(\frac{1}{\tan^2\beta}+1\right)} \tag{A.46}$$

and, when referred to the impeller outlet $\phi = \bar{w}_{r,2}/(\omega r_2)$

$$Z_p = \frac{\pi}{\phi\left(\frac{1}{\tan^2\beta_2}+1\right)} \tag{A.47}$$

For safety, as in $\tan\beta_2' > \tan\beta_2$, we can set the minimum number of blades such that:

$$Z_p \geq \frac{\pi}{\phi\left(\frac{1}{\tan^2\beta_2'}+1\right)} \tag{A.48}$$

A.4 FLOW-AREA RELATIONSHIPS FOR ISENTROPIC FLOW OF AN IDEAL GAS

The steady, one-dimensional continuity equation in differential form is:

$$\frac{d\rho}{\rho}+\frac{dc}{c}+\frac{dA}{A}=0 \tag{A.49}$$

To relate changes in density and velocity, we use the inviscid one-dimensional momentum equation:

$$dp+\rho c\,dc=0 \tag{A.50}$$

together with the isentropic relation:

$$dp=a^2\,d\rho \tag{A.51}$$

where a is the speed of sound. Eliminating dp between (A.50) and (A.51) gives:

$$\frac{d\rho}{\rho}=-\frac{c^2}{a^2}\frac{dc}{c}=-\mathrm{Ma}^2\frac{dc}{c} \tag{A.52}$$

Substituting (A.52) into the continuity equation (A.49) yields the standard area–velocity relation:

$$\frac{dA}{A}=(\mathrm{Ma}^2-1)\frac{dc}{c} \tag{A.53}$$

This expression implies that:

- For subsonic flow ($\mathrm{Ma} < 1$), an increase in area ($dA>0$) leads to a decrease in c.
- For supersonic flow ($\mathrm{Ma}>1$), an increase in area results in an increase in c.
- At sonic conditions, the Mach number reaches 1, which indicates a critical point.

When the flow is sonic, any differential change in area dA leads to no change in flow properties unless there is also an infinitesimal change in velocity. Mathematically, this means:

$$\frac{dA}{dx}=0 \quad \text{at Ma}=1$$

Thus, the sonic condition can only exist at a stationary point of the area profile, which could be a maximum or a minimum. For physical reasons, we discard the maximum-area scenario. In fact, in an expanding duct, the subsonic flow further slows with an increase in area, making it physically impossible to accelerate to sonic or supersonic speeds. Conversely, a converging duct leading to a minimum area naturally accelerates the flow as it narrows. This geometry allows the subsonic flow to reach Mach 1 in the throat. Only after the flow reaches sonic speed can it expand in a diverging section to achieve supersonic speeds.

For supersonic flow to decelerate to subsonic speeds (e.g. in a diffuser), it must also pass through a minimum-area throat. This property ensures that the sonic condition serves as a reversible boundary for accelerating or decelerating compressible flows.

It is essential to note that while the sonic condition in the throat (Ma = 1) is a necessary condition for the transition between subsonic and supersonic flows, it is not sufficient on its own. For the flow to truly transition to supersonic speeds, the pressure at the channel discharge must be compatible with fully isentropic expansion (where the flow goes from subsonic to supersonic) in the divergent section of the nozzle.

In other words, sufficient conditions for achieving supersonic flow include:

- **Pressure compatibility**. The discharge pressure must be lower than the critical pressure for the flow to expand isentropically in the diverging section.
- **Isentropic expansion**. The flow must undergo a fully isentropic expansion in the divergent section to reach supersonic speeds effectively.

Without these conditions, although the flow can reach Mach 1 at the throat, it may not continue to accelerate to supersonic speeds in the divergent section.

The second Hugoniot relationship is for the density change and is obtained as follows. The speed of sound is given by:

$$a = \sqrt{\frac{\gamma p}{\rho}} \tag{A.54}$$

Differentiating this expression gives:

$$da = \frac{1}{2\sqrt{\frac{\gamma p}{\rho}}}\left(\frac{\gamma dp}{\rho} - \gamma p\frac{d\rho}{\rho^2}\right) \rightarrow da = \frac{1}{2a}\left(\frac{\gamma dp}{\rho} - \gamma p\frac{d\rho}{\rho^2}\right) \tag{A.55}$$

Substituting da into equation A.51 gives:

$$\frac{d\rho}{\rho} + 2\frac{da}{a} = \frac{dp}{p} \tag{A.56}$$

$$\frac{d\rho}{\rho} + 2\cdot\frac{1}{2a}\left(\frac{\gamma dp}{\rho} - \gamma p\frac{d\rho}{\rho^2}\right) = \frac{dp}{p} \tag{A.57}$$

$$\frac{d\rho}{\rho} + \frac{1}{a}\left(\frac{\gamma dp}{\rho} - \gamma p\frac{d\rho}{\rho^2}\right) = \frac{dp}{p} \tag{A.58}$$

Rearranging gives:

$$\frac{d\rho}{\rho} = \frac{dp}{p} - \frac{\gamma}{\rho a}\left(\frac{dp}{\rho} - p\frac{d\rho}{\rho^2}\right) \tag{A.59}$$

After simplification, we find:

$$\frac{d\rho}{\rho} = -\left(1 - \frac{1}{\mathrm{Ma}^2}\right)\frac{dA}{A} \tag{A.60}$$

The third Hugoniot equation computes the pressure change. Using the density-change relationship derived earlier:

$$\frac{d\rho}{\rho} + 2\frac{da}{a} = \frac{dp}{p} \tag{A.61}$$

Substituting the expression for $\frac{d\rho}{\rho}$ gives:

$$-\left(1 - \frac{1}{\mathrm{Ma}^2}\right)\frac{dA}{A} + 2\frac{da}{a} = \frac{dp}{p} \tag{A.62}$$

Substituting the relationship for da gives:

$$-\left(1 - \frac{1}{\mathrm{Ma}^2}\right)\frac{dA}{A} + 2\left(\frac{\gamma dp}{\rho} - \gamma p\frac{d\rho}{\rho^2}\right) = \frac{dp}{p} \tag{A.63}$$

Rearranging yields:

$$\frac{dp}{p} = -\gamma\left(1 - \frac{1}{\mathrm{Ma}^2}\right)\frac{dA}{A} \tag{A.64}$$

A.5 ANALYTICAL CALCULATION OF RADIAL INFLOW TURBINE MAP

The equation for π_e is:

$$\pi_e = \left[1 - (\gamma - 1)\frac{\mathrm{Ma}_u}{\tan\alpha_2}\cdot k \cdot \dot{m}_c\left(1 + \frac{\gamma+1}{2}\mathrm{Ma}_2^2\right)^{\frac{1}{\gamma-1}}\right]^{\frac{1-\gamma}{\gamma\eta_{\mathrm{pol,T}}}} \tag{A.65}$$

Substituting Ma_2^2 from:

$$\mathrm{Ma}_2^2 = \left(\frac{\mathrm{Ma}_u}{\sin\alpha_2}\cdot\frac{\tan\beta_2\cdot\tan\alpha_2}{\tan\beta_2 + \tan\alpha_2}\right)^2 \tag{A.66}$$

wherein:

$$\tan^2\beta_2 = \frac{\mathrm{Ma}_{2,0}^2 \sin^2\alpha_2}{(\mathrm{Ma}_u - \mathrm{Ma}_{2,0}\cot\alpha_2)^2} \tag{A.67}$$

Substitute $\tan^2\beta_2$ into equation A.68:

$$\mathrm{Ma}_2^2 = \left(\frac{\mathrm{Ma}_u}{\sin\alpha_2} \cdot \frac{\frac{\mathrm{Ma}_{2,0}^2 \sin^2\alpha_2}{(\mathrm{Ma}_u - \mathrm{Ma}_{2,0}\cot\alpha_2)^2} \cdot \tan\alpha_2}{\frac{\mathrm{Ma}_{2,0}^2 \sin^2\alpha_2}{(\mathrm{Ma}_u - \mathrm{Ma}_{2,0}\cot\alpha_2)^2} + \tan^2\alpha_2} \right)^2 \tag{A.68}$$

From:

$$\mathrm{Ma}_{2,0} = \left(\frac{\mathrm{Ma}_2^2}{1 + \frac{\gamma+1}{2}\mathrm{Ma}_2^2} \right)^{\frac{1}{2}} \tag{A.69}$$

Substitute this into equation A.69:

$$\tan^2\beta_2 = \frac{\frac{\mathrm{Ma}_2^2 \sin^2\alpha_2}{1 + \frac{\gamma+1}{2}\mathrm{Ma}_2^2}}{\left(\mathrm{Ma}_u - \frac{\mathrm{Ma}_2^2 \cot\alpha_2}{\sqrt{1 + \frac{\gamma+1}{2}\mathrm{Ma}_2^2}} \right)^2} \tag{A.70}$$

After substituting equation A.72 into equation A.68, we simplify iteratively to eliminate β_2. After algebraic reductions:

$$\mathrm{Ma}_2^2 = \frac{\mathrm{Ma}_u^2}{1 + \frac{\gamma+1}{2}\mathrm{Ma}_u^2} \tag{A.71}$$

Substitute equation A.73 into equation A.67:

$$\pi_e = \left[1 - (\gamma - 1)\frac{\mathrm{Ma}_u}{\tan\alpha_2} \cdot k \cdot \dot{m}_c \cdot \left(1 + \frac{\gamma+1}{2} \cdot \frac{\mathrm{Ma}_u^2}{1 + \frac{\gamma+1}{2}\mathrm{Ma}_u^2} \right)^{\frac{1}{\gamma-1}} \right]^{\frac{1-\gamma}{\gamma e \eta_{\mathrm{pol,T}}}} \tag{A.72}$$

Simplify the term inside the parentheses:

$$1 + \frac{\gamma+1}{2} \cdot \frac{\mathrm{Ma}_u^2}{1 + \frac{\gamma+1}{2}\mathrm{Ma}_u^2} = 1 \tag{A.73}$$

Thus, the final expression becomes:

$$\pi_e = \left[1 - (\gamma - 1)\frac{\mathrm{Ma}_u}{\tan\alpha_2} \cdot k \cdot \dot{m}_c \right]^{\frac{1-\gamma}{\gamma \eta_{\mathrm{pol,T}}}} \tag{A.74}$$

To solve for β_2 as a function of $k \cdot \dot{m}_c$ and Ma_u, let us start with:

$$\tan\beta_2 = \frac{k \cdot \dot{m}_c \left(1 + \frac{\gamma+1}{2}\mathrm{Ma}_2^2 \right)^{\frac{1}{\gamma-1}}}{\mathrm{Ma}_u - \mathrm{Ma}_{2,0}\cot\alpha_2} \tag{A.75}$$

$$\mathrm{Ma}_2 = \frac{\mathrm{Ma}_u}{\sin\alpha_2}\left(\frac{\tan\beta_2 \cdot \tan\alpha_2}{\tan\beta_2 + \tan\alpha_2}\right) \tag{A.76}$$

$$\mathrm{Ma}_{2,0}\sin\alpha_2 = (\mathrm{Ma}_u - \mathrm{Ma}_{2,0}\cot\alpha_2)\tan\beta_2 \tag{A.77}$$

Substituting equation A.78 into equation A.77, we write:

$$\mathrm{Ma}_2^2 = \frac{\mathrm{Ma}_u^2}{\sin^2\alpha_2}\cdot\left(\frac{\tan\beta_2\tan\alpha_2}{\tan\beta_2 + \tan\alpha_2}\right)^2 \tag{A.78}$$

Substituting this into equation A.77 gives:

$$\tan\beta_2 = \frac{k\cdot\dot{m}_c\left(1+\frac{\gamma+1}{2}\cdot\frac{\mathrm{Ma}_u^2}{\sin^2\alpha_2}\cdot\left(\frac{\tan\beta_2\tan\alpha_2}{\tan\beta_2+\tan\alpha_2}\right)^2\right)^{\frac{1}{\gamma-1}}}{\mathrm{Ma}_u - \mathrm{Ma}_{2,0}\cot\alpha_2} \tag{A.79}$$

Rearrange equation A.79 to solve for $\mathrm{Ma}_{2,0}$:

$$\mathrm{Ma}_{2,0} = \frac{\mathrm{Ma}_u\tan\beta_2}{\sin\alpha_2 + \cot\alpha_2\tan\beta_2} \tag{A.80}$$

Substitute equation A.82 into equation A.81:

$$\tan\beta_2 = \frac{k\cdot\dot{m}_c\left(1+\frac{\gamma+1}{2}\cdot\frac{\mathrm{Ma}_u^2}{\sin^2\alpha_2}\cdot\left(\frac{\tan\beta_2\tan\alpha_2}{\tan\beta_2+\tan\alpha_2}\right)^2\right)^{\frac{1}{\gamma-1}}}{\mathrm{Ma}_u - \frac{\mathrm{Ma}_u\tan\beta_2\cdot\cot\alpha_2}{\sin\alpha_2+\cot\alpha_2\tan\beta_2}} \tag{A.81}$$

Finally, simplify further:

$$\tan\beta_2(\sin\alpha_2 + \cot\alpha_2\tan\beta_2) = k\cdot\dot{m}_c\cdot\sin\alpha_2 \cdot\left(1+\frac{\gamma+1}{2}\cdot\frac{\mathrm{Ma}_u^2}{\sin^2\alpha_2}\cdot\left(\frac{\tan\beta_2\tan\alpha_2}{\tan\beta_2+\tan\alpha_2}\right)^2\right)^{\frac{1}{\gamma-1}} \tag{A.82}$$

equation A.84 is the implicit equation for $\tan\beta_2$. Solving explicitly for β_2 generally requires numerical methods.

Index

For Product Safety Concerns and Information please contact our EU representative GPSR@taylorandfrancis.com Taylor & Francis Verlag GmbH, Kaufingerstraße 24, 80331 München, Germany